Healthy Cities

Evelyne de Leeuw • Jean Simos
Editors

Healthy Cities

The Theory, Policy, and Practice of Value-Based Urban Planning

Springer

Editors
Evelyne de Leeuw
Centre for Health Equity Training
Research & Evaluation (CHETRE)
University of New South Wales
South Western Sydney Local Health District
Ingham Institute
Liverpool, NSW, Australia

Jean Simos
Institute of Global Health
University of Geneva
Campus Biotech
Geneva, Switzerland

ISBN 978-1-4939-6692-9 ISBN 978-1-4939-6694-3 (eBook)
DOI 10.1007/978-1-4939-6694-3

Library of Congress Control Number: 2016956250

Printed on acid-free paper

This Springer imprint is published by Springer Nature
The registered company is Springer Science+Business Media LLC
The registered company address is: 233 Spring Street, New York, NY 10013, U.S.A.

Foreword: Healthy Cities Is Political and It Works

Over my 30-year international public health career, there have been countless reminders to our partners and stakeholders that health is not only the absence of disease, that health is everybody's business, and that health is essential for social and sustainable development. At WHO, we had hoped that our increasingly strong and compelling evidence on the determinants of health and the root causes of health inequalities would catapult us to unstoppable action, but the reality proved otherwise.

Science alone and preaching to the converted do not make change happen. Today we finally speak of health as a political choice. It is about the kind of society we want to live in. It is about the values that we wish to crown our visions and policies. Embracing health and equity as a whole-of-government goal is the premise for an intersectoral frame of cooperation where everyone addresses the question: what can you do for health and what health can do for you?

A minister of health is de facto in a less influential position to mobilize intersectoral action for health than a mayor who is inspired by the vision and the values of a healthy city. The WHO Healthy Cities movement draws its success from this realization. The conceptual 'software' of a Healthy City aligned perfectly with the local government 'hardware' and its strategic and convening power across sectors and civil society. This was a project that soon became a movement: An international initiative that provided to city leaders the opportunity, the legitimacy and the space to experiment with and learn from some of the most challenging concepts and ideas of modern public health.

It has been a fascinating experience to witness step by step throughout the last 27 years, since the launch of Healthy Cities in Europe, how ideas such as equity, community participation and creating physical and social environments supportive to health found fertile ground in such a diverse range of cities. Healthy Cities unleashed the creative powers of cities in unprecedented ways. Capturing this unique experience was always a challenge. The same applies to making the case that 'Healthy Cities' *works*. Scientific controversy about evaluating interventions of the type of Healthy Cities has been going on for many years.

We in the European Region of WHO made Healthy City evaluation a priority. It was never to be an easy or comprehensive exercise because of the lack of adequate resources, but nevertheless each of the 5-year phases of the Healthy Cities movement was concluded with an evaluation which provided a wealth of information and became the source of key guidance and know-how products. I should like to congratulate the editors of this amazing publication, Evelyne de Leeuw and Jean Simos. This book brings together the knowledge and inspiration from Healthy Cities from around the globe. Healthy Cities embedded our hopes and dreams for a new public health movement that was inspired by Health for All and the Ottawa Charter. Healthy Cities works and I am sure you will feel the same reading this timely and rich publication.

Agis D. Tsouros
Institute for Global Health Innovation
Imperial College, London, UK

Acknowledgements

Recognizing and honouring people that have had a role in the development and publication of a work like this easily finds itself wading through the syrup of cliché and platitude. To paraphrase the ominous Agricola Tacitus and US President John F. Kennedy, 'Success has a hundred fathers, and failure is an orphan' (Simpson and Speake 2008a). Another American political leader, Hillary Clinton, used a Nigerian (Igbo and Yoruba) proverb to indicate that 'It takes a whole village to bring up a child' (Simpson and Speake 2008b).

So let's stick into it.

In a long lineage of ancestors, Healthy Cities has one grandfather: Leonard (Len) Duhl. Without his wisdom, kindness and role as the proverbial maven (from the Hebrew *mevin* (מבין), meaning 'one who understands', and related to the *binah* (בינה), which denotes understanding or wisdom in general), Evelyne would never have found herself on the Healthy Cities path at all. When she was studying at the University of California at Berkeley, her academic advisor, John Ratcliffe, advised her to explore her strange, European, 'out-of-the-box' ideas about public health with Len, who happily took her in to his classes and potluck dinner at his exotic and eclectic Berkeley Hills home—and this is where she met characters like Réal Lacombe and Trevor Hancock. Len—deep gratitude.

Trevor, with John Ashton, can legitimately be labelled one of the hundred fathers of Healthy Cities, although they both happened to have a few Uncles and Aunties (in the Indigenous sense—respected elders in the public health community) who enabled the development of what is now a global phenomenon. Two regional directors of the European office, Leo Kaprio and Jo Asvall, and a director general of WHO, Halfdan Mahler, recognized the sign of the times and were willing to allow people like Trevor, John, Kathryn Dean and Keith Barnard to explore the limits of what was acceptable. We really feel small on the shoulders of these giants.

The real catalyst was Auntie/Mummy Ilona Kickbusch. Ilona—you skillfully and elegantly pushed, and continue to push, the edge of reason. When Evelyne, in 1984, wanted to visit the Copenhagen office you started her on a journey through health and political science. This involved stops in Ottawa and other global destinations, but you convinced Evelyne that local health was the thing to pursue. You remain the best.

Trevor Hancock is the relentless visionary without compromise. With such an infallible moral compass, it is no wonder that many people, with Evelyne and Jean, appreciate him as a thought leader. This leadership does not limit itself to Healthy Cities. Through his futurism, green credentials (Trevor co-founded the Canadian Greens) and deep concern for individual and planetary well-being, he has touched the lives of thousands if not millions—of many species. Trevor—thanks.

Ron Draper, another brilliant Canadian, brought inspiration and wisdom to the management of Healthy Cities. Without his first creative and careful research and development nudges, the project would never have become a movement. We are sure he is somewhere on the Mount Olympus of health looking down on us and smiling his wry grin. Ron—you were unique.

Michel O'Neill really started this book, and both Jean and Evelyne are happy to see that retirement for Michel has meant greater productivity and happiness. His spirit pervades this book. We wanted to make him co-author of virtually every chapter, but he gracefully declined. Merci beaucoup Michel, et espérons que Springer va bientôt publier ce livre en français!

Réal Lacombe left us too soon, in early 2016. He quietly passed away, suddenly, from standing right in the middle of life. That life truly was shared with many Canadians in the Réseau québécois des villes et villages en santé; Jean and Evelyne consider themselves honorary Québécois and we grieve with the family.

Evelyne and Jean would like to thank the warm, happy and welcoming team in Copenhagen, including Jill Farrington, Charles Price, Premila Webster, Erlinda Petersen (who also left us too early), Bente Drachmann and of course the steadfast Connie Petersen.

Evelyne would like to thank generations of students and colleagues who helped pursue her visions of Healthy Cities evidence, including Marco Strik, Marleen Goumans, Loes Polman, Gretha van der West, Kristel Logghe, Sonja Danen, Angeli Blankers, Esmée Kolthof, Mathil Sanders, Ellis Abbema, Julie Salgado, Shumei Wang, Tim Milewa, Marjan Hoeijmakers, Ellis Abbema, Matt Commers, Thomas Skovgaard, Lucy Spanswick and Nicola Palmer.

Agis Tsouros was the European Healthy Cities force to be reckoned with for nearly three decades. At some personal risk to career, sanity and bliss he fought repeatedly for the survival of the project in a highly charged European political environment. Well done, Agis, we owe you!

Jean thanks the Francophone Network (in particular Annette Sabouraud, Christiane Gosset, Martine Bantuelle, late Eric Giroult, Pascal Thébault, late Réal Lacombe, Louis Poirier, Julie Levesque, Denis Lapointe, Michel O'Neill, Charlotte Marchandise-Franquet, Patricia Saraux-Salaün, Natacha Litzistorf, Souhail Latrèche, Alexandre Bédat, Abdelhamid Haroun), his Swiss team for their ongoing support (in particular Derek Christie, Nicola Cantoreggi and Thierno Diallo), colleagues and friends from Greece (in particular Daizy Papathanassopoulou and Sotiris Papaspyropoulos) and Hawa Senkoro, from WHO-Libreville.

Of course this book could not have happened without the dozens of authors and co-authors. We thank you, in spite of the nagging and whinnying about deadlines—in both directions!

We are all most grateful to political leaders of Healthy Cities around the world (including the late Slobodan Lang—Zagreb; Antonio de Blasio—Pecs; and Nicolo Gianotti—Milan, who moved from Healthy City practice into politics, back into practice and activism, and beyond their city into the glocal) who have supported and challenged us.

Margaret Johnson, the Book Doctor—we think your surgical technique may have drawn considerable blood, but here's a patient that turned out unrecognizably healthy! Thanks for your relentless criticism and support.

Thanks also to Janet Kim and Khristine Queja of Springer who have been (too) understanding of the many delays and pitfalls in putting this book together.

And finally a great big debt of gratitude to the lovely and loving women in our lives, Lynne and Béatrice. You put up with a lot of our silliness, and we can only hope that the friendships we made will last and grow.

Evelyne de Leeuw
Jean Simos

References

Simpson, J., & Speake, J. (2008a). *The Oxford dictionary of proverbs* (5th ed.). Retrieved from http://www.oxfordreference.com/oso/viewentry/10.1093$002facref$002f9780199539536.001.0001$002facref-9780199539536-e-2160#.

Simpson, J., & Speake, J. (2008b). *The Oxford dictionary of proverbs* (5th ed.) Retrieved from http://www.oxfordreference.com/view/10.1093/acref/9780199539536.001.0001/acref-9780199539536-e-2379?rskey=S02J4S&result=1.

Contents

Contributors

Ama de-Graft Aikins is Professor of Social Psychology at the Regional Institute for Population Studies, University of Ghana (UG). Her research focuses on experiences and representations of chronic physical and mental illnesses and on Africa's chronic non-communicable disease (NCD) burden. Her current collaborative projects—with colleagues from UG, University of Amsterdam, New York University and London School of Hygiene and Tropical Medicine—include diabetes and obesity perceptions among Ghanaians in Ghana and Europe, food beliefs and practices among Ghanaians in Ghana and the US, child and adult mental health care, and community-based cardiovascular disease (CVD) and mental health interventions in rural and urban Ghana. She is involved in NCD advocacy and has consulted for health organizations including the Ghana Health Service (GHS) and the World Health Organization (WHO).

Mohammad Assai is medical doctor M.B.B.S. with M.P.H. degree from Dundee University of Scotland, and currently holds the position of Coordinator Integrated Service Delivery in the WHO—Eastern Mediterranean region since 2012. He held the position of Regional Adviser of Community Based Initiatives from 2004 to 2012 where he gained a lot of experience in community organization and mobilization for health and social actions. He played a substantial role in promoting Healthy City Programme in the Eastern Mediterranean Region and worked closely with WHO Center of Health Development in Kobe (Japan) in designing and implementing Urban Health Equity Assessment and Response Tool (Urban HEART). He worked in Pakistan as Medical Officer for Primary Health Care from 2000 to 2004. Before joining WHO, he had a position of Director General, Primary Health Care, in his country (Iran). He is the author of many papers and also co-author of a WONCA book on *The Contribution of Family Medicine to Improving Health System.*

Ai Chaobang John Ashton is a writer and blogger who holds an M.Sc. in Sustainability, Development and Peace from the United Nations University—Institute of Sustainability and Peace (UNU-ISP) in Tokyo, Japan. He is the author of the Chaobang's Travels blog at http://www.aichaobang.blogspot.com as well as a Master's thesis on sexual diversity politics at the United Nations. His research and writing interests are broad and extensive, with a frequent focus on history, gender and alienation among other socio-political and cultural concerns.

Clem Bezold is Chairman and Senior Futurist of the Institute for Alternative Futures (IAF). He is a leading developer of foresight techniques, particularly scenarios. He has worked with the largest government agencies, corporations and non-profit organizations to enhance their foresight. At IAF and through its work he has evolved a particular approach—Aspirational Futures—that considers likely challenging and visionary scenarios and aids organizations to better create the future they prefer. In the public sector he has lead IAF's futures work with several US government agencies (GAO, GSA, Defense, Justice, the VA, DHHS, CDC, and NIH). He has consulted with the World Health Organization in Geneva and its European Office (WHO/EURO) as well as the Pan American Health Organization (PAHO) on enhancing their ability to use foresight in their operations, particularly in support of WHO's vision of Health for all.

Antonio de Blasio is the Coordinator of the Hungarian national network of Healthy Cities (founded in 1992), and former Project Coordinator of the Pécs Healthy Cities Project for 25 years. Originally a teacher of English, he was involved in the Healthy Cities Project as an interpreter in Pécs during the preparation of the project. He became Project Coordinator in Pécs when the City of Pécs joined the WHO European Healthy Cities Project as a founding member in 1987. He has been WHO temporary advisor in several occasions on various topics and in a number of missions and trainings. He was an elected member of the local government (1990–1994, 1998–2006), the county government (1994–1998), and the European Parliament (2006–2008). His main research areas are health impact assessment in the local decision-making system, health development planning and health communication.

Lucero Rodriguez is a surgeon, graduated from UNAM. She has a Master of Clinical Sciences in Family Medicine from the University of Western Ontario, Canada. She was a University professor. She has held numerous positions related to health promotion and education in the Mexico Ministry of Health. Currently she is the Director of Healthy Settings Strategies and Development in Mexico's General Directorate of Health. She has published in numerous national and international journals as well as developed many educational materials.

Karina Cimmino is graduated in Education Sciences with a specialization in Health Education and a master's degree in Public Health. She has been working in health promotion field for more than 20 years, designing, developing and evaluating health promotion intervention programmes (working for regional, national, provincial and local levels); teaching and training professionals, doing research; and producing educational materials. Moreover, she has been working in health promotion in several countries such as Argentina, the United States, Italy, the Philippines and the Latin American Region. She works as independent consultant in Health Promotion for international and national organizations. She is a member of the Program of Social Sciences and Health at FLACSO (Latin American School of Social Sciences) and works on different projects to promote the rights of children and women in the health system. She coordinates the external evaluation team of The National Healthy Communities and Municipalities Project in Argentina (PAHO Argentina—National Ministry of Health).

Ana Rivière-Cinnamond A public health expert with more than 14 years of experience, Ana Rivière-Cinnamond holds an M.Sc. from the LSE and a Ph.D. from the LSHTM. She has worked for the United Nations since 2002 (FAO and WHO) in different regions of the world, focusing mainly on a systemic-transdisciplinary approach to the prevention and progressive control of communicable, emerging, zoonotic and transboundary diseases such as H5N1, H1N1, plague, dengue, cholera, FMD (etc.), and on funding strategies for public health care service delivery, insurance mechanisms, compensation schemes, and conducting socio-economic impact evaluations using quantitative and qualitative methods. She has provided in-country technical assistance to more than 24 countries across four different regions of the world and collaborated with several international organizations, including the WBG, OIE, the ILRI and IFPRI. She is currently the Regional EcoHealth Advisor for the Pan American Health Organization in Lima, Peru.

Carole Clavier (Ph.D. in political science, Université de Rennes 1—France) is an assistant professor in the Department of Political Science at Université du Québec à Montréal (Canada). Her research interests are the study of urban and regional public policies in relation to public health issues—or healthy public policy; the contribution of the policy studies to health promotion research; the practices of actors involved in intersectoral programmes and policies; and comparative policies. Carole currently leads, together with France Gagnon, a research on active transportation policies in Montreal and Toronto (CIHR—Heart and Stroke Foundation, 2014–2016). She recently co-edited the book *Health Promotion and the Policy Process* together with Evelyne de Leeuw (Oxford University Press, 2013).

Jason Corburn is associate professor in the School of Public Health and Department of City and Regional Planning, and Director of the Institute of Urban and Regional Development and the Center for Global Healthy Cites at the University of California, Berkeley. Corburn's research explores how urban poverty influences population health, the relationships between urban places, stress and health inequalities, urban climate change, and the health impacts of inclusive community development for informal settlements in cities of Africa, Asia and Latin America. He is a leading practitioner of urban health impact assessment (HIA), community-based 'citizen science' and slum upgrading. He is the author of award winning books including *Street Science: Community Knowledge and Environmental Health Justice* (MIT Press, 2005); *Toward the Healthy City* (MIT Press, 2009); *Healthy City Planning* (Routledge, 2013) and the editor of *Healthy Cities* (Routledge, 2015) and *Slum Health: From the Cell to the Street* (UC Press, 2016).

Samar ElFeky is a medical doctor qualified in Reproductive, Maternal and Child Health from Cairo University and learned Public Health in Liverpool University. She got many training certificates on maternal and child health, reproductive health, quality of health care services, accreditation and health system management from national and international universities. She has experience working in WHO/EMRO in Primary and Community Health Care and WHO country office of Somalia in Health System and Maternal and Child Health Care. She participated in introducing Healthy City Programme in some countries of the region including Lebanon, Jordan, Kuwait, Sharjah and Sudan. In addition, she participated in implementing Urban Health Equity Assessment and Response Tool (Urban HEART) in Egypt, Morocco, Jordan, Sudan and Tunisia. She has many publications in the field of planning effective communication strategies, community health workers, home health care for elderly and community-based disaster risk reduction. Currently she works as Technical Officer Research Policy and Development (RPD) in Regional Office of Eastern Mediterranean Region (EMRO).

Paola Pliego is a Communication Specialist with experience in Social Marketing, Social Media and Health Promotion. She worked at the Ministry of Health of Mexico at the General Directorate of Health Promotion, participating in several projects with the aim of promoting healthy lifestyles and wellness. She collaborated in the training of health personnel on the following topics: health marketing, social media, and risk communication. She worked with different associations to promote healthy lifestyles by developing programmes, events, activities, campaigns and methods to inform about benefits of having healthy habits, regular check-up, and timely diagnostics in order to prevent diseases. Paola is co-author of the 'Manual for the use of social media in health emergencies'. Currently she is working as Communication Director at Federal Court for Fiscal and Administrative Justice.

Cris Franceschini is the Executive Director of the Center for Studies, Research and Documentation on Healthy Cities and Municipalities (CEPEDOC), a research centre linked to the University of Sao Paulo, Brazil. She has a B.A. in anthropology from the University of Maryland and a Master's in international health from Johns Hopkins University. She is currently a doctoral student at the School of Public Health of the University of Sao Paulo. Her work experience includes working for the Pan American Health Organization and the World Bank.

Ronald Gould is a M.R.Pharm.S., Diploma Mental Health, member of the Pharmaceutical Society of Great Britain, of the General Pharmaceutical Council (2011–2016), and advisor to the Local Pharmaceutical Committee. He was a

Community and Mental Health Pharmacist to the former Liverpool Primary Care Trust (2001–2013, now the Liverpool Clinical Care Commissioning Group). Lord Mayor of Liverpool in 2003–2004, he has been a Honorary Alderman of the City of Liverpool, a Cabinet Member, Liverpool City Council, responsible for Adult and Social Care and Political Leader with special responsibility for WHO and Healthy Cities (1998–2008), member of the information panel of the National Institute of Clinical Health and Excellence issuing guidance on a huge variety of topics from Dementia to the use of Statins and forming legislation on Smoking in Public Places. He introduced the idea of Smoke Free Liverpool (2004) and was the lead politician for this groundbreaking public health initiative which altered the landscape of Public Health in the United Kingdom forever (2008).

Geoff Green is professor emeritus of urban policy at the Centre for Health and Social Care Research, Sheffield Hallam University, in the UK and chair of the WHO Strategic Advisory Group on Global Age-Friendly Cities and Communities. He has coordinated evaluation of the WHO European Healthy Cities Network and advises WHO on Age-Friendly cities and city health development planning. His professional life is at the interface of local government and academia, working strategically in the Chief Executive Departments of four cities of the United Kingdom and for the universities of Sheffield, Oxford, Birmingham and now Sheffield Hallam. He coordinated 20 university research programmes on the 'upstream' determinants of public health for local governments, health authorities and the UK Government. This primary research has now been supplanted by modelling impacts using a realist synthesis methodology.

Trevor Hancock is a public health physician and health promotion consultant and a Professor of Public Health and Senior Scholar at the University of Victoria. As one of the pioneers of the now-global Healthy Cities and Communities movement, he co-authored the original background paper for WHO Europe's Healthy Cities initiative with Prof. Len Duhl in 1986. Over the past 30 years, he has worked as a consultant on Healthy Cities/Communities for local communities, municipal, provincial and national governments, health care organizations, NGOs and the World Health Organization. In recent years he has been a member of the Knowledge Network on Urban Settings (part of the WHO Commission on the Social Determinants of Health), a member of the Advisory Board for the Urban HEART project of the WHO Kobe Centre for Health and Development, and a member of the Global Research Network on Urban Health Equity. He was awarded Honorary Fellowship in the UK's Faculty of Public Health in 2015.

Abdelhamid Haroun holds the position of Deputy Mayor of the city of El Khroub and he is currently leading the planning and management of the territory of the city. He is a member of the scientific committee of the Community Health Centre, of the Hospital Board of Directors of El Khroub and of S2D (Rennes). After getting the general practitioner's degree in public health in 1987, he made a professional career into the care sector. He had several training courses mainly related to the environment and city management and participated as director of the Community Health Centre to promote health and environment programmes and the fight against waterborne diseases. In 2003, he led the Departmental Directorate of Health TIPASA, while participating in various seminars on environmental health at the University of Constantine. He has several participations in the international level, among others at the International Congress of Healthy Cities in Athens (1998), and as representative of Algeria to the WHO Africa 52nd session in Brazzaville (2001).

Claire Henderson-Wilson is a Senior Lecturer in health promotion and ecological health within the School of Health and Social Development at Deakin University, Melbourne, Australia. Claire has a background in family health and consumer studies, and prior to commencing her Ph.D. (which investigated the links between nature access and health for inner city high-rise residents), she worked in an Adult Training and Support Service for adults with a mild to moderate intellectually disability. She has 12 years' experience in people-environment research using mixed methods. Her research investigates links between housing and health, families and health and the human health benefits of contact with nature. Claire is a key member of the 'Health Nature and Sustainability Research Group', located within the School of Health and Social Development, which investigates links between health, nature contact, and sustainability issues. She is also a member of several professional organizations including Ecohealth and the International Association for People-Environment Studies and is author of numerous publications.

Gwyn Jolley is a (now retired) researcher with over 20 years' experience in research and evaluation in public health, community health and health promotion. Gwyn was employed as a senior researcher at Flinders University, South Australia, and was a chief investigator on a number of grant-funded research projects. Gwyn has many years' association with Healthy Cities, being past chair of Healthy Cities Onkaparinga and currently deputy chair. She has facilitated workshops and presented at Australian and international conferences, including the Global Alliance for Healthy Cities. Gwyn is the co-convenor of the Primary Health Care Special Interest Group of the Public Health Association of Australia. Gwyn's research interests include community empowerment, healthy eating, and qualitative and mixed methods evaluation. She has published extensively in peer-reviewed journals and produced numerous reports and resources. Gwyn has a master's degree in primary health care and completed her Ph.D. in 2013.

Irene Karanja is the former Executive Director of Muungano Support Trust (MuST), a Kenyan NGO that acted as the technical arm of the Kenyan slum dwellers federation, Muungano wa Wanavijiji. Before Directing MuST, I. Karanja established the Research and Advocacy unit of the Pamoja Trust, a community development organization in Kenya. She is a specialist in participatory research, community organizing and capacity building of the urban poor. She has experience working in Kenya, Ghana, Liberia, Malawi, Uganda, South Africa and Tanzania.

Jean de Dieu Konongo has a Public Health degree from the University Marien Ngouabi (Brazzaville—Congo), a postgraduate certificate in statistical epidemiological and applied methods in Public Health from the National School of Public Health of the Free University of Brussels (Belgium), and a specialization certificate of génie sanitaire de l'Ecole Inter-Etats des Ingénieurs d'Equipement Rural (Ouagadougou, Burkina Faso) and from the Ecole Polytechnique Fédérale of Lausanne (Federal Institute of Technology—Switzerland). He is a sanitary engineer, epidemiologist, public health expert, former employee of the Ministry of Health of Congo and of WHO, with high responsibilities within these institutions, and national and international consultant (Congo, RDC, Guinea-Conakry and Burundi). He is author or co-author of several scientific publications. He contributed to set up Healthy Cities programmes in Brazzaville and Pointe Noire and the specific programmes 'Healthy Markets' and 'Healthy Schools'.

Réal Lacombe started his medical practice in 1973 as a family physician in Abitibi-Témiscamingue, a remote and underserved region of Quebec (Canada). Six years later he became its first Regional Public Health Director. He had to deal with PH issues mainly related to dominant mining and lumber industries, major environmental pollution, poor worker's health and safety conditions but also their impacts on the health of communities. After a MPH at UC Berkeley in 1985, he came back home and in 1987 he started the first official Healthy City Project in America, in Rouyn-Noranda, his hometown. Then he moved to Quebec City to set up what is still the Quebec Network of Healthy Cities and Towns. Over his 13 years in Quebec City, he was involved actively in a large number of activities at the national and international level, mostly at the request of WHO or other HC networks. In 2001, he returned to his region to resume his PH Regional Directorship function.

Ljubica Latinovic is a native of Belgrade, Serbia, who has lived in Mexico since 1999, and a medical doctor who graduated from the Faculty of Medicine, University of Belgrade, Serbia. She has a Master's in Management of Health Organizations from the University la Salle, Mexico City, with several courses and training in public health, international health and diplomacy, and risk communication. For the last 10 years, she worked in various positions in the Ministry of Health's General Directorate of Health Promotion as Assistant Director for Social Marketing, Healthy Environments and Capacity Building of Health Personnel. She has worked on international projects and as a member of various international working groups on risk communication as well as social determinants of health and public policy. She has been a health promotion professor at the National Institute of Public Health, and a speaker at various global conferences as well as facilitator of the workshops on social marketing, risk communication and social determinants of health for the public and private sectors. She has published articles and book chapters relevant to health promotion issues.

Evelyne de Leeuw has worked and studied in universities in the Netherlands, the USA, El Salvador, Canada, Denmark and Australia. She holds degrees from Maastricht University and the University of California at Berkeley. Her Ph.D. (1989) investigated the feasibility of developing and implementing Healthy Public Policy at the national level, and analysing that this is virtually impossible she found that whole-of-government health (policy) approaches would be more easily achieved at the local level—hence a commitment to Healthy Cities since the mid-1980s. Her quest for the development and application of health political science was sidelined routinely—by being elected Secretary General of the Association of Schools of Public Health in the European Region (ASPHER) (1992–1998), Editor-in-Chief of Health Promotion International (since 2010), foundation director of health promotion research and public health teaching institutions (Syddansk Universitet, 2000–2005, Deakin University 2008–2013), and recreational writing outings. Currently she enjoys being professor and director of the Centre for Health Equity Training Research and Evaluation (CHETRE), based at the University of New South Wales, South Western Sydney Local Health District, and the Ingham Institute.

Vivian Lin has more than 30 years' experience in public health, with a variety of roles in policy and programme development, health services planning, research and teaching. She is Professor of Public Health at La Trobe University, in Melbourne, and currently on leave to serve as Director of Health Systems for the World Health Organization in the Western Pacific Regional Office. She is the author of several leading textbooks in Australia as well as in China. Vivian has also worked at senior levels in health policy in several Australian jurisdictions. She has also consulted widely for the World Bank, UK Department for International Development, AusAID as well as the WHO and served on multiple academic, government and community boards.

Jack Makau is the Director of Shack/Slum Dwellers International (SDI)—Kenya. He was previously research and communications director of Muungano Support Trust and Pamoja Trust, leading NGOs in Kenya that support the urban poor to reclaim their rights and negotiate for improved living conditions, among other issues. Makau is a global expert in slum upgrading, designing planning and policy solutions that prioritize the urban poor and working to ensure the voices of youth are prioritized in African city development. Mr. Makau is based in Nairobi, Kenya, but works with SDI affiliates across the world.

Erik Martin is a public health academic with experience in researching tobacco control, policy, global health and healthy settings. His doctoral degree explored the implementation of the Framework Convention on Tobacco Control, the first World Health Organization (WHO) treaty, in the Pacific Islands. This utilized a qualitative and multidisciplinary approach drawing from health promotion and political science. Erik has undertaken consulting work for the WHO's Healthy Islands recognition scheme since 2012 and has been teaching public health to medical students at Deakin University's School of Medicine in Geelong, Australia, since 2010. Erik is the author of numerous publications in peer-reviewed journals and has presented at numerous conferences on tobacco control, health promotion and medical education. He also holds a B.AppSci. (Health Promotion).

Sandra McCarthy is an Honorary Senior Research Fellow at the centre for Health Initiatives (CHI) at the University of Wollongong. Sandra is a global practitioner in local government, public health and community development. Sandra has post-graduate qualifications in Education and a Master of Public Health. Sandra is an Emeritus Mayor, International Leading Pioneer Expert AFHC (WHO) and Chair of the Australian chapter of Healthy Cities. In addition to her vast community planning expertise, as Chair of the Australian chapter of Healthy Cities she has been successful in merging the policy efforts of local government and communities and translating this knowledge in advising and mentoring many international cities with health planning.

Raul Mercer is a paediatrician and public health practitioner with a master's degree from the University of Wisconsin at Madison. He worked in the areas of Maternal and Child Health in Argentina and as a consultant for PAHO. Since 2000, Raul has moved into the areas of women's health, child development, rights and gender. He is the Coordinator of the Program of Social Sciences and Health at FLACSO (Latin American School of Social Sciences) and Academic Coordinator of the 'CIPS' (International Course of Health Promotion). Raul is a researcher at CISAP (Center for Research in Population Health) located at the Durand Hospital in Buenos Aires City. FLACSO promotes the rights of children and women in the health system through the 'Te Escucho' (I hear you) project. Raul is involved with professional organizations dealing with child social issues on health and development, like the International Society of Social Pediatrics (ISSOP), the Committee of Social Pediatrics of ALAPE (Latin American Association of Pediatrics) and ACEI (Association of Child Education International). He is the author of numerous publications and contributions to books.

Nastaran Keshavarz Mohammadi is an associate professor in health promotion at Shahid Beheshti University of Medical Sciences (Iran). She has about 20 years of experience in teaching and research in Iran, Australia and for a short period of time in WHO-Geneva. Her competencies are in public health, complexity science, health promoting settings, empowerment and community-based programmes and community participation. Her Ph.D. at the University of Sydney on application of complexity science in health promotion practice opened a new approach in health promotion. She has served as temporary adviser to WHO in several occasions, have been commissioned research by WHO/HQ and WHO office in Tehran on settings health promotion and is a national adviser of health promotion to the ministry of health of Iran. She is the author of several papers, and books including *Complexity Science, Schools and Health*. She is also an editorial board member of *Health Promotion International* Journal.

Zsuzsanna Nagy has been involved in the WHO European Healthy Cities Project since 1999 as the project assistant of the Healthy City Foundation of Pécs (Hungary) and the Hungarian Speaking Association of Healthy Cities. She is the Project

Coordinator of the Healthy Cities Project in Pécs since 2014. She has participated in various health development projects on different themes—on the European, national and local level as well, e.g. childhood obesity, health impact assessment, occupational health, city health development planning and physical activity programmes for elderly people. Her main research areas are health development planning and effects of television on the eating habits of children.

Françoise Belemel Naissem is graduated from the Faculty of Applied Sciences, University of N'Djamena. She has 11 years of experience as sanitary engineer at the Ministry of health in Chad and she is the local focal point for Wash, WHO project against cholera. She is a health promotion trainer through environmental health and ATPC (Assainissement Piloté par la Communauté) and was also trained to PBF in Cotonou, Benin, 2015. From 2013 to 2015 she was the coordinator of 'Safety for street food' (Salubrité des aliments de la rue), a project funded by FAO. Since 2010, she is the head of the Environmental Hygiene and Sanitation division in the Ministry of Health (Chad).

Jonas Naissem was born in 1969 in N'Djamena (Chad) and is a graduate in modern literature of the University of Chad (now University of N'Djamena) and in library science of the School of Librarians, Archivists and documentalists (EBAD), University Cheikh Anta Diop, Dakar (Senegal). He has also a diploma in computer sciences and communication of the University of N'Djamena. He is for 16 years in charge of health promotion, communication and advocacy at the local WHO office in Chad, where he is also the Focal Point for Environmental Health. He advises for the last 10 years the Ministry of Health in Chad on topics such as environmental health, Healthy Cities, WASH/cholera prevention, food safety, communication and social mobilization, community involvement, and struggle against Ebola virus.

Keiko Nakamura is a public health scholar with over 25 years of experiences with Healthy Cities in the Asia-Pacific region. Dr. Nakamura is Head of the Department of Global Health Entrepreneurship, Tokyo Medical and Dental University (TMDU). She has researched health and environment in urban areas, with particular focus on determinants of health, healthy cities and urban policy, use of information technology for health promotion, trades in health services, and cooperation for health development. She has worked with the WHO Healthy Cities Programs since the early 1991 and is strongly committed to the Alliance for Healthy Cities, an international network working closely with WHO. She is in charge of TMDU's Disease Prevention Global Leader Program, where young professionals from public and private institutions worldwide become leaders in global health policy and community health. She has published extensively in peer-reviewed journals and holds an M.D. and a Ph.D. in Medicine.

Tyler Norris is an entrepreneur and founder of over a dozen social ventures. His three decades of service in the public, private and non-profit sectors have focused on population health, community vitality, and equitable prosperity. Currently, he serves as Vice President, Total Health Partnerships at Kaiser Permanente, where he helps lead the implementation of 'anchor institution' strategies, applying all KP assets to measurably improve population health and community well-being. Tyler is a trustee of Naropa University and serves on advisory bodies for the Convergence Partnership, Transportation for America, Active Living by Design, and Samueli Institute—and is active with Social Venture Network, BALLE and SOCAP. As an early leader in the US healthy/sustainable communities movement, Tyler has worked in over 400 communities and with scores of organizations in the United States and globally. Previously he served as founding board chair of IP3, a technology venture that powers www.CommunityCommons.org and GIS data mapping platforms for health philanthropies and governmental agencies. He was founding CEO of Community Initiatives Inc.; convener of Advancing the Movement; founding co-chair of the W.K. Kellogg Foundation's National Leadership Alliance; and head coach of the YMCA's Pioneering Healthier Communities initiative. He is an avid mountain biker, backcountry skier and pilot, and now resides in Oakland, CA. In January 2017 Tyler became the first Chief Executive of the new Institute for Mental Health and Wellness in the USA.

Michel O'Neill is Professor Emeritus of the Université Laval Faculty of Nursing, in Quebec City, Canada, where he was previously on the faculty for 30 years up to September 2011. Since then he offers his expertise as trainer and consultant and has also developed a keen interest about the sociology of religious phenomena on which he is currently writing books. Michel, who received his Ph.D. in Sociology from Boston University, has worked in different capacities in the field of community/public/population health since the beginning of the 1970s: researcher in a public health regional agency, teacher, academic researcher, consultant and activist. His fields of interest are mainly the historical and political dimensions of public health and health promotion, notably around the Healthy Cities movement in Quebec and globally. He has published extensively on these topics, on which he has also given courses and presentations all over the world.

Dionysia Papathanasopoulou is a Sociologist and Social Worker. She is the National Coordinator of Hellenic Healthy Cities Network and since 2013 Chair of the Advisory Committee of WHO Europe for the Programme's National Networks. From 1982 to 2010, Dasy worked in the City of Athens as head of different departments in charge with the Social Sector. She was involved in the Healthy Cities Programme since 1993 and was a founder member of the Programme in Athens and its first Coordinator, from 1993 to 2010. She was also founder member of the foundation of Hellenic Healthy Cities Network and member of its Board of Directors from 1994 to 2010. She has made numerous missions as a WHO temporary advisor.

Ann Pederson became the Director of Population Health Promotion at BC Women's Hospital + Health Centre in Vancouver, Canada, in 2014 after 18 years at the BC Centre of Excellence for Women's Health, most recently as Co-Director. Ann's career in health promotion started in traffic safety education but has focused on women's health, gender and health, and health care reform for two decades. Ann recently completed a doctorate exploring the potential of health promotion interventions to support gender equity at the University of British Columbia, having completed a Master of Science in Community Health at the University of Toronto in 1989. She serves on the Board of the Canadian Public Health Association and reviews for several journals, including *Health Promotion International*. Ann recently co-edited *Making it Better: Gender-Transformative Health Promotion* and *Women's Health: Intersections of Policy, Research, and Practice*. She is currently working with Irv Rootman and Katherine Frohlich on a fourth edition of *Health Promotion in Canada*.

Fran Perkins is a health promotion and public health specialist with 25 years of experience in the broader public health field. In the 1980s she was part of the team that developed and implemented a new conceptual approach to innovative Public Health change in Toronto and the genesis of the Healthy City movement. She co-developed the 'Mandala of Health' with Trevor Hancock. This conceptual model had become a basic in many educational settings and texts. Fran was Executive Director of The City of Toronto Healthy City Office for 5 years. This group worked across all City Department to address the determinants of health for the citizens of Toronto. At the International Unit at the Centre for Health Promotion, University of Toronto, she worked in 11 counties concentrating on programme development and education in health promotion. Presently she concentrates on advocating for and implementing the Housing First model in Toronto.

Gloria Libia Polanía Aguillón is a dentist, specialized in Social Security Health Institutions Management and First Level Health Care Management. She has experience in middle and general management of health services, specifically in State Social Enterprise, and has been responsible for clinical management and local public health services, with social and community participation. This has required creating a balance between the management of human talent, the provision of health services with high quality standards, networking services delivery, inter-agency coordination, social responsibility and financial sustainability. The main results she has achieved include maintaining the accreditation of the Pablo VI health institution, having the Hospital classified as a low fiscal and financial risk institution in 2012 and as an institution without risk in 2014, and growing the offices and health services for the community of Bosa and the city of Bogota D.C.

Marilyn Rice is Chief Executive Officer (CEO) of merci (Marilyn E Rice Consulting International), an international consulting firm that advises on promoting equity in health and development. Until June 2011 she was the Senior Advisor in Health Promotion and Coordinator of the Urban Health and Determinants of Health Team for the Pan American Health Organization (PAHO/WHO). With over 40 years of experience working in multi-cultural, multi-lingual environments, she is known globally for her work in community development, social mobilization, health promotion and education, participatory research and evaluation, and the emerging field of urban health. Serving on boards of numerous international, non-profit and professional organizations, as well as on editorial boards of numerous peer review journals, she has published extensively in her fields of expertise and about her experiences in countries throughout the world. She continues to conduct workshops, courses and conferences in any of four languages (English, French, Spanish and Portuguese). She holds degrees in Sociology, English, Public Health and Health Education.

Jose Osney Velandia Rodriguez is a philosopher and is currently a master's degree candidate in Bioethics. He has 6 years of experience in the management of public health projects for the Hospital Pablo VI Bosa, which is part of Bogotá's District Health Network. During this time, he worked as the focal point for health policies management. Later on, he coordinated the Health Equity Analysis Project, which implemented the Urban HEART (Urban Health Equity Assessment and Response Tool) methodology, and he was in charge of leading an urban health network. He participated in an expert consultation, organized by the WHO Centre for Health Development in Kobe, Japan, regarding the Urban HEART methodology. He contributed to the Local Health Diagnostic publication through the social participation of the seventh district of Bosa between 2011 and 2012.

Irving Rootman has worked in the field of Health Promotion for more than 35 years as a researcher, research manager, programme manager, professor, consultant, volunteer and event planner. He is currently an Adjunct Professor in the School of Public Health and Social Policy at the University of Victoria and a Visiting Professor in the Department of Gerontology at Simon Fraser University. He is an Honorary Lifetime member of the Canadian Public Health Association and the Ontario Public Health Association and Chair of the Steering Committee for the British Columbia Health Literacy Networks. He has many publications in Health Promotion, Health Literacy and Mental Health. One of his recent publications was the third edition of Health Promotion in Canada for which he was the managing editor and an author. He has a Ph.D. in sociology from Yale University (1970) and an Honorary Doctor of Laws degree from the University of Victoria (2014).

Ronice Franco de Sá is a researcher on Health and Development, Health Promotion and Healthy Cities. She has been working for several years in Northeast Brazil developing social technologies with local managers and people from the Pernambuco Network of Healthy Municipalities. A member of the Executive Board of the International Union for Health Promotion and Education (IUHPE), Dr. Ronice Franco de Sá is also a professor at the Federal University of Pernambuco in Qualitative Research, Health Promotion, Qualitative Evaluation, and Research Seminars. As a member of the Coordination Committee of the Thematic Group on Health Promotion and Sustainable Development of the Brazilian Association of Collective Health, she actively participated in the revision of the Brazilian Health Promotion National Policy. Author of several books and articles about this theme, she is also a transcendental meditator and scholar on consciousness, spirituality, and happiness.

Annette Sabouraud is a psychologist, with professional experience in a hospital environment (neurology, paediatrics) and community activities in the areas, such as creation of a daily centre for the elderly and the handicapped. Elected at the municipal level, she assumed from 1983 to 1995 the responsibility of the Deputy Mayor in charge of the Health of the City of Rennes, the first in France. Present at the Lisbon meeting and at the Ottawa Conference in 1986, she engaged her city in the WHO Healthy City Project. She initiated the creation of the French Healthy Cities Network, which became in 1990 an association that she presided over until 1996. She was also the initiator of the creation of the Francophone Healthy Cities Network and of the WHO Collaborating Centre for Francophone Healthy Cities, which became in 2000 the International Association for Health and Sustainable Development (S2D) and which she presided over until 2010.

Patricia Saraux-Salaün is a general practitioner, geriatrician, addictologist and tropical medicine expert. She has 20 years of medical practice in Brest with social public in great difficulty. She was also the Healthy City Project leader, Department of Public Health of the city of Brest (Brittany, France), for 12 years. In 2011 she moved to the City of Nantes and became the city's head of the Department of Public Health, Direction Générale Déléguée à la Cohésion Sociale et aux Territoires, and again in charge of the local Healthy City programme. Nantes is from the beginning a member of the French Network of Healthy Cities (since 2014, it assumes the vice-presidency of this national network) and since 2012 a member of the WHO European Network of Healthy Cities. She is also teaching at the University of Brest—Department of GP, Faculty of Medicine.

Jean Simos became, after his doctorate thesis, head of a research group on environmental assessment at the Federal Institute of Technology (EPFL), Lausanne, Switzerland. From 1992 to 2000 he worked as scientific advisor for the Directorate of Public Health, Geneva, Switzerland, and from 2000 to 2007 he became its Deputy Director, in charge of environmental health and health promotion. Since 2007, he is head of the research group on environmental health, Institute of Global Health, University of Geneva. He was involved in the WHO Healthy Cities European Network from 1994 to 2007 as a local coordinator and since 2009 as a WHO advisor. He is also president of the WHO Collaborating Centre for francophone Healthy Cities, Rennes, France. He took on several missions around the world for WHO, World Bank and Swiss Development & Cooperation Agency as an environmental health expert, particularly in link with health promotion, Healthy Cities, health impact assessment, and multiple criteria decision aid (ELECTRE methods). He is invited professor at the University of Liege, Belgium, since 2011, was invited lecturer in University Paris-Dauphine in 2015, and is a member of the High Council for Public Health, France, since 2011.

Anna Stevenson is a public health specialist leading the Health in All Policies team at Canterbury District Health Board, New Zealand. She graduated from Auckland Medical School, did her general practice training in the South Island and has completed postgraduate degrees in epidemiology and public health from the Universities of London (UK) and Dunedin (NZ). Anna has worked in clinical practice in rural and urban New Zealand and in several developing world countries. In 2006 she project managed the largest policy level health impact assessment to date in New Zealand on the Greater Christchurch Urban Development Strategy. Her work over the last decade has concentrated on implementing a Health in All Policies approach at local and regional government level in Canterbury. She has particularly focused on transport and urban planning as significant determinants of health—this experience has been useful in the recovery phases following a series of devastating earthquakes in the Canterbury region which began in late 2010 and have not fully settled at the time of writing.

Julia Taylor was one of the first non-medical students to graduate from the Medical Faculty in the University of Liverpool with a Master's in Public Health. She made a career in Private Dental Practice, Health Promotion and Public Health. She was appointed Director of Liverpool Healthy Cities initiative in 1993 until retirement in 2011. She is a holder of the Duncan Medal for the outstanding contribution to Public Health in Liverpool. She was chair of the first WHO Healthy Cities Advisory Committee, awarded Honorary Fellow of the Faculty of Public Health, presented work across Europe, at the UN, and keynote speaker at the first Healthy Cities Conference in China, Hong Kong and Macau. A Magistrate for 22 years, she assumed the Chair of Governors of a Special School for children with moderate learning difficulties, behavioural problems and autistic spectrum conditions for the past 28 years. She is continuing to keep fit dancing and organizing events for a local Walking Group.

Mardie Townsend is an Honorary Associate Professor in the School of Health and Social Development at Deakin University, having retired from full-time academic work in 2014. Coming from a background in environmental sociology, for the past 15 years Mardie has researched and taught in the health promotion/public health area, where she has led a programme of research exploring the human health/environment interface, focusing particularly on contact with nature as a source of health promotion and healing, and on the impacts of sustainability or its absence on human health. Mardie has received in excess of $1 million in research funding and has 66 manuscripts (26 first author) published or in press. She holds a B.Soc.Sci (with Distinction) and a Ph.D. in environmental sociology. Her 'Healthy Parks, Healthy People' research is widely recognized internationally, and her research into the motivations for and benefits of environmental volunteering has extended across several states of Australia and to the United Kingdom.

Nina Wallerstein Professor of Public Health; Director, Center for Participatory Research (cpr.unm.edu), University of New Mexico, has been developing community-based participatory research (CBPR) and empowerment/Freirian interventions for over 30 years. Among over a hundred publications are *Community-Based Participatory Research (CBPR) for Health*, 2nd edition, 2008, and *Problem-Posing at Work: A Popular Educator's Guide*. She has worked in North American and Latin American contexts, in participatory evaluation of healthy municipalities; in family, youth, and women's health intervention research; and with tribal partners to support culturally centred research in New Mexico and the USA. To improve the science of CBPR and reduce health inequities, she is researching best partnering practices associated with health outcomes and developing a partnership evaluation/reflection toolkit. She has collaboratively produced with Latin American colleagues a train-the-trainer Empowerment, Participatory Research and Health Promotion curriculum (initially sponsored by Pan American Health Organization), available in Spanish, Portuguese and English.

Helen Wilding is a policy and development practitioner and a scholar. Since 2009, she has been coordinating partnership working for well-being and health in Newcastle upon Tyne, UK—a post which incorporates the role of Healthy Cities Coordinator for this designated city and a lead role in the coordination of the UK Healthy Cities Network. Prior to 2009, she worked in the fields of drugs and informal carers, always with a focus on collaborative governance and facilitating cross-cutting organizational and practice changes. Alongside her paid work, Helen studies part-time. She holds an M.B.A. and an M.Sc. in Systems Thinking in Practice both awarded by The Open University, UK. She is currently studying for a Ph.D. in Public Health at Lancaster University, UK. Her research focus is local government policy work practice and its development, inspired by her interest in health in all policies at an urban level.

Part I
Healthy Cities in History

Chapter 1
Cities and Health from the Neolithic to the Anthropocene

Evelyne de Leeuw

Introduction: Urban and Human Development

Human settlements are pivotal for the human, social and ecological development of our planet. Initially, people started to band together for reasons such as a secure food supply, but safety and security also extended to protection against man-made and environmental threats, including tribal violence and joint efforts to deal with (and exploit) seasonal events like floods.

Early human settlements in Mesopotamia and Egypt were driven by agricultural development around rivers, but with the accumulation of human and natural capital these settlements could also engage in trade with other settlements and with people who led nomadic or semi-nomadic lifestyles. This led to specialization of human capacity and development, and the need for particular trades unique to the geographical uniqueness of the settlement (Diamond 1998). Trade led to the emergence of reliable travel routes that in turn had to be protected and supplied by further settlements, and cities started to develop around the confluence of natural features that allowed for the movement of people and goods (rivers, valleys, and harbours and roads).

With urbanization came institutionalization: implicit and explicit rules about the shape of cities and how to behave in them, as individuals and as communities. An early urban planning directive for the ruler's complex in the Assyrian city of Larsa suggested appropriate locations of storage facilities for food. Governance arrangements in the city-states of ancient Greece led to the precursors of democratic systems. This is not to say that urban development necessarily leads to democratic rights for all—full participation of people in the affairs of their city has still not been accomplished in many parts of the world (e.g. Hardoy and Satterthwaite 2014).

E. de Leeuw (✉)
Centre for Health Equity Training, Research & Evaluation (CHETRE),
University of New South Wales, South Western Sydney Local Health District,
Ingham Institute, Locked Bag 7103, Liverpool, BC NSW 1871, Australia
e-mail: e.deleeuw@unsw.edu.au

E. de Leeuw, J. Simos (eds.), *Healthy Cities*,
DOI 10.1007/978-1-4939-6694-3_1

Slaves were not eligible to engage with early iterations of democratic decision-making in ancient Greece, nor were women. In most parts of the world female suffrage took another 2000 years to be achieved. In Australia, electoral equality between Aboriginal Australians and newcomers (colonization by the British started in 1786) was not established until the *Commonwealth Electoral Amendment Act* of 1983, which made voting compulsory for the Indigenous population as it was for other Australians. But for cities around the world with large slum or 'informal' areas—favelas, barrios or shanty towns—participation of individuals and communities in the processes that shape their environment remains a challenge precisely because of the informal nature of their conurbations.

Cities shape our lives, and our lives shape cities. To what extent is the same true for our health?

The Neolithic Demographic Transition: Health Shifts

Archaeological records show that during the Neolithic era (from ~10,000 to ~4500–2000 BC, depending on the geographical area), a hitherto relatively stable human population started to grow rapidly, displaying much higher birth and death rates than previously. Paleoepidemiology suggests that before this Neolithic demographic transition, human populations thrived in relatively good health but did not expand in size. After the transition populations experienced a far higher burden of disease, offset by an even greater increase in birth rates and possibly female fertility. McKeown (2014; based on an original statement published in 1979) attributes these changes to greater food security, but current interpretation of the archaeological record (Bocquet-Appel and Bar-Yosef 2008) suggests a more complex interplay between a number of factors claiming a more direct relationship between demographic growth during the Neolithic transition and the settlement of hunter-gatherer populations. They suggest that the fertility explosion can be attributed only in part to the emerging farming economy during the Neolithic age, and may be better explained by factors more directly associated with sedentism: that is, the process of people settling permanently in one place. The suggestion is that the limitation of the need to be constantly mobile increased resources, allowing Neolithic women to bear and successfully deliver more children. New resources included secure access to good food and nutrition, but more generally could be attributed to improved energy balance at population level, for with security and sedentism came opportunities to invest energy beyond mere survival.

The Neolithic demographic transition meant that more people began living sedentary lifestyles in a more permanent way—and in so doing they changed the etiological and pathological ecosystem: that is, the complex set of relationships among living resources, habitats and residents of an area, and the ways and means by which they cause and transmit health, disease, disability or infirmity. The determinants of health in early human civilizations were generally associated with seasonal and environmental conditions and the greater mobility of human groups following favourable climate conditions. Once these groups started to inhabit more stable and permanent dwellings, the types and impacts of pathogens became different; they were, in a way, more 'man-made' (or part of the Anthropocene, as we will elucidate below). In permanent settlements, as we have described above, social determinants of health in addition to shifts in environmental determinants (associated with building materials, proximity and species density) cause different aetiologies and stimulate different pathogens.

When We Settle: Determinants of Health, Disease and Death

Armelagos (2009) argues that Omran's (1971) theory of epidemiological transitions should extend further into the past. Epidemiological transition theory has been described as a useful heuristic that models the impact of various disease types on demographic change and population viability and longevity, but without the predictive and operational precision that a 'real' theory should display (Mackenbach 1994).

Armelagos adds another dimension to his criticism. He quotes Thomas Hobbes who described our paleolithic ancestors as living in 'continual fear' with 'a danger of violent death and a life that was … solitary, poor, nasty, brutish, and short' (*Leviathan*, I xiii 9), but the archaeological record strongly suggests that they were relatively healthy and well nourished. The first epidemiological transition would have been associated with the Neolithic demographic transition in which more sedentary populations began to be exposed to an increase in infectious diseases (Table 1.1).

Indications of the relatively good health of pre-Neolithic populations were provided, among others, by Bengmark (2000) who found that 'the diet of our forefathers contained at least a billion times more non-pathogenic health-promoting bacteria' and that in contemporary developing countries humans' commensal (health supporting) gut flora weighs 2 kg, while in developed countries it is less than 1.3 kg. Empirical evidence is added by Hershkovitz and Gopher (2008) who compared archaeological dietary inferences from two populations in the Levant. This finding, for better or for worse, has inspired the current fad for the 'paleo diet' (Frassetto et al. 2009).

Among archaeologists, paleo-epidemiologists and anthropologists studying social transitions between forager and sedentary societies, a debate still rages as to what precisely caused the Neolithic demographic transition (cf. Caldwell and Caldwell 2003), but there is no doubt that the shift towards lasting and denser settlements that occurred around the world from 6000 to 10,000 years ago had a great impact on patterns of morbidity, mortality and aetiology. The presence of a greater diversity of species, spatially and temporally more continuous, meant that humans were exposed on a more constant basis to many pathogens, with fewer changes to their occurrence and impact. Changes (in the forager way of life also stemming from shifting habitats and ecological variation, e.g. McMichael 2001) became merely seasonal: for example, autumn and winter influenza, and the rainy season bringing floods and water-borne diseases (Cohen 1989).

Table 1.1 Causes of death and disease in Palaeolithic and Neolithic eras

Palaeolithic		Neolithic	
Vectors/pathways	Diseases/causes of death	Vectors/pathways	Diseases/causes of death
Patriarchy	Anthropoktony (war, homicide, infanticide, etc.)	Faecal contamination	Hookworm
			Ascarids
Wild animals and their parasites	Rabies	Rats, mice, ticks, fleas, mosquitoes, lice	Rubella
Other zoonoses	Tuberculosis	Domesticated animals	Cholera
	Brucellosis	Irrigation	Smallpox
	Plague		Mumps
	Tularaemia		Measles
	Leptospirosis		Chickenpox (wrongly)
	Chagas disease		Salmonellosis
	Yellow fever		Yellow fever
	Encephalitis		Typhoid
	Rickettsiosis		Poliomyelitis
	Chickenpox		Malaria
	Herpes		Influenza
	Leprosy		Common cold
	Staphylococcal and streptococcal infection		
	Haemorrhagic fever		
	Anthrax		
	Gangrene		
	Botulism		
	Tetanus		
	Encephalitis		
	Trypanosomiasis		

Compiled from Caldwell and Caldwell (2003)

This differential exposure to pathogens did not go unnoticed by the rulers of the cities that emerged in the Levant, in Mesopotamia and in the Indus valley. In the latter around 4000 years ago, Possehl (2002:105) reports, elaborate domestic water management systems clearly suggest a plan to store water safely and dispose of it after use. In the Indus urban settlements of Mohenjo-daro, Harappa, Nausharo and Lothal, this happened not just in individual dwellings but at a collective level through sewer systems, and the evidence suggests that human waste and faeces were collected and disposed of separately and deliberately for sanitary reasons.

Other health dimensions have been identified in ancient Babylonia, particularly in the royal palaces of Ur and Larsa, where food storage facilities were determined to be about 5000 years old (Crawford 2007:91). Goetze (1950) reports cuneiform tablets in which sesame inventories from food stores are listed—which qualifies as one of the earliest attempts to establish food security.

The great cities in the cradle of humankind continued to grow and thrive, leading to the next great shift: increased sophistication in the governance and management of city states—either loosely connected to each other as in the Ancient Greek civilization, or elaborate continent-spanning bureaucracies as in the Persian Achaemenid empire (e.g. Farazmand 1998).

This is, therefore, the situation in which we can place writings in the Hippocratic tradition. The worldview espoused by the Hippocratic physicians valued an assessment of the social and physical environment as much as individual dream reading and analyses of bodily fluids. Driven by a strong sense of empiricism, they identified connections between the location and architecture of dwellings and the classes of disease that were likely to afflict their occupants. The location and quality of dwellings, and their physical and social environments, has been integral to architectural paradigms from the days of Marcus Vitruvius Pollio (also known as Vitruvius) (c. 75 BC–15 BC). He is part of the popular lexicon in the phrase 'Vitruvian man'—more correctly, *Le proporzioni del corpo umano secondo Vitruvio* (the proportions of the human body according to Vitruvius) drawn by Leonardo da Vinci around 1490. This iconic perspective suggests ideal proportions of the human body, as well as a measure of its relation to the broader environment.

Proportions and balance are key to Vitruvian thinking. Vitruvius is famous for asserting in *De Architectura* that a structure must exhibit the three qualities of *firmitas, utilitas, venustas*—that is, it must be solid, useful, beautiful. These are sometimes termed the Vitruvian virtues or the Vitruvian Triad, and still drive good design and architecture; for instance, in the Design Quality Indicator tool (www.dqi.org.uk), which argues that good design leads to better health and well-being, and brings economic, ecological and sustainability benefits as well as better governance. Vitruvius is also hands-on in stressing the need to test environmental quality before establishing new communities. To test the soil, for instance, animals were left to graze and then slaughtered, and their livers were inspected to gauge healthfulness (Porter 1999:19).

In practical terms, the Vitruvian legacy lives in contemporary architecture through the heritage of Andrea Palladio (Venice, 1508–1580). His renaissance concept of *commoditas* resembles *utilitas* (utility) but is generally framed as the idea that the vision of utility is embraced in a balanced and aesthetically pleasing manner (Alexander 2013, in Basta and Moroni 2013). Boschi and Pagliughi (2002) review the Vitruvian and Palladian perspectives on these matters in relation to perspectives of health and well-being (Table 1.2).

Hippocrates, Vitrivius and Palladio may be seen as proponents of a rational approach to place-based health, but religion and health practice go hand in hand. Cities, as focal points of religious adoration and practice, made for a connection between settlement, cleanliness and godliness. Paul Unschuld (1985a, 1985b:264) traces such a gaze back to ancient China, where 'Chinese doctors under the Chou Dynasty (1122–250 BC) linked physical health to moral well-being and spiritual serenity, which led to cosmic harmony'. Porter (1999:13) finds the same aspiration

Table 1.2 Vitruvius' and Palladio's concepts of utility, durability and beauty (based on Boschi and Pagliughi 2002)

Dimension	Vitruvius	Palladio
Utilitas/commoditas and location	First comes the choice of a very healthy site if our design of private houses are to be correct we must at the onset take note of the countries and climates in which they are built	Great care … not to build near those waters … with regard the wholesomeness of the air
Utilitas/commoditas and ventilation, thermal comfort, health	Different rooms require different exposures … in libraries with southern exposures books are ruined by worm and dampness… summer dining rooms to the north … it makes the use of the rooms both healthy and agreeable	An edifice may be esteemed commodious, when every part or member stands in its due place and fit situation, neither above or below its dignity and use; or when the loggia's, halls, chambers cellars and granaries are conveniently disposed, and in their proper places
Utilitas/commoditas and social-well being	After settling the position of the rooms … we must consider the principles on which should be constructed those apartments … to suit different classes of persons	The habitation for the master ought to be made with regard to his family and condition
Utilitas/commoditas and economy	Proper management of materials and of site … thrifty balancing of cost and common sense in the construction work … plan for different kinds of dwellings suitable for ordinary people, great wealth	That one may learn, by little and little, to lay aside the strange abuses, the barbarous inventions, the superfluous expenses
Durability	When foundations are carried down to solid ground ….	Foundations strong and solid…
Beauty	Members are in due proportion. Proportion is a correspondence among the measures of the members if the entire work and the whole to a certain part selected as standard…	Will result from the form and correspondence of the whole, with respect to the several parts, of the parts with regard to each other, and of these again to the whole; that the structure may appear an entire and complete body, wherein each member agrees with the other, and all the necessary to compose what you intend to form

to achieve balance in ancient Egypt, where full health is seen to exist in a space between temporal and spiritual existence. In Babylonian, Assyrian and Hebrew codes, the injunction was to achieve 'purity before God'. Jewish (and later, Muslim) doctors continued to influence public health thought through these perspectives well into the Middle Ages.

In fact, Porter finds that organized *urban* health care provision was first witnessed under Muslim rule, when Caliph Harun al-Rashid and Caliph al-Muktadir in the ninth and tenth centuries established hospitals in Baghdad. In the Christian world the provision of health care on a charity basis became the source of political power later in the Middle Ages, where 'providing welfare for the sick-poor may have been one route to power; controlling their ability to spread disease was another' (1999:23). The connection between health and urban planning has therefore always existed, implicitly or explicitly. Common practices in the European Middle Ages included expelling people who were suspected of carrying fatal infectious diseases (e.g. bubonic plague or the Black Death in the mid-fourteenth century) outside the city limits, as well as excluding certain population groups such as Jews (often on the basis of 'moral' arguments) from urban living. These practices consolidated a strong sense of the need for governance arrangements in urban health. In central Europe this led, for instance, to the establishment of sanitation squads during epidemic events. This tradition inspired early scholars of public health like Johann Peter Frank (1788) to develop encyclopaedic instructions for the creation of a *Medizinische Polizey* (medical police).

A strong belief emerged in the Enlightenment that certain types of urban planning would be more beneficial to health (and the prevention of disease) than others. Maneglier (1990) describes how Voltaire, in the same era as Johann Peter Frank, complained about the markets of Paris, 'established in narrow streets, showing off their filthiness, spreading infection and causing continuing disorders'. It took another century before Louis-Napoleon Bonaparte (elected president of the French Republic in 1848 before becoming emperor Napoleon III through an odd, almost reverse, coup d'état) committed to improving the lot of the Parisian working classes through better urban planning. The original work of French hygienist and early sociologist Louis-René Villermé, looking at occupational health hazards, had been instrumental in the French 'mouvement hygiéniste' to legislate for the 'Première Loi sur l'urbanisme interdisant la location des logements insalubres' (1850; the 'First Urban Planning Act Prohibiting the Rent of Substandard Housing'). This—almost literally—paved the way for Louis-Napoleon to commission George Eugene Haussmann (1809–1891) to *aérer, unifier, et embellir* (provide air, unify, and beautify) the great city. The grands boulevards radiating through the city have become emblematic of Parisian charm, but Haussmann clearly had health in mind in designing the infrastructure; in his memoirs he wrote:

> The underground galleries are an organ of the great city, functioning like an organ of the human body, without seeing the light of day; clean and fresh water, light and heat circulate like the various fluids whose movement and maintenance serves the life of the body; the secretions are taken away mysteriously and don't disturb the good functioning of the city and without spoiling its beautiful exterior. (De Moncan and Heurteux 2002).

In Britain, similar reflexions led to a surge in urban and public health development. Edwin Chadwick founded, and gained support from, the Health of Towns Association in 1844 (Ashton and Ubido 1991). In 1875, Sir Benjamin Ward-Richardson presented his vision of *Hygeia: A City of Health* to the Social Science Association in Brighton, UK (Cassedy 1962). In his address he marked the passing of the recent Public Health Act by describing a utopian city incorporating concepts like clean air, public transport, small community-based hospitals, community homes for the aged and the insane, occupational health and safety, the absence of tobacco and alcohol, and many other advances. His ideas were taken up by others, notably Ebenezer Howard, who proposed and developed the first 'garden cities' in Britain in the 1890s—a movement still alive today through New Towns, sustainable communities or transition town projects (Alexander 2009). One would think that the longevity of such initiatives would have moved them to the mainstream of public health and urban planning ideas in the twentieth century, but the reality is that throughout that century they remained marginal in the face of Cartesian, reductionist and structural-biomedical approaches to urban health.

One year after the presentation of *Hygeia*, Pasteur and Koch marked the birth of the public health era that has been variously labelled 'germ theory' or the 'biomedical' wave (Davies et al. 2014; Kickbusch 2007). This perspective has great potency—rather than having to look at a multitude of abstract and interrelated 'big picture' concepts, ill health can be dealt with by eliminating a single organism (or such is the belief—Birn (2011) has shown in her analysis of smallpox eradication that even vaccination campaigns are complex social constructions). A similar competitive slant has been identified in the area of urban planning (Deelstra 1985): from visionary sociocultural perspectives in the nineteenth century, the urban planning perspective shifted to a structural-physical model in which architectural hardware is all that matters in planning.

All the World Was Conquered by Reductionism, or Was It?

As an echo of nineteenth century sociocultural ideals (or more likely—with great twenty-first century foresight) US psychiatrist Leonard Duhl initiated a long-range programme development in the National Institute of Mental Health in 1955. His vision was to explore, across disciplines, the impact of the physical environment on human behaviour. From 1956, with zoologist John B. Calhoun, he organized conferences on social and physical environmental variables as determinants of mental health. The first 17 cross-disciplinary participants preferred to call themselves the Space Cadets—where their space was not *outer* space (these were the days of the space race between the Soviet Union and the USA) but *urban* space. The Space Cadets grew to several hundred participants (Martin 2014). The conversations they had across the country, and most notably at the 39th Annual Meeting of the American Orthopsychiatric Association in 1962, led Duhl to compile and edit a book of pivotal importance to Healthy Cities.

In this, *The Urban Condition—People and Policy in the Metropolis* (Duhl 1963), an eclectic, innovative and remarkably funny group of thirty men and two women discuss health, housing, ethology, violence, mental health, pathology, planning and matters that have only re-emerged since the turn of the century: climate change, systems thinking and complexity in the urban environment. The Space Cadets contributing to this volume agree that the city must be viewed as a system in which physical infrastructure (which Deevey in 1963 dubbed 'urbs') and its people and their capabilities (which he called 'civitas') constantly interact, dynamically and often unpredictably. Duhl himself, in his introduction to the book, states in no uncertain terms that looking at the parts of the city is possible and often understandable, but in order to make sense of the connection between health and urban dynamics one must see its complexity as an ecological whole.

The psychiatrist Duhl moved into academia at the University of California (Berkeley) to take up a chair in public health and planning, from where he continued to refine his humanistic and ecological perspective on urban development through teaching and research into community health, entrepreneurship for change and city dynamics. His work appeared on centre stage during the 1984 conference in Toronto that celebrated the Lalonde Report and declared the Canadian city's ambition to become a Healthy City (see Trevor Hancock's account of this event in Chap. 4 of this book). Duhl presented a keynote speech, and Ilona Kickbusch of the European office of WHO saw the potential of Healthy Cities to advance innovation for the new public health.

In one of the foundation documents of the WHO European Healthy Cities project, Hancock and Duhl justified the ecological, inclusive and dynamically complex view of urban health:

> Some question the city's ability to initiate and implement health initiatives in the face of a variety of problems that include deterioration of the physical environment, poverty, unemployment, economic stagnation, homelessness, hunger, family violence, and crime and youth alienation. In some respects, cities may be seen as the potential or actual 'victims' of national and international policies—most spectacularly in connection with the threat of nuclear annihilation, more mundanely as a result of social, economic, immigration and other policies. On the other hand, others point to the many real strengths of the city. For instance, it is in the city that the greatest variety of skills, resources and talents are available; it is the city, Jane Jacobs argues, that is the economic powerhouse of nations; it is in the cities that invention, the arts and other forms of creativity have traditionally flourished. Furthermore, city governments are often the closest level of government to people that have the mandate, the authority and the administrative resources needed to bring together the wide variety of skills and resources needed for a multi-sectoral approach to health. (1986)

Many global city networks have flourished since the early 1980s, and Davies (2015) lists others besides Healthy Cities:

- Just Cities
- Green Towns and Cities
- Sustainable Cities
- Transition Towns and Ecodistricts
- Winter Cities
- Resilient Cities

- Creative Cities
- Knowledge Cities
- Safe Cities and Communities
- Festive Cities
- Slow Cities

In addition, we know of international networks of local governments coming together as Happy Cities (Montgomery 2013), Smart Cities (Caragliu et al. 2011), Child-friendly Cities (Riggio 2002), and Age-friendly Cities (WHO 2007), Conscious Cities (Palti 2016) and Inclusive Cities (UN-HABITAT 2010). All of these show the importance of human settlements for (social) change, and the search for local as well as global opportunities to make cities better places for all their inhabitants and all their other users.

Hancock and Duhl ask a quintessential question which once again leads to an inclusive, ecological and dynamic answer:

> What do we then mean when we talk of 'the city'? As Phillips and LeGates (1981, 1982) point out, the concept of the city does not inspire consensus: there is no precise definition of the word 'city' that social scientists or anyone else can agree upon. At one level, the city is a collection of buildings and roads and their associated transportation, communication, water and sewage systems—the hard infrastructure. However, this is more a description of an archaeological site than of a city. Clearly, a city is more than simply bricks and mortar. A city has a life of its own, it has a soul, a spirit, a personality … Historically, the city may have begun as a centre of trade, because of its strategic siting on a main transportation route, or because of the religious and symbolic nature of the site. Cities frequently had considerable religious symbolism, what Lynch (1981) refers to as the "cosmic city", one that negates time, decay, death and fearful chaos. Such a city was based upon order, stability and dominance, with religion and faith as its core. This religious and spiritual significance is seldom found today, its place being taken, perhaps, by Mammon and his banking headquarters. But the role of spiritual or mythic symbol remains an important part of what a city is. (1986)

On a perhaps quintessential aside, it seems important to note that in many cities around the world the largest economic sector is not banking, but the health or medical–industrial complex. Some authors, like Clerc and Stern (2004) or Foucault (1983) go so far as to claim that modern medicine has taken over the role that faith and religion played in bygone times. This creates an interesting perspective for a Healthy City: whereas it is relatively easy to attribute all sorts of 'bad' (e.g. poverty, new public management and inequity) to 'Mammon and his temples', the health care sector—in spite of many critical perspectives—is held in high moral regard across societies and communities. But as Hancock and Duhl (1986) argue, it becomes clear that it is not (only) the health care sector and medical–industrial complex that create urban health, but rather the complex interplay between all the infrastructure and all the people in the cityscape:

> Phillips and LeGates (1981) suggest that the common elements that describe a city are permanent residents, a large population living at high density and a heterogeneous population. However, there are no criteria delimiting how large, how dense or how varied. To an economist, a city is a place 'where the local inhabitants satisfy an economically substantial part of their daily wants in the local market' (Weber 1999, originally published 1921). To an anthropologist, it may be that a city exists 'only where there are cultural ingredients considered essential to urban life—the fine arts, exact science and, in particular, writing'. (Philips and LeGates 1981:83).

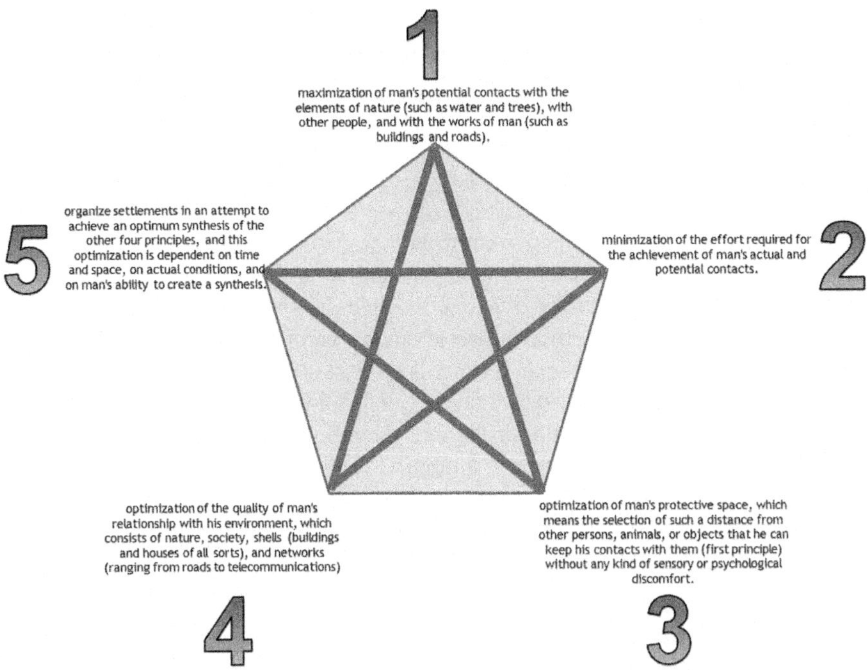

Fig. 1.1 Five principles that guide human settlements (Doxiadis 1970)

A sociologist, of course, would focus upon the interactions between the inhabitants of the city, and for her or him that would be the city. It is these interactions that Duhl terms the 'soft infrastructure':

> It is the geography and history of the city, with its varied populations, their immigration patterns and cultures and their art, music and poetry that governs the city. These characteristics and events, interacting with the still broader context of region, state, nation and world in all their manifestations, determine how people are born, live and die. The laws, taxes, regulations, business practices and the availability of infrastructure emerge out of the political struggles of all these groups and people.

Hancock and Duhl turn to Konstantinos Doxiadis' *ekistics* (a term coined by himself—'the science of human settlement'). Doxiadis devised a scholarly approach to framing the drivers of settlement and urbanization. The establishment of settlements and cities is driven by five principles (Fig. 1.1) that constantly interact.

Hancock and Duhl (1986) consider that 'according to Doxiadis, our image of the city has gone through a number of phases in the past four decades, from buildings to transportation, then to society and now to nature and energy':

> '...we know that in fact people all over the world suffer from much more complex situations than these fashionable attitudes would have us believe, we must not allow passing fashions and incomplete diagnoses to divert us from understanding the real problems or basic diseases of human settlements and their causes' (Doxiadis 1977:50).

We believe that if we are to come to grips with the complex reality of the city, it is perhaps best understood holistically, as an organic, living system, partly organism, partly ecosystem. As an organism it is composed of a number of subsystems—arteries to transport materials and nutrition, nerves to carry messages, an excretory system, a respiratory system—and like an organism it must learn from its mistakes, adapt to and cope with change, repair itself, and communicate and exchange with its fellows. As an ecosystem—'a functioning interaction system of living organisms and their effective environment, physical, biological and cultural' (Berry and Kasarda 1977:16)—it is composed of a variety of competing and cooperating groups in a state of dynamic balance. Its strength lies in its diversity, its interdependence, in the efficient use of energy and the continuing recycling of material (Lynch 1981). Of course, the city is really a social ecosystem, and a direct analogy to a biological ecosystem may be misleading. But clearly the city is an important subsystem level in understanding the human ecosystem, and the vision of the city as a human ecosystem is a popular one (Burton 1982; Deelstra 1985; Duhl 1963; Lynch 1981).

According to Kevin Lynch, a noted urban designer and theorist,

> The fundamental good is the continuous development of the individual or the small group and their culture … a settlement is good which enhances the continuity of a culture and the survival of its people, increases the sense of connection in time and space, and permits or spurs individual growth: development, within continuity, via openness and connection. (1981:116)

He goes on to discuss five dimensions of good city form. Four of these (sense, fit, access and control) refer to the need for the settlement to be clearly perceived and put in a context of time and space; the extent to which a settlement's spatial pattern matches the customary behaviour of its inhabitants; the extent to which people are enabled to reach other people, resources, activities, services, information or places that they might want access to; and the extent to which those who use, work or reside in a space control the use of and access to the space.

However, important as several of these are to health (perhaps particularly access and control), it is Lynch's first dimension—vitality—that is of particular interest. It is also the dimension of city form which, in Lynch's view, 'comes as close to being a pure public good as any on our list, since health and survival are values very widely held' (1981:125). Vitality, as understood by Lynch, means that the settlement supports the vital functions and meets the biological requirements of human beings. He suggests three main features relating to human health and well-being, and two features relating more generally to environmental/ecological well-being:

- Sustenance: 'An adequate supply of food, energy, water and air and a proper disposal of wastes'
- Safety: 'A good settlement is one in which hazards, poisons and diseases are absent or controlled and the fear of encountering them is low. It is a physically secure environment'
- Consonance: 'The spatial environment should be consonant with the basic biological structure of the human being'. By this Lynch means that it supports natural

rhythms of sleeping and waking, provides optimum sensory input (not overloading and not boring), promotes exercise, and controls the harmful effects of light, noise and indoor air pollution.

Lynch's other two aspects of vitality are concerned with

- How well the environment provides for the health and genetic diversity of species economically useful to man.
- The present and future stability of the total ecological community.

From Lynch we may take the concepts of provision of basic human needs, a physically safe and clean environment, biological connectedness, ecological sustainability, access, mobility, control, and the extent to which the settlement's form permits—or even encourages—health-enhancing behaviour. Doxiadis adds:

> Out of all definitions of the goal of the city which we have inherited, I think that there is only one that is valid for all human societies and this is the Aristotelian one: to make the citizen happy and safe.

Hancock and Duhl observe that 'Doxiadis went on to expand upon that definition to develop his own goal for the city: to make the citizens happy and safe and help them in their human development' (1986:6). Doxiadis suggests that there are five human needs that the city must satisfy (cf. Fig. 1.1):

- The maximization of potential contacts with other people, resources, etc.
- The minimization of effort in making those contacts (measured as energy, time and cost). In their attempts to maximize potential contacts, humans try to bring things closer to them.
- Optimization of protective space, the attempt to create a balance between bringing people and objects close and feeling crowded or threatened.
- Optimization of relationships with other elements of the system (nature, society, buildings and communication networks).
- Striking a balance among the four principles listed here. On the basis of these principles, Doxiadis suggests five things that humans demand in their cities:
  - Freedom to move (so as to maximize potential contacts)
  - Safety
  - A quality of life which satisfies their aspirations
  - Human contacts
  - Creativity and human development

As Doxiadis sees it, the city has to assist individuals in responding to challenges and developing to the fullest extent possible by bringing people 'closer together to benefit from their contacts, but at the same time to form a proper structure that can keep them sufficiently far apart, so that the exposure to and the danger from each other is minimized' (1977:87). From him we may take the concepts of human development, maximum contacts, freedom to move, safety, efficiency, human contacts and quality of life.

Another interesting urban planning theorist is Malcolm Fitzpatrick, who has attempted to combine the science of ecology, the values of public health and the methodology of urban planning in deriving his four criteria for evaluating community design (Fitzpatrick 1978), which are:

- Minimize intrusion into the natural state. Since health is dependent upon maintaining an ecological equilibrium and the dynamic natural state can adjust more easily to small changes than to large ones, it is important that cities minimize the alteration of their ecosystem.
- Maximize variety. The availability of choice, together with the ability to choose (which requires participation and opportunity), will promote adaptability and prevent monotony and boredom.
- Close the system as much as possible. In ecological terms, we want a closed loop system, so that its outputs become its inputs, thus making the city as self-perpetuating and sustainable as possible.
- Maintain an optimum balance between population and resource use.

From Fitzpatrick we may take the concepts of variety, ecological sustainability and participation.

Another important planner who has developed a framework for thinking about the 'health of the city is Hans Blumenfeld (cited in University of Waterloo 1984). He suggests six parameters for the functions of a city:

- A place to make a living, and for living
- Accessibility and transportation
- The environment
- Relationship of the physical with the social environment
- Privacy and neighbourliness
- Flexibility

Eighteen criteria are listed. Among the most prominent and recurring themes are variety and choice, accessibility and mobility, safety, ecological preservation/conservation, housing quality, community, and continuity and identity (connectedness). Although planners such as Lynch talk of space, design, ecosystem and other physical or biological attributes, they are of course fully aware that the city is more than bricks and mortar. A great deal of work has been done in the field of urban sociology that may help us to understand what it is that makes a good city from a sociological point of view, recognizing that 'good' depends on who you are, and how you are affected'.

Based on their profound understanding of urban and human systems, and on field trips they undertook in the early 1980s, Hancock and Duhl (1986) add their final layer of understanding to what a Healthy City is all about. Good cities

- Have a common 'gameboard' where everyone comes together to make decisions by a commonly accepted set of rules.
- Are multidimensional, yet succeed in relating the various parts to each other.

- Are homogeneous and heterogeneous at the same time (the dominant culture accepts new cultures without engulfing them, and is enriched by them).
- Have an extensive and redundant network of formal and informal communication linkages, both among its own people and with the outside world.
- Can adapt to change, cope with breakdown, repair themselves, and learn from their own experience and that of other cities.
- Have a commonly accepted mythology about themselves, in terms of a sense of history, and image of the city as it is today and a vision of what the city should be in the future.

Combining these insights, and connecting them to historical and developmental parameters and definitions (e.g. of health and salutogenesis), the fathers of Healthy Cities propose that a Healthy City is a city

> that is continually creating and improving those physical and social environments and expanding those community resources which enable people to mutually support each other in performing all the functions of life and in developing to their maximum potential

and strives to provide the following qualities to its people and infrastructure:

- A clean, safe, high quality physical environment (including housing quality)
- An ecosystem which is stable now and sustainable in the long term
- A strong, mutually supportive and non-exploitative community
- A high degree of public participation in and control over the decisions affecting one's life, health and well-being
- The meeting of basic needs (food, water, shelter, income, safety, work) for all the city's people
- Access to a wide variety of experiences and resources with the possibility of multiple contacts, interaction and communication
- A diverse, vital and innovative city economy
- Encouragement of connectedness with the past, with the cultural and biological heritage and with other groups and individuals
- A city form that is compatible with and enhances the above parameters and behaviour
- An optimum level of appropriate public health and sick care services accessible to all
- High health status (both high positive health status and low disease status)

The Anthropocene and Healthy Cities

We started this chapter—and book—arguing that Healthy Cities in a way emerged in the late Neolithic. This archaeological and physical anthropological characterization of the era coincides with the Holocene—the geological and climatic epoch that is classified as the current interglacial period in Earth's evolution that started ~11,000 years ago (Walker et al. 2009). And where the Neolithic (or Stone Age) evolved into a Bronze and Iron Age (and currently arguably the Plastic or Quantum Age) the Holocene, some argue, is being transcended by the Anthropocene.

A proposal to formalize this term is being developed by the Anthropocene Working Group of the International Commission on Stratigraphy. These geologists characterize the—tentative—period as the current one, started in the early 1800s, in which

> many geologically significant conditions and processes are profoundly altered by human activities. These include changes in:
>
> - erosion and sediment transport associated with a variety of anthropogenic processes, including colonization, agriculture, urbanization and global warming.
> - the chemical composition of the atmosphere, oceans and soils, with significant anthropogenic perturbations of the cycles of elements such as carbon, nitrogen, phosphorus and various metals.
> - environmental conditions generated by these perturbations; these include global warming, ocean acidification and spreading oceanic 'dead zones'.
> - the biosphere both on land and in the sea, as a result of habitat loss, predation, species invasions and the physical and chemical changes noted above. (International Commission on Stratigraphy 2016)

The idea of the Anthropocene, even without formally being adopted as a geological epoch, is gaining traction outside the earth sciences. Hancock (2015), in continuing an eco-social paradigm of (public) health, argues that detrimental human impact on the planet in the Anthropocene must be countered with sane, participatory and sustainability-driven actions, research and policy. There seems to be a fair bit of doom and gloom about the ways in which humanity since the dawn of the Anthropocene has systemically compromised the future of the planet and the health and well-being of coming generations (Steffen et al. 2007; Whitmee et al. 2015). But integral to human development in the epoch have not just been exploitation, deterioration, pollution and degradation: the human spirit has also led to creativity, boundless thought, recognition of complexity, and fun (de Leeuw 2011). The 'hardware' deficits of the Anthropocene may well be offset by the 'software' of human ingenuity and flexibility.

We are not—just—talking about technological innovation and inventions to compensate for general planetary destruction (e.g. through elaborate industrial processes for carbon capture to mitigate the consequences of climate change). Derickson and MacKinnon (2015), for instance, argue for a feminist praxis of anthropocenic justice and resourcefulness which is being tried and tested with communities in, for instance, West Atlanta. Buck (2015) takes the argument even further: she sees opportunities for a charming—and healthful, imaginative, positive and expansive—Anthropocene. She does so because the emerging evidence on 'fear framing' (i.e. risk-focused appeals to motivate public support to act for, notably, better climate change policies) show these perspectives do not work, or work less well than positive frames and scenarios aiming at betterment and gain (Moser and Dilling 2011). This is an honoured mantra in health promotion, where healthy choices are to be made the easier (or more fun) choices (Kickbusch et al. 2005).

Buck paints four scenarios:

- The *Futuristically Ancient*, in which cunning 'rewilding' (balanced reintroduction of beasts and monsters—predators—in relatively monocultural ecosystems) will yield greater biodiversity and ecosystem resilience
- *Art and Craft*, where human creativity shapes biophilic cities that are constantly evolving, highly varied urban-diversity-ecodesign-politics environments that generate new and more opportunities for an expanded ethic, activities, attitudes, knowledge, institutions, and governance for all
- *Connection and Care*, aiming at holistic and integrative systems of planetary farming honouring the human scale and Indigenous knowledge systems in which human, community and creation are one
- *Convergences and Distributed Systems*, where peer-to-peer, distributed, open-source and rhizomatic connections shape natural and information landscapes that are directly connected to the human experience. Bruno Latour (2015) has typified this as the ultimate dimension of people connected to the Anthropocene: the ability to feel, anticipate and control consequences.

Healthy Cities, we feel, are quintessential vehicles to create a charming Anthropocene, an epoch in which community and creativity work together to shape institutions and governance arrangements that benefit equitable health and well-being for all in the new natural habitat of humanity: *urbs* and *civitas*, the city and its people.

This Book

Healthy Cities, particularly in Europe, has been described often as a visionary movement. It drew on the evidence and foresight of seers like Duhl, Hancock, Kickbusch and Tsouros. We believe that the vision was so strong (and enabled the movement) because it resonated with deeply held human values that urbanites and their institutions can easily identify with. In most cities, (health) inequity is at your doorstep, or around the corner in your gutter. Solidarity is what happens at train stations and in farmers' markets. Sustainability is seen in gardens and parks. Admittedly urban planning toward *Truman Show*-like gated communities does effectively block out misery and filth (Cunningham 2005) and community action is easily compromised by a 'smile and you're happy' subterfuge (Ehrenreich 2010), but that is precisely why urbanites identify with, or at least have emotional connections with, efforts to strengthen communities, effective and healthy transport, etc.

Discussions about this book started about a decade ago, in meetings between European and Canadian Healthy Cities aficionados, organized by the Centre Québécois Collaborateur de l'OMS pour le Développement des Villes et Villages en Santé (The Quebec WHO Collaborating Centre for Healthy Cities and Villages Development) where Michel O'Neill and Evelyne de Leeuw saw human values—not planning checklists or scholarly definitions—as the key to the existence and success of this international initiative. Values are an essential expression of the

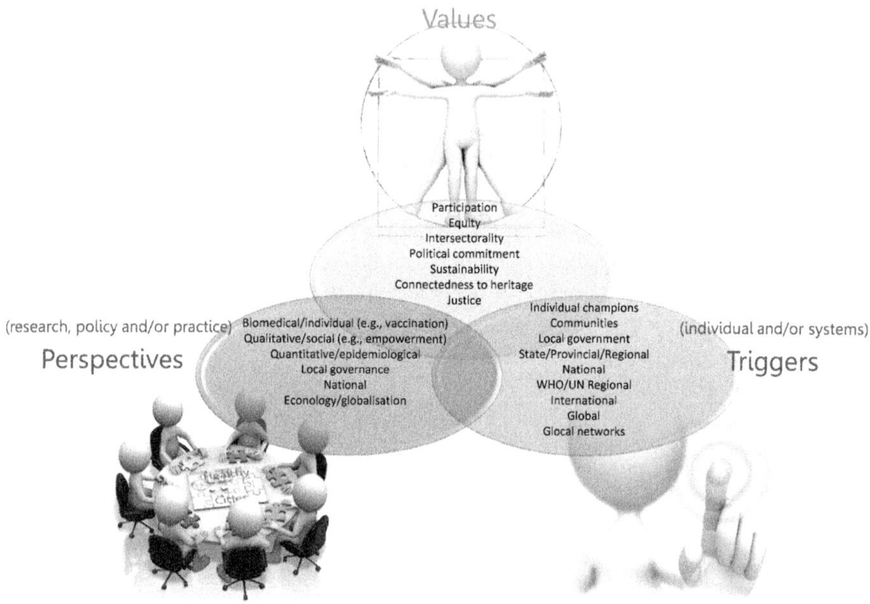

Fig. 1.2 The conceptual design of this book

Anthropocene: the redwood tree or the quoll do not hold values or aspire to them. Values are what matter in Healthy Cities, and the successful advocates of the effort are all firmly grounded in a deeply profound ethical value-base—one that pans out in their personal lives and actions. We believed, however, that values are the key ingredient but not the reason for getting cooking. For the generation of a proper argument in a collection of essays we needed to identify the triggers, the catalysts, that urge communities and (political) institutions to act on those values. Those triggers, we recognized, are always contextual. They are created, shaped and morphed by opportunity, serendipity and sheer coincidence. If Healthy Cities were 'easy' they would be ubiquitous, and they aren't (yet): why? How do values, triggers and opportunities interact to create, in the words of Buck (2015), that charming Anthropocene?

We invited 'friends and family of the movement' (as Agis Tsouros used to call them) to reflect on this conceptual development (Fig. 1.2) and structured the book accordingly.

We asked our colleagues to reflect on their passion in Healthy Cities and to identify the values that pervaded their experiences in being scholarly and activist in the area. As value prompts, we mentioned participation, equity, intersectorality, political commitment, sustainability, connectedness to heritage, and justice—but many proposed other values and value systems. We also asked them to identify the 'triggers' that led to the birth of the Healthy City (or Healthy City network or experience) that they reflect on. We suggested that a Healthy City could have been triggered by individual champions (we have already mentioned the likes of Ilona

Kickbusch, Trevor Hancock and John Ashton), but also by a fortuitous confluence of actors and factors that might or might not include communities, local government, governments at a higher level, or a range of glocal networks. Finally we told our authors to not restrict themselves to a standard 'scienc-ey' view of the world in which the standard repertoire of the biomedical model is dominant. We wanted to harvest the richness of narrative, epidemiology, governance, evaluation and assessment, storytelling, etc. The result is a highly diverse, possibly slightly rambling, but absolutely exciting collection of chapters.

There are three parts to this work. In Part I ('Healthy Cities in History') we look at what happened to create such a fertile soil for the emergence of a range of Healthy Cities initiatives in the 1980s. Jason Corburn makes an argument that twenty-first-century urbanization necessarily must lead to new forms of urban governance and the relations of all those people and their institutions in the urban context. Following this, Ann Pederson and Irv Rootman take up the 'values' argument and show that Canada in the 1970s and 1980s was deeply rooted in a commitment to do things differently, precisely because of a strong value base. Trevor Hancock describes the events that created the serendipity of getting the right people at the right places—for a series of daring meetings in the second half of the 1980s that were the catalysts for new ways of thinking about urban health.

In Part II ('Regional World Perspectives') we present a global inventory of Healthy City and Community efforts. We asked regional leads/editors to apply our conceptual gaze (Fig. 1.2) and invited city case studies from their region. The process and general perspective are outlined in the first chapter of this section by Evelyne de Leeuw. Jean Simos then kicks off with Anglophone and Francophone African Healthy Cities. They display an enormous variety in triggers (e.g. overwhelming pollution or population pressures) and values (e.g. a very strong sense of community). Nastaran Keshavarz Mohammadi has joined forces with colleagues from the 'Eastern Mediterranean' (and Islamic world) to describe a similar diversity in Oman, Iran and Egypt. Marilyn Rice, with many co-authors from across the continent, contributes an exceptional piece from Latin America where hundreds, if not thousands, of Healthy Cities and Communities thrive. Trevor Hancock, Tyler Norris, Réal Lacombe and Fran Perkins present an inventory of North American towns and cities—which sometimes seem to have been struggling with the values but always resiliently re-emerge. Jean Simos takes the lead on the European account of six veteran cities of the network. Keiko Nakamura assembles three strongly value-inspired case studies from Asia and puts them in the context of the thriving Alliance for Healthy Cities, and Evelyne de Leeuw compiles material from Oceania (Australia, New Zealand and the Pacific)—not just towns and cities, but also reflecting on the firm commitment to values of sustainability and community that are found in the network of Healthy Islands in the Pacific.

Part III looks forward to 'Grand Healthy City Challenges and Perspectives'. Healthy City evaluations have happened throughout the life of the movement, importantly in Europe, but under different guises in other parts of the world as well. Carole Clavier and Michel O'Neill use this evidence to compile evidence on the role of community participation in influencing local policy. Mardie Townsend and Claire

Henderson-Wilson accumulate evidence on urban ecosystems and the important role of accessibility of green space for health, wealth and resilience in urban environments. De Leeuw and Lin review what is known about urban health governance and how connected networks of community values and institutional values and commitments make Healthy City initiatives stronger and more resilient to change. Jean Simos reviews the role of Health (Equity) Impact Assessment as a critical tool, not just for describing risk and impact, but for maintaining and building the very fabric of Healthy City communities, policies and practices. Trevor Hancock and Clem Bezold describe how communities and their political representatives can be enabled to think proactively about the future of their (urban) environment. Evelyne de Leeuw and Geoff Green, who have both been instrumental in shaping the European Healthy City evaluation context and deliverables for about 30 years, describe how the theoretical and methodological context for developing and implementing such challenging multi-national research programmes has evolved.

As outlined above, the arguments in this book centre around values. In the conceptualizations around the development of this work, both with Michel O'Neill and Jean Simos, it was almost implicit that there is an 'old style' 'Urban Health' that is important and that should continue to be embraced, but that 'Healthy Cities' do something extraordinary that the old style has not accomplished—it added a strong grounding in human and development values. Values are, of course, closely connected and often aligned to politics. In the last chapter Agis Tsouros reflects on the political nature of Healthy Cities—and he frames very well what this book is all about: if we want to advance the health of people that live in cities we *must* be political about it. Describing what the occurrence and distribution of health in cities is, and documenting the range of interventions that might resolve or exacerbate health challenges is important but 'old style' (as illustrated in World Health Organization and UN Habitat 2016). Shaping urban (health) politics is an entirely different, and far more pressing issue.

Reflecting on the arguments in this book we cover many of these values. There is one domain, however, where we have not ventured with any significance. This relates to the quality of urban planning and design, and the role of aesthetics in health experiences and outcomes. Beauty, although in the eye of the beholder, clearly is a factor in the determinants of individual and population health opportunity and development. There is a growing body of work that explores the aesthetics of urban planning and its relation to health (e.g. Loder 2014 and Ghertner 2015), and above we illustrated that this is one of the foundations of Healthy City thinking, from the early dawn of humanity through the Classics. But this book does not systematically review the qualities of urban design for health, as compiled by and validated by Ewing and Clemente (2013)—Table 1.3. This is one area where the value base of Healthy Cities needs to connect more significantly with that of Healthy Urban Planning (e.g. Barton and Tsourou 2013; Grant 2015)

Table 1.3 Fifty-one perceptual qualities of the built environment

Adaptability	Expectancy	Ornateness
Distinctiveness	Mystery	Upkeep
Intricacy	Territoriality	Continuity
Richness	Comfort	**Imageability**
Ambiguity	Focality	Prospect
Diversity	Naturalness	Variety
Legibility	Texture	Contrast
Sensuousness	Compatibility	Intelligibility
Centrality	Formality	Refuge
Dominance	Novelty	Visibility
Linkage	**Transparency**	Deflection
Singularity	Complementarity	Interest
Clarity	**Human scale**	Regularity
Enclosure	Openness	Vividness
Meaning	Unity	Depth
Spaciousness	**Complexity**	Intimacy
Coherence	Identifiability	Rhythm

Qualities in Courier font have been tested, with those in bold validated, and those in italics not passing the test (adapted from Box 1.2 in Ewing and Clemente 2013)

A Positioning Note

What *is* a Healthy City? And who *needs* a Healthy City?

Sometimes the real story and the true answers are not being related in the official literature and it seems important to position ourselves in our experience. In this chapter, and in Part I of this book, we have outlined the contours of Healthy Cities. We have, with the help of Len Duhl and Trevor Hancock, tentatively defined and characterized the Healthy Cities idea. But it is important to realize that we may be blindsided. The fact that cities around the world identify as Just, Happy, Sustainable, Winter, Resilient, Creative, Knowledge, Child-friendly, Festive, Slow, Smart, Conscious, Inclusive, Age-friendly Green Transition Towns and Ecodistricts, and Safe Cities and Communities, shows one thing: health may not be the all-important consideration for local government. And it isn't, really, for Healthy Cities either.

This may come as a surprise. It did so, very much, when we engaged in the very first review of European Healthy Cities (Draper et al. 1993). One of the questions we asked (and we thought this was a mere validation check …) was 'Why did you join the network?' The responses—which we thought better not to include in the official WHO document in full at the time—were interesting:

- 'As we are the home of the National School of Public Health we thought it impossible *not* to join'.
- 'We needed to create political momentum to invest in sustainable public transport and we thought that Healthy Cities would give us significant clout'.

- 'Being a Healthy City would enable us to bring back the old grandeur of our city as a spa town and would bring more tourists'.
- 'We wanted to show off our efforts at urban renewal internationally—Healthy Cities seemed a good forum'.
- 'We were very happy to be able to join a pan-European city network'.

I remember a conversation I subsequently had with Ron Draper. I was very disappointed that the majority of cities had not claimed they wanted to improve the health of their populations; that none of these reasons argued for the development of health policy; that the official rhetoric of determinants of health was absent from coal-face European Healthy Cities considerations. Thoughtful as ever, Draper said, 'I think they all have committed to investing in the determinants of health and well-being. They may not label it that way. But they do commit. And that is a major achievement'.

There is more than one lesson in this anecdote. Healthy Cities embrace diversity—they do not exclude a biomedical model of health in favour of a social model. They are not hung up on ownership of their programme by the health care sector (e.g. de Leeuw 2015). Goumans (1998), in fact, showed that for Dutch and British Healthy Cities ownership can be shared between the education sector, police, parks and recreation, and many more—very much beyond health *care*. Healthy Cities are archetypical for Health in All Policies development (see Chap. 17). For many professionals in the health (care) silo this generates ambivalence: how can you make health if it is not under the control of the health sector? In fact, health *is* created outside the health sector, and many Healthy Cities are prime examples of how that

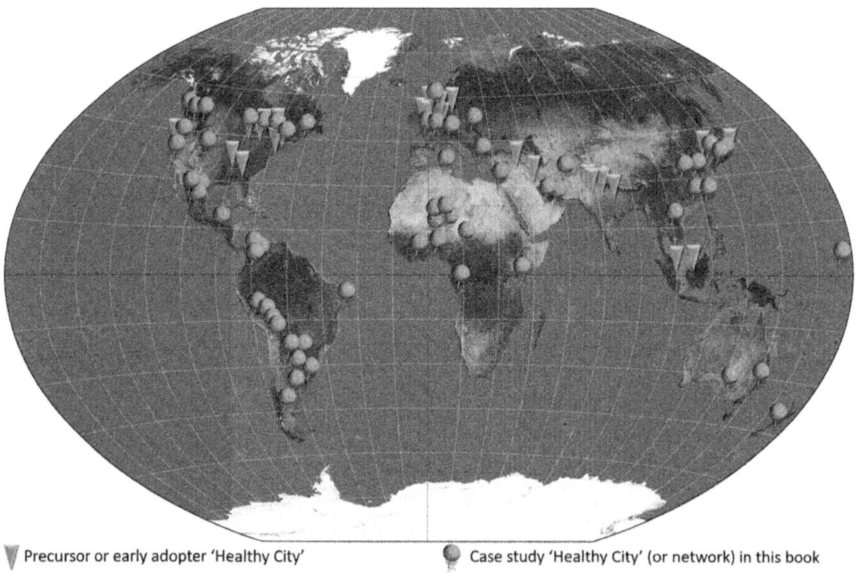

Fig. 1.3 Cities mentioned in this book

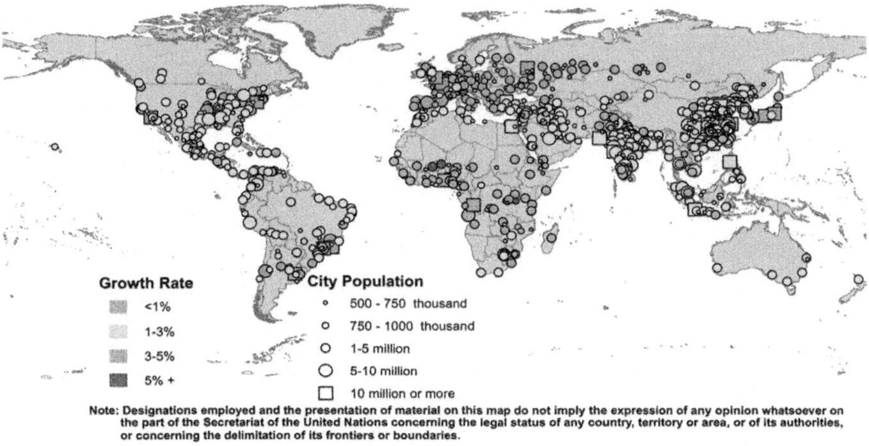

Fig. 1.4 Urbanization rates. Overview map of global urban growth. Source: UN World Urbanization Prospects Report, 2014 Update

works and can be managed: not through ever-increasing (social) epidemiological data collections and analyses, but through a much broader understanding of what creates pleasant, liveable urban environments (Crammond and Carey 2016). This is what this book is about.

For that very reason it is also interesting to develop a quick geography of Healthy Cities. In Fig. 1.3 we show the location of Healthy Cities in history mentioned in this book, and Healthy Cities mentioned as case studies. We see that Healthy City precursors in the early and mid-Holocene are found in the cradles of urbanization—the Middle East and the Valley of the Indus. The Healthy Cities ancestors during the industrialization era were European and US East Coast, whereas the small group of early adopters before the movement even was a project can be found in Canada, Britain and the Far East.

A superficial assessment of Fig. 1.3 would suggest that the current global spread of Healthy Cities (and our documentation of them) is fairly evenly distributed, and aligns with general patterns of urbanization.

Figure 1.4, however, shows that this is only partly true. There are very few cities in the sparsely populated tundras of Russia, Kazakhstan and Mongolia. The saying that most Canadians live within 50 km of the United States seems to be holding, and there are—fortunately perhaps—not too many cities in the deserts of Africa and Australia, or the jungles of Amazonia and Melanesia. But Fig. 1.4 shows that cities around the African equator are growing faster than their American and Asian counterparts. The Middle East, the Indian subcontinent and the Eastern half of China continue to see a slow but steady urbanization growth. Indonesia alone is

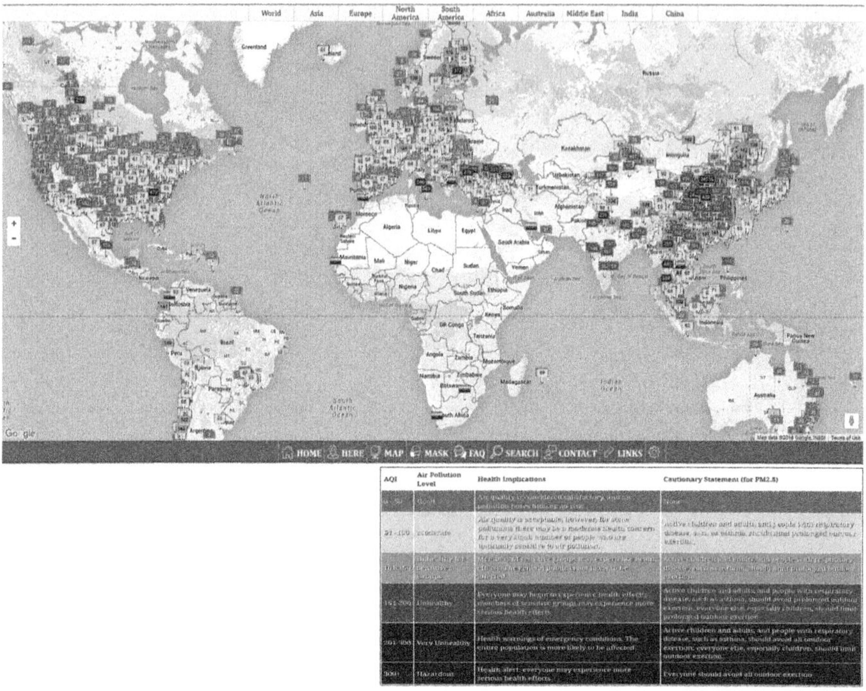

Fig. 1.5 The real-time World Air Quality Index, developed and maintained by a China-based social enterprise at www.aqicn.org. World data per city on 24 March 2016

showing greater urbanization than all its neighbours combined. Some of these patterns have not been reflected in this book. After initial successful efforts at establishing Healthy Cities on the subcontinent (for instance, in Cox's Bazaar in BanglaDesh—see Burton 1999 and Harpham et al. 2001), we have been unable to identify any urban environment in India, Pakistan, BanglaDesh, Myanmar or Indonesia that even remotely embraces the remit of this book—and we are quite certain that this is the fault of the editors, as we have heard rumours of Healthy City projects in Surabaya and Mandalay. Our investigative powers to track these down, however, failed.

The proverbial alien arriving from a galaxy far far away, looking down on our urbanizing world, may perhaps think that these lands (the Indian subcontinent, Eastern China, Indonesia) do not require Healthy Cities. Perhaps in their modern evolution they have not made the mistakes of the crowded American and European cities of the nineteenth century industrialization era. Alas—Fig. 1.5 shows that they, more than any other local government area, need some 'Healthy, Just, Happy, Sustainable, Winter, Resilient, Creative, Knowledge, Child-friendly, Festive, Slow, Smart, Conscious, Inclusive, Age-friendly Green Transition

Towns and EcoDistricts, and Safe Cities and Communities'-thinking. They are in the red-to-maroon zones of urban air pollution (with the two top tiers being hazardous).

This book is not for the cities in Fig. 1.3. It is for the cities on the wrong side of the track in Fig. 1.5. And we hope that in a second edition of this book they will have been inspired to take centre stage.

References

Alexander, A. (2009). *Britain's new towns: Garden cities to sustainable communities*. London: Routledge.

Alexander, E. R. (2013). Values in planning and design: A process perspective on ethics in forming the built environment. In C. Basta & S. Moroni (Eds.), *Ethics, design and planning of the built environment*. New York: Springer. Chapter 1.

Armelagos, G. J. (2009). The paleolithic disease-scape, the hygiene hypothesis, and the second epidemiological transition. In G. A. W. Rook (Ed.), *The hygiene hypothesis and Darwinian medicine* (pp. 29–43). Basel: Birkhäuser.

Ashton, J., & Ubido, J. (1991). The healthy city and the ecological idea. *Social History of Medicine, 4*(1), 173–180.

Barton, H., & Tsourou, C. (2013). *Healthy urban planning*. London: Routledge.

Basta, C., & Moroni, S. (Eds.). (2013). *Ethics, design and planning of the built environment*. New York: Springer.

Bengmark, S. (2000). Bacteria for optimal health. *Nutrition, 16*(7), 611–615.

Berry, B., & Kasarda, J. D. (1977). *Contemporary urban ecology*. New York: MacMillan.

Birn, A. E. (2011). Small (pox) success? *Ciência & Saúde Coletiva, 16*(2), 591–597.

Bocquet-Appel, J. P., & Bar-Yosef, O. (Eds.). (2008). *The neolithic demographic transition and its consequences*. Dordrecht, The Netherlands: Springer.

Boschi, N., & Pagliughi, L. M. (2002). Quality of life: Meditations on people and architecture. In *Proceedings of Indoor Air 2002*. Santa Cruz.

Buck, H. J. (2015). On the possibilities of a charming Anthropocene. *Annals of the Association of American Geographers, 105*(2), 369–377.

Burton, I. (1982). A human ecological approach to cities. In R. White, & I. Burton (Eds.), *Ecoville: Urbanization in the context of eco-development*. Institute for Environmental Studies, University of Toronto.

Burton, S. (1999). Evaluation of healthy city projects: Stakeholder analysis of two projects in Bangladesh. *Environment and Urbanization, 11*(1), 41–52.

Caldwell, J. C., & Caldwell, B. K. (2003). Was there a neolithic mortality crisis? *Journal of Population Research, 20*(2), 153–168.

Caragliu, A., Del Bo, C., & Nijkamp, P. (2011). Smart cities in Europe. *Journal of urban technology, 18*(2), 65–82.

Cassedy, J. H. (1962). Hygeia: A mid-Victorian dream of a city of health. *Journal of the History of Medicine and Allied Sciences, 17*(2), 217–228.

Clerc, O., & Stern, R. (2004). *Modern medicine: The new world religion: How beliefs secretly influence medical dogma & practices*. Fawnskin: Personhood Press.

Cohen, M. N. (1989). *Health and the rise of civilization*. New Haven: Yale University Press.

Crammond, B. R., & Carey, G. (2016) Policy change for the social determinants of health: The strange irrelevance of social epidemiology. *Evidence & Policy: A Journal of Research, Debate and Practice*. doi:10.1332/174426416X14538920677201.

Crawford, H. (2007). Architecture in the old Babylonian period. In G. Leick (Ed.), *The Babylonian world*. New York: Routledge.
Cunningham, D. A. (2005). A theme park built for one: The new Urbanism vs. Disney design in the Truman show. *Critical Survey, 17*(1), 109–130.
Davies, W. K. D. (Ed.). (2015). *Theme cities: Solutions for urban problems*. New York: Springer.
Davies, S. C., Winpenny, E., Ball, S., Fowler, T., Rubin, J., & Nolte, E. (2014). For debate: A new wave in public health improvement. *The Lancet, 384*(9957), 1889–1895.
de Leeuw, E. (2011). Theory and policy innovation for health: Where has the creativity and fun gone? *Health Promotion International, 26*(1), 1–3.
de Leeuw, E. (2015). Intersectoral action, policy and governance in European Healthy Cities. *Public health Panorama, 1*(2), 175–182.
De Moncan, P., & Heurteux, C. (2002). *Le Paris d'Haussmann*. Paris: Les éditions du Mécène.
Deelstra, T. (1985) *Urban development & mental health: A planners view*. In Paper presented at the 'Health in Towns' conference. Strasbourg: Council of Europe, December 3–5, 1985.
Deevey, E. S. (1963). General and urban ecology. In L. J. Duhl (Ed.), *The urban condition—People and policy in the metropolis* (pp. 20–32). New York: Basic Books. Chapter 3.
Derickson, K. D., & MacKinnon, D. (2015). Toward an interim politics of resourcefulness for the Anthropocene. *Annals of the Association of American Geographers, 105*(2), 304–312.
Diamond, J. M. (1998). *Guns, germs, and steel: A short history of everybody for the last 13,000 years*. London: Random House.
Doxiadis, C. A. (1970). Ekistics, the science of human settlements. *Science, 170*(3956), 393–404.
Doxiadis, C. A. (1977). *Ecology and ekistics*. Boulder: Westview Press.
Draper, R., Curtice, L., Hooper, J., & Goumans, M. (1993). *Review of the first five years: WHO Healthy Cities Project (1987–1992)* (Document EUR/ICP/HSC 644). Copenhagen: WHO Regional Office for Europe.
Duhl, L. J. (Ed.). (1963). *The urban condition—People and policy in the metropolis*. New York: Basic Books.
Ehrenreich, B. (2010). *Smile or die: How positive thinking fooled America and the world*. London: Granta books.
Ewing, R., & Clemente, O. (2013). *Measuring urban design. Metrics for livable places*. Washington, DC: Island Press.
Farazmand, A. (1998). Administration of the Persian achaemenid world-state empire: Implications for modern public administration. *International Journal of Public Administration, 21*(1), 25–86.
Fitzpatrick, M. (1978). *Environmental health planning: Community development based on environmental & health precepts*. Cambridge: Ballinger.
Foucault, M. (1983). *Naissance de la Clinique. Une archéologie du regard médical*. Paris: Presses Universitaires de France.
Frank, J. P. (1788). *System einer vollständigen medicinischen Poliezy*. Solwan.
Frassetto, L. A., Schloetter, M., Mietus-Synder, M., Morris, R. C., & Sebastian, A. (2009). Metabolic and physiologic improvements from consuming a paleolithic, hunter-gatherer type diet. *European Journal of Clinical Nutrition, 63*(8), 947–955.
Ghertner, D. A. (2015). *Rule by aesthetics: World-class city making in Delhi*. New York: Oxford University Press.
Goetze, A. (1950). Sin-Iddinam of Larsa. New tablets from his reign. *Journal of Cuneiform Studies, 4*, 83–118.
Goumans, M. J. (1998). *Innovations in a fuzzy domain: Healthy cities and (health) policy development in the Netherlands and the United Kingdom* (Doctoral dissertation). Maastricht University.
Grant, M. (2015). European healthy city network phase V: Patterns emerging for healthy urban planning. *Health Promotion International, 30*(Suppl 1), i54–i70.
Hancock, T. (2015). Population health promotion 2.0: An eco-social approach to public health in the Anthropocene. *Canadian Journal of Public Health, 106*(4), E252.

Hancock, T., & Duhl, L. (1986). *Promoting health in the urban context* (WHO Healthy Cities Papers No. 1). Copenhagen: FADL.
Hardoy, J. E., & Satterthwaite, D. (2014). *Squatter citizen: Life in the urban third world*. New York: Routledge.
Harpham, T., Burton, S., & Blue, I. (2001). Healthy city projects in developing countries: The first evaluation. *Health Promotion International, 16*(2), 111–125.
Hershkovitz, I., & Gopher, A. (2008). Demographic, biological and cultural aspects of the neolithic revolution: A view from the southern Levant. In J. P. Bocquet-Appel & O. Bar-Yosef (Eds.), *The Neolithic demographic transition and its consequences*. Dordrecht, The Netherlands: Springer.
International Commission on Stratigraphy. (2016). *Subcommission on quaternary stratigraphy. Working Group on the Anthropocene*. Retrieved February 1, 2016, from http://quaternary.stratigraphy.org/workinggroups/anthropocene/.
Kickbusch, I. (2007). Health governance: The health society. In D. V. McQueen, I. Kickbusch, & L. Potvin (Eds.), *Health and modernity: The role of theory in health promotion* (pp. 144–161). New York: Springer.
Kickbusch, I., Maag, D., & Saan, H. (2005) *Enabling healthy choices in modern health societies*. In Eighth European Health Forum, Bad Gastein, Austria, 5–8 October 2005.
Latour, B. (2015). Telling friends from foes in the time of the Anthropocene. In C. Hamilton, C. Bonneuil, & F. Gemenne (Eds.), *The Anthropocene and the global environmental crisis: Rethinking modernity in a new epoch* (pp. 145–155). Abingdon: Routledge.
Loder, A. (2014). 'There's a meadow outside my workplace': A phenomenological exploration of aesthetics and green roofs in Chicago and Toronto. *Landscape and Urban Planning, 126*, 94–106.
Lynch, K. (1981). *A theory of good city form*. Cambridge: MIT Press.
Mackenbach, J. P. (1994). The epidemiologic transition theory. *Journal of Epidemiology and Community Health, 48*(4), 329–331.
Maneglier, H. (1990). *Paris impérial: La vie quotidienne sous le Second Empire*. Paris: Armand Colin.
Martin, L. J. (2014). *Space cadets and rat utopias. The appendix—Futures of the past 2(3) August 28, 2014*. Retrieved November 6, 2015, from http://theappendix.net/issues/2014/7/space-cadets-and-rat-utopias.
McKeown, T. (2014). *The role of medicine: Dream, mirage, or nemesis?* Princeton: Princeton University Press.
McMichael, A. J. (2001). *Human frontiers, environments and disease: Past patterns, uncertain futures*. Cambridge: Cambridge University Press.
Montgomery, C. (2013). *Happy city: Transforming our lives through urban design*. New York: Macmillan.
Moser, S. C., & Dilling, L. (2011). *Communicating climate change: Closing the science-action gap. The oxford handbook of climate change and society* (pp. 161–174). Oxford: Oxford University Press.
Omran, A. R. (1971). The epidemiologic transition: A theory of the epidemiology of population change. *The Milbank Memorial Fund Quarterly, 49*(4), 509–538.
Palti, I. (2016). Defining conscious cities. *Journal of Urban Design and Mental Health, 1*(1), 1.
Philips, E. B., & LeGates, R. T. (1981). *City lights: An introduction to urban studies*. New York: Oxford University Press.
Porter, D. (1999). *Health, civilization, and the state: A history of public health from ancient to modern times*. New York: Psychology Press.
Possehl, G. L. (2002). *The Indus civilization: A contemporary perspective*. Lanham: Rowman Altamira.
Riggio, E. (2002). Child friendly cities: Good governance in the best interests of the child. *Environment and Urbanization, 14*(2), 45–58.

Steffen, W., Crutzen, P. J., & McNeill, J. R. (2007). The Anthropocene: Are humans now overwhelming the great forces of nature. *AMBIO: A Journal of the Human Environment, 36*(8), 614–621.

UN-HABITAT. (2010). *State of the world's cities 2010/2011: Bridging the urban divide*. London: Earthscan.

University of Waterloo. (1984). *Technological futures & human settlements* (Principal investigator, Len Genier). Waterloo, ON.

Unschuld, P. U. (1985a). *Medicine in China: A history of ideas*. Berkeley: University of California.

Unschuld, P. U. (1985b). Die traditionelle chinesische Medizin im 20. Jahrhundert: Überlebenskampf und Legitimationsstrategien. *Medizinhistorisches Journal,* (H. 3), 263–269

Walker, M., Johnsen, S., Rasmussen, S. O., Popp, T., Steffensen, J. P., Gibbard, P., … Cwynar, L. C. (2009) Formal definition and dating of the GSSP (Global Stratotype Section and Point) for the base of the Holocene using the Greenland NGRIP ice core, and selected auxiliary records. *Journal of Quaternary Science, 24*(1), 3–17.

Weber, M. (1999) *Schriften und Reden XXII. Wirtschaft und Gesellschaft. Die Wirtschaft und die gesellschaftlichen Ordnungen und Mächte*. Nachlaß.: Teilband 5: Die Stadt (Vol. 5). Mohr Siebeck.

Whitmee, S., Haines, A., Beyrer, C., Boltz, F., Capon, A. G., de Souza Dias, B. F., … Horton, R. (2015). Safeguarding human health in the Anthropocene epoch: report of The Rockefeller Foundation—Lancet Commission on planetary health. *The Lancet, 386*(10007), 1973–2028.

World Health Organization. (2007). *Global age-friendly cities: A guide*. Geneva: World Health Organization.

World Health Organization, & UN Habitat. (2016). *Urban health. Global report on urban health: Equitable, healthier cities for sustainable development*. Geneva: World Health Organization.

Chapter 2
Equitable and Healthy City Planning: Towards Healthy Urban Governance in the Century of the City

Jason Corburn

Introduction

City planning is a discipline and professional practice that includes political processes, institutions, and discourses that generate policies, rules, and physical plans that shape where we live, learn, work, and play in cities and towns. Yet city planning is much more than rules and regulations concerning land use and the built environment; planners regularly make discretionary decisions that shape the implementation of formal rules, can provide greater public access to (or stymie) forums for democratic decision-making, and interpret national and even global ideologies about urbanization and development. In these ways, city planning helps structure the distribution of social, physical, and economic 'goods and bads' that influence human health and explain persistent urban health inequities. In other words, city planning acts as a structural determinant of health through its formal and informal institutions, its micro and macro politics, and how these intersect with our day-to-day activities, from access to employment and food to the qualities of our neighbourhoods and housing to the allocation and distribution of social and health care services.

The twenty-first century is the century of the city; the United Nations Population Reference Bureau predicts that by 2050, 70 % of the world's population will be living in metropolitan areas. Thus, city planning must be understood as part and parcel of global health. In the nineteenth century, city planning emerged as a modern profession and discipline with close ties to public health initiatives, including tenement housing reforms, the construction of urban water supply and sewerage systems, and the design of parks and playgrounds. While having similar visions of the equitable and healthy city, the work of professionals in each field diverged throughout the

J. Corburn (✉)
University of California, Berkeley, Center for Global Healthy Cities.
School of Public Health. Berkeley, CA 94720, USA
e-mail: jcorburn@berkeley.edu

E. de Leeuw, J. Simos (eds.), *Healthy Cities*,
DOI 10.1007/978-1-4939-6694-3_2

twentieth century, which has contributed to the persistence of health inequities for many poor and minority urban populations. Efforts in the early years of the twenty-first century aimed at reconnecting city planning and public health have been limited and largely focused on a narrow framing of the intersection of the fields, such as whether and how the built environment might change behaviours to increase physical activity and related health outcomes. The built environment and health framing tends to ignore the politics of planning and, as I will highlight in this chapter, important lessons from each field's modern history. Throughout this chapter, I identify at least three lessons from each field's modern history that present challenges for reconnection related to the structural determinants of urban health inequities, including an overemphasis on physical changes for improving social conditions, scientific rationality, and professionalization and fragmentation of the disciplines. Building on the historic review and these themes, the chapter suggests a set of reconnection strategies and practices for moving towards 'healthy and equitable city planning' that will require continued critical engagement with the histories of the fields along with new issues and problem framings, investigative and analytic techniques, and inclusive and deliberative public processes that together can generate new norms, discourses, and practices for greater health equity. In short, healthy city planning will require new commitments to inject health and social justice into *urban governance*. Healthy urban governance will mean that the decision-making processes and institutions that shape places are altered to focus on equity, and that new decisions ensure a more equitable distribution of the positive physical and social characteristics of places that promote urban health.

Early Connections Between City Planning and Health, 1840s–1890s

In the 1820s, French epidemiologist Louis René Villermé noted that the wealthier the Parisian neighbourhood, or arrondissement, the lower was the mortality rate and likelihood of illness. By 1842, Edwin Chadwick had built on Villermé's work and published the *Report on the Sanitary Conditions of the Labouring Population in Great Britain*, documenting that the 'gentry and professional' classes lived longer than 'labourers and artisans' (1842:87). Miasma—filth or dirty air—was the leading theory of disease causation, and sanitary commissions were created in European and American cities to clean up urban environments with the hope of arresting infectious disease epidemics (Duffy 1990). Sanitary engineers tended to address urban health issues by employing new technologies to remove waste by, for instance, piping it away from cities into rivers and oceans (Melosi 2000). When removing the miasma did not seem to reduce disease, the sick were removed from society. *Contagion*, the belief in the direct passage of poison from one person to another, led to large quarantines of immigrants and justified state-sponsored interventions in the economy.

Research and practice linking planning and public health during this era included the sanitary survey, park and playground planning, and the work of settlement houses. Intensive spatial and social surveys of urban environments were the epidemiologic methods of this era, such as the sanitary maps of Bethnal Green and Liverpool prepared by Chadwick and the 1878 Memphis sanitary survey that described every street, structure, and individual lot within the city to determine the environmental conditions that might 'breed' diseases (Peterson 1979:90). Progressive Era park and playground movements advocated for urban play spaces next to schools so that gymnasiums, reading rooms, and baths could all be used for children's recreation, literacy, and hygiene.

The women of Hull House in Chicago, influenced by the burgeoning Chicago School of Sociology that initiated the study of neighbourhood effects on well-being, worked with residents to document unsanitary neighbourhood and workplace conditions and advocated on behalf of immigrants new social policies (Hull House Residents 1895). The researchers at Hull House, particularly Florence Kelley, borrowed methods from public health researchers in England at the time, namely Charles Booth, to conduct some of the first community health surveys that included neighbourhood mapping of both non-infectious and infectious diseases in the United States.

By the end of the nineteenth century, many urban health interventions focused on physical removal of both 'environmental miasmas'—garbage, wastewater, air pollution, and the like—and 'undesirable and sick' people. For sanitarians, the local solution to pollution was removal and dilution, and downstream environmental health impacts were often ignored (Tarr 1996). Housing reformers were split between those advocating physical improvements, such as bathrooms, ventilation, and fire escapes, and those calling for the construction of safe and affordable housing. Many housing and sanitary reforms were grounded in the belief that technological and design innovations could not only improve living conditions but also make 'immoral' slum-dwellers more orderly and healthy (Fairfield 1994).

Germ Theory and Planning 'the City scientific': 1900s–1940s

By the turn of the new century, it was well known in public health that both miasma and contagion failed to explain certain aspects of urban health, such as why, with ubiquitous filth, epidemics occurred only some times and in some places. By now, the driving ideology in public health had shifted to *germ theory*, which stated that microbes were the specific agents causing infectious disease (Susser and Stein 2009). Medical treatment and disease management began to replace strategies of physically removing harms, and public health shifted towards laboratory research and interventions aimed at eliminating bacteria, such as vaccinations and chlorination of municipal drinking water supplies. Separate municipal bureaucracies emerged during this era to manage these urban improvement schemes, such as water service delivery infrastructure, sewage systems, public housing, parks and recreation, school-based health, and others, resulting in greater fragmentation and segmentation of work in public health and city planning (Peterson 1979).

Public Health, Social Justice, and the Emergence of American City Planning

Social inequalities kept the fields of planning and public health somewhat linked. In 1906 W.E.B. Du Bois published *The Health and Physique of the Negro American* and used data from census reports, vital statistics and insurance company records to generate some of the first documentation of the poor health status of African-Americans in both southern and northern cities in comparison to white Americans. Du Bois questioned whether improvements in science, technology, housing, and other services were benefiting all Americans and concluded that racial health disparities were a result of social conditions—not eugenic ideas of inherent racial inferiority that were commonly accepted at the time (Du Bois 1906).

As American planners prepared for their first national conference in the early years of the twentieth century, Benjamin Marsh and his group, the New York Committee on the Congestion of Population (CCP), argued that city planning ought to embrace more explicitly a social justice agenda. In a 1908 article entitled *City Planning in Justice to the Working Population*, Marsh argued that American planning ought to make 'the right of the citizen to leisure, to health, to care in sickness, to work under normal conditions, and to live under conditions which will not impair his health or efficiency' a 'foresight' rather than an afterthought (1908:1514). Fredrick Law Olmsted Jr., the first president of the American City Planning Institute and its conference, responded by defining the field as primarily concerned with the means of 'circulation', the distribution of public space and the development of private land (Olmsted 1910:69). As Peterson (2003:249) noted, 'while Marsh battled for justice, Olmsted focused more on planning itself, especially ways to build it up as a technically effective field'. The comprehensive, rational, and aesthetically focused 'City Beautiful' ideas of Olmsted took hold, while the human health and social justice concerns of Marsh and the CCP largely fell off the agenda. By the fifth National City Planning conference in 1913, entitled 'The City Scientific', American planning had defined itself largely as a technocratic profession aimed at designing efficient cities (Fairfield 1994).

The Biomedical Model, Housing Policies, and the Fractured City: 1950s–1990s

By mid-century the driving theory in public health would shift again to a biomedical model of disease (Susser and Stein 2009). This model combined laboratory science with a focus on individuals and attributed morbidity and mortality to molecular-level pathogens brought about by lifestyles, behaviours, hereditary biology, or genetics. The biomedical model provided public health with explanations for emerging chronic diseases such as heart disease, and shifted research and interventions to personal health 'risk factors' such as smoking, diet, and exercise. At the

same time, New Deal programmes helped usher in the era of the 'bureaucratic city', where a new set of impersonal public institutions, staffed by newly credentialled professionals, helped to establish disciplinary boundaries and further separate planning from public health (Duhl and Sanchez 1999).

A series of Federal Housing Administration policies in the USA helped to fracture poor urban communities of colour both physically and socially in the 1940s and 1950s by refusing to insure mortgages on older homes and razing poor communities of colour under the guise of urban renewal. Planners and public health practitioners often justified the labelling of an area as blighted and thus subject to demolition by using healthy housing and neighbourhood guidelines recommended in the 1948 American Public Health Association publication, *Planning the Neighborhood*, which offered design guidelines for 'healthy' new neighbourhoods, but no guidance to *improve* existing housing (APHA 1948). The result was widespread displacement of the urban poor and policies that acted to shut out people of colour from the health-promoting benefits of suburban living, such as home ownership, capital accumulation, access to better-funded schools, and participation in the growing suburban economy.

Urban Activism and Rejection of Rational Planning

Social movements in the 1950s and 1960s pushed back against the displacement and inequities perpetuated by federal housing and other urban policies. Social movements helped pass important legislation such as Medicaid and civil rights laws that would improve the health of all populations, but especially of the urban poor and people of colour. During this time planning and urban policymakers, supported by federal War on Poverty programmes, turned back to improving neighbourhoods, and the Office of Economic Opportunity supported a national network of community health centres the first two of which were located in Columbia Point, Massachusetts, and Mound Bayou, Mississippi. The community health centre model focused on providing both immediate care and preventive services for the poor, instead of treating people and sending them back to the living and working conditions that had made them sick in the first place. According to Jack Geiger, a physician and leader of the health centre movement, community health centres aimed to take a holistic view of health by 'writing prescriptions for the building blocks of health' such as food, rent, jobs, and sanitation (Geiger 2005).

By the mid-1980s academics, government agencies, and the World Health Organization came together to create the Healthy Cities programme in order to reinvigorate the linkages between city planning and public health (Duhl and Sanchez 1999). Focused primarily in Europe, the programme enrolled cities to draft 'healthy city plans' and share examples of the opportunities and barriers for planning more healthy cities (Barton et al. 2009). By the 1990s public health researchers began to reconceptualize explanations for the distribution of disease across populations in order to explain health disparities, energizing the field of social epidemiology

(Krieger 2011). By the early twenty-first century place inequities such as residential segregation, urban divestment, and environmental injustices began to be seen by epidemiologists as linked social determinants of health that might help explain distributions of death and disease across different population groups and places and issues needing urgent policy attention.

Towards Healthy and Equitable Urban Governance

As this brief review has suggested, reconnecting city planning and public health to address structural inequities must be attentive to the political—not just the scientific, technical, and physical—design challenges of healthy urban governance, which demands more than just governmental reform and includes such political processes as identifying and framing new policy issues, generating appropriate standards of evidence, constituting some social actors as 'experts', adjudicating scientific uncertainty and different knowledge claims, securing public accountability for decisions, and implementing and monitoring policies. Governance practices are, in short, the rules, norms, and processes for exercising power over collective actions and, when inattentive to social inequalities, often result in unequal population health outcomes by upholding existing distributions of resources like political power, wealth, and knowledge (Corburn 2009).

The key challenge for reconnecting planning and public health in the twenty-first century is to learn from each field's history and jointly develop strategies that address the root causes of poor health, not just to devise interventions aimed at specific diseases or individual behaviours. Grappling with the political challenges of reconnecting the fields demands that, for instance, planners experiment with new institutional designs that can handle cross-disciplinary conflicts over political power, social justice, and health values, such as when a state or private sector-sponsored development project clashes with the health objectives of a local community. Planners will also need to question critically the adequacy of existing norms and institutions that help determine how practitioners use or abuse power, respond to or even resist market forces, work to empower some groups and dis-empower others, promote multi-party decision-making, or simply rationalize decisions already made.

Health Impact Assessment: Preventing, Not Displacing, Urban Health Risks

The first lesson from the histories of planning and public health for healthy urban governance is that removal and displacement of 'problems' does not necessarily improve urban health and may leave the most vulnerable groups and places worse off. Health impact assessment (HIA) is one example of a precautionary practice that can help reconnect planning and public health and prevent problem displacement. Now widely used for healthy urban planning in Europe, HIA is both an analytic tool and political process that might bring together the built and social environmental factors that influence urban health. The San Francisco Department of Public Health has used HIA to assess a proposed living wage ordinance, new housing proposals and rezoning plans. During one HIA process, called the Eastern Neighborhoods Community Health Impact Assessment, the San Francisco Department of Public Health organized a participatory planning process that included over 40 governmental and non-governmental organizations for evaluating the positive and negative human health impacts of a proposed rezoning plan. During the process, stakeholders collaboratively defined the elements of a healthy place, discussed how land use does or does not influence these elements, and investigated how a set of rezoning proposals and potential alternatives might influence the health of the largely low-income, immigrant communities of the Mission and South of Market neighbourhoods in San Francisco (Corburn 2009).

One outcome of the Eastern Neighborhoods Community Health Impact Assessment process was the Healthy Development Measurement Tool, an analytic method outlining the broad social indicators, land use development goals, and quantitative and qualitative data for healthy urban development (now available at http://www.sfindicatorproject.org). While no panacea, health impact assessment is beginning to bring together government agencies, advocacy organizations, and researchers, groups that rarely work together, suggesting that it may be a process for breaking-down disciplinary boundaries and other institutional barriers impeding efforts to reconnect planning and public health (Corburn 2009).

Integrated, Spatial Planning for Returning Inmates

Healthy and equitable city planning must also address the urban 'epidemic' of incarceration and prisoner re-entry. American jail populations are disproportionately young, urban, African-American and Latino men—the same groups that have the poorest health. Incarceration has created a planning and public health challenge by spatially concentrating the removal of young men from both families and the workforce and the social stress that accompanies their return to their neighbourhood. The constant cycle of incarceration and re-entry in American cities in particular has brought the health issues of prisons into neighbourhoods, including infectious disease, addiction, mental health problems, and routine physical violence. In addition, returning inmates face homelessness, family evictions from public housing, denial of food stamps, terminated Medicaid benefits, and regular workplace discrimination (Freudenberg et al. 2005; Rich et al. 2011).

In New York City, a neighbourhood-focused reintegration project called the Community Reintegration Network is working to address the strains on public safety, community health, family stability, and municipal budgets that come from neighbourhood concentrations of former inmates. As a partner in this coalition, the

Vera Institute of Justice launched Project Greenlight, which prepares inmates for release and reintegration by matching them with programmes and organizations in their home community, including supportive and special needs housing, drug treatment programmes, job training, and health clinics (Brown and Campell 2005). While these initiatives aim to build one-stop locations for housing, job training, and social and health services, city planners have a unique opportunity to reconnect with the social justice roots of the profession by participating in community-based re-entry programmes, and might offer their knowledge of spatial and social programming to help reduce recidivism. In the twenty-first century, incarceration and re-entry must become planning and public health issues so that, for instance, municipal funds are redirected to provide the place-based housing, education, employment, and social services necessary to reduce the spatial impacts of incarceration.

New Institutions for Healthy and Equitable City Planning

A second set of related lessons from the histories of urban planning and public health is that an over-reliance on technological solutions and physical designs, without accompanying institutional change, fails to protect the most vulnerable population groups. Healthy city governance will depend, in part, on robust community-based institutions. The histories of planning and public health suggest that, for instance, economic development in the absence of community-based institutions may fail to improve living conditions because local institutions help ensure that the benefits of development are distributed to meet local needs. For example, a coalition of community-based organizations called the Figueroa Corridor Coalition successfully negotiated a community benefit agreement with the Los Angeles Arena Land Company, a private developer, over the Staples Center Phase II project in downtown Los Angeles. The community benefit agreement is a legally binding agreement guaranteeing that the developer include affordable housing and public amenities such as new parks, and that the new commercial establishments hire local residents at a living wage (Gross et al. 2002). Reminiscent of struggles for early twentieth century workplace and neighbourhood improvements, organized labour unions are acting as integral partners in shaping these agreements, recognizing that their members increasingly come from low-income communities and are working in service-sector jobs.

Tempering Professional Models with Lay Expertise

A final lesson from the histories of urban planning and public health is that if what local people know, how they move through the world, and their subjective experiences with illness and their environment are ignored, interventions ultimately fail. Local knowledge, including the experiences and narratives shared by populations living with persistent hazardous exposures, chronic diseases, and social marginalization, is a valuable form of expertise that can improve scientific analyses, enhance the relevance of health-promoting interventions, and enrich the democratic character of public decisions (Corburn 2005). Drawing from the lessons of Progressive Era neighbourhood surveys and the neighbourhood health centre movement, twenty-first century planning and public health might re-embrace local knowledge by promoting and supporting networks of community health workers (CHWs).

Often called *promotoras de salud*, CHWs are frontline lay health outreach workers who organize neighbourhood residents around health issues and provide health education, basic disease screening and translation and interpretation services. CHWs often act as the bridge-builders between poor, minority and immigrant communities and professional health services and institutions. Since CHWs live in the places within which they work, they have a keen awareness of local culture and practices, and often have lived experience of how macro-social structures impact on the daily lives of local residents. Ultimately, CHWs tap into local knowledge to improve health by helping to build community networks, formal associations and other connections between socially dissimilar persons or groups that can be crucial both for securing immediate health-promoting resources and for organizing long-term policy advocacy coalitions.

An asthma prevention and management programme in Brooklyn, New York, relies on CHWs to organize residents and articulate the barriers asthmatics and their families face in managing this disease. The Community Health Educator (CHE) team at an organization called *El Puente* (www.elpuente.us) works in the Latino neighbourhoods of south Williamsburg, Bushwick, and Bedford-Stuyvesant, performing home visits and holding community-wide focus groups to assist families to manage asthma. During a series of focus group meetings, the CHE team heard from local women that many asthmatic families were turning to home and herbal remedies in place of physician-prescribed medications (Ledogar et al. 2000). When asked why, the women, many of whom were Spanish speakers and Dominican immigrants, said that time-tested herbal treatments acted as a familiar practice in a world where asthma caused unparalleled social and economic disruption to families and communities by forcing children to miss school, adults to miss work, and families to fall further into poverty. The information from residents about the use of home remedies and the social and cultural role they play in immigrants' lives was translated by CHWs into new interventions aimed at addressing the multiple dimensions of asthma (Corburn 2005). Since community and experiential knowledge can ensure interventions are contextually and culturally relevant, healthy twenty-first century planning must learn from and with local experts.

Planning and Public Health for the Twenty-First Century

This chapter has aimed both to critically review the histories of city planning and public health and, by drawing lessons from this review, to suggest strategies to reconnect these two fields to confront twenty-first century challenges in promoting urban health equity. The contemporary challenges to achieving this are daunting: global spread of disease, trans-boundary environmental pollution, burgeoning urban and slum populations, and increasing health disparities mirroring widening class inequalities. As momentum for the reconnection effort builds, reflected in recent journal articles, conferences, and government and foundation efforts, the lessons from each fields' historic missteps ought to be given closer scrutiny.

The recommendations offered here should also be viewed in a comparative perspective, since they reflect experiments in 'healthy urban planning' from around the world. For instance, the World Health Organization's City Action Group on Healthy Urban Planning (www.euro.who.int/document/e82657.pdf) and the United Kingdom Office of the Deputy Prime Minister's Creating Sustainable Communities initiative (www.odpm.gov.uk) both principally focus on how to reverse social exclusion, and inequality more generally, through the design of new collaborative governance schemes, the development of state regulations, and the building of non-governmental capacity. In the United Kingdom, London and Merseyside are using HIA to address planning and development decisions (Barnes and Scott-Samuel 2002). Efforts to upgrade slums in developing countries, including strategies for achieving the Sustainable Development Goals, frequently call for the inclusion of local knowledge and the integration of urban planning with health equity (WHO/UN-HABITAT 2010; https://sustainabledevelopment.un.org/index.php?menu=1252). Finally, one of the most successful campaigns to reduce AIDS in Haiti, the poorest country with the highest rate of HIV infection, was designed around networks of community health workers (Farmer 1999).

Contemporary efforts to reconnect planning and public health can learn from the past to understand how current trends gained resonance and what alternative futures are possible. Alternative paradigms can be both practical and socially just, as the examples here suggest; however, since many of the recommendations are offered as frameworks, not specific guidance, more work needs to be done to evaluate which practices might work best in specific cultural and political contexts. As efforts to reconnect urban planning and public health move forward, a historic perspective is necessary to effectively re-engage with city planning as a structural determinant of health.

References

American Public Health Association (APHA) Committee on the Hygiene of Housing. (1948). *Planning the neighborhood: Standards for healthful housing*. Chicago: Public Administration Service.

Barnes, R., & Scott-Samuel, A. (2002). Health impact assessment and inequalities. *Pan American Journal of Public Health, 11*(5/6), 449–453.

Barton, H., Grant, M., Mitcham, C., & Tsourou, C. (2009). Healthy urban planning in European cities. *Health Promotion International, 24*, i91–i99.

Brown, B., & Campell, R. (2005). *Smoothing the path from prison to home*. New York: Vera Institute of Justice. Retrieved February 23, 2009, from www.vera.org/publication_pdf/319_590.pdf.

Chadwick, E. (1842, May). *Report on the sanitary condition of the labouring population and on the means of its improvement*. London. Retrieved August 22, 2012, from www.deltaomega.org/ChadwickClassic.pdf.
Corburn, J. (2005). *Street science: Community knowledge and environmental health justice*. Cambridge: MIT Press.
Corburn, J. (2009). *Toward the healthy city: People, places and the politics of planning*. Cambridge: MIT Press.
Du Bois, W. E. B. (1906). *The health and physique of the Negro American: Report of a social study made under the direction of Atlanta University*. Retrieved July 23, 2012, from http://www.archive.org/details/healthphysiqueof00dubo.
Duffy, J. (1990). *The sanitarians: A history of American public health*. Chicago: University of Illinois Press.
Duhl, L. J., & Sanchez, A. K. (1999). *Healthy cities and the city planning process*. Retrieved November 11, 2012, from www.who.dk/document/e67843.pdf.
Fairfield, J. D. (1994). The scientific management of urban space: Professional city planning and the legacy of progressive reform. *Journal of Urban History, 20*, 179–204.
Farmer, P. (1999). *Infections and inequalities: The modern plagues*. Berkeley: University of California Press.
Freudenberg, N., Daniels, J., Crum, M., Perkins, T., & Richie, B. E. (2005). Coming home from jail: The social and health consequences of community reentry for women, male adolescents, and their families and communities. *American Journal of Public Health, 95*(10), 1725–1736.
Geiger, J. (2005). The unsteady march. *Perspectives in Biology and Medicine, 48*(1), 1–9.
Gross, J., LeRoy, G., & Janis-Aparicio, M. (2002). *Community benefit agreements: Making development projects accountable*. Washington, DC: Good Jobs First and the California Public Subsidies Project.
Krieger, N. (2011). *Epidemiology and the people's health: Theory and context*. Cambridge: Oxford University Press.
Ledogar, R., Penchaszadeh, A., & Garden-Acosta, L. (2000). Asthma and Latino cultures: Different prevalence reported among groups sharing the same environment. *American Journal of Public Health, 90*, 929–935.
Marsh, B. (1908). City planning in justice to the working population. *Charities and the Commons, 19*, 1514–1518.
Melosi, M. (2000). *The sanitary city: Urban infrastructure in America from colonial times to the present*. Baltimore: Johns Hopkins University Press.
Olmsted, F. L., Jr. (1910). City planning: An introductory address. *American Civic Association, 2*(4), 1–30.
Peterson, J. (1979). The impact of sanitary reform upon American urban planning, 1840–1890. *Journal of Social History, 13*, 83–103.
Peterson, J. (2003). *The birth of city planning in the United States, 1840–1917*. Baltimore: Johns Hopkins University Press.
Residents, H. H. (1895). *Hull House maps and papers*. New York: Thomas Y. Crowell.
Rich, J. D., Wakeman, S. E., & Dickman, S. L. (2011). Medicine and the epidemic of incarceration in the United States. *New England Journal of Medicine, 364*(22), 2081–2083.
Susser, M., & Stein, Z. (2009). *Eras in epidemiology: The evolution of ideas*. Oxford: Oxford University Press.
Tarr, J. A. (1996). *The search for the ultimate sink: Urban pollution in historical perspective*. Akron, OH: University of Akron Press.
World Health Organization (WHO), & United Nations Human Settlements Programme (UN-HABITAT). (2010). *Hidden cities: Unmasking and overcoming health inequities in urban settings*. Kobe: World Health Organization Centre for Health Development. Retrieved January 16, 2011, from http://www.hiddencities.org/report.html.

Chapter 3
From Health Care to the Promotion of Health: Establishing the Conditions for Healthy Communities in Canada

Ann Pederson and Irving Rootman

Introduction

In Canada we claim the Healthy Cities movement as part of our health promotion legacy, although two of its early instigators were an American (Leonard Duhl) and a transplanted Englishman (Trevor Hancock). A number of conditions in Canada in the 1970s and 1980s laid the groundwork for the articulation of the Healthy Cities concept, such as the country's post-war universal health insurance scheme, the establishment of the Long Range Health Planning Branch within the federal Health and Welfare bureaucracy, and a decentralized, local public health system. These factors created a context of innovation and leadership that contributed to the eventual hosting of the Beyond Health Care conference in Toronto and a 1-day workshop on the healthy city in 1984; but without the practical, political and discursive context that gave rise to the field of health promotion itself, starting with the release in 1974 of *A New Perspective on the Health of Canadians* (consistently referred to worldwide as the Lalonde Report), the workshop and the global movement it inspired would not likely have occurred. Thus, the emergence of the Healthy Cities movement is intimately tied up with the articulation of the concept of health promotion as explicitly concerned with the contributions of lifestyle and environments to health in Canada and the World Health Organization (WHO), and its alignment with the organization of public health as a municipal responsibility in Canada.

In this chapter we pay particular attention to the federal government activities that generated the Lalonde Report and its successor, *Achieving Health for All* (Epp 1986) over a decade later. We rely heavily on historical analyses prepared by McKay

A. Pederson
BC Women's Hospital + Health Centre, Vancouver, BC, Canada
e-mail: apederson@cw.bc.ca

I. Rootman (✉)
University of Victoria, Victoria, BC, Canada
e-mail: irootman@telus.net

E. de Leeuw, J. Simos (eds.), *Healthy Cities*,
DOI 10.1007/978-1-4939-6694-3_3

(McKay 2000) and Glouberman (Glouberman 2001), among others (Hancock et al. 1985; Pederson et al. 1994; Raphael 2008a), as well as our own engagement with the field of health promotion—for we are partly to blame for Canada's fame in relation to health promotion. For nearly three decades we have reflected on the emergence, development, decline and renewal of health promotion in Canada (e.g. O'Neill et al. 2007; Pederson et al. 1994; Pederson et al. 2005; Rootman et al. 2012). As researchers, advocates and practitioners, we have contributed to the substance of a discourse of health promotion in and about Canada, and been agents in its organization, application and transformation. From our unique vantage point as 'inside agitators' (Eisenstein 1996), we think that an interesting confluence of events and actors produced the conditions in Canada that helped establish the conditions for the Healthy Cities programme.

Although Canada was not the first or only country to grapple with rising health care costs, an epidemiological transition and various social movements in the 1960s and 1970s, it was nevertheless one of the first to formally challenge the view that health care was the central resource for producing health. Its core framework, the health field concept, recognized the limits of what formal health care could do to create health and identified lifestyle, biological endowment and the physical environment as three other contributory factors—early forerunners of the contemporary frameworks of the 'determinants of health' (Baum 2008). Having developed the health field concept, Canada was uniquely positioned to act with the European Regional Office of WHO in Copenhagen to articulate its emerging framework for health promotion.

If it ever was, Canada is no longer the country with the strongest network of health promotion research, policy and practice. Decades of restructuring, cost containment and shifts in political culture, combined with the emergence of the competing discourse of 'population health', have all contributed to a context in which health promotion, and projects such as the Healthy Cities/Healthy Communities initiative, has been severely curtailed (Raphael 2008b). However, three elements of health promotion: a social view of health, a commitment to social justice through equity, and support for balancing individual and collective responsibility for health, have outlived the particular political and organizational structures established to support them, illustrating the remarkable resilience of core health promotion concepts and principles. As Kickbusch (2007) suggested, health promotion is more like a rhizome than a tree. The continuing emergence of shoots of support for health promotion and small-scale Healthy Communities networks in this country suggest that there continues to be an important alternative discourse and practice, spawning related initiatives such as Age-friendly Cities (World Health Organization 2007), Smoke-free Cities (World Health Organization 2011) and Active Cities or Communities (see http://www.triple-ac.net/).

The Significance of the Lalonde Report

McKay (2000) has argued that a set of unique conditions and actors contributed to the development of the Lalonde Report. Based on interviews with individuals involved in its preparation, as well as the contents of official records from the House of Commons (the elected parliament), she suggested that Marc Lalonde, as Canadian Minister of Health and Welfare, released *A New Perspective on the Health of Canadians* as a green paper intended to challenge the conventional wisdom of the time regarding health. Specifically, it was intended to challenge the notion that the main determinant of health was health care, by suggesting that lifestyle, environment and human biology were also important, if not more important (see Fig. 3.1). This idea grew out of the work of the British epidemiologist Thomas McKeown (Mckeown et al. 1972), who argued that 'one had to go beyond standard medical services to improve the health of the population' (Glouberman 2001:10). The 'McKeown thesis' was that improved social conditions rather than medical or public

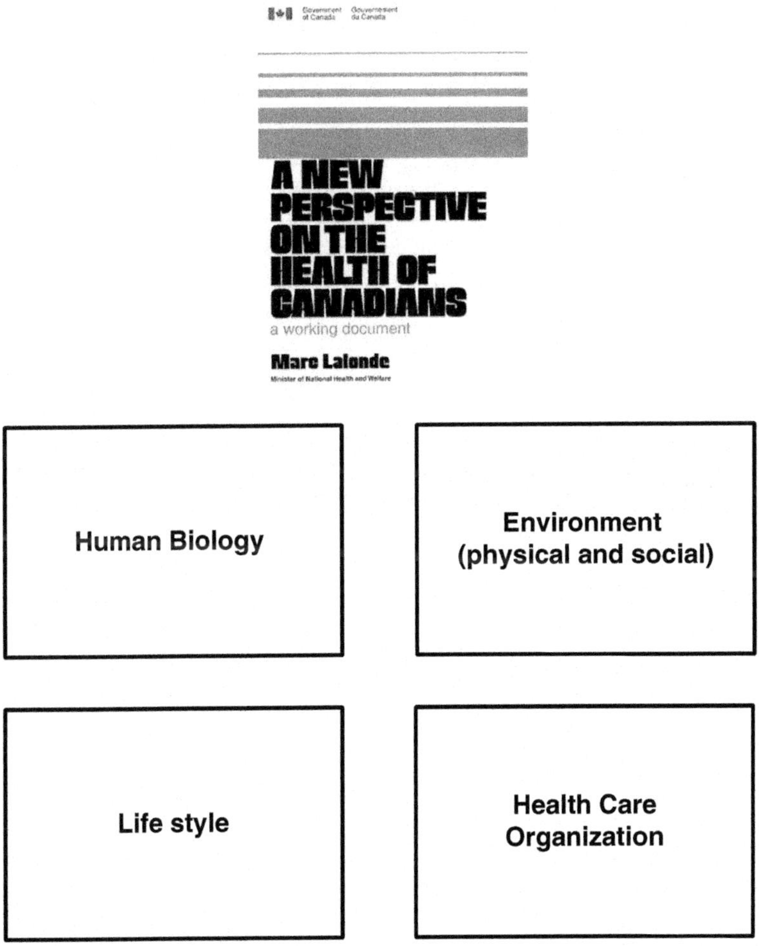

Fig. 3.1 Elements of the Health Field Concept Source: Lalonde, M. (1974). A new perspective on the health of Canadians (p. 31). Ottawa: Government of Canada

health interventions were responsible for the dramatic reductions in mortality in the eighteenth and nineteenth centuries (Bynum 2008; Link and Phelan 2001). Although still occasionally debated (see Colgrove 2002), this thesis emphasizes that improvements in nutrition, followed by improvements in sanitation—both of which preceded modern medical interventions such as antibiotics—led to the reductions in mortality observed. This understanding that social conditions are fundamental to the health of the population forms the central argument of the Health Field Concept.

In Canada, green papers, following British tradition, are not formal policy statements or legislative direction but ‘think pieces’ or working documents intended for discussion. As such, the Lalonde Report did not reflect official government policy when it was released on May 1, 1974. McKay (2000) suggested two reasons why an informal green paper was produced by the Long-Range Planning Branch of Health and Welfare Canada, which had been established only a few years earlier, in 1971. First, she suggested, a shift in policy focus was significant. Medicare, Canada’s medical insurance scheme, had been the central element of health policy in the post-war era through the 1960s; however, by the early 1970s there was growing concern among politicians and policymakers whether governments were receiving value for money, given that ‘the health status of Canadians did not appear to have significantly improved’ (McKay 2000 cited in Glouberman 2001:10). Second, rapid and significant inflation was affecting health care costs, and the federal government was committed to funding 50 % of provincial health insurance schemes and operating costs although under the constitution, initially framed by the *Constitution Act* 1867 and revised in successive documents, the federal government directly controls neither health services nor their costs: this is the domain of the provinces. However, provincial governments too were finding it difficult to contain the costs of Medicare in the 1970s and anticipated that these costs would continue to increase because of population growth and ageing, among other factors (McKay 2000). Thus, there were increasing worries ‘about the sustainability of the health care system and about the roles of the federal and provincial governments with respect to health’ (Glouberman 2001:10).

In this context, Hubert (Bert) Laframboise was asked to lead an effort within Health and Welfare Canada to generate some new ideas as the director-general of the Long-Range Planning Branch of the Department. In this capacity he assembled a team of talented policy planners to think about the future of health in Canada in order to inform the Federal government. The team embraced McKeown’s ideas ‘because they helped explain the current situation, opened the door to a broader vision of the future, and presented fresh policy options for improving the health of Canadians’ (Glouberman 2001:10).

Working from the basic ideas of McKeown, the team developed the ‘Health Field Concept’ and a working paper, *A New Perspective on the Health of Canadians*, under the Minister’s name. In addition to containing a description of the health field concept, the report proposed two objectives: the first to reduce mental and physical health hazards for those parts of the population whose risks were high; the second to improve accessibility to good mental and physical health care for those whose current access was unsatisfactory (Lalonde 1974:66). The report outlined five strategies: health promotion, regulatory, research, health care efficiency and goal setting, and listed a total of 74 possible courses of action aligned with the five strategies, one of which was health promotion. Although nothing specifically targeted cities and the term 'community' was only mentioned a few times in very specific contexts, such as in relation to community facilities and helping aged people stay in their communities. On the other hand, a number of sectors, including 'communities', were mentioned in relation to awareness-raising activities 'for the reduction of self-imposed and environmental health risks', as in business, trade unions, governments, voluntary associations, action groups, communities, professions and teachers (Lalonde 1974:67). Municipal governments *were* mentioned in a list of participants in decisions affecting health (Lalonde 1974:63).

The Lalonde Report identified several courses of action focusing on disadvantaged or marginalized groups such as native peoples, the handicapped, the aged and the economically deprived (in the language of the day) which, along with the two objectives noted above, suggest at least an implicit concern for equity. The green paper also committed the Government of Canada to work with others in implementing the two objectives, and named specific actions that it could undertake. It is clear, then, that the Lalonde Report implied a commitment to several of the 'values' that have come to be associated with the Healthy Cities/Communities movement but it was not explicit about the connection.

The report was explicit, however, with regard to the commitment to pursuing a health promotion strategy, albeit a rather narrow one 'aimed at informing, influencing, and assisting both individuals and organizations so that they will accept more responsibility and be more active in matters affecting mental and physical health' (Lalonde 1974:66). Interestingly, when the report was presented to parliament, some members of the opposition argued that 'stress on lifestyle was primarily a way for the government to reduce funding for Medicare by blaming the victims of disease' (Glouberman 2001:11), a criticism that was subsequently reiterated by the professional community in their critiques of health promotion (e.g. Labonte and Penfold 1981) and which continues to be a tension in the field today (Raphael 2003).

Follow-Up to the Lalonde Report

Following its initial release, the Lalonde Report was distributed slowly but widely among government bodies, policymakers and health professionals in Canada and beyond. Marc Lalonde and others in Health and Welfare Canada gave speeches at various conferences and meetings, including the 1974 conference of the Canadian Public Health Association. Despite the lack of a programme of systematic marketing, the Lalonde Report, especially the health field concept, became very well known in Canada and internationally among governments, policymakers and health professionals.

Ivan Illich, a well-known Austrian philosopher, mentioned the Lalonde Report on the first page of his book, *Limits to Medicine: Medical Nemesis: The Expropriation of Health* (Illich 1976). Released shortly after the Report, *Limits to Medicine* praised it as the first government document to adopt the health field concept as public policy and propose strategies and an action plan to implement it (Lalonde 2002:5). Other public health academics too wrote positive reviews, and Halfdan Mahler, then Director-General of WHO, endorsed the report (Lalonde 2002:2). As a result of these positive reviews and the fact that the health field concept addressed many of the issues that resonated with governments and policymakers in other countries, a number of countries adopted similar perspectives in their own health policy statements, including the US Surgeon-General's 1979 *Healthy People* report (Califano 1979; Green and Fielding 2011).

This positive international reception led to Health and Welfare Canada taking action on the report. Initially it established a small 'focal point' on health promotion, followed in 1978 by a Health Promotion Directorate consisting of about 130 people drawn from various parts of the Health and Welfare department. The largest of these was the Non-Medical Use of Drugs Directorate, whose director-general, Ron Draper, was appointed as the new directorate's director-general. Coincidentally, the staff that came from the Non-Medical Use of Drugs Directorate, especially those from the regional offices, were committed to various social movements of the time such as the 'women's movement' and the so-called 'self-help movement' (R. Stern, personal communication, July 29, 2013).

The Early Years of the Federal Health Promotion Directorate (1978–1986)

One of the actions when the Directorate was created was to establish its structure. It was decided by senior management that it would consist of four divisions: programmes, programme resources, programme promotion and regional services. Each contained a number of units such as health promotion studies in the programme resources division. In the case of regional services, these units were regional offices created from the ones that existed in the Non-Medical Use of Drugs Directorate. They proved to be particularly important in terms of developing relationships with regional and local organizations and individuals involved in health promotion and in supporting innovative initiatives funded by the Health Promotion Contribution Program established in 1980, modelled on the programme that had operated in the Non-Medical Use of Drugs Directorate. For example, this funding programme provided support to community-based organizations and groups to engage in a wide range of activities under the umbrella of 'health promotion', many of which were community development initiatives.

The directorate took its early cues from the Lalonde Report itself and emphasized the lifestyle components of the health field concept. This quickly came under attack by critics of the 'Lifestyle' approach such as Ron Labonté and Susan Penfold (Labonte and Penfold 1981) who identified concerns about 'blaming the victim'

(Ryan 1971) expressed by opposition members of the parliament, mentioned above. Reflecting on this critique a few years later, Labonté noted that it 'essentially incorporated the knowledge challenges of social movements as they might apply to health' (Labonte 1994:80).

Although the directorate continued down the 'lifestyle' track, senior management became somewhat more open to other avenues of action, and regional offices continued their practice of listening to the various communities that they were working with, many of which were interested in other approaches, particularly community-based ones. They also funded a number of non-lifestyle projects though the Health Promotion Contribution Program and encouraged and supported the development of others, including many that were community-based and that may have been precursors to 'healthy community' projects.

All this activity attracted the attention of Ilona Kickbusch, who was appointed Regional Advisor for Health Education for the European Office of WHO in the early 1980s. Early in her mandate, Kickbusch visited Canada with Rob Anderson, a medical sociologist, to view and assess what was happening and its implications for WHO/EURO. During this trip they began to establish working relationships with Canadians such as Ron Draper, Trevor Hancock (then a public health officer with the City of Toronto Department of Public Health) and Gerry Dafoe (then CEO of the Canadian Public Health Association). This led to WHO/EURO's participation in the *Beyond Health Care* Conference in Toronto in 1984, the Healthy Cities workshop that followed and ultimately a partnership between WHO/EURO, Health and Welfare Canada, and the Canadian Public Health Association to host the first international conference on health promotion in Ottawa in 1986.

The partnership with the European office of WHO also led the Health Promotion Directorate to rethink its mandate and emphasis along the lines being developed by WHO/EURO. Under the leadership of Ron Draper, this rethinking came to fruition in the lead-up to the Ottawa conference when Draper and Rita Stern, director of the western regional office of the Health Promotion Directorate, drafted a discussion paper on health promotion that reflected many of the concepts contained in the WHO/EURO paper, 'Concepts and Principles of Health Promotion' (WHO-EURO 1984) as well as some other concepts such as 'knowledge-development' that subsequently proved to be very powerful in shaping the field of health promotion in Canada. The Draper–Stern paper formed the foundation for a subsequent policy paper entitled 'Achieving Health for All: A Framework for Health Promotion'. Drafted by Barbara Naegele, a staff member of the health promotion branch of Health and Welfare Canada, it was released at the conference by Jake Epp, Minister of Health and Welfare Canada at the time (Epp 1986), without approval of Cabinet or Parliament. As suggested by its title, it centred on a framework for health promotion (see Fig. 3.2). Among other things, the framework emphasized 'equity', both in its overarching 'aim' for health promotion ('Achieving Health for All') and as one of three health challenges (reducing inequities). It also included 'healthy environments' as a 'mechanism' for health promotion along with 'fostering public participation', 'strengthening community health' and 'coordinating healthy public policy' (Epp 1986). Most of these concepts were also elements of the *Ottawa*

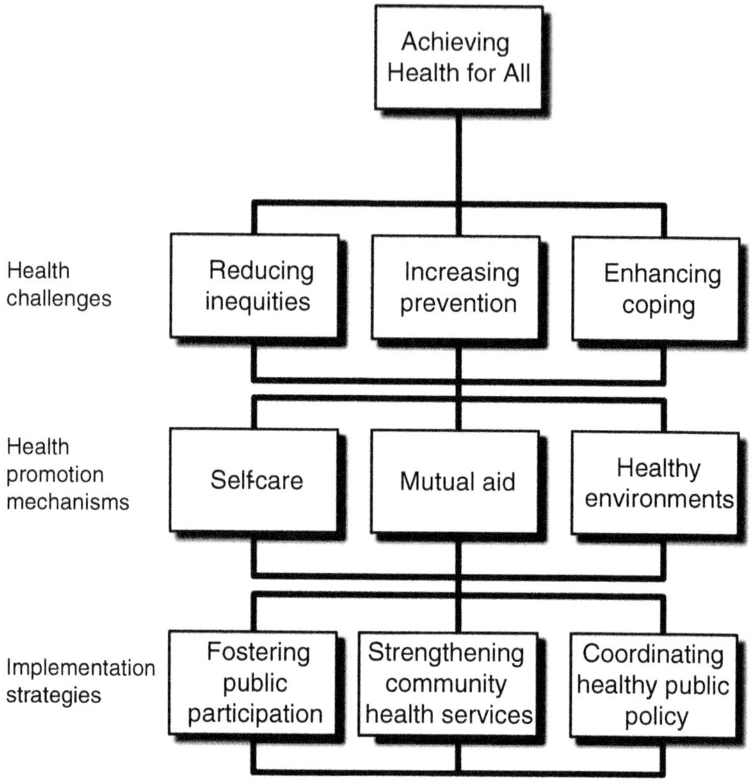

Fig. 3.2 A framework for Health Promotion, Epp 1986

Charter for Health Promotion (World Health Organization 1986) which was also released at the Conference.

From an international point of view, the Ottawa Charter was the main outcome of the conference. Although drafted by a small group led by Ilona Kickbusch and based on the WHO/EURO discussion paper on the concepts and principles of health promotion, it also incorporated suggestions from conference participants, most of whom were from developed countries. As mentioned, the Charter contained some of the ideas in the Epp document, including the same definition of health promotion, drawn from the 'Concepts and Principles' paper (e.g. health promotion is 'the *process* of enabling people to increase control over, and to improve, their health' (WHO-EURO 1984)). However, the Charter also had elements that differed from the Epp document. For example, its prerequisites for health included peace, shelter, education, food, income, a stable ecosystem, sustainable resources, social justice and equity; some of these were not mentioned in the Canadian document. It also proposed five action areas (building healthy public policy, creating supportive environments, strengthening community action, developing personal skills and reorienting health services) that were mentioned in the Epp Report, but in some cases used different language. The Epp document was also unique in its presentation of the health promotion framework noted above. Despite these differences, the two documents endorsed the same core values and approaches to health promotion.

Follow-Up to the Conference

In contrast to the low-key dissemination of the Lalonde Report, Health and Welfare Canada undertook a systematic social marketing programme to promote the Epp Report to stakeholders in the health field. The process included a series of workshops across the country to inform health and related practitioners of the content of the report, especially the health promotion framework.

The workshops were also intended to inform audiences of the findings of the 1985 National Health Promotion Survey that provided some of the empirical support for the Epp Report. This survey was also the basis for a series of 'active health reports' (Health and Welfare Canada 1986a–f) that were distributed to media and practitioners interested in health and health promotion. A series of meetings specifically intended to inform provincial policymakers about the report and the department's interests, and to elicit provincial reactions, was also held; in most cases, the provincial response was relatively positive.

Shortly after this cross-country tour, the Health Promotion Directorate launched a 'knowledge development' project, led by its Health Promotion Studies Unit, in which academics from across Canada were commissioned to conduct literature reviews on all the components of the EPP Framework. These reviews were summarized (Rootman 1989), and a discussion paper prepared (Health and Welfare Canada 1988) and used as the basis for another cross-country set of workshops, organized by the people who had done the reviews. The results of the workshops were summarized at a national workshop on health promotion knowledge development, in Ottawa. A call to action based on the workshop was drafted and disseminated (Health and Welfare Canada 1989) and, among other things, recommended that Health and Welfare Canada support the establishment of university-based centres for health promotion research across the country.

During the time that the knowledge development project took place, the Health Promotion Directorate funded two projects directly related to Healthy Communities, the preferred term in Canada for the Healthy Cities project because it accommodated the many municipalities in Canada that were not technically cities. The first project was co-sponsored by the Canadian Public Health Association and the Federation of Canadian Municipalities: Healthy Communities in Canada was based at the Canadian Institute of Planners as a 'deliberate strategy to move the project beyond the health sector' (Hancock 1993:13). The second project was called Strengthening Community Health, to reflect the similarly named cell in the Epp Report, and was based at and managed by the Canadian Public Health Association from its Ottawa office. Its aim was to 'assist communities in the process of defining the health issues that affect them most directly, and in organizing to take action on these issues' (Hoffman 1994:123), an aim that certainly was consistent with 'healthy cities' efforts. According to Ken Hoffman, Assistant Program Director of the Strengthening Community Health project, the project 'had its roots' in both the Ottawa Charter and the Epp Report (Hoffman 1994:124). Although funding for both these projects ended in 1991, according to Trevor Hancock the National Healthy Communities Project 'was successful in stimulating a number of provincial projects (…) as a direct result of which over 200 cities, towns and villages [were] involved in the Canadian healthy communities network' two years later (Hancock 1993:14).

During this same period, the Health Promotion Directorate received some increase in funds for its Health Promotion Contribution Program (although not as much as requested). This enabled the regional offices to provide support to regional and local projects related to Healthy Communities. National funding for research also increased during this period as a result of the knowledge development project, particularly in support for six university-based health promotion research centres which were funded by the Department's National Health Research Development Program in partnership with the Social Sciences and Humanities Research Council. Funding for health promotion research was also available from other sources, such as universities that established health promotion research centres with their own funds and provincial governments that supported the research and knowledge development activities of these centres, many of which were focused on supporting local action on health promotion, encouraging participatory action research and building the capacity of the field in health promotion.

In the second edition of *Health Promotion in Canada*, we suggested that the period between 1986 and 1994 could be seen as the golden era for health promotion in Canada (O'Neill et al. 2007:8). Many innovative initiatives were conducted at the national, regional and local levels, including many 'healthy community' projects.

Rise of Population Health

While health promotion ideas, as reflected in the Epp Report and the Ottawa Charter, were taking hold among health and other practitioners and some governments in Canada during this time, another set of related ideas developed mainly by economists and epidemiologists began to attract the interest of governments and health policymakers. These ideas, which came to be called 'population health', were developed by The Canadian Institute for Advanced Research (CIAR), established in 1982 by Dr Fraser Mustard, a prominent physician and former dean of the medical school at McMaster University. Specifically, they were developed by the CIAR's Population Health Program set up in 1987, 1 year after the first international conference on health promotion. One of its first products was a paper by two health economists, Bob Evans and Greg Stoddart (1990), proposing a new conceptual framework for looking at health, building on the Health Field Concept (see Fig. 3.2) but differing in a number of ways. For one, it split the environment into two categories, social and physical, and added others including prosperity, with links to the social and physical environment as well as to well-being. It also differentiated between health and disease.

This framework and other work and advocacy by the CIAR proved very successful in persuading policymakers, researchers and others to focus on the determinants of health, which were mentioned in both the Epp Report and the Ottawa Charter. It also helped to stimulate and guide a whole programme of synthesis of knowledge and research on a number of related topics: the impact of early childhood experiences on biological development, long-term health and social development; the importance of social and economic gradients; the impact of lack of control over work; and the importance of social supports in buffering stress.

Lavada Pinder, the Director General of the Health Promotion Directorate at the time, described the emergence of population health within policy circles:

> Population Health gained currency by providing a coherent set of analyses and priorities. The provinces of Quebec, British Columbia, Manitoba, Ontario and Saskatchewan were the first to take up the notion of population health. And their influence on colleagues in Federal, Provincial and Territorial meetings gained support from the federal government. By 1992, the population health framework won the attention of the Federal, Provincial and Territorial Conference of Deputy Ministers of Health which created an Advisory Committee on Population Health (ACPH) to reflect this new focus. (Pinder 2007:94)

One of the first products of this committee was a report that identified the following determinants of health: income and social status; social support networks; education; employment and working conditions; physical environments; biology and genetic endowment; personal health practices and coping skills; healthy child development; and health services (subsequently, and significantly, Health Canada added culture and gender to this list of determinants). It also outlined three strategic directions: strengthening public understanding of the determinants of health; building understanding of the determinants of health among sectors outside of health; and developing comprehensive intersectoral population health initiatives (Federal, Provincial and Territorial Committee on Population Health 1994).

According to Pinder,

> There is no doubt this document was pivotal in the federal government move from health promotion to the 'the population health approach'. With the exception of one reference to the Lalonde Report (1974), it made no mention of 20 years of work in health promotion. (2007:95)

The Evans and Stoddart paper was reprinted in a book entitled *Why Are Some People Healthy and Others Not*? (Evans et al. 1994). Although released at the same time as the first edition of *Health Promotion in Canada* (Pederson et al. 1994), it garnered much more interest from policymakers and certain kinds of researcher in Canada and internationally than did our book, which was mainly known and used in health promotion circles. Ironically, *Why Are Some People Healthy and Others Not* contributed to the decline of health promotion in government circles in Canada.

At the beginning of 1995, in an article reflecting on the twentieth anniversary of the Lalonde Report, Lavada Pinder remarked, 'I'm sure there will still be for many years to come—a health promotion directorate in the federal government, with offices in the regions and funds for communities' (1994–1995:2). Ironically, the Health Promotion Directorate was renamed the 'Population Health Directorate' a few months later, indicating a change of perspective, and the Health Promotion Contribution programme changed its terms of reference and ultimately was eliminated altogether (Pinder 2007:95). These actions significantly diminished federal investment in health promotion and community action in particular, weakening the Healthy Communities movement.

Critiques of the Population Health approach quickly began to appear. One of the earliest was by Ron Labonté who, as noted above, was also a critic of the 'Lifestyle Approach' to health promotion. Although he suggested there were commonalities between Health Promotion and Population Health, he suggested that there were a number of key differences: health promotion questions economic rationalism, whereas population health espouses it; health promotion draws from a range of social sciences, whereas population health draws heavily from economics; health promotion tends to be 'bottom-up', whereas population health is 'top-down'; and health promotion has a 'community focus', whereas population health has a 'population focus' (Labonte 1995:26). Although this characterization of the differences between the two fields has been challenged (e.g. Frank 1995), it can be argued that the last difference in particular had significant implications for Healthy City/Healthy Community initiatives and may have contributed to the elimination of the Health Promotion Contribution Fund and the downplaying of community initiatives in favour of higher level policy initiatives by governments in Canada that we are still waiting to see come to fruition.

In spite of the eclipse of health promotion by population health in government circles, as we indicated in the second edition of *Health Promotion in Canada* (O'Neill et al. 1997), health promotion continued to develop in some areas, particularly with respect to knowledge development. For example, the federally funded university-based research centres, as well as those funded through provincial and other resources, continued to develop individually and collectively through the 1990s and into the first decade of the twenty-first century. To some degree these research enterprises replaced the national leadership of the federal government in health promotion (Rootman et al. 2007)—possibly because of their freedom to pursue topics of interest and the availability of multiple funding streams for academic research.

The Canadian Consortium for Health Promotion Research, a coalition of all university-based centres for health promotion research established in 1996 as well as others funded though other sources, offered leadership to the field until it dissolved shortly after the global conference of the International Union of Health Promotion and Health Education in Vancouver in 2007. Since that time a number of the centres have dissolved, including the two that led the Consortium; currently some of the surviving centres are experiencing funding and other difficulties.

To date, no other bodies have stepped forward to offer pan-Canadian leadership in the field of health promotion although a new pan-Canadian body consisting mainly of health promotion practitioners is currently under development and may offer this leadership.

Although the Healthy Communities movement has been significantly affected by the decline of health promotion in Canada over the last couple of decades, there are still many active healthy community initiatives with relatively strong provincial networks, particularly in Quebec and British Columbia, supported by provincial and local, and some external funding, including some outside the health sector for such initiatives as age-friendly cities, smoke-free cities and active cities.

Discussion

The Lalonde Report was the trigger for most of the developments described in this chapter, including the Epp Report and the Ottawa Charter, which each contributed to other developments. With regard to the Lalonde Report, the Canadian Policy Research Network carried out a study in the late 1990s that considered its impact on health policy. Among other things, the study looked at its possible impact on health attitudes and behaviours. Following an analysis of relevant data, Shalom Glouberman concluded that 'We eat more nutritious foods, smoke less, wear seatbelts, and so on. We are no longer tolerant of smoking in others; we think of a much higher frequency of exercise as normal'. However, he conceded that 'these changes were not by themselves proof of the impact of the report' as they may have been co-incidental and not causal. On the other hand, he noted that the Lalonde Report 'declared their relevance and preceded their occurrence' and suggested that it 'marked the beginning of [a] transition period'. In addition, he expressed the view that 'Whether the Lalonde Report was an important agent for change or merely a harbinger of change that was already afoot, it is nonetheless widely seen as a key health policy document both in Canada and the rest of the world' (2001:9).

Glouberman suggested three critical achievements of the Lalonde Report:

- It recognized that health was a complex concept that went beyond medical care.
- It launched health promotion and prophesied a change in attitudes to health.
- Ideas were linked through a framework to policy recommendations.

In discussing each of these achievements, he identified a number of concrete developments, some of which have already been mentioned in this chapter. For example, he mentioned the list of determinants of health adopted by Health Canada (Nickoloff and Health Canada 1996). Similarly, he mentioned 'the emergence of health promotion as an instrument of government policy in Canada and elsewhere' (Glouberman 2001:12) and noted that the Lalonde Report 'connected innovative ideas about health to a new platform for analysis and evaluation … and to policy recommendations using those tools' (Glouberman 2001:12–13). In that regard, he might also have mentioned that it led to the use of frameworks in health policymaking such as the Epp Report and the Evans-Stoddart Framework to convey complex policy ideas or elaborate on earlier conceptualizations. Related to this latter point,

Glouberman noted two main shortcomings of the Lalonde Report, that 'an overemphasis on lifestyle missed the importance of the social environment' and that the report failed 'to grasp the strong interactions among the four quadrants' or components (Human Biology; Environment; Lifestyle; Health Care Organization) which to some extent were subsequently addressed both by health promotion and population health.

Interestingly, Lalonde, in reflecting on the report 27 years after its publication, noted the importance of the 'interrelationship between the four components' (2002:2). He also mentioned that the ideas in the report had found their way into two important government documents on the Canadian health care system two decades later, one by a federal senate committee (Kirby 2003) and another by a royal commission (Romanow 2002), and that the report itself was cited in many other reports at all levels of government in Canada and around the world. Although, it is extremely difficult to attribute causality to the Lalonde Report, without a doubt it has been a powerful influence on health policies and actions internationally.

Having said all this, however, it would be fair to say that it had very little direct impact on communities in Canada or on the Healthy Communities movement that began to develop about 10 years after its release, although it did lead to the establishment of the Health Promotion Directorate at the national level and perhaps to the establishment of similar units at the provincial/territorial level in Canada. For example, in the chapter on Manitoba in the first edition of *Health Promotion in Canada*, the author noted that the Lalonde Report 'was a driving force behind the formation of the Manitoba Health Promotion Directorate' (English 1974:197). Such units at both national and provincial/territorial levels provided support in terms of advice and funding to local communities and ultimately provided support to the Healthy Communities movement itself, both nationally and provincially (see Pederson et al. 1994, section on Provincial Perspectives).

To the best of our knowledge, there has not been an assessment similar to the one by Glouberman on the *Lalonde Report* of either the Epp Report or the *Ottawa Charter for Health Promotion*; however, based in part on our own knowledge as well as materials in the first and second editions of *Health Promotion in Canada*, it is possible to make some reasonable assessment of their relative impacts.

The Epp Report appears to have had very little effect on policy at the federal level, although the Health Promotion Directorate was granted some additional funding and the actual preparation of the report resulted in increased communication and collaboration between the two parts of the health promotion and services branch. One example of this was the publication of a report on mental health promotion produced by the health services branch and patterned after the Epp Report (Health and Welfare Canada 1988). Another policy-related outcome at the federal level was the knowledge development project, which led to some policy-related research on health promotion in Canada by a number of the health promotion research centres as well as by the Canadian Consortium for Health Promotion Research (See Rootman et al. 2007).

At the provincial/territorial level, the Epp Report encouraged organizations and people working in health promotion to seek policy changes. For example, in the

Saskatchewan chapter of the first edition of *Health Promotion in Canada*, the author noted that 'the Framework was invaluable as an expression of the aspirations of many individuals committed to reducing inequities and struggling for policies that would create more supportive environments for health' (Feather 1994:189). In addition, there were a number of policy reports produced at provincial level that made reference to the Epp Report (Pederson et al. 1994:153–311). The Health Promotion Directorate provided support for such policy efforts as well as offering general support and funding for work in health promotion following the release of the Epp Report.

The Epp Report also led the Health Promotion Directorate to support the two national projects mentioned earlier (i.e. Strengthening Community Health and Canadian Healthy Communities) as well as local and regional Healthy Community efforts through regional offices and the Health Promotion Contribution Fund until funding ended. The national projects also provided support for projects at local and provincial levels until their funding ended in 1991; thus, the Epp Report directly led to the launching of a number of processes that supported the development of Healthy Communities as well as health promotion in general. Its influence began to wane in 1989 when Epp was moved to another ministry and his successor failed to endorse the report; however, the ideas have not totally disappeared. For example, the Canadian Strategy on Aging recently incorporated the three mechanisms (supportive environments, mutual aid and self-care) into its strategic plan (Edwards and Mawani 2006).

Interestingly, the withdrawal of formal support for the Epp Report opened the doors to the Ottawa Charter, which had been ignored by the Department of Health and Welfare in favour of the Epp Report but which subsequently became the key document guiding health promotion practitioners in Canada until the present time.

The Ottawa Charter did not have an impact on the development of the infrastructure for health promotion in Canada; nor did it have an impact on government policies, at federal level at least. Instead, its impact has been in terms of providing a vision and a clear set of values for the field, linking Canadians with others of like mind around the world. The Ottawa Charter also provided a framework for organizing work in Canada and elsewhere through the articulation of the five action areas. In addition, the release of the charter at the same time as the Epp Report 'suddenly provided a powerful boost to the credibility of health promotion' (Altman and Martin 1994:156), stimulating a great deal of discussion and action among practitioners and researchers in health and related fields and helping to create the 'professional movement' of health promotion mentioned earlier, which is still active as indicated by the title of the well-attended workshop (Health Promotion in Canada: Alive and Kicking) that we, along with others across Canada, organized at the 2013 Conference of the Canadian Public Health Association (Rootman et al. 2010) as well as several followup workshops since then.

The Ottawa Charter did have a significant impact on the development of the Healthy Communities movement in Canada; indeed, Trevor Hancock has explained that 'The Healthy City concept has evolved in parallel and as an application of health promotion' and that 'right from its inception the project was seen as a vehicle for health promotion' (1993:7).

Conclusion

For about 20 years (1974–1994), Canada was deeply, visibly engaged in health promotion. Twenty-two years after that, we think that the for the most part the Lalonde Report and the Ottawa Charter (World Health Organization 1986) have both stood the test of time and have been of significant influence worldwide. The Healthy Cities concept, which developed in part out of the health promotion context in Toronto, Canada's largest city, seems also to have been an important organizing mechanism that continues to expand its reach through burgeoning interest in urban health. Now that the majority of the world's population lives in cities, debate over how best to support the health of urban residents and the urban environment will increase.

Although the Lalonde and Epp Reports and the Ottawa Charter may have been important as triggers for action very little would have happened without a convergence of people, structures and resources that helped to create conditions for action. For example, without Marc Lalonde and Jake Epp, two government ministers who were willing to defend the work of the people in their departments who produced the reports, the reports themselves might never have come to the attention of practitioners across the country. Similarly, if the Lalonde Report had not been praised by people like Ivan Illich as well as by governments in other countries, the government of Canada might not have established the Health Promotion Directorate and Ron Draper might not have been given the support that he needed to encourage innovation. Without the decentralized structure of the regional offices of the Directorate, many of the healthy community initiatives might not have taken root. And without the resources allocated by federal research funding agencies, university-based health promotion research centres might not have been established and so would not have been able to provide support to community efforts to promote health.

However, it is also essential to understand that in Canada public health has a long history at the local level of government. The city of Toronto, in particular, has long been a central figure in urban health, supporting as it does the largest urban region of the country. Although a detailed history of the City of Toronto Department of Public Health is beyond the scope of this discussion, that leaders such as Trevor Hancock and Fran Perkins found opportunities within it at critical stages in their careers warrants greater analysis of how public health infrastructure in Canada supports innovation and action.

Perhaps it is worth repeating here, as we close, that we are not convinced that health promotion was ever a social movement per se: rather, we understand that much of what has been labelled as health promotion in Canada have been bureaucratic and technocratic activities designed to reduce health care costs by increasing preventive activities and fostering individual responsibility for health. Nevertheless, we have welcomed the 'legitimate social and professional space to challenge a health agenda focused on the consumption of services and care rather than on the production of health' (Kickbusch 1994:8). We recognize and applaud the ways that health promotion has functioned in and through the state to articulate a social view of health and support the importance of looking beyond the health care sector for the origins of health and the solutions to health inequities. We were long ago convinced by McKeown (1976) and others that health arises through the conditions of living (e.g. Commission on Social Determinants of Health 2008) and that those conditions include but extend beyond the formal health care system in any country.

Postscript

As we put the finishing touches to this chapter, health promotion in Canada has come under attack by some of its potential allies for its failure to advance the social determinants of its health agenda. Specifically Ted Schrecker, a respected scientist of the Institute of Population Health at the University of Ottawa, published a paper in which he argues that

> Health promotion initiatives that neglect structural influences on health and reinforce or legitimize neoliberal attributions of responsibility are not just irrelevant to the health equity agenda advanced by the WHO Commission; they are actively destructive of that agenda and its commitment to social justice. If such initiatives cannot be reinvented quickly and comprehensively, it is worth considering whether equity would be best served by shutting them down.

In addition to 'shutting them down' he suggests that

> In the Canadian context ... reinvention [of health promotion] will require that health promoters acquire a higher level of methodological self-consciousness; recognize the congruence of their current focus with the presumptions of neo-liberalism; and rethink that focus in a way that explicitly confronts neoliberalism and articulates the alternatives at the levels of policy formation and quotidian practice. Such reinvention is possible, but implausible. (Schrecker 2013)

While we agree that health promotion efforts need to be re-imagined so as to not reproduce health inequities, we think that Schrecker's pessimism may be unjustified. Many health promotion initiatives in Canada, including those carried out under the aegis of the Healthy Community movement, do address the health equity agenda of the WHO Commission on the Social Determinants of Health, as do current training and research programmes in health promotion. For example, work conducted through the British Columbia Centre of Excellence for Women's Health over the past few years has spawned a lively discussion about how to ensure that health promotion initiatives support not only women's health and their social and economic status. As we heed Schrecker's advice, we also need to be careful that we do not inadvertently destroy the good work in health promotion in general and in Healthy Communities that is currently under way.

References

Baum, F. (2008). The commission on the social determinants of health: Reinventing health promotion for the twenty-first century? *Critical Public Health, 18*(4), 457–466.

Bynum, B. (2008). The art of medicine: The McKeown thesis. *The Lancet, 371*, 644–645.

Califano, J. (1979). *Healthy people: The surgeon general's report on health promotion and disease prevention*. Washington, DC: United States Public Health Service, Office of the Surgeon General.

Colgrove, J. (2002). The McKeown thesis: A historical controversy and its enduring influence. *American Journal of Public Health, 92*(5), 725–729.
Commission on Social Determinants of Health. (2008). *Closing the gap in a generation: Health equity through action on the social determinants of health.* Final Report of the Commission Social Determinants of Health, Geneva.
Eisenstein, H. (1996). *Inside agitators: Australian femocrats and the state.* Philadelphia, PA: Temple University Press.
Epp, J. (1986). *Achieving health for all: A framework for health promotion.* Ottawa: Health and Welfare Canada.
Evans, R. G., Barer, M. L., & Marmor, T. R. (Eds.). (1994). *Why are some people healthy and others not? The determinants of health of populations.* New York: A. de Gruyter.
Frank, J. W. (1995). Why "population health"? *Canadian Journal of Public Health, 86*, 162–164.
Glouberman, S. (2001). *Towards a new perspective on health policy.* Ottawa: Canadian Policy Research Networks.
Green, L. W., & Fielding, J. (2011). The U.S. healthy people initiative: Its genesis and its sustainability. *Annual Review of Public Health, 32*(1), 451–470. doi:10.1146/annurev-publhealth-031210-101148.
Hancock, T. (1993). The evolution, impact and significance of the health cities/healthy communities movement. *Journal of Public Health Policy, 14*(1), 5–18. doi:10.2307/3342823.
Hancock, T., Robertson, J., Buck, C., Carlson, R., Emery, F., Oberlander, P., Hellberg, H. (1985). Beyond health care: proceedings of a conference on healthy public policy. *Canadian Journal of Public Health, 76*, 9–104.
Health and Welfare Canada. (1988). *Priorities and strategies for research to promote the health of Canadians: A discussion paper.* Ottawa: Health Services and Promotion Branch.
Health and Welfare Canada. (1989). *Knowledge development for health promotion: A call to action.* Ottawa: Health Services and Promotion Branch.
Hoffman, K. (1994). The strengthening community health program: Lessons for community development. In A. Pederson, M. O'Neil, & I. Rootman (Eds.), *Health promotion in Canada: Provincial, national and international perspectives* (pp. 123–138). Toronto: W.B. Saunders.
Illich, I. (1976). *Limits to medicine: Medical nemesis: The expropriation of health.* Toronto: McClelland and Stewart.
Kickbusch, I. (1994). Introduction: Tell me a story. In A. Pederson, M. O'Neill, & I. Rootman (Eds.), *Health promotion in Canada: Provincial, national and international perspectives* (pp. 8–17). Toronto: W. B. Saunders Canada.
Kickbusch, I. (2007). Health promotion: Not a tree but a rhizome. In M. O'Neill, A. Pederson, S. Dupere, & I. Rootman (Eds.), *Health promotion in Canada: Critical perspectives* (pp. 363–366). Toronto: Canadian Scholars' Press.
Labonte, R. (1994). Death of program, birth of metaphor: The development of health promotion in Canada. In A. Pederson, M. O'Neill, & I. Rootman (Eds.), *Health promotion in Canada: Provincial, national and international perspectives* (pp. 72–90). Toronto: W.B. Saunders.
Labonte, R., & Penfold, S. (1981). Canadian perspectives in health promotion: A critique. *Health Education, 19*(3), 4–9.
Lalonde, M. (1974). *A new perspective on the health of Canadians: A working document.* Ottawa: Ministry of Supply and Services Canada.
Lalonde, M. (2002). A new perspective on the health of Canadians revised. *The Health Promotion Exchange, Special Edition* (Winter), 4–6.
Link, B. G., & Phelan, J. C. (2001). McKeown and the idea that social conditions are fundamental causes of disease. *American Journal of Public Health, 92*(5), 730–732.
McKay, L. (2000). *Making the Lalonde report.* Ottawa: Canadian Policy Research Networks.
McKeown, T. (1976). *The role of medicine: Dream, mirage, or nemesis?* London: Nuffield Provincial Hospitals Trust.
Mckeown, T., Brown, R. G., & Record, R. G. (1972). An interpretation of the modern rise of population in Europe. *Population Studies, 26*(3), 345–382.

O'Neill, M., Pederson, A., Dupéré, S., & Rootman, I. (Eds.). (2007). *Health promotion in Canada: Critical perspectives* (2nd ed.). Toronto, ON: Canadian Scholars' Press.

Pederson, A., O'Neill, M., & Rootman, I. (Eds.). (1994). *Health promotion in Canada: Provincial, national and international perspectives*. Toronto: W.B. Saunders.

Pederson, A., Rootman, I., & O'Neill, M. (2005). Health promotion in Canada: Back to the past or towards a promising future? In A. Scriven & S. Garman (Eds.), *Promoting health: Global perspectives* (pp. 255–265). Houndsmills: Palgrave Macmillan.

Pinder, L. (2007). The federal role in health promotion: Under the radar. In M. O'Neill, A. Pederson, S. Dupéré, & I. Rootman (Eds.), *Health promotion in Canada: Critical perspectives* (2nd ed., pp. 92–105). Toronto: Canadian Scholars' Press.

Raphael, D. (2003). Barriers to addressing the societal determinants of health: Public health units and poverty in Ontario, Canada. *Health Promotion International, 18*(4), 397–405. doi:10.1093/heapro/dag411.

Raphael, D. (2008a). Getting serious about the social determinants of health: New directions for public health workers. *Promotion & Education, 15*(3), 15–20.

Raphael, D. (2008b). Grasping at straws: A recent history of health promotion in Canada. *Critical Public Health, 18*(4), 483–495.

Rootman, I. (1989). Knowledge for health promotion: A summary of Canadian literature reviews. *Health Promotion International, 4*(1), 67–72.

Rootman, I., Dupéré, S., Pederson, A., & O'Neill, M. (Eds.). (2012). *Health promotion in Canada: Critical perspectives on practice*. Toronto: Canadian Scholars Press.

Rootman, I., Jackson, S. F., & Hills, M. (2007). Developing knowledge for health promotion. In M. O'Neill, A. Pederson, S. Dupéré, & I. Rootman (Eds.), *Health promotion in Canada: Critical perspectives* (2nd ed., pp. 123–138). Toronto: Canadian Scholars' Press.

Rootman, I., Warren, R., and Catlin, G. (2010). Canada's health promotion survey as a milestone in public health research. *Canadian Journal of Public Health, 101*(6), 436–438.

Ryan, W. (1971). *Blaming the victim*. New York: Pantheon Books.

WHO-EURO. (1984). *Health promotion: A discussion document on the concept and principles*. Copenhagen: WHO-European Regional Office.

World Health Organization. (1986). Ottawa charter for health promotion. *Canadian Journal of Public Health, 77*(6), 425–430.

World Health Organization. (2007). *Global age-friendly cities: A guide*. Retrieved from http://www.who.int/ageing/publications/Global_age_friendly_cities_Guide_English.pdf.

World Health Organization. (2011). *Making cities smoke-free*. Geneva: World Health Organization.

Chapter 4
Healthy Cities Emerge: Toronto–Ottawa–Copenhagen

Trevor Hancock

The concept of a healthy city is hardly a new one, as was discussed in the lead-up to this chapter. The currents of modern urban public health that began in the industrializing cities of England and spread across Europe and the United States in the mid- to late nineteenth century soon came to Canada. Toronto, which was to become one of the key sources of the modern incarnation of Healthy Cities, had been there before, becoming a public health leader in the early twentieth century. Since we stand on the shoulders of giants and are the product of our history, an understanding of Toronto's public health history helps us understand, at least in part, why and how it came to be such a centre of innovation in the 1980s.

Toronto: The Early Days

The settlement that was to become Toronto began as a French trading post on the north shore of Lake Ontario in the 1750s, but was abandoned after the British defeated the French in 1760. In 1793 Governor Simcoe chose the small settlement that had sprung up as the capital of Upper Canada and named it York. The settlement became Toronto when it was incorporated as a city in 1834, at which point it had a population of less than 10,000.[1] One of the first things the new City did was appoint a Board of Health to deal with the epidemics of cholera and typhoid that swept up from Quebec and Montreal each summer with each wave of new immigrants from Europe; but the board was temporary, and once the emergency was over

[1] http://en.wikipedia.org/wiki/Old_Toronto

T. Hancock (✉)
Professor and Senior Scholar, School of Public Health and Social Policy, University of Victoria, Victoria BC, Canada
e-mail: thancock@uvic.ca

E. de Leeuw, J. Simos (eds.), *Healthy Cities*,
DOI 10.1007/978-1-4939-6694-3_4

it was disbanded; not until 1884, after the passage of Ontario's Public Health Act, did the city establish a permanent Board of Health.

Around that time it was estimated that the average age of death for males was around 25 years, and 27 for females, mainly because of the shockingly high infant mortality rate (which distorts the average age of death): in 1908, the rate was 193 per 1000 live births (Board of Health 1989:28)—in other words, almost one in five children did not see their first birthday!

Not surprisingly, health was a major concern: of 36 municipal departments established in Toronto between 1834 and 1989, 16 (44 %) were found to have health concerns as contributors to their establishment and in ten of these, health had a major impact on their growth (Severs 1989). Examples include the Abattoir, City Engineer, Housing Authority and Public Welfare. Clearly, the City's leaders recognized that health was a responsibility of many different departments and not just the Department of Public Health.

Canada: The City Healthy (1909–1920)

> It is not so much the city beautiful as the city healthy that we want for Canada. (Dr. Charles Hodgett, Commission of Conservation 1915)

In the early twentieth century Toronto was blessed by a far-sighted Medical Officer of Health at a time when there was a national interest in better health in Canada's cities. The combination was fortuitous. In 1909 the Canadian government, following the lead of President Roosevelt in the United States, established a Commission of Conservation—what today we would call a commission of sustainable development, except that their conception of conservation was broader than our modern-day notion of sustainability:

> There are two important factors in the question of national conservation, the physical and the vital. The former relates to the protecting of our land, our forests, our minerals, our waters, our sunlight, our fresh air; the latter to the prevention of diseases, to health and to the prolongation of life. (Commission of Conservation 1912)

This statement was part of the report of the Public Health Committee of the Commission by Dr. Charles Hodgett, the Commission's advisor on public health and a former secretary of the Ontario Board of Health. He added,

> In housing and town planning we are dealing with most of the former [conservation of natural resources] and all of the latter [conservation of vital resources].

The roots of modern urban planning in Canada can be traced to the Public Health Committee of the Commission of Conservation. Because of its interest in health in towns, the Public Health Committee recommended in 1913 that a national conference on housing and town planning be held and that Thomas Adams be invited. Thomas Adams had been Secretary to the first Garden City Company at Letchworth in England and was a renowned advocate and practitioner of town planning, serving as the first President of the British Town Planning Institute. In 1914, the Commission of Conservation secured his services from the British government (Armstrong 1959). In the

5 years that he spent in Canada as the Commission's town planning advisor, from 1914 to 1919, Adams revised the Commission's model town planning bill and had a hand in preparing town planning bills in most provinces, prepared a pioneer document on rural planning and development, consulted with some 40 local councils, wrote most of the material for a quarterly bulletin called *Conservation of Life* put out by the Commission, helped to organize the Civic Improvement League, and in 1919 was elected as the first president of the Town Planning Institute of Canada (Armstrong 1959).

The Healthiest of Large Cities: Toronto, 1910–1929[2]

Coincident with the work of the Commission of Conservation, Dr. Charles Hasting became the Medical Officer of Health for the City of Toronto in 1910,[3] a position he held until 1929. The early twentieth century was a time of great social and political reform, and during that period he developed the Department to the point that it became one of the foremost public health departments in the world.

In July 1915, *MacLean's Magazine* in Toronto published an article entitled 'Saving lives on the wholesale plan: How Toronto has been made the healthiest of large cities' (large cities being those with more than 350,000 residents). The article compared the health status of Toronto with that of similar large cities in Great Britain and the United States and declared that in terms of mortality Toronto came first, followed by Milwaukee, New York, Chicago, and then the large British cities. This pre-eminence in health, *MacLean's* stated, was due to 'an awakening, a man and several years of varied and strenuous activity'. The awakening

> was in the municipal air all over the continent of North America. New light had broken on the subject of keeping people well by the cityfull or the townfull. New machinery to prevent

> **The Social Determinants of Health as Seen in 1915**
> Every nation that permits its people to remain under the fetters of preventable disease, and permits social conditions to exist that make it impossible for them to be fed, clothed and housed … and that endorses a wage that does not afford a sufficient revenue for the home, a revenue that will make possible the development of a sound mind and body, is trampling a primary principle of democracy under its feet.
>
> *Charles Hastings, Presidential address, American Public Health Association, 1915 (quoted in MacDougall 1990:27)*

---

[2] For an in-depth history of the City of Toronto Department of Public Health 1883–1983 see MacDougall 1990.

[3] He was also one of the founders of the Canadian Public Health Association (MacDougall 1990 p. 26) and worked with Hodgett and the Local Council of Women to start school medical inspections, both in 1910 (MacDougall 1990:186).

> people from becoming ill was coming into being from the Pacific to the Atlantic. A new idea of health and health work had taken shape, a more economic and democratic and twentieth century idea. (Yorke 1915)

This new idea consisted of conservation of the human asset, an idea analogous to conserving vital resources as proposed by the Commission of Conservation, was encapsulated in the legend in the masthead of the Toronto Department of Public Health's monthly newsletter:

> With a well organised Department of Public Health, a municipality may have as much health as it is willing to pay for.

The man to undertake this was Dr. Hastings, a well-respected member of the medical fraternity in Toronto who became attracted to public health somewhat late in life in response to the death of one of his daughters from tuberculosis contracted after drinking infected milk. Not surprisingly, one of his first initiatives was to ensure that the city's entire milk supply was pasteurized. In addition, Toronto began chlorinating its drinking water in 1910—one of the first cities in the world to do so—and by 1915 was also chlorinating its sewage and filtering its water.

Under Hastings' leadership, the City's Public Health Department became one of the most respected in the world. Its work expanded to include industrial hygiene, social welfare, housing, school health and municipal housekeeping, which taught women the principles of hygiene and sanitation, diet and nutrition, child rearing and care of the sick (Board of Health 1989). It grew tenfold under Hastings' leadership, with the nursing staff increasing dramatically from two when he became Medical Officer of Health in 1910, to 110 when he retired 19 years later, by when the total staff was nearly 500 strong (MacDougall 1990:31). Also during his tenure the city established an abattoir, a housing commission and a street cleaning commission (Board of Health 1989).

The results were impressive: the mortality rate from typhoid fever dropped 90 % between 1910 and 1915; the last case of smallpox was recorded in 1932; and the last diphtheria death was in 1934 (Board of Health 1989). Through the leadership of one person, the political commitment of the City Council and the support of the community, Toronto had achieved a commendably high level of public health.

A New Perspective on the Health of Canadians

Ironically, the end of Hastings' tenure in 1929 also marked the beginning of the ascendancy of the medical model in health care. With the advent of antibiotics in the 1930s, as well as the development of improved anaesthesia, surgery, diagnostic technologies and pharmaceuticals prior to, during and after the Second World War, the pendulum began to swing towards health care for the individual and away from prevention for the community. The advent of the National Health Service in the United Kingdom in 1948 was somewhat matched in Canada by the federal government's Hospital Construction Grants that same year.[4] This was followed by a federal

[4] In Canada health care is a provincial responsibility under the Constitution, and the federal government can only offer funding as an incentive for change, having no legislative power in health care.

initiative for hospital insurance in 1958 and medical insurance in 1968. By 1972 all provinces had bought into the programme.

It was perhaps no coincidence that 2 years later, in 1974, the federal government produced the Lalonde Report (Health and Welfare Canada 1974). The government had little on its plate by way of an agenda in the health sector, and did little other than hand over money to the provinces and take what political credit they could. Marc Lalonde, the federal Minister of Health and Welfare at the time, consciously or not, in effect opened up whole new fields of health policy for the federal government. The Report identified four health fields—human biology, health care, lifestyle and environment—and famously noted:

> the traditional view of equating the level of health in Canada with the availability of physicians and hospitals is inadequate ... there is little doubt that future improvements in the level of health of Canadians lie mainly in improving the environment, moderating self-imposed risks and adding to our knowledge of human biology. (p. 18).

Suddenly there were three new fields within which the federal government could be active, and where it arguably could take on a greater role and more power: environment, lifestyle and human biology. The Lalonde Report's fame, justifiably, lies in the fact that it was the first government report anywhere to acknowledge that health care is not the most important determinant of health, and to turn the spotlight on broader societal factors—although, regrettably, in the years that followed the focus was placed mainly on lifestyle, personal behaviour and personal responsibility. It was followed by similar reports in the United Kingdom, the United States and Sweden, all of which gave credit to the Lalonde Report (Hancock 1986a), and laid the groundwork for Canada's emergence as a world leader in health promotion in the 1980s, and in population health in the 1990s.

It was in this context of innovation in Canadian thinking about health that the Board of Health in Toronto revived and reassumed the mantle of Charles Hastings.

Public Health in the 1980s: A municipal Lalonde Report

In Toronto, as elsewhere in the 1970s, reform was in the air. In Toronto it led to the election in 1972 of a reform-minded Council headed by David Crombie. Because the Council appointed the Board of Health, this led to the appointment of reformers to the Board, where they soon found themselves in conflict with the traditional and rather conservative leadership of the Department of Public Health. The conflict crystallized around the issue of a lead smelter operating in the city and pollution of the local environment, resulting in high lead levels in children. The new reformers on the Board of Health felt that the Department's leadership was insufficiently responsive to the needs of the citizens, too unaware of the emerging science of environmental health science and too reluctant to take strong action (see MacDougall

1990, Chap. 14, pp. 274–290 for a detailed discussion of this issue, of the report, and of the early stages of implementation up to 1983).

The upshot was that the Board decided that the Department needed to be reformed, to pay more attention to new thinking about health and new, more democratic ways of working. In 1976 the Board established the Health Planning Steering Committee and in 1978, after extensive community consultation, staff input and expert advice, including a report on modern-day public health from the University of Toronto's School of Hygiene, they tabled their report, *Public Health in the 1980s* (Board of Health 1978).

This report was as important for municipal public health as was the Lalonde Report at the national level. It redefined good public health practice, calling for a much greater emphasis on upon data-based planning, health education, health promotion, community involvement and advocacy (Hancock 1984), and proposed the creation of a new group to do this: and in 1979 the 16-member Health Advocacy Unit was created.[5]

I was hired as the health planner in this unit in early 1980. I was responsible for leading the public health planning process for the city, and that entailed the development of a new model, a new understanding of health, as well as the development of a mission statement. If we were to deal with public health in a new way, we needed a new understanding of what health is and where it comes from, and we needed to communicate and share this not only within the Department but across the city government and beyond.

The new model—The Mandala of Health (Hancock 1985; Hancock and Perkins 1985)—was developed with my good friend Fran Perkins, who a decade later would be the Director of Toronto's Healthy City Office but at the time was a senior public health nurse in the Department.

The mission statement, which was adopted in 1982, was developed in collaboration with the Department's leadership and the Board of Health, stated:

> The mission of the City of Toronto Department of Public Health is to make Toronto the healthiest city in North America (Hancock 1986b)

The development of this statement was not without its moments of humour. At one point there was a discussion as to whether such a bold statement was perhaps too bold, not very Canadian; the issue was resolved when I suggested we could aim to be the fifth healthiest, or the tenth, or whatever we chose—so where did we want to be? But the adoption of this statement brought with it some interesting questions (see text box), in which can be seen the seeds of the development of the Healthy City concept, which was to be further explored in 1984.

---

[5] For a full description of the work of the Health Advocacy Unit, see Hancock 1984.

Challenges to Becoming the Healthiest City in North America
First, what is health?
... second ... how are we to measure the City's health?
... third, obviously, the Department is not solely responsible for the health of the city. Therefore the Department must work closely with and support any individual or group whose purpose is to improve or maintain the health of the citizens of Toronto.
... fourth ... [this] led to the recognition that if the Department is to make Toronto the healthiest City in North America, it is crucial that the Department of Public Health become a leading force in the community, whether that community be the City of Toronto and its many neighbourhoods and ethnic groups, the wider community of Ontario and Canada and their political leaders, or the public health, environment and social justice movements locally, provincially, nationally or even internationally.
(Hancock 1986b)

Beyond Health Care: A Working Conference on Healthy Public Policy

1984 was a good year for hosting a major conference, being the sesquicentennial (150th anniversary) of the City of Toronto, the centennial of the Board of Health, the seventy-fifth anniversary of the founding of the Canadian Public Health Association and the tenth anniversary of the Lalonde Report—so there was money and resources available from all of these. Based on the recognition that the main determinants of health lay beyond the health care system, the topic of this 1984 conference was healthy public policy, a term coined in 1982 (Hancock 1982), but borrowing in particular from the concept of health-promoting policy (Milio 1981). The conference explored five themes: healthy people, healthy work, healthy communities, a healthy nation and a healthy world; and brought together a stellar group of internationally renowned speakers including Fred Emery, John McKnight, Nancy Milio, Peter Draper and one of the worlds leading alternative futurists, James Robertson (Beyond Health Care 1985).

But a parallel event had even greater impact. Given that the City and its Board of Health were the major sponsor and main host of the event, and given the Department's mission statement, it was felt important to have a strong focus on the city and to offer an event that was more locally focused and relevant to members of the public. Thus a day-long workshop was held immediately after the conference, called Healthy Toronto 2000. It marked the kick-off of what would be a 5-year process culminating in the report *Healthy Toronto 2000*, which would be the successor to *Public Health in the 1980s* but focused less on the Department of Public Health and more on the whole city.

The keynote speaker at the workshop was Len Duhl, Professor of Public Health and Urban Planning at the University of California, Berkeley, and one of the first people to begin to stitch public health and urban planning back together in his 1963 edited book *The Urban Condition* (Duhl 1963). Also in attendance was Ilona Kickbusch, Health Promotion Officer for WHO Europe, as the guest of Ron Draper, the Director-General of Health Canada's Health Promotion Directorate and one of the key supporters of the Beyond Health Care conference; the price for his support was that Ilona Kickbusch be invited (she had already visited the Health Advocacy Unit with Ron a year or so earlier as she began to establish links with the health promotion movement in Canada, a partnership that was to culminate in November 1986 with the *Ottawa Charter for Health Promotion* (WHO 1986)—but that is another story).

What Ilona saw in the *Healthy Toronto 2000* workshop was a key strategy that would take the concept of health promotion that she was developing with Ron Draper and others out on to the streets of Europe and make it concrete. She went back to Europe to convene the first meeting of a planning team for WHO Europe's Healthy Cities Project.

Healthy Cities in Europe, 1986

In January 1986, Ilona Kickbusch gathered a small group together at WHO Europe's headquarters in Copenhagen to begin planning a Healthy Cities project.[6] The plan was for a small group project of four to six cities, and Constantino Sakellarides (who despite his Greek name was from Lisbon) offered to host a first meeting in Lisbon in March 1986, while John Ashton took on the task of coordinating the project from Liverpool.

Len Duhl and I offered to write a background paper because, as it happened, I was going from Copenhagen to Berkeley for a 6-week stint as a Visiting Scholar, at Len's invitation. So within those 6 weeks we put together the first WHO Healthy City paper (Hancock and Duhl 1988), which included the definition that is still widely used today, and a set of 11 characteristics of a healthy city that we gleaned from the literature.

The definition was:

> A healthy city is one that is continually creating and improving those physical and social environments and strengthening those community resources which enable people to mutually support each other in performing all the functions of life and achieving their maximum potential.

It is worth taking a bit of time to explore some of the key concepts within this definition.

1. It is a definition of a process, not a status; a Healthy City is one that is continually doing a number of things to become more healthy—the implication being that this is a process of constant striving. It is worth noting, in this connection,

[6] The list of participants can be found in the Preface of Ashton 1992.

that my favourite definition of a goal is one that was used in the late 1970s by the US Public Health Service: It is a timeless statement of aspiration; it is something always beyond our reach, something we are always striving for but will never attain.

2. While developed some months before the Ottawa Charter for Health Promotion was finalized at the First International Conference on Health Promotion (November 1986), it not surprisingly contains many elements of the Charter that were then under development by Ilona Kickbusch, Ron Draper and others, including the concepts of creating supportive environments and strengthening community.
3. The aim is broader than health: it is about functioning well in normal daily life, and about human development and human potential. The focus is not in fact the city itself, but the people who live there.

As to the characteristics, they too are worth some discussion. The 11 characteristics we identified are:

1. A clean, safe, high quality physical environment (including housing quality).
2. An ecosystem which is stable now and sustainable in the long term.
3. A strong, mutually supportive and non-exploitative community.
4. A high degree of public participation in and control over the decisions affecting one's life, health and well-being.
5. The meeting of basic needs (food, water, shelter, income, safety, work) for all the city's people.
6. Access to a wide variety of experiences and resources with the possibility of multiple contacts, interaction and communication.
7. A diverse, vital and innovative city economy.
8. Encouragement of connectedness with the past, with the cultural and biological heritage and with other groups and individuals.
9. A city form that is compatible with and enhances the above parameters and behaviours.
10. An optimum level of appropriate public health and sick care services accessible to all.
11. High health status (both high positive health status and low disease status).

To begin with, we recognized both the physical environment (today I would call it the built environment) and (in wording that prefigured the wording in the Ottawa Charter) an ecosystem which is stable now and sustainable in the long term. We also called for a strong but non-exploitative community, a qualifier that came out of a conversation I had with Aaron Antonovsky, who had created the concept of coherence (Antonovsky 1979). I had asked him if we could apply the concept to a community rather than just to individuals, and he cautioned that Nazi Germany would have scored high on a community coherence index. So when we said non-exploitative, we meant that a strong community would not exploit (or worse) its weaker members, its weaker neighbours (including neighbours halfway around the world) or the natural ecosystems of which it and other communities are a part.

In calling for a high degree of participation we were again picking up on a key emerging theme in health promotion, that of empowerment. Similarly we picked up on another key emerging theme; the basic determinants of health—what the Ottawa Charter called prerequisites—peace, food, shelter, education and income. In addition, we stressed what the literature shows: that healthy people need variety, diversity and stimulation, ideas that are captured in the Liveable Cities movement. Arguably, it is partly this need that makes cities attractive to people and brings them there—that and the need for an income, which is the product of a good economy in the city.

One point that has perhaps not attracted enough attention in the world of health promotion is the importance of connections to history, culture, art and nature, all of which help us to establish a sense of place. There has been a growing interest in the role of the arts in health promotion in recent years, and Richard Louv's book *Last Child in the Woods* (2005) brought us the concept of nature deficit disorder and a growing focus on the health benefits of contact with nature; how we get nature to people in the city is an important issue again, as it was 150 years ago.

We also bring together the need for good urban planning that enables and supports all the preceding characteristics, and public health services, two of the bedrocks of a healthy city. It was no mistake that we placed public health before health care: prevention should always come before treatment; and we refer to it as 'sick care' services because health care is important when you are sick, but is not a particularly important contributor to population health itself. The final point brings us back to where we started; it is not mainly about health status per se, but about the processes that lead to better health status, which is why we listed health status itself last.

It did not take long for the power of the idea to become apparent. Instead of the four or six cities we had anticipated, 21 cities from across Europe sent representatives to the Lisbon meeting, and the project took off. It was aided, undoubtedly, by its very close links to the Ottawa Charter for Health Promotion, being an exemplar of the settings approach that the Charter identified as an important strategy for health promotion. This was reinforced at the Second International Conference on Health Promotion in Adelaide in early 1988, where the focus was on healthy public policy and where Healthy Cities became an example of healthy public policy at the local level (Evers et al. 1990).

Some sense of the rapidity with which the idea of Healthy Cities spread around the world can be gained from the contents of the first book about the concept (Ashton 1992). It contains chapters about developments in Canada, the United States, Australia and developing countries, as well as case studies from nine European cities, Toronto, two US states and three Australian communities—and at that point, the project was barely 6 years old!

In 1915, as already noted, *MacLean's Magazine* had touted Toronto's success in becoming the healthiest of large cities, attributing that success to an awakening, a man, and several years of varied and strenuous activity. Fast forward a bit more than 75 years and we can say the same, only this time it's an awakening, a woman, and several years of varied and strenuous activity. The awakening was health promotion, with its roots in the Lalonde Report and the Ottawa Charter; the woman was Ilona Kickbusch, and the several years of varied and strenuous activity are recorded in part in Ashton (1992). But this time the impact was not just in one large city in Canada, but in an urbanizing world.

References

Antonovsky, A. (1979). *Health, stress and coping*. San Francisco: Jossey-Bass.

Armstrong, A. (1959). Thomas Adams and the Commission of Conservation. *Plan Canada, 1*(1), 14–32.

Ashton, J. (Ed.). (1992). *Healthy cities*. Milton Keynes: Open University Press.

Board of Health. (1978). *Public health in the 1980s*. Toronto: Local Board of Health, City of Toronto.

Board of Health. (1989). *Healthy Toronto 2000*. Toronto: Local Board of Health, City of Toronto.

Care, B. H. (1985). Proceedings of a working conference on healthy public policy. *Canadian Journal of Public Health, 76*(Suppl. 1), 1–104.

Commission of Conservation. (1912). *Annual report* (p. 148). Ottawa: Commission of Conservation.

Commission of Conservation. (1915). *Annual report* (p. 270H). Ottawa: Commission of Conservation.

Duhl, L. J. (1963). *The urban condition: People and policy in the metropolis*. New York: Simon & Schuster.

Evers, A., Farrant, W., & Trojan, A. (1990). *Healthy public policy at the local level*. Campus: Frankfurt.

Hancock, T. (1982). Beyond health care. *The Futurist, 16*(4), 4–13.

Hancock, T. (1984). Health as a social and political issue: Toronto's Health Advocacy Unit. In D. P. Lumsden (Ed.), *Community mental health action*. Ottawa: Canadian Public Health Association.

Hancock, T. (1985). The Mandala of health: A model of the human ecosystem. *Family and Community Health, 8*(3), 1–10.

Hancock, T. (1986a). Lalonde and beyond: Looking back at "A New Perspective on the Health of Canadians". *Health Promotion, 1*(1), 93–100. (Also published as Efter Lalonde in *Vigor* (Swedish National Board of Health), August 1987)

Hancock, T. (1986b). Public health planning in the city of Toronto. Part 1. Conceptual planning. *Canadian Journal of Public Health, 77*, 180–184.

Hancock, T., & Duhl, L. (1988). *Healthy cities: Promoting health in the urban context* (WHO Healthy Cities Paper #1) Copenhagen: FADL. (Originally published in 1986 by WHO Europe, Copenhagen)

Hancock, T., & Perkins, F. (1985). The Mandala of health—A conceptual model and teaching tool. *Health Education, 24*(1), 8–10.

Health and Welfare Canada. (1974). *A new perspective on the health of Canadians* (Lalonde Report). Ottawa: Health & Welfare Canada.

Louv, R. (2005). *Last child in the woods: Saving our children from nature-deficit disorder*. Chapel Hill: Algonquin Books.

MacDougall, H. (1990). *Activists and advocates: Torontos Health Department 1883–9183*. Toronto: Dundurn Press.

Milio, N. (1981). *Promoting health through public*. Philadelphia: F.A. Davis.

Severs, A. (1989). Public health as a catalyst to the growth of municipal departments in early Toronto. *Canadian Journal of Public Health, 80*(4), 291–294.

WHO. (1986). *The Ottawa charter for health promotion*. Copenhagen: WHO Europe.

Yorke, K. M. (1915). Saving lives on wholesale plan. *MacLeans Magazine, 28*(12).

Chapter 5
Healthy Cities Move to Maturity

Evelyne de Leeuw and Jean Simos

In Part I of this book we have taken a look at the antecedents and birth of Healthy Cities. Since the dawn of human settlement 'health' has been an integral part of the greater vision that some have had for urban planning and resilience. 'Health' here is deliberately put in quotation marks, as throughout history our understanding of the idea of health, and its causes and consequences, has shifted. From the inevitable consequences of, for instance, 'the Eye of God' and other spirit explanations, through concepts of balances in body (e.g. Hippocrates) and the greater cosmos (e.g. Yin and Yang), views of the importance of hygiene, and the grand searches for pathological agents that cause disease, we now seem to have arrived at a new understanding that health is multi-dimensional.

The many dimensions of health play out at individual and community levels—for instance, how individuals in their social context are conscious of the (un)healthy choices they can make, and the extent to which they are internally and externally enabled to act on that awareness. But the multi-dimensional nature of health also attains, at the institutional level, the character of a powerful, highly professionalized medical-industrial complex that operates networks of health care diagnostic and treatment facilities within strict legal and governance frameworks. At virtually all levels of government and governance this medical-industrial complex can be considered an important economic and political influence. In many cities and regions around the world—particularly in OECD countries—the local health care industry is

E. de Leeuw (✉)
Centre for Health Equity Training, Research & Evaluation (CHETRE), University of New South Wales, South Western Sydney Local Health District, Ingham Institute, Locked Bag 7103, Liverpool, BC NSW 1871, Australia
e-mail: e.deleeuw@unsw.edu.au

J. Simos (✉)
Institute of Global Health, University of Geneva, Campus Biotech, Geneva, Switzerland
e-mail: jean.simos@unige.ch

E. de Leeuw, J. Simos (eds.), *Healthy Cities*,
DOI 10.1007/978-1-4939-6694-3_5

the largest employer. At the same time there is a strong emergence of a sane, humane and ecological perspective on health, which may not necessarily align very well with the (often implicit) value system of the medical-industrial complex. Fortunately, the medical dominance that has been so pervasive in and beyond both the medical sector and matters of state between the 1930s and 1970s may now be waning (at least, so the authoritative Evan Willis (2006) claims). Regrettably, this is attributed to the rise of another phenomenon that some have labelled evil, or at least not conducive to achieving health and health equity: neo-liberalism (e.g. Bambra 2011).

A Project Becomes a Movement

Even with the decline of medical dominance, the legacy of the biomedical model is strong (Diallo et al. 2015). In an early evaluation of Healthy Cities efforts, we heard from local coordinators and politicians that it was not easy to sell the complex ecological and strongly value-based Healthy Cities message. '*We say public health, they hear medicine; we say health, they hear disease!*' was a complaint that we came across more than once. The consequence of this tension, for many, was that a more activist approach to advocacy for a different health perspective was necessary. The early champions of a Healthy City programme (Len Duhl, Trevor Hancock, John Ashton and Ilona Kickbusch) clearly embraced this type of activism (see Ashton 1990; Ashton and Kickbusch 1986), a call that has been sustained over time when seeking more and better activist scholarship and scholarly activism (de Leeuw 2012a).

In this context it comes as no surprise that one of the first review documents of the accomplishments of the European Healthy Cities endeavour announced that the project had become a movement (Tsouros 1991). A validation of this position can be found in the fact that this publication has had many reprints of thousands of copies, has been translated into at least ten other languages (and published as a glossy, richly illustrated work across Europe) and even in the scholarly world has been cited prolifically. Indeed, Healthy Cities has become a global movement, in practical and in conceptual terms.

Starting with the latter, we take our bearings from Social Movement Theory (McAdam et al. 1996)—or rather, the growing body of theorizing about the new types of social movement that embrace values and cultural perspectives more than the action for full suffrage, economic equality and the advancement of particular groups in society. The key for such value-based movements is that they identify political opportunity to advance, for instance, greater solidarity, equity, sustainability or empowerment. In the first chapter of this book we claim that these values set Healthy Cities apart from other global urban initiatives. Healthy Cities, as value advocates, can be comfortably labelled a 'health social movement' or HSM (Brown et al. 2004). Health social movements

> challenge medical policy and politics, belief systems, research, and practice … They include formal and informal organizations, supporters, networks of cooperation, and media. In pursuit of their objectives, HSMs may challenge political power, professional authority, and existing definitions of personal and collective identity (Brown et al. 2004:16–17).

Admittedly, the conceptualizations of social movements focus predominantly on the coalescing of 'communities', 'lay people', 'those affected', 'clients' and 'consumers' (and other ways in which those groups can be labelled—an Australian study claimed a sizeable portion of the population would be happy with 'mate' (Lloyd et al. 2001)). The idea of Healthy Cities as a movement could (and should?) be criticized as coopting those in power who could be considered guilty of maintaining the status quo that the movement would challenge. The explicit membership and responsibility of local politicians (be they councillors or mayors) in designated European Healthy Cities would not augur well for the espousal of radical change, one could say, as outlined in the original vision for the project. The Eleven Qualities (see Chap. 1; Hancock and Duhl 1986) could easily be reconstructed as a revolutionary manifesto.

But we contend that it is precisely the combination of a relatively radical agenda, advocated and promoted by an authoritative organization like WHO's European office over a long period of time, with all local stakeholders (including communities, practitioners and political actors and activists) involved that sustained the vision and viability of a project that truly did become a movement. Is this true around the world?

The Many Faces of a Healthy City

Through the World Health Organization

Since their 'official' launch in 1986 Healthy Cities has evolved and grown naturally and in great diversity. There has been general support for the initiative by WHO in its different regions, but this support has come in different shapes, directions, and times. Admittedly, many local healthy city and community initiatives have had very little to do with the vision of a bureaucratic international agency such as WHO; other interests and commitments stimulated local action, for instance concern about sustainability spurred by things like the Brundtland report and the first Rio conference (the Earth Summit—in Lafferty and Eckerberg 2013), and the Québécois network of Villes et Villages en Santé initially developed wholly autonomously from endeavours by the Pan American Health Organization (O'Neill et al. 1990). But the strongest effort at directing Healthy City development seems to be in the European region where the movement has been working with cities and national networks continuously over nearly 30 years. This has been facilitated by a sustained personnel presence in the WHO Office and an enthusiastic and committed network of city representatives and academic supporters. The European Healthy City approach is strictly codified, yet recognizes unique diversity between cities. The programme is urban government-based and requires formal designation through the submission of documentation on deliverables, processes and governance arrangements, endorsed by a formal political statement (usually a council decision). One of the commitments each city expresses is to take action on the development of a national network of Healthy Cities. Designated Healthy Cities become a resource hub for other cities in a member state. WHO EURO Healthy Cities work in phases of approximately 5 years (thus transcending the 'regular' WHO budget cycle), during which predetermined strategic and thematic priorities need to be addressed. Currently the network operates in Phase VI. Each of the phases has been evaluated, with increasing sophistication (below). The 'unity in diversity' approach in the designation process means that although all European Healthy Cities sign up to the same parameters and values, the operational aspects of their implementation can be wildly diverse. In a comparative evaluation of five UK and five Dutch Healthy Cities, Goumans (1998) found that Healthy Cities can be, and are being, driven and championed by many sectors, including education, social work, police, infrastructure, transport, and parks and recreation. In most but not all the health sector is represented.

Other WHO regions have expressed commitment to the Healthy City idea, too. The Pan American Health Office has consistently nurtured and supported local government action for health (e.g. through resource packs for mayors and other local politicians) but has not engaged in the formalization of networking. The Western Pacific Regional Office (WPRO) has always had a number of member states where Healthy Cities is high on the agenda, initially with Australia, Japan and Malaysia (Kuching) a consistent presence. Since the turn of the century this region has formally acknowledged the need to focus on urbanization and health, and guidance documentation and a strategic programme to this effect have been developed. WPRO is not proactive in its approach of 'designating' cities, but through its networking efforts many cities across the region have signed up as a 'Healthy City', for instance in Cambodia, Mongolia, and Laos, with an abundance of cities in the Republic of Korea. However, there is no formal programme that establishes monitored relations between local governments and WPRO. The WHO Eastern Mediterranean Region has over the last 5 years revitalized a fledgling Healthy Cities and Villages programme by building on active programmes, notably in the Gulf and Iran. Again, no formal designation, support or recognition has been developed. AFRO has always been active in Healthy Cities and has been supported by a global Francophone network (Europe and Quebec) in developing capacities for Healthy Cities in its French-speaking countries. After some promising initiatives in SEARO (e.g. Cox's Bazar in Bangladesh), the Healthy Cities and urbanization argument seems to have disappeared from the region's agenda, but in other member states (notably Indonesia) many Healthy City-type efforts continue. Globally, WHO headquarters supported Healthy Cities with significance, engaging in World Health Day 1996 ('Healthy Cities for a Better Life'). After this the emphasis shifted to the WHO Kobe Centre (WKC), which used to have urban health as one of its official remits.

WKC develops tools in collaboration with health authorities: for instance, Urban HEART (the Urban Health Equity Assessment and Response Tool as described by Kumaresan et al. 2010) is offered to (local) governments to identify and reduce health inequities at the city level. The tool has been applied widely and is reported on in the peer-reviewed literature, notably that from Iran and the Philippines.

Urban health has positioned itself firmly on the agenda of many national and international organizations. After the first joint publication of a volume called *Hidden Cities* (WHO 2010), the WHO Kobe Centre and UN Habitat published an update on healthier and more equitable cities for sustainable development (WHO 2016). But as we argue in this book, a commitment to urban health is not the same as 'Healthy Cities'. Part I makes abundantly clear that Healthy Cities is not about measurement and epidemiology (alone), but about political commitment and activism (cf. de Leeuw 2012a, in response to Rydin et al. 2012).

Through National Government Schemes

Local governments do not need WHO support to develop and implement urban health activities, but in many countries legislative schemes have taken such ideas as the social determinants of health and Healthy Cities as an inspiration. The Republic of Korea is one such country, with a proliferation of dozens of Healthy Cities, often associated with local civil society and academia (Nam et al. 2011); there is an emerging evaluation and reporting tradition there that follows earlier surges in Japan (e.g. Takano 2003). Another example is the Australian state of Victoria, where a Public Health and Well-Being Act exists that is inspired by evidence of social determinants, the Ottawa Charter and Healthy Cities thinking (see de Leeuw 2012b and 'Case study: Environments for health policy research - Environments for Health (E4H) policy effectiveness' in de Leeuw et al. 2014a)—but the hundreds of Victorian municipalities that produce the required health plans do not identify themselves as Healthy Cities. Even more interestingly, perhaps, Healthy Cities are not necessarily driven by, or accountable to, local government. Two of the longest-running Healthy Cities in the world (Kiama/Illawarra in New South Wales and Noarlunga/Onkaparinga in the state of South Australia) are run by civil society foundations which may—or may not—connect to an official government health agenda: both are represented in this book.

This situation has been recognized by several international NGOs, the most active of which seems to be the Alliance for Healthy Cities with a membership of hundreds in Asia and Oceania. Although the Alliance keeps a repository of case studies and members' health policies, and organizes regular scientific conferences, there is a dearth of systematic and peer-reviewed evaluation from this rich resource. Professor Keiko Nakamura, a long-standing champion of Healthy Cities and the Alliance, contributes a chapter (Chap. 11) on the Asian region in Part II of this book.

Healthy City Evaluations

Since the mid-1980s there has been a proliferation of Healthy City initiatives around the world, with local governments and civil societies often adopting an eclectic mix of objectives and emphases. In some cases a comprehensive, value-driven and long-term perspective is taken (for instance, by designated cities in the European Region of WHO, and some of the European National Networks). Such an approach lends itself to larger and more systematic evaluation efforts, e.g. Harpham et al. (2001), Kegler et al. (2000), Werna and Harpham (1995), and Werna et al. (2014).

In other cases Healthy Cities have wanted to demonstrate the outcomes of their particular emphases, e.g. on oral health in Brazil (Moysés et al. 2006). Choosing evaluation parameters turns out to be a political endeavour (O'Neill and Simard 2006). Not all urban health interventions should be labelled as Healthy City initiatives—they should at the very least demonstrate an engagement with broader social determinants of health discourse, recognize profound social and community involvement, and be driven by a comprehensive set of values such as equity, sustainability and participation. Ideally they need to be grounded in the Eleven Qualities Hancock and Duhl (1986) outlined, and to acknowledge Healthy Cities as a process rather than an intervention. For instance, an urban-based evaluation of an intervention that aims to reduce the incidence of obesity among children through behavioural change cannot be considered a Healthy City effort; at the same time, a sanitation programme to improve access to safe water that recognizes the critical role of the local community, women's groups and good governance may well be a Healthy City (or village) initiative even when the label 'Healthy City' has not been applied.

The Effectiveness of the Application of Principles and Values

Healthy Cities are not necessarily interventions: they are value systems and socio-cultural 'gazes' that may lead to opportunities for better health and the reduction of health inequity. A number of evaluations have looked at how such a value system and gaze has come about, and what its influence is on the potential and capacity of the local level to develop novel programmes. Barton and Grant (2013), Baum et al. (2006), Burton (1999), de Leeuw et al. (2014a, 2014b, 2015c, 2015a), Donchin et al. (2006), Hall et al. (2010), Kim (2012), Plümer et al. (2010) and Twiss et al. (2000) look at systems-wide and value-based drivers of Healthy Cities. They all find that with careful preparation and conscientious consultation with all stakeholders (including communities and industry, and across government sectors) Healthy Cities create conditions for better governance, enhanced intersectoral action and opportunities for Health in All Policies, and for greater resilience of particular programmes facing resource or political turmoil. Lipp et al. (2013) also find that such a systems perspective creates wider and stronger partnerships for health. Farrington et al. (2015) show that enhanced systems parameters, and broader and tighter partnerships, lead to more effective city-based programmes that aim to facilitate healthier lifestyle choices.

The European Office of WHO has consistently sponsored and facilitated these types of evaluation of its networks, which consistently show that taking the time to set the stage in adopting principled values for good governance, community engagement, sustainability and a drive to equity (Ritsatakis 2009 2013; Ritsatakis et al. 2015) yields results.

Focused Interventions from a Healthy City Platform

The pointy end of Healthy City initiatives is, and should be, to improve the health of individuals and communities and to reduce any health inequities. Such an approach boils down to the identification of health threats, disease patterns and effective interventions. Nam et al. (2011) show that the strong value base of Healthy Cities in the Republic of Korea created opportunities for increased tobacco taxation, the proceeds from which were pumped back into health promotion programmes. The resulting reduction in tobacco consumption is consistent with similar research findings elsewhere, but here we can uniquely attribute the success to the Healthy City vision.

The emerging evidence is strong that embracing general Healthy City values at any level of government or governance yields positive health and well-being effects, although this evidence as yet is a bit far and wide. Kiyu et al. (2006) evaluate the development and effectiveness of Healthy Villages in Malaysia: a clear and comprehensive vision, engaging the community with respect, leads to sustained outcomes. Nohara et al. (2015) and Mohan et al. (2014) take a similar approach to diabetes and non-communicable diseases in India and find that new IT technologies may enhance and sustain village-based efforts. Coghlan et al. (2014) look at child wasting in Lao PDR and find sustained health effects of taking a broad intervention and policy perspective. Fanzo (2014) applies a Healthy Village perspective to matters of food security in the Global South and makes similar findings. Masud Rana et al. (2009) investigate quality of life of the elderly in BanglaDeshi villages to the same effect. Even in remote communities in higher income countries this approach works: Tomlin et al. (2012) show that a comprehensive Healthy Village-type programme promotes the health of rural and remote children in Canada. Suntayakorn and Rojjanasrirat (2013) demonstrate its effectiveness in the area of stroke reduction in Thailand, and Acharya et al. (2015) in the improvement of maternal and child health in rural India. At the other end of the spectrum, Healthy Communities has been found to improve the health and well-being of rural elderly in China (Wang et al. 2014).

Introduction to Regions: Part II of This Book

Our initial intention was to have a chapter per WHO region, but reality showed this was not such a good idea. For instance, to gather together in a single chapter all the richness and diversity of Healthy Cities experiences in North and Latin America would under-emphasize their richness and would unbalance the structure of the book. Eventually we opted for a mixed solution, where geographic clustering and cultural proximity both played a role. In Chap. 1 we outlined our logic for compiling the practical material that is collected in Part II: in particular we sought to demonstrate the enormous, value-based, complexity and diversity of Healthy Cities around the world. We believe, thanks to many colleagues who have contributed to this book, that we have succeeded in achieving that objective. We are less happy—as we also mentioned in Chap. 1—that we have been unable to identify Healthy City projects and descriptions in some parts of the world with the highest rates of urbanization.

All these various examples are presented by their protagonists, the local officials who initiated them (and some of whom are still in office), allowing them a wide margin of expression so that they can emphasize what seems to them, in hindsight, most relevant.

We kick off Part II with Africa. The cradle of humanity presents many different faces. It is desert dry and jungle lush. It has in the last few hundred years been dominated by colonial powers, with Anglophone, Francophone and Lusophone legacies. A massively complex tribal and language network covers the continent, not to speak of a range of religious allegiances. After an introduction to the general urbanization and perspectives of Africa, we list case studies that present Healthy Cities implementation across this diversity: Sahelian type (N'Djamena, Healthy Cities in Mali), Tropical (Pointe-Noire), the common concern about healthy markets (Togo) and informal settlements (Mathare), specific problems of deprived urban African communities (Ga Mashie in Accra), and cultural contexts in North and Arabic Africa (El Khroub).

Although the Eastern Mediterranean region is a mere geopolitical construct (with a superficially unifying notion of Islam as the common denominator), we asked our colleagues to reflect on Healthy City cases around the (Persian or Arabian—we do not wish to align with any controversial naming preference) Gulf and the remainder of the WHO region. In Chap. 7 initiatives are described at national levels in, for instance, Iran and Egypt, with local illustrations that centre predominantly around community actions for liveability and sanitation.

Latin America provides the by far longest and most comprehensive chapter (Chap. 8) in this book. The authors of this piece ground themselves firmly in historical antecedents to local health action and show that there has been an enduring legacy of community action and solidarity among the peoples of this part of the world. Our friends from the region describe national programmes and concrete local examples from Argentina, Colombia, Brazil, Peru and Guatemala. Their focus—as we suggested in our invitation—is both on the value base for Healthy Cities and on their governance structures, and they provide ample evidence that these initiatives make a difference, locally and globally.

The North American chapter (Chap. 9) (and in particular both Anglophone and Francophone Canada and the United States—unfortunately we have not been able to illustrate the highly successful local health programmes across Mexico) argues that, rather than resting on the geopolitical urban entity that is addressed in other Healthy Cities, the success of the movement in that part of the world can be attributed to grounding initiatives in local communities, which may translate into political commitment but also may not. The authors, for that reason, tend to focus on national contexts for community action.

For the European chapter (Chap. 10), the geographical diversity—which is, incidentally, one of the criteria of the WHO European Network of Healthy Cities (ENHC)—presents cities from North (Liverpool), West (Rennes and Nantes), South (Athens), East (Pécs) and Central Europe (Geneva). However, there is a particularly good opportunity in this chapter to describe more in detail various institutional forms that influence Healthy Cities structures: they may be attached directly under the mayor's leadership (Athens), located within a municipal administration department (Rennes, Nantes) or another administrative level than the municipality, or even outsourced to a foundation (Pécs). As well as this, innovative actions that could be applied in most cities are presented, such as smoke-free initiatives (Liverpool), the green fork label or the future's scenario workshop method (Geneva). It was important from this perspective to select as examples cities that had at least 15 consecutive years as Healthy Cities, in order to benefit from the necessary reflection. An extended introduction to the chapter describes and analyses ENHC, the only network operated directly by a WHO Regional Office.

In the Asia-Pacific region we find an enormous diversity of initiatives, and Keiko Nakamura in her chapter (Chap. 11) describes in countries as varied as Cambodia, Japan, China and Mongolia, the evolution and foundations of Healthy Cities as very early adoptions of the principles of European Healthy Cities. The case studies represent a fascinating and intriguing mix, combining 'advanced' healthy urban planning approaches with 'traditional' urban environmental health and community action initiatives.

In the Oceania chapter (Chap. 12) we have compiled four perspectives from New Zealand, Australia and the Pacific. Anna Stevenson from Christchurch describes how Healthy Cities thinking brought creativity and resilience to the earthquake-ravaged city. Veteran Healthy City operators from New South Wales and South Australia describe two very different but highly successful Healthy Cities of amazing longevity despite the challenges they have faced over time from adverse state and federal health policies. And finally, in this chapter we also briefly describe some of the most urbanized places on earth, in localities where space is at a premium and sprawl is virtually impossible: islands. We reflect on the challenges of Small Island Developing States and the unique opportunities they have to choose a Healthy City—or rather Healthy Island—path.

Conclusion

This chapter is a prelude to Part II of this book, in which we present an enormously wide array of Healthy City cases and contexts. Based on the historical accounts of Part I, we have described the emerging evidence base for Healthy Cities but acknowledge that their global nature and sheer diversity does not necessarily make it easy to determine whether a Healthy City 'works'. The cases in Part II show that at a very practical and narrative level Healthy Cities most certainly work, and in Chap. 13 we will attempt to wrap up the strong emerging evidence and identify which challenges remain. These comprise Part III of this work.

References

Acharya, A., Lalwani, T., Dutta, R., Rajaratnam, J. K., Ruducha, J., Varkey, L. C., … Bernson, J. (2015). Evaluating a large-scale community-based intervention to improve pregnancy and newborn health among the rural poor in India. *American Journal of Public Health, 105*(1), 144–152

Ashton, J. (1990). Urban lifestyle and public health in the city. *Journal of the Royal Statistical Society Series D (The Statistician), 39*(2), 147–156.

Ashton, J., & Kickbusch, I. (1986). *Healthy cities—Action strategies for health promotion, first healthy cities brochure*. Copenhagen: World Health Organization Regional Office for Europe.

Bambra, C. (2011). Health inequalities and welfare state regimes: Theoretical insights on a public health 'puzzle'. *Journal of Epidemiology and Community Health, 65*(9), 740–745.

Barton, H., & Grant, M. (2013). Urban planning for healthy cities: A review of the progress of the European Healthy Cities Programme. *Journal of Urban Health: Bulletin of the New York Academy of Medicine, 90*(Suppl 1), 129–141.

Baum, F., Jolley, G., Hicks, R., Saint, K., & Parker, S. (2006). What makes for sustainable healthy cities initiatives?—A review of the evidence from Noarlunga, Australia after 18 years. *Health Promotion International, 21*(4), 259–265.

Brown, P., Zavestoski, S., McCormick, S., Mayer, B., Morello-Frosch, R., & Gasior Altman, R. (2004). Embodied health movements: New approaches to social movements in health. *Sociology of Health & Illness, 26*(1), 50–80.

Burton, S. (1999). Evaluation of healthy city projects: Stakeholder analysis of two projects in Bangladesh. *Environment and Urbanization, 11*(1), 41–52.

Coghlan, B., Toole, M. J., Chanlivong, N., Kounnavong, S., Vongsaiya, K., & Renzaho, A. (2014). The impact on child wasting of a capacity building project implemented by community and district health staff in rural Lao PDR. *Asia Pacific Journal of Clinical Nutrition, 23*(1), 105.

de Leeuw, E. (2012a). Healthy cities deserve better. *The Lancet, 380*(9850), 1306–1307.

de Leeuw, E. (2012b). Do healthy cities work? A logic of method for assessing impact and outcome of healthy cities. *Journal of Urban Health, 89*(2), 217–231.

de Leeuw, E., Clavier, C., & Breton, E. (2014a). Health policy–Why research it and how: Health political science. *Health Research Policy and Systems, 12*(1), 55.

de Leeuw, E., Tsouros, A. D., Dyakova, M., & Green, G. (2014b). *Healthy cities, promoting health and equity-evidence for local policy and practice*. Copenhagen: WHO Regional Office for Europe.

de Leeuw, E., Green, G., Spanswick, L., & Palmer, N. (2015a). Policymaking in European healthy cities. *Health Promotion International, 30*(suppl 1), i18–i31 de Leeuw, E., Green, G., Tsouros, A., Dyakova, M., Farrington, J., Faskunger, J., … Lease, H. (2015b). Healthy Cities Phase V evaluation: Further synthesizing realism. *Health promotion international, 30*(Suppl 1), i118–i125.

de Leeuw, E., Kickbusch, I., Palmer, N., & Spanswick, L. (2015c). European Healthy Cities come to terms with health network governance. *Health Promotion International, 30*(suppl 1), i32–i44

de Leeuw, E., & Skovgaard, T. (2005). Utility-driven evidence for healthy cities: Problems with evidence generation and application. *Social Science & Medicine, 61*(6), 1331–1341.

Diallo, T., Cantoreggi, N., Simos, J., Christie, D. (submitted in December 2015). Health is multisectoral, but is it practiced as such? A case study in Switzerland. *Public Health.*

Donchin, M., Shemesh, A. A., Horowitz, P., & Daoud, N. (2006). Implementation of the Healthy Cities' principles and strategies: An evaluation of the Israel Healthy Cities network. *Health Promotion International, 21*(4), 266–273.

Fanzo, J. (2014). Strengthening the engagement of food and health systems to improve nutrition security: Synthesis and overview of approaches to address malnutrition. *Global Food Security, 3*(3), 183–192.

Farrington, J. L., Faskunger, J., & Mackiewicz, K. (2015). Evaluation of risk factor reduction in a European City Network. *Health Promotion International, 30*(suppl 1), i86–i98.

Goumans, M. J. (1998) *Innovations in a fuzzy domain: Healthy cities and (health) policy development in the Netherlands and the United Kingdom* (Doctoral dissertation). Maastricht University.

Hall, C., Davies, J. K., & Sherriff, N. (2010). Health in the urban environment: A qualitative review of the Brighton and Hove WHO Healthy City Program. *Journal of Urban Health, 87*(1), 8–28.

Hancock, T., & Duhl, L. (1986). *Promoting health in the urban context* (WHO Healthy Cities Papers No. 1). Copenhagen: FADL Publishers.

Harpham, T., Burton, S., & Blue, I. (2001). Healthy city projects in developing countries: The first evaluation. *Health Promotion International, 16*(2), 111–125.

Kegler, M. C., Twiss, J. M., & Look, V. (2000). Assessing community change at multiple levels: The genesis of an evaluation framework for the California Healthy Cities Project. *Health Education & Behavior, 27*(6), 760–779.

Kim, J. (2012) *Principles of healthy city evaluation*. 保健 敎 育·建康 增進 學會 誌 第 29 卷 4 號, 2012.10, 67–81.

Kiyu, A., Steinkuehler, A. A., Hashim, J., Hall, J., Lee, P. F., & Taylor, R. (2006). Evaluation of the healthy village program in Kapit district, Sarawak, Malaysia. *Health Promotion International, 21*(1), 13–18.

Kumaresan, J., Prasad, A., Alwan, A., & Ishikawa, N. (2010). Promoting health equity in cities through evidence-based action. *Journal of Urban Health, 87*(5), 727–773.

Lafferty, W. M., & Eckerberg, K. (2013). *From the Earth Summit to Local Agenda 21: Working towards sustainable development* (Vol. 12). London: Routledge.

Lipp, A., Winters, T., & de Leeuw, E. (2013). Evaluation of partnership working in cities in phase IV of the WHO Healthy Cities Network. *Journal of Urban Health, 90*(1), 37–51.

Lloyd, C., King, R., Bassett, H., Sandland, S., & Savige, G. (2001). Patient, client or consumer? A survey of preferred terms. *Australasian Psychiatry, 9*(4), 321–324.

Masud Rana, A. K. M., Wahlin, Å., Stålsby Lundborg, C., & Nahar Kabir, Z. (2009). Impact of health education on health-related quality of life among elderly persons: Results from a community-based intervention study in rural Bangladesh. *Health Promotion International, 24*(1), 36–45.

McAdam, D., McCarthy, J. D., & Zald, M. N. (Eds.). (1996). *Comparative perspectives on social movements: Political opportunities, mobilizing structures, and cultural framings*. Cambridge: Cambridge University Press.

Mohan, V., Prathiba, V., & Pradeepa, R. (2014). Tele-diabetology to screen for diabetes and associated complications in Rural India: The Chunampet Rural Diabetes Prevention Project Model. *Journal of Diabetes Science and Technology, 8*(2), 256–261.

Moysés, S. J., Moysés, S. T., McCarthy, M., & Sheiham, A. (2006). Intra-urban differentials in child dental trauma in relation to healthy cities policies in Curitiba, Brazil. *Health & Place, 12*(1), 48–64.

Nam, E. W., De Leuw, E., Moon, J. Y., Nayu, I., Dorjsuren, B., & Park, M. B. (2011). Sustainable funding of health initiatives in Wonju, Republic of Korea via a tobacco consumption tax. *Health Promotion International, 26*(4), 457–464.

Nohara, Y., Kai, E., Ghosh, P. P., Islam, R., Ahmed, A., Kuroda, M., … Nakashima, N. (2015). Health checkup and telemedical intervention program for preventive medicine in developing countries: Verification study. *Journal of Medical Internet Research, 17*(1), e2.

O'Neill, M., Cardinal, L., Fortin, J. P., & Groleau, G. (1990). La naissance du réseau québécois de Villes et de villages en santé. *Recherches sociographiques, 31*(3), 405–418.

O'Neill, M., & Simard, P. (2006). Choosing indicators to evaluate Healthy Cities projects: A political task? *Health Promotion International, 21*(2), 145–152.

Plümer, K. D., Kennedy, L., & Trojan, A. (2010). Evaluating the implementation of the WHO Healthy Cities Programme across Germany (1999–2002). *Health Promotion International, 25*(3), 342–354. doi:10.1093/heapro/daq028.

Ritsatakis, A. (2009). Equity and social determinants of health at a city level. *Health Promotion International, 24*(suppl 1), i81–i90.

Ritsatakis, A. (2013). Equity and the social determinants of health in European cities. *Journal of Urban Health, 90*(1), 92–104.

Ritsatakis, A., Ostergren, P. O., & Webster, P. (2015). Tackling the social determinants of inequalities in health during Phase V of the Healthy Cities Project in Europe. *Health Promotion International, 30*(suppl 1), i45–i53.

Rydin, Y., Bleahu, A., Davies, M., Dávila, J. D., Friel, S., De Grandis, G., … Lai, K. M. (2012). Shaping cities for health: complexity and the planning of urban environments in the 21st century. *The Lancet, 379*(9831), 2079–2108.

Suntayakorn, C., & Rojjanasrirat, W. (2013). Effects of an empowerment program on health behaviors and stroke risk reduction of at-risk group in Thai rural community. *Journal of the Medical Association of Thailand=Chotmaihet thangphaet, 96*(9), 1229–1238.

Takano, T. (Ed.). (2003). *Healthy cities and urban policy research*. London: Routledge.

Tomlin, D., Naylor, P. J., McKay, H., Zorzi, A., Mitchell, M., & Panagiotopoulos, C. (2012). The impact of Action Schools! BC on the health of Aboriginal children and youth living in rural and remote communities in British Columbia. *International Journal of Circumpolar Health, 71*, 17999.

Tsouros, A. D. (1991). *World Health Organization Healthy Cities Project: A project becomes a movement* (Review of progress 1987 to 1990). Milano: Sogess.

Twiss, J. M., Duma, S., Look, V., Shaffer, G. S., & Watkins, A. C. (2000). Twelve years and counting: California's experience with a statewide Healthy Cities and Community program. *Public Health Reports, 115*(2–3), 125.

Wang, J., Chen, C. Y., Lai, L. J., Chen, M. L., & Chen, M. Y. (2014). The effectiveness of a community-based health promotion program for rural elders: A quasi-experimental design. *Applied Nursing Research, 27*(3), 181–185.

Werna, E., & Harpham, T. (1995). The evaluation of healthy city projects in developing countries. *Habitat International, 19*(4), 629–641.

Werna, E., Harpham, T., Blue, I., & Goldstein, G. (2014). *Healthy city projects in developing countries: An international approach to local problems*. London: Routledge.

Willis, E. (2006). Introduction: Taking stock of medical dominance. *Health Sociology Review, 15*(5), 421–431.

World Health Organization. (2010). *Hidden cities: Unmasking and overcoming health inequities in urban settings*. Geneva: World Health Organization, UN Habitat.

World Health Organization. (2016). *Global report on urban health: Equitable, healthier cities for sustainable development*. Geneva: World Health Organization, UN Habitat.

Part II
Regional World Perspectives

Chapter 6
Healthy Cities in Africa: A Continent of Difference

Jean Simos, Françoise Belemel Naissem, Jonas Naissem, Fatoumata Maïga Sokona, Jean de Dieu Konongo, Amidou Sani, Jason Corburn, Irene Karanja, Jack Makau, Ama de-Graft Aikins, and Abdelhamid Haroun

Introduction

Africa is one of the least urbanized continents. Less than 40 % of the population (413 million of over one billion) lived in urban areas in 2011. This situation is changing rapidly, and in 2050 around 70 % of the continent's population is expected to be urban. This extremely fast pace means that the gap between Africa and other continents is closing rapidly—the world average was 52.1 % in 2011 (see Fig. 6.1).

Considerable regional differences can be observed between still relatively rural East Africa (23.7 %), highly urbanized Southern Africa (58.7 %), and the Northern Africa region (51.1 %), which is close to the world average (51.5 % urbanization in 2011). There are also big differences between countries, from urbanization at 17 % in Ethiopia to over 50 % in Côte d'Ivoire and in Northern African states as a whole (UN-Habitat 2014a, b).

J. Simos (✉)
Institute of Global Health, University of Geneva, Campus Biotech, Geneva, Switzerland
e-mail: jean.simos@unige.ch

F.B. Naissem
Chef de Division Hygiène du Milieu et Assainissement, Direction de la Santé Environnementale et de Promotion de la Santé, Ministry of Public Health, N'Djamena, Chad

J. Naissem
Conseiller National Chargé de l'Information et de la Santé (NPO/HIP) & Point Focal Santé/Environnement (PHE), WHO Local Office in Chad

F. Maïga Sokona
Conseiller Santé & Environnement, WHO Local Office in Mali

J.d.D. Konongo
independant consultant, former WHO Environmental Health Officer, Brazzaville, Republic of Congo

E. de Leeuw, J. Simos (eds.), *Healthy Cities*,
DOI 10.1007/978-1-4939-6694-3_6

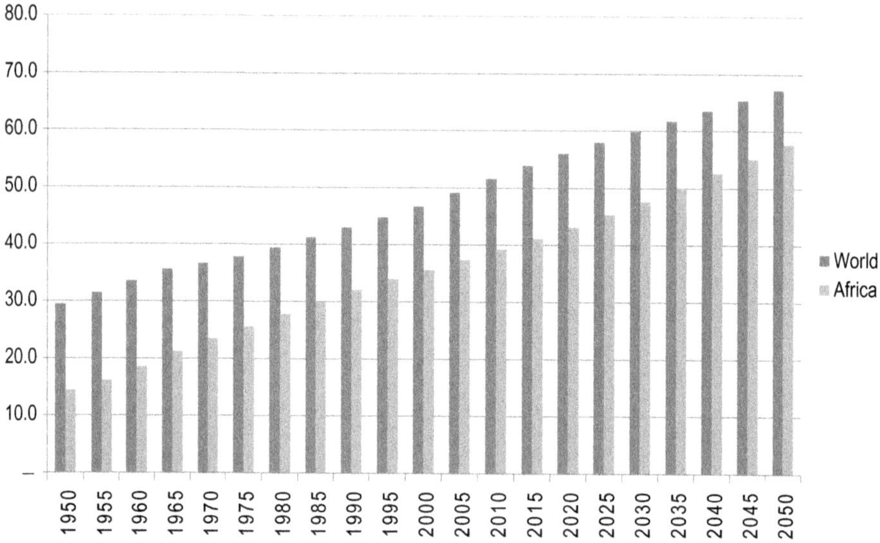

Fig. 6.1 World Urbanization Prospects (Source: *World Urbanization Prospects, 2011*)

African cities can be summarily categorized as follows (Chenal 2013):

- North African cities, most of which have a long history (several were founded more than 2000 years ago).
- Cities around the Gulf of Guinea, many of them the largest on the continent.
- Cities in Southern Africa, developed under British or Dutch influence.

Many of the largest cities are ports, and developed looking out to sea rather than towards the interior of the continent. This is clearly a heritage of colonization: most

A. Sani
Division de l'Assainissement et de l'Hygiène du Milieu (DAHM), Direction des Soins de Santé Primaires (DSSP), Ministry of Health, Togo

J. Corburn
Director of the Institute of Urban and Regional Development and the Center for Global Healthy Cites at the University of California, Berkeley, CA, USA

I. Karanja
former Executive Director of Muungano Support Trust (MuST), Nairobi, Kenya

J. Makau
Director of Shack/Slum Dwellers International (SDI), Nairobi, Kenya

A.d.-G. Aikins
Professor of Social Psychology at the Regional Institute for Population Studies, University of Ghana, Accra, Ghana

A. Haroun
Centre de Santé Communautaire, City of El Khroub, Algeria

of the railways reaching between these cities and the interior were to convey raw materials to the coast and on to Europe (Durand-Lasserve & Selod 2009).

The typical colonial city had a dual identity, with a business-oriented centre for the elite surrounded by poorer areas. These originally had strict boundaries between the two types of area, made possible by the fact that the state (usually a European state) owned all the land. Such cities developed gradually as each plot was mapped out and rented or sold by the government. An important aspect of contemporary African cities is the tension between modern and traditional property of land, the former being based on written land titles and the latter on customary tenure by local residents. In general, urban land is owned by the state and allocated to occupants who enjoy use-rights. Vacant urban lands are owned by the state (USAID 2010). This has important implications regarding the ability of cities to manage the current influx of new residents who are displaced from rural areas through a combination of population growth, falling agricultural yields, increased insecurity, and a general lack of perspectives for the future, and who have no access to land on the free market (Durand-Lasserve 2005). In 2010, 57 % of the urban population of Africa lived in cities with fewer than half a million inhabitants and 26 % in cities with one to five million inhabitants. In these larger cities urbanization has increased the most. Such fast-track urbanization process brings with it a number of problems. Informal settlements becomes the norm; there is inevitable environmental degradation, and social and spatial segregation lead to territorial fragmentation, inappropriate or ineffective public policies, insufficient and outdated infrastructure, and inequitably shared economic growth, whose dividends remain the property of a chosen few (Chenal 2013).

Local and national governments are often unable to address the new needs generated by rapid urban growth: although responsible for providing infrastructure and services, they often do not have the resources to deliver what is needed. Poor infrastructure does not deter new urban migrants: even when demand for urban infrastructure outstrips what can be made available in the short term, whatever can be provided may be an improvement on the services available in rural areas. Infrastructure in cities, even when it is insufficient to meet demand, is often far more available than in the country (Bart and Bonvallot 2002). Local participatory democracy is sometimes able to supplement insufficient budgets at neighbourhood level.

Some decentralization processes that are occurring in the same timeframe as urbanization may help achieve better local governance, but very few African countries have launched a true devolution process.

The informal sector is an important part of the African economy today. It has its positive aspects because it can help redistribute riches, or even contribute to managing a city or neighbourhood. Informal activities are practised not only by poor people but also by higher social classes.

It may seem paradoxical that in some countries the wealthiest people can almost do without a legal framework, while the poorer sectors of society push for the legal and physical protection provided by the rule of law (Chenal 2013).

Among the challenges specific to Africa is an extremely strong population growth rate (250 million more inhabitants are expected within the next 20 years) accompanied by rapid urbanization and no guarantee of economic growth. This situation is very different from what occurred during the industrial development of Europe. Such rapid growth leads to another key challenge: urban sprawl. The continent's 250 million new inhabitants will mainly be moving towards cities without the financial or physical ability to live in their centres, which are still the domain of the privileged. Rather than increasing urban density, these newcomers tend to develop informal settlements on the outskirts, to the detriment of rural activities and natural landscapes (Myers 2011).

The fundamental problems surrounding sanitation, waste disposal and the supply of clean water are compounded by problems linked to transportation. This is because, in a rapidly growing city, all populations—and especially the most impoverished and marginalized—need to travel more frequently and over greater distances. Those living in informal settlements on city outskirts, on land poor in roads and public transportation services, will have to invest disproportionately in travel. Mobility may become an important issue, not only due to its environmental effects (air pollution, noise, congestion, accident risks) but also regarding social effects such as spatial segregation and exclusion.

Population growth, rapid urbanization, urban sprawl and increased demand for transport all lead to increased greenhouse gas emissions. These factors are therefore among the drivers of climate change, which will affect African cities in ways that are difficult to evaluate today.

Healthy Cities can be set up in such complex, diverse and often very challenging settings, and the Healthy Cities Programme can help populations and governments respond to these challenges and prepare for the future.

Healthy Cities in Chad: The Case of the City of N'Djamena[1]

Introduction

Situated in the central north of the African continent, Chad is one of the poorest nations in the world, ranking 183rd out of 187 countries on the Human Development Index established by UNDP in 2011. Poverty in Chad is both monetary and non-monetary. According to a study of 'household consumption and the informal sector in Chad' (ECOSIT3), published in 2011, a little less than half of the population of the country (46.7 %) is in a state of monetary poverty considered extreme: this means below a UNPD-defined poverty level of 652 CFA francs (approximately USD 1.30 per day) in 2012. The proportion of the population under the poverty level was higher in 2003, at around 55 %. Rural–urban differences remain high, with respectively 59 % and 25 % of the population under the poverty level (INSEED, ECOSIT3, 2011).

[1]Françoise Belemel Naissem and Jonas Naissem, with the support of Jean Simos.

Chad nevertheless has high economic potential, as yet insufficiently tapped. The socio-economic situation is partly due to the landlocked position of the country, a highly unequal distribution of rain both throughout the year and geographically, environmental degradation, weakness in communications and transport infrastructure, lack of qualified human resources, bad governance, an ongoing HIV epidemic, and other epidemic and endemic diseases. Health coverage remains unsatisfactory and varies considerably between different parts of the country, despite recent efforts by government agencies. Public spending on health (4.72 % of the national budget) is far below what is usual in other developing countries. Health problems are compounded by a poor communications infrastructure and the weak purchasing power of the population, which all limit access to health services.

Climatic variations complicate the provision of health services. Chad is one of the Sahel[2] region's richest countries as regards water supply, and during the rainy season inhabitants of the flood plains find it extremely difficult to access health services at the time when transmissible diseases such as malaria are at a peak. The incidence of many insect-borne diseases diminishes as one moves away from sources of water (the distribution of parasitic diseases and the hydrographic map of the country overlap considerably).

The administrative organization of the country is heavily centralized in the capital, N'Djamena, where the major decision centres and most social and health services can be found. However, Chad opted in 1996 for a decentralized structure, and the decentralization and devolution of services to outer regions has begun.

The economy of the country is based on the primary sector. Agriculture, including stock-raising, counts for 40 % of GDP. Extractive industries, mainly oil and gas, have become important sources of export earnings in recent years.

Water and Sanitation

Less than one-third of the population has access to drinking water (31.3 %, of which 11.2 % have access to tap water and 20.1 % use public wells).

Sewage and drainage systems are far from satisfactory. Most of the population lives in unhygienic conditions, and some 71.3 % have no sanitation (Ministry for Economic Development and Planning, 2000). Only 24.6 % have an adequate sewage system and only 0.2 % modern (flushing) toilets. This situation is associated with the presence of diseases such as diarrhoea and polio, and parasitic diseases, and to the country's high levels of morbidity and mortality.

Household waste disposal is another intractable problem (see Tables 6.1 and 6.2). Infrastructure for waste disposal is not available yet, so waste is left to decompose in the open, which during the wet season is unpleasant and highly unhygienic.

[2]The Sahel is the ecoclimatic and biogeographic zone of transition in Africa between the Sahara Desert to the north and the Sudanian Savanna to the south.

Table 6.1 Production of waste in the city of N'Djamena

	Solid waste production with sand		Solid waste production without sand	
Year	Tonnes/day	m^3/day	Tonnes/day	m^3/day
2000	600	634,364	340	364,723
2005	752	800,011	428	461,467
2010	931	999,591	533	578,994

Table 6.2 Production of waste on the markets of N'Djamena

	Waste produced					
	With sand	Without sand	With sand	Without sand	With sand	Without sand
Year	kg/year	kg/year	m^3/year	m^3/year	m^3/year	m^3/year
2000	16,889	12,312	55	40	14,600	20,075
2005	22,601	16,411	74	53	19,345	27,010
2010	30,246	21,959	98	71	25,915	35,770

The Urban System in Chad

This system has four main characteristics:

- A polarized and slow urbanization process
  The proportion of the population living in urban areas increased from 9 % in 1970 to 23.8 % in 2000 and 42 % in 2014. This is relatively slow, with an urbanization rate of around 3 % per year over 2010–2015. N'Djamena is the only sizeable city, with slightly more than one million inhabitants, in a country whose total population is 11,500,000; Chad's other towns are far smaller. Almost all of them lie in the southern and central areas of the country, where they have grown thanks to an influx from the north (some of them nomadic), who have been fleeing drought and ethnic strife over the past 20 years. Sociopolitical crises lie at the root of many population movements in Chad.

- The dominant role of N'Djamena
  The capital city concentrates political and administrative power, as well as economic decision centres. Some 78.5 % of its population works in the services sector, and the city offers half of the salaried jobs available nationally.

- Need to boost the decentralization process
  The pace of institutional reform was accelerated in August 2004 when 18 administrative regions were created (see Fig. 6.2). Decentralization is a transversal objective, which has the potential to enable and stimulate planning activities across the country. However, regions and local communities (whether urban or rural) have not yet succeeded in implementing their own development agendas.

- Need for infrastructure to manage internal migration processes
  The arrival of migrants in urban centres puts infrastructure under pressure at many levels: housing, health, education, transport, water and sanitation. Local

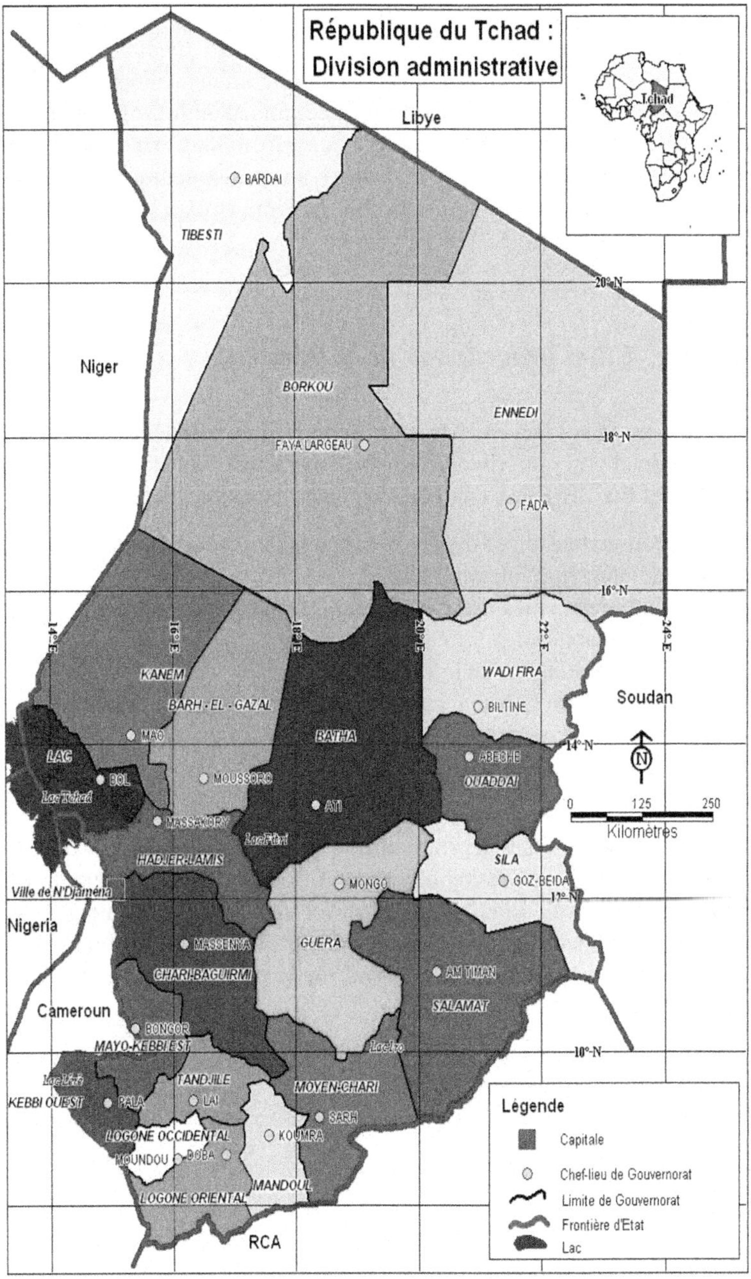

Fig. 6.2 Administrative map of Chad

and national government officials need to be able to plan ahead and set up construction programmes and projects for which funding (and therefore multi-lateral support) is needed, ideally enabling urban centres to increase their 'absorption capacity' in the face of the continued influx of populations and goods. An example is Abéché (the fourth largest city in Chad, with an estimated population of 76,492 in July 2014), where an infrastructure development programme, the installation of tap water, set up with German support has met with a great deal of success.

The Healthy Cities programme in N'Djamena

The economic profile of this city has been dominated by petroleum extraction and an active commercial port on a river leading to Lake Chad. The demographic increase since the end of the 1990s has compounded environment degradation such as:

- A spontaneous extension of the city through the creation of informal settlements with no accompanying infrastructure.
- High population densities in swampy areas, some of which have been officially declared non-inhabitable.
- Destruction of existing urban infrastructure.
- Waste dumps close to housing, causing precarious hygienic conditions.
- Presence of mosquitos, rats and other pests, some of which carry disease.
- Increased risk of faecal contamination.

In July 2000, the Municipality of N'Djamena and the Ministry for Public Health set up a Healthy Cities programme following a workshop which brought together public and private stakeholders, civil society, and representatives of the smaller administrative units: boroughs (*arrondissements*), neighbourhoods (*quartiers*) and blocks (*carrés*). This workshop prioritized 13 topics for urgent attention, all linked to quality of life and the precarious environmental health situation in the city:

- Household waste disposal
- Food safety
- Good governance
- Institutional support
- The proliferation of plastic bags
- Access to healthcare services
- Noise
- Rainwater collection and improvement of run-off water networks
- Transport
- Housing and town planning
- Coastal and riverside development
- Sanitation systems
- Drinking water supply

Participants in this workshop issued recommendations, in particular to set up a Health Plan for N'Djamena which was to evolve into a Healthy Cities action plan; this plan was adopted in January 2012. It is constructed around 3 core objectives and 33 actions, the concrete measures associated with each core objective. The plan is founded on the mutual engagement of over 10 different ministries, including Environment, Water, Public Works, Transport, Housing and Urban Planning, Justice, the Interior, Economic Development, and Social Services.

- Waste disposal for solids, liquids and biomedical waste. This core objective has two main orientations: to ensure the efficient management of household, industrial and biomedical waste, and to empower local populations to adopt basic hygiene practices and adhere to the rules regarding sanitation.
- Access to clean and healthy water supplies.
- Follow-up of the action plan implementation/evaluation.

Results, Progress and Challenges

Housing is an underdeveloped sector and faces constraints linked to the lack of a policy encouraging and funding social housing, antiquated legal propriety laws (1967), and construction rules (1970) which are not all supported by the local population and whose enforcement is time-consuming.

Regarding drinking water supply, the current objective is to ensure coverage of 70 % of the urban population in the short term. The strategy is to provide local populations with the necessary infrastructure and empower them to perform the management and maintenance services themselves.

Significant challenges exist regarding sanitation. The second general population survey (*Recensement général de la population et de l'habitat*—RGPH.2) showed that most homes and people in Chad (74 %) still use the outdoors rather than toilets. In 1993, over three-quarters of households used the outdoors for their basic bodily functions; consequently there are frequent diarrhoeal diseases and high child mortality.

Over the country as a whole, 70.9 % of households dispose of their waste in the immediate vicinity. A further fifth (19.5 %) leave it by the roadside. Nationally, only 2.2 % use modern waste disposable methods (garbage bags or municipal containers). The proportion 'using other methods' is 4.7 %, according to the 2013–2015 National Development Plan document, published in April 2013.

With the roll-out of community-managed sanitation in urban and semi-urban settings, new projects have evolved, building on the success of previous attempts:

- Creation of a wastewater master plan for N'Djamena
- Development of studies for the creation of a wastewater collection network and treatment plant for a priority area (equivalent to 150,000 inhabitants)
- Creation of wastewater master plans in four major urban centres, and contributions to sanitation through the construction of toilets, latrines and washbasins

- Extension of a project to provide running water provision in N'Djamena (4 km/year), N'Djari and Diguel
- Projected construction of water towers in Chagoua, N'Djari and N'Djamena
- Training and empowerment of communities furnished with latrines
- Construction of improved latrines in public buildings and medical centres throughout the country
- Improvement of existing latrines
- Establishment of a household waste collection system

Conclusion

The access rate to basic sanitation, which was at 7 % of the population in 2000, increased to 12 % in 2010. The objective of 35 % coverage in 2015, one of the country's Millennium Development Goals, was impossible to reach, despite substantial efforts made by national authorities.

Adding to its purely environmental challenges, the capital city is exposed to strong internal migration pressure, coupled with low productivity (both of workforce and capital) and weak infrastructure (energy, water, transport, communications), compounded by low private sector investment in construction. The situation tends towards a positive evolution but a fragile one, because of the strong constraints facing N'Djamena.

The setting up of the Healthy Cities programme therefore constitutes a positive dynamic for the capital and other urban settings throughout Chad, with a view to improving the urban environment and the health of urban populations.

Implementing the Healthy Cities Initiative: The Malian Experience[3]

Introduction

Mali is a large country in the heart of the Sahel region in West Africa; around 65 % of its surface is desert or arid territory.

The epidemiological profile of Mali is dominated by infectious diseases, some of them parasitic, as well as nutrition-related problems. The latest Demography and Health Survey (*EDS IV*, initials from the French) and National Health Information System (*SNIS*) show that malaria is the first cause of morbidity with 37.5 % of cases, followed by lower respiratory tract infections (9.3 %) and diarrhoea (5.4 %). This last figure does not include cholera, which continues to affect the country: over the

[3] Fatoumata Sokona Maïga, with the support of Jean Simos.

past 20 years, Mali has had eight cholera epidemics resulting in 21,559 cases and 3014 deaths.

Demographic growth, insufficient sanitation (provision of drinking water and disposal of waste water), combined with inappropriate hygiene practices on the part of the population, explain to a great extent the high levels of water-related diseases including diarrhoea.

The Healthy Cities Approach in Mali: History and Priorities

Confronted with a challenging public health profile, Mali adopted the Healthy Cities approach following the World Health Day call to action issued by WHO on 7 April 2003 under the motto 'Healthy Cities for a better lifestyle'. Given its specific needs related to healthy environments and food safety, Healthy Cities in Mali concentrated on living environments, with a two-pronged approach focusing on open-air markets and schools, dubbed 'Healthy Markets' and 'Healthy Schools'. Such priorities are common to several other countries in the region and WHO Africa has delivered considerable methodological and scientific support for them.

Investigations within schools including an evaluation of the level of access to water and sanitation showed that almost all the 12 investigated schools had traces of defecation around toilets, which had been closed because of the lack of water (see Photo 6.1). Ten of these schools (83.3 %) were storing drinking water in non-hygienic jars, inside classrooms that were heavily overcrowded (see Photos 6.2 and 6.3).

An investigation of open-air food markets in four cities, Bamako, Sikasso, San and Koutiala, led to the following observations:

- Access to drinking water is very difficult (no source of water inside the market)
- Provision of toilets is insufficient
- Exposition of food to flies is a serious concern
- Hygiene regarding meat is unsatisfactory across the product cycle (transport, packaging, handling, conservation after the meat is sold, etc.; see Photos 6.4 and 6.5)

Similar observations were made in all markets.

Implementation and Results

The Healthy Markets concept was introduced into several countries in the African region, including Mali, in order to improve coordination between stakeholders, leading to slow but more importantly lasting improvements in food markets. As well as enhancing infrastructure and essential services, there is a strong emphasis on changing the behaviour of the community in and around the marketplace, including food producers, sales personnel and the consumers themselves. Healthy Markets is a unique opportunity for educating and informing key stakeholders on their roles in

Photo 6.1 school latrines abandoned for lack of water

Photo 6.2 Storage of drinkable water at school

Photo 6.3 Classroom

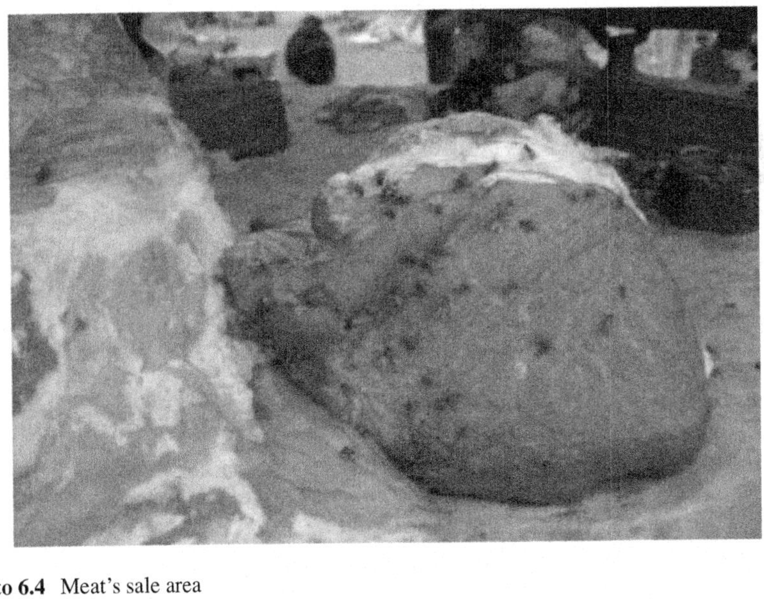

Photo 6.4 Meat's sale area

Photo 6.5 Water shortage in poultry cleaning area

the delivery of healthy food: and approach with the potential to foster hygienic behaviours at all stages of the food chain.

Implementation of the Healthy Cities initiative in Mali has led to several results. At the institutional level of municipalities,

- The city of Bamako joined the Healthy Cities community, a Healthy City action plan was created, and a round table discussion to mobilize resources was organized with city officials.
- A Healthy City action plan was drawn up for the cities of Sikasso and Ségou.
- A competition was organized for 'Cleaner Homes' and another for 'Healthy Schools'.

The Healthy Schools initiative is a strategy to improve hygiene within schools, through the following actions:

- Guarantee access and quality of social and sanitary services for schoolchildren, students and children below school age.
- Improve medical surveillance of schoolchildren and children below school age.
- Involve teachers in health education and in the improvement of living conditions within the school.
- Improve knowledge and practices regarding hygiene, sanitation, parasites, environmental protection, prevention of diseases linked to micro-nutriments, malnutrition, and diseases linked to insufficient hygiene.
- Promote health in a healthy and attractive environment within the school.

The following results were achieved:

- Introduction of lessons and collective exercises for hand washing at school (see Photos 6.6 and 6.7).
- Organization of a 'Clean Schools' competition.
- Construction of infrastructure for drinking water, hygiene and sanitation.

Photo 6.6 School washbasin

Photo 6.7 Celebration at school of Hand Washing Day

Photo 6.8 Device for drinking water storage and waste management

Actions in Markets

After a preliminary analysis of the situation, a planning workshop was organized bringing together all stakeholders, including the board of market retailers, women's groups, municipal and administrative authorities.

The implementation of the action plan was facilitated by the availability of resources at local level (support fund for market retailers, municipal authorities, market management board). Results included improved drinking water storage and waste management (see Photo 6.8); better protection of areas where meat is stored and sold (see Photo 6.9); and wide sharing of the five basic practices for food safety (see Photo 6.10):

- General cleanliness, including hand-washing
- Separation of cooked and raw foodstuffs
- Sufficient cooking of cooked foodstuffs
- Storage of foodstuffs at appropriate temperatures
- Provision of drinking water and safe products

Obstacles

The three main obstacles to the roll-out of Healthy Cities in Mali were an insufficient mobilization of financial resources, insufficient monitoring and lack of commitment from municipal authorities.

Photo 6.9 Protected area where meat is stored and sold

Photo 6.10 Display of the five basic practices for food safety

Conclusion

Beyond aspects linked to health and the improvement of living conditions, the implementation of a Healthy Cities approach through Healthy Markets and Healthy Schools has enabled a strong focus on problems related to hygiene, sanitation and access to drinkable water. The planning and implementation of the various action plans have had a positive impact on living conditions in schools and markets. The empowerment of local stakeholders can be observed through the mobilization of internal resources within the municipalities that took part in the suggested activities.

The Healthy Cities Programme in the City of Pointe-Noire (Republic of Congo-Brazzaville)[4]

Introduction and Context

Pointe-Noire is the economic capital of the Republic of Congo-Brazzaville. It has undergone a continuous demographic explosion since the end of the 1990s. From 500,000 inhabitants in 1977, its population increased to around one million in 2013.[5] The city, built on a swamp area, has developed in an uncontrolled fashion. Town planning has been overwhelmed since a massive influx of people fleeing civil unrest at the end of the 1990s. The search for land has led populations to settle in *non aedificandi* (not for construction) areas where they are exposed to health risks linked to environmental degradation and natural disasters.

Various water and sanitation master plans based on traditional sewage and rainwater collectors did not materialize because of the financial costs of the technology as well as the lack of reliable management and evaluation services. At sunset the air is visibly polluted (see Photo 6.11), the result of a flourishing petroleum industry and ever-increasing numbers of second-hand cars in a dense traffic grid, not to mention the proliferation of uncontrolled waste dumping sites across the city (see Photo 6.12). Local government is unable to deliver the services that would improve the living conditions of the population because of insufficiently trained personnel, insufficient financial resources, and a chronic lack of tools and technical equipment.

Drinking water coverage has been estimated at around 68.9 %,[6] which is low. People depend on well water distributed by the *Société Nationale de Distribution*

[4] Jean de Dieu Bienvenu Konongo, with the support of Jean Simos.

[5] Congo Terminal: Pointe Noire in brief. http://www.congo-terminal.net/index.php?page=pointe-noire-en-bref.

[6] Ministère du plan, de l'aménagement du territoire et de l'intégration économique. Centre national de la statistique et des études économiques: Enquête congolaise auprès des ménages pour l'évaluation de la pauvreté (Ministry for planning and economic integration. National centre for statistics and economic studies: Congolese household study for the evaluation of poverty), 2005.

Photo 6.11 Heavily air-polluted Charles-De-Gaulle Avenue

Photo 6.12 Illicit rubbish in a Pointe-Noire neighbourhood

d'Eau, SNDE (National Society for Water Distribution), but also from wells often not managed and on rainwater gathered by equipment which is rarely disinfected or washed. The SNDE network does not cover areas outside the city centre. Most of the wells are located downstream of and too close to latrines. Over a large area of the city, the water table lies close to the surface, which creates favourable breeding conditions for malaria, diarrhoeal diseases, skin diseases, acute respiratory conditions, typhoid fever, Shigellosis etc. Since 1998 Pointe-Noire has been exposed several times to cholera. Polio struck the city in 2010, with 589 cases of acute flaccid paralysis and 189 deaths, a lethality rate of over 32 %.[7]

Establishing the Healthy Cities Programme Locally

In 2004, with the help of WHO, the mayoral office at Pointe-Noire decided to set up a Healthy Cities programme locally with a view to promoting municipal policies favourable to health and sustainable development. This was seen as a useful instrument for guiding city policy regarding public health. The Departmental and Municipal Council of Pointe-Noire[8] formally voted on 14 October 2004 to institute this programme. A focal point and an intersectoral committee were set up.

In 2007 Pointe-Noire drew up a 2007–2011 Healthy Cities 5-year action plan, founded on an intersectoral and participatory approach. Various public-sector and private-sector participants played an important role in drawing up the plan, as did the population as a whole. The methodology was to undertake together an analysis of the environmental and public health situation of the city, to determine needs and to prioritize the suggested actions. This plan was approved by the city council before the end of 2007 and used within the framework of the struggle against the cholera epidemic which hit the city between November 2007 and June 2008, because managing epidemics was one of the activities specified in the Healthy Cities 2007–2011 action plan. Two specific projects were created with neighbourhood committees, food market committees and school representatives from the most severely impacted areas: 'Healthy Schools' (*écoles-santé*) in the KM4 and Mbota Carlos neighbourhoods, and 'Healthy Markets' (*marché-santé*) in the Mbota Carlos neighbourhood.

Results

Under the programme, two state schools were able to obtain new infrastructure for the provision of drinking water and sanitation. New latrine-type toilets linked to ventilated multi-cabin ditches were built, as well as an impluvium linked to a water

---

[7] OMS Congo : Poster épidémie de poliomyélite (WHO Congo: Poster of the polio epidemic). Pointe Noire, 2010.

[8] Pointe-Noire is a department as well as being a municipality.

distribution system. The Mbota Carlos market was renovated, with the addition of new market stands, toilets and taps for drinking water.

Access roads were paid for by the mayor's office of Pointe-Noire. Some were tar sealed while others were paved. The Department of environmental hygiene received tools and anti-pest products such as wheelbarrows, spades, shovels, rakes, hoes, machetes, buckets, backpack sprayers, vaporizers, fumigators, disinfectants and pesticides. After several waves of diarrhoeal diseases including cholera, and to prepare to fight these epidemics, the Department of Environmental Hygiene also set up a system to monitor water quality in the city.

The development of uncontrolled fishing activities along the Atlantic coast led to the establishment of informal settlements along the beach. The beach had lost its recreational character and was choked with all types of waste produced by the fishing community. As part of its Healthy Cities development plan, the mayoral office had the fishing village moved; a traditional craft village with a potential for tourism was erected in its place.

Challenges and How to Face Them

In 2010, the person responsible for the focal point of the Healthy Cities programme departed, and commitments that had been made up to then were not met. The programme, which already was failing to evolve in a predictable way, came to a standstill. Making the most of this vacuum, a French-registered NGO began to carry out a series of actions under the umbrella of the local Healthy Cities programme, although its workshops and meetings to further projects had little in common with the municipality's programme. To re-boot the Healthy Cities programme, the mayoral office again asked WHO for support. This led to a rapid evaluation of the local programme, to identify strong points and challenges. A review of the 2007–2011 action plan was carried out. A new *rapporteur* (coordinator) for the Healthy Cities focal point was designated through a municipal by-law voted on 17 June 2013, and the intersectoral programme committee was reinforced, thanks to four neighbourhood leaders plus representatives of the police and of the major oil and gas companies active in the area. The terms of reference of the intersectoral committee were revised as a consequence of these changes.

Perspectives

The new action plans covers 2013–2016 and is presently being rolled out. An outbreak of rabies in dogs occurred at the beginning of 2014 in Pointe-Noire, and the intersectoral committee was involved in the actions carried out against this epidemic. The *rapporteur* ensured the coordination of the activities. With the support of the Healthy Cities programme the following actions have been carried out:

- Reinforcement of epidemiological surveillance
- Creation of an anti-rabies centre
- Acquisition of anti-rabies vaccines and serum, for humans and for animals, and of equipment for catching animals
- Acquisition of a dog pound truck
- Provision of a site for building a clinic for surgery, including pre- and post-operative care

Conclusion

The Healthy Cities programme in Pointe-Noire is moving forward, like many other projects, thanks to an iterative trial-and-error process where it constantly has to adapt to the local situation. It is vulnerable to the complexities of the political situation of the country, the economic difficulties and, especially, the problems faced by local civil servants as they attempt to manage urban spaces. To overcome these challenges, the mayoral office of Pointe-Noire, through its Healthy Cities programme, has sought to enforce an ambitious health promotion policy, recognize its intersectoral nature, and develop programmes with the full cooperation of neighbourhood residents, and of major oil and gas companies active in the region.[9] The creation of environments conducive to health is ongoing, and focuses on healthy schools, healthy markets and healthy neighbourhoods.

The Process of Implementing Healthy Markets in Togo, 2009–2013[10]

Introduction

At the behest of WHO, Togo decided to engage in the Healthy Cities process by introducing its flagship project, Healthy Markets. For the pilot phase of this project, the markets of Vogan, Kétao, and Sokodé came under consideration, but the selection committee, after carefully weighing several criteria, in particular sanitary emergency (more specifically cholera cases outbreak), finally decided on the market of Sokodé.

[9] The company Total E&P Congo, together with the mayoral office, a neighbourhood NGO and the residents of the four Loandjili neighbourhoods, are setting up a pilot project entitled 'Clean Neighbourhood' with the objective of ensuring the collective management of domestic waste collection and the construction of a waste separation facility able to produce compost.

[10] Amidou Sani, with the support of Jean Simos.

Local Context

Sokodé, which literally means 'place of tasty meat' in the Cotocoli or Tem languages, lies 350 km north of Lomé and is the second city of Togo. It is crossed on a north–south axis by the Lomé-Dapaong main road, the backbone of the country, and on an east–west axis by another major road leading from Benin to Ghana.

From around 29,000 inhabitants in 1970, the population reached 113,000 in 2013. The average household is eight people. The Tem are the major ethnic group, followed by the two groups of Kabyè-Naoudéba-Lamba and Adja-Ewé. Islam is the main religion.

The gross activity level (number of working people over the total population) is 43 %. The job market is dominated by transport, retail and arts and crafts. Women mainly carry on retail. The economy of Sokodé relies to a great extent on direct money transfers from their young people who work overseas.

The market, in the centre of Sokodé, is home to some 1700 retailers. The number of customers per day is around 10,000 according to a 2013 estimate (*Rapport 2013 SRAHM*). Products for sale include fish, meat, poultry, fruit, vegetables, cereals, spices, prepared foods, preserves and other industrialized products.

Physical Environment and Infrastructure

Buildings

The floor of the market was in good condition, although the roofs were the worse for wear. The type of construction, the materials used and the general state of the buildings was satisfactory. Poor access roads to the market and the absence of a parking lot made the market difficult to access, and increased the risk of road accidents and of food contamination.

Occupation of the Market by Different Categories of Retailers

The occupation plan was not adequate and risked leading to cross-contamination of food products.

Working Conditions

The shelves and stalls did not meet norms. Lighting and ventilation were inadequate. There was a high risk of pollution through dust and smoke from mopeds and other vehicles. Unpleasant smells were noticeable in various places within the market.

Maintenance Services

At a general level, the market was not managed on a regular basis for things such as the provision of running water, waste water management, building maintenance or infrastructure for rainwater run-off.

Operations and Services

Running Water

There were altogether three taps for the whole market. Water from these taps was sold. Very often, the quality of the water thus collected and intended for consumption was compromised by being transported without care and in dirty and uncovered vessels (contamination through particles in the air) and deposited on the ground (see Photo 6.13).

Health Services

The Sokodé regional hospital was not far away, so it was easy, at least geographically, for market retailers to access health services.

Photo 6.13 Tap water transportation uncovered vessels

Health Education

There was no official education campaign, although the regional service for hygiene and sanitation imparts health information when it organizes flash controls on the market.

Places for Resting and Eating

The market had no spaces for resting or eating. Retailers had to eat their meals at their stands, which does not correspond to adequate hygienic conditions.

Fire Prevention

Although electric equipment and the electric grid conformed to national standards, some practices were potentially very risky. The city of Sokodé has no fire brigade, so a fire could have very serious consequences.

Security

The market had no fence around it. Security was ensured by patrols.

Health at Work

The level of hygiene of the working environment was not satisfactory. Garbage could be seen lying around, and sweeping and cleaning occurred only rarely.

Toilets

Access to toilets was difficult for both women and men. A latrine with four stalls split between men and women, and with paying access, catered for the more than 10,000 people at the market. A look inside the facility showed insufficient maintenance, and more especially inappropriate use of the facility.

Waste Management

The majority of the waste produced in the market is similar to household waste, to which is added liquid waste from butcher's stalls and restaurants, and a few dangerous wastes such as used batteries and dyes. The mayoral office had set up five dustbins for household waste and had entrusted the NGO MAD—*Maison d'Assistance aux Déshérités* with sweeping and garbage pre-collection; but this NGO is limited by its insufficient technical and financial capacity, and especially because of the lack of designated areas for the intermediate stockpiling of waste. As a consequence all available spaces, streets, and especially ditches and gutters are treated like garbage dumps. Furthermore, damaging behaviour has been observed in the population, some members of which are satisfied with the current management of solid waste or are actually hostile to any campaign for cleanliness in the market.

Insects and Other Harmful Animals

Such creatures are present and there is no campaign to control or eradicate them.

Drainage System

Waste water is not properly evacuated towards the outlets. The pipes are often obstructed by household waste, which hampers drainage and gives rise to unpleasant smells and a host of flies and mosquitoes. Market users have to throw their waste water onto the pavements or streets.

Chemical Security

No chemicals are stored or sold in this market.

Food Hygiene

In Togo there is a set of rules to ensure food safety in markets, and a national action plan for food safety has been set up. There are frequent inspections by the services responsible for food safety; however, supporting by-laws are not available, which impedes the surveillance mechanism as well as any quality control of foodstuffs in the market.

Personal

There is no wearing of appropriate clothing or headgear. Hand washing after going to the toilet or after working with raw meat or any other potentially contaminated product is rarely bothered with.

Handling of Foodstuffs

Most of the sources of food are trustworthy, but transport conditions are rarely adequate. In particular:

- Stocking places along the cold chain are not optimal.
- Processes for cooking and preparing food are inadequate.
- Foodstuffs which require particular temperature conditions tend not to be well conserved.
- Packaging of foodstuffs is often inadequate.

Cleanliness of Stalls and Tools

The environment where food is prepared and sold is not hygienic. Cake shops and cafeterias do not respect minimal hygienic norms. Some products for immediate consumption (drinking water, fruit juice, locally produced drinks) are left uncovered or exposed to the air.

Placement of Foodstuffs

There is a risk of contamination due to dust, flies and other insects. Nor are foodstuffs protected from excessive manipulation by potential buyers (see Photo 6.14).

Photo 6.14 Uncovered and exposed food

Consumer Education

There are no campaigns advocating food hygiene or hygiene in general, neither in Togo or Sokodé.

Setting Up a Healthy Market in Sokodé

An action plan aiming at addressing each of the identified problems was set up and implemented in 2010, with a budget of around 133 million CFA francs,[11] distributed in the following way: 1.7 % for market management and services, 86.4 % for infrastructure and environment, and 11.9 % for food safety. WHO was the main financial partner.

A workshop for all stakeholders reviewed progress halfway through the project in June 2013, and undertook some modifications to the action plan. By that time the following actions had already been implemented:

- Creation of tools and materials for advocacy
- Advocacy with municipality leaders and key market stakeholders in Sokodé around the importance of the project and of food safety in general
- Capacity-building for outreach workers and hygiene officers to be able to take basic measures necessary to improve the safety of food sold in the market and in the streets
- Mass advocacy and door-to-door campaigns by trained outreach workers in markets, schools and homes
- Investment in equipment for waste management in the market

It was decided to empower market retailers in the Sokodé main market and in surrounding food markets, among which was the smaller market in Komah. A new action plan was therefore drawn up with a budget of around 135 million CFA francs,[12] 7.1 % for market management and services, 71.8 % for infrastructure and environment, and 21.1 % for food safety.

Results

Strengths

- Today the Healthy Markets project is a reality in Sokodé.
- The engagement of technical services and of stakeholders in the project is strong.

[11] Approximately 203,000 EUR or 244,000 USD, at the 1 June 2010 exchange rate.

[12] Approximately 206,000 EUR or 268,000 USD at the 1 June 2013 exchange rate.

- The activities of the project were reviewed after an operation time of 2.5 years.
- WHO, the main financial partner of this project, has lived up to expectations and has already invested in several activities to a total of over 8 million CFA francs.
- The Sokodé municipality has also invested in and carried out several activities at a total value of over 7 million CFA francs.

Challenges

- Several activities listed in the 2010 action plan were not implemented due to inadequate financial means.
- It is necessary to look for other financial partners and strengthen advocacy.

Perspectives

- With support from WHO, the municipality is to organize a round table to mobilize resources for implementation of the action plan.
- It is intended to strengthen the ongoing monitoring of the action plan and the follow-up of activities by the committee.
- It is also intended to strengthen advocacy for resource mobilization, engaging the media in the process.

Obstacles

- Implementation of some of the planned activities requires considerable funds.
- The mayoral office must make choices in the roll-out and practical usefulness of some of the planned activities (e.g. reorganizing the distribution of spaces for market stands).
- There is not enough free space inside or around the market to accommodate certain facilities.

Conclusion

The Healthy Markets project in the town of Sokodé is a reality thanks to the strong commitment of all stakeholders and partners. Important positive knock-on effects are to be expected for local populations and consumers. Work on the project is ongoing, although several planned activities have not yet been carried out because of funding inadequacies.

Planning for Urban Health in Mathare, Nairobi, Kenya[13]

In Nairobi, Kenya, close to 65 % of the population live in informal settlements, or slums, on about ten per cent of the city's land area. Children under five living in Nairobi's slums are almost three times more likely to die than their counterparts in the rest of the city. Women in Nairobi's slums experience disproportionate health burdens compared with men. For example, over a quarter of all women and girls in Nairobi's slums reported an episode of diarrhoea in the past month, compared with about a fifth for all Kenyans. Women in these slums have a 38 % greater prevalence of HIV/AIDS than men. Adding to their disease and care burdens, over 36 % of slum-dwelling women report being physically forced to have sex, and over one-third report being sexually abused.

Many of these problems are the result of a combination of dangerous physical environment (i.e. crowded housing, pathogens in open sewerage, indoor air pollution from lack of electricity), social inequities (discrimination in the workplace, disproportionately high food, water and toilet costs, insecure housing/threats of eviction) and structural violence (lack of health care services, rampant sexual violence against women, underinvestment by government in security and basic infrastructure). These related health determinants in an informal settlement like Mathare have contributed to the slum dwellers responding to co-produce health improvements—by negotiating with the state for immediate needs, partnering with outsiders to develop long-term improvement strategies, and working to redefine relationships between the urban poor and development institutions that have for too long only viewed slums and their residents as in need of treatment and professional care—in a totally passive attitude.

Mathare is an informal settlement located about 6 km from the city centre and comprising 13 different villages: Mashimoni, Mabatini, Village No. 10, Village 2, Kosovo, 3A, 3B, 3C, 4A, 4B, Gitathuru, Kiamutisya and Kwa Kariuki. Mathare sits in a river valley, a former quarry, and its steep slopes and river frontage present a host of environmental risks to residents. Mathare has a history of resident organizing and activism, and some of its Kikuyu[14] residents were active in the struggle for independence (1950s and 1960s); later, in the 1970s and 1980s, they formed cooperatives to buy land and build housing. However, ethnic tensions and state neglect, along with an influx of new residents, divided residents and contributed to widespread violence in the 1990s and control of much of the local economy by the Mungiki gang in the early 2000s. Post-election violence erupted in Mathare in 2007, and civil society organizations have played a role in re-establishing peace, organizing residents and helping them advocate for essential, life-promoting basic services.

One civil society organization, Muungano wa Wanavijiji (the Kenyan Federation of Slum-dwellers), began organizing residents in Mathare in 2002. A member of

[13] Jason Corburn, Irene Karanja and Jack Makau.

[14] The largest ethnic group in Kenya.

Shack/Slum Dwellers International (SDI), a network of urban poor federations active in 34 nations and headquartered in Cape Town (www.sdinet.org), Muungano has over 160,000 active members from hundreds of different slums. Muungano and its sister organization, Muungano Support Trust (the research and training arm of the group, mustkenya.or.ke), use daily micro-savings to organize residents. This organization then helps 'make the invisible visible' through household enumerations and mapping. This establishes a base upon which local groups can prioritize needs and advocate for resources and support to deliver services and improve living conditions, such as building new housing, providing water and sanitary infrastructure, engaging youth in employment training, or establishing a childcare cooperative.

In 2008, the United Nations Environment Programme and the Government of Kenya launched the Nairobi River Cleanup Programme, intended to improve water quality in 'the place of cold waters' (the meaning of Nairobi). The first task was to establish a thirty metre 'riparian buffer' on each side of all rivers in the city, declaring these zones off-limits to any human activity. Residents of Mathare were particularly threatened by this move, as an estimated 50,000 lived within the riparian buffer and faced eviction. In an effort to 'encourage' slum dwellers to vacate their homes along the river, the Nairobi Water and Sewer Company shut off the water supply to Mathare. Within hours Mathare residents were organizing and Muungano representatives demanded that the city turn the water back on. After a week Muungano organized a march to Nairobi City Hall to protest this brutal treatment, requesting a meeting with the City Council at the same time. The direct action and face-to-face negotiations proved a temporary success: the city agreed to turn the water back on with the condition that Mathare residents develop a plan to stop polluting the river.

Muungano representatives organized a series of community meetings to discuss their options and reached out to their international network through SDI for help. The University of Nairobi, Center for Urban innovations (http://www.centreforurbaninnovations.com/) and the University of California, Berkeley, Center for Global Healthy Cities (www.healthycities.berkeley.edu and http://nairobistudio.blogspot.com/) partnered with Mathare residents and civil society groups over the next 5 years to develop integrated upgrade plans with goals that emerged from community-led negotiations:

- Limited or no displacement of existing residents while accommodating for growth
- Plans that built upon existing community assets and strengths
- Environmental protection with, not at, the expense of, human development
- Infrastructure planning as an entry-point to address other related social inequalities
- All work to have meaningful community participation and ownership
- Addressing immediate needs as a means to influence policy at municipal and national levels

A series of village-scale plans were drafted with residents by 2009. One plan in the village Kosovo detailed a new piped water service for the entire village and, for the first time in a Nairobi slum, in-home water taps. Most slum water projects provided service to community yard taps or kiosks, which were often taken over by local gangs or cartels who charged exorbitantly high prices for their use. This often forced slum dwellers to tap into a pipe and siphon off water, contaminating the water supply, reducing pressure, and creating a system of 'spaghetti pipes' haphazardly strewn across the community. The new project, called the Kosovo Water Project, was endorsed by community members, financed by the Athi Water Board, and constructed and managed by local NGOs, including Muungano (http://wstf.go.ke/index.php?option=com_docman&task=doc_download&gid=13&Itemid=496). By 2014, over 20,000 Mathare residents were benefiting from the improved water service (http://www.upfi.info/projects/mathare-kosovo-water-connections/).

Not satisfied with boutique, village-scale upgrading, Muungano and its academic partners immediately began research to organize a Mathare Zonal Plan. This focused on integrated improvement strategies for the entire Mathare valley so that co-benefits could be realized from upgrading livelihoods and economic opportunities, infrastructure, and other social services such as schools and childcare, housing and human health. After over a year of collaborative research by slum dwellers, civil society and academic partners, the Mathare Zonal Plan was completed in 2012 (http://www.sdinet.org/media/upload/documents/Mathare_Zonal_Plan_25_06_2012_low_res-2.pdf).

Over the next year, a series of events in Kenya and East Africa showcased the Mathare Zonal Plan, building interest among slum dwellers and decision makers across the region. The Kenyan Ministry of Lands and Housing eventually endorsed it, and its sanitation proposal is being used by the World Bank to upgrade the municipal sewer in Mathare (http://www.mustkenya.or.ke/index.php/blog/10-qiuck-links/26-sanitation).

As importantly, the co-production process with Mathare residents and civil organizations enhanced their standing in the eyes of government policymakers and international organizations. The Mathare Zonal Plan data and report helped put civil society partners in leadership positions in a government-sponsored process drafting a new Kenya Slum Upgrading and Prevention Policy and a Nairobi Master Plan, both in 2013. International organizations such as Canada's IDRC and Sweden's SIDA reached out to Mathare organizations to lead new violence prevention projects and build a network of urban slum dweller groups in Kenya. As a result of the Mathare Zonal Plan process, which helped stabilize the community and reduce violence by engaging young people and others in a visionary project, investments in health services, transport, housing, urban agriculture, recreation space and other social determinants of health are increasing. This is an ongoing process, and residents, NGOs and academics continue to be partners co-producing innovative strategies to improve health in Mathare, other informal settlements in Nairobi, and the Kenyan urban poor more generally.

Operationalizing the 'Healthy City' in Africa: A Case Study of Accra's Urban Poor[15]

Introduction

The fast structural and cultural rise of Ghana's capital, Accra, has had an impact on the life and health of its inhabitants. Like all African cities, Accra's urban poor experience an extremely complex set of burdens. I focus on the health of Ga Mashie, one of Accra's oldest indigenous Ga communities and one of the poorest, to examine the ways in which this community uses the city to various individual and communal ends to address health problems, and what this means for operationalizing the concept of 'healthy cities' for deprived urban African communities. I choose Ga Mashie for two reasons. First, Accra was conferred Millennium City status as part of the Millennium Cities Initiative established by the Earth Institute, Columbia University and development economist Jeffery Sachs to support sub-Saharan African cities to eradicate poverty and attain Millennium Development Goals (MDGs) (AMA-UN HABITAT 2011). This project has since initiated plans to develop basic infrastructure, housing and commercial space to create a 'healthier and more equitable environment' for Ga Mashie. The community therefore is situated concretely within global discussions on healthy cities. Second, since 2010 Ga Mashie has been the site of longitudinal multidisciplinary field research conducted by the Regional Institute for Population Studies (RIPS), University of Ghana, with the aim of studying inequities in the health and welfare of the community, and to sensitize local and regional stakeholders to urban poverty and health issues. This case study draws from data gathered by the RIPS chronic non-communicable disease (NCD) working group on representations and experiences of health, illness and NCDs in the community (Afrifa-Anane et al. 2014; Awuah et al. 2014; de-Graft Aikins et al. 2014a; b; Peterson et al. 2014). The focus on NCD in Ga Mashie is driven by current evidence that it kills more adult men and women in Africa than in other parts of the world; and that poor, urban African communities are disproportionately affected (de-Graft Aikins et al. 2010).

The Biography and Demography of Ga Mashie

Ga Mashie is referred to as Old Accra as it existed before Accra became the capital of Gold Coast in 1877. By the late sixteenth century Ga Mashie was well known as a powerful trading Ga township and home to immigrant non-Ga ethnic groups (Quayson 2014). By the seventeenth century the structure of the township had been established along lines of kinship (indigenous and immigrant), occupation (fishing, fish processing, salt making) and trade, into seven quarters (*akutsei* in the Ga language): three of these quarters were located in Jamestown (Akanmaji, Sempe, Alata) and four in Usshertown (Abola, Gbese, Asere, Otublohum) (Fig. 6.3).

[15] Ama de-Graft Aikins.

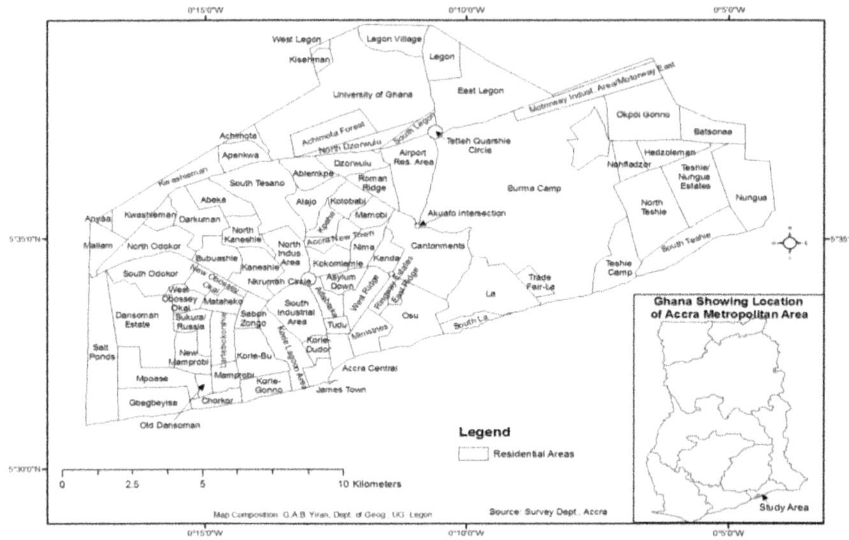

Fig. 6.3 Map of Accra

Like other Ga townships along the coast of Accra, the history of Ga Mashie is intricately linked to the history of European colonial conquest, as the British, Dutch and Danes conducted their political, economic and social affairs from various forts and castles situated along the Gold Coast. After the abolition of slavery in the nineteenth century, freed slaves from Brazil settled in the area; their descendants are the Tabon of Ga Mashie. To multidisciplinary scholars of Accra, therefore, Ga Mashie and its twin towns of Jamestown and Usshertown became urban, global and multicultural long before other parts of Accra and the Gold Coast (Quayson 2014).

Today Ga Mashie is situated in prime real estate surrounded by major institutions such as the houses of parliament, law courts, government ministries, the financial district and the Korle-Bu Teaching Hospital. It has been categorized a heritage site and is a major tourist destination and space for artistic expression, with communities like Jamestown garnering postmodern labels such as 'the Soho of Accra' (New African 2013). This vibrant social, cultural and political identity coexists with the experience of chronic intergenerational poverty.

The current population of Ga Mashie is about 30,000 (10,000 for Jamestown and 20,000 for Ussher Town). Ga Mashie occupies the lowest income class in the Accra metropolitan area. The average monthly household income is GHC126.13 (USD 78.83), a tenth of the average monthly household income of the first income class (AMA-UN-HABITAT 2011). The main economic activities remain fishing and petty trading. About three-quarters of the population have attained junior high school (or middle school) education and above; however, there is higher male enrolment in school and the quality of education is low, with a dominance of poorly resourced public schools. The environment is heavily built up, housing structures are old and weak, households are overcrowded with members ranging from 6 to 20, and sanitation remains a serious problem, especially when flooding occurs during the rainy seasons.[16] There is a strong presence of civil society organizations including the Ga Mashie Development Agency (GAMADA), which focus on addressing the developmental needs of the community. Targeted members receive cash transfers from the Livelihood Empowerment Against Poverty (LEAP) programme, the government's flagship social protection initiative. Further government investments have been made to improve roads, pavements and drainage in recent years. The Earth Institute's Millennium City Challenge project of creating a 'healthier and more equitable environment' for Ga Mashie is thus situated within a history of externally driven development interventions *for the community*.

The Health Status of Ga Mashie

Ga Mashie, like other urban poor African communities, experiences a double burden of infectious diseases associated with poverty and chronic non-communicable diseases (de-Graft Aikins et al. 2014a). Infectious diseases like malaria and diarrhoea persist and outbreaks of cholera recur; e.g. in 2014, 20,000 people were affected in Accra. These conditions co-exist with chronic non-communicable diseases like hypertension, diabetes, stroke and common mental disorders. The overall prevalence of hypertension in the community is 28.3 %, higher among men (31.0 %) than women (25.6 %) (Awuah et al. 2014). The prevalence of diabetes is 5.7 %, slightly lower among men (5.6 %) than women (5.7 %). Just over 30 % of the study population are either overweight or obese, with higher rates in women than in men. Self-reported psychological distress is substantial: 55 % of 'very high' psychological distress was recorded among a cohort of 952 individuals interviewed, with the Kessler scale (Peterson et al. 2014). Old age, low income, low education and limited social participation are major predictors of subjective experience of psychological distress.

Everyday Healthcare Practices in Ga Mashie

Community members know the range of prevalent infectious and chronic conditions in their community but lack comprehensive knowledge of the causes and outcomes of these conditions. For example, we administered a cardiovascular disease

[16] Accra has two rainy seasons, one beginning in March, and peaking in May or June, the other in September/October.

knowledge scale to approximately 250 individuals in households, churches, mosques and community centres. While most respondents knew of hypertension, stroke and diabetes and some knew people living with these conditions, they lacked knowledge of important risk factors such as the relationship between the consumption of high salt, high fat foods and alcohol and conditions such as hypertension and diabetes, and on the importance of lifelong pharmacological treatment for cardiovascular disease (de-Graft Aikins et al. 2014a).

There are important associations between socio-economic context, gender, and cardiovascular disease risk and health outcomes in the community. For example, dietary practices are poor: consumption of fruits and vegetables is low, consumption of processed fats is high, and snacking on processed foods is common. Poor eating practices are associated with an obesogenic environment with a dominant presence of fast foods and processed foods on most community streets. Women experience both undernutrition and over-nutrition: they are less likely than men to have three meals a day and more likely to be overweight and obese. These differences are mediated by livelihood strategies: women traders, unlike men fishers, are likely to work late into the night and eat their meals late.

We examined hypertension prevalence, awareness and treatment among 714 individuals (Awuah et al. 2014). Among respondents who had hypertension (just under 30 %), only 7.4 % were aware of their condition, only 4 % were on antihypertensive medication, and only 3.5 % of hypertensive individuals had adequate blood pressure (BP) control (BP < 140/90 mmHg). Knowledge of hypertension and diabetes symptoms was poor particularly among women. The level of awareness and treatment of hypertension was lower in men than in women (3.1 % and 1.3 % for men and 11.9 % and 6.5 % for women, respectively) and the rate of control was higher among women than men (5.0 % and 2.1 %, respectively).

The majority of health maintenance activities—eating, drinking, exercising—occur within the community, but illness management practices move beyond community boundaries. Ga Mashie is served by one public polyclinic, two private clinics and six community pharmacies; the Korle-Bu Teaching Hospital is situated within walking distance (Fig. 6.4). The community is also served by alternative health services provided by herbalists, shrines, churches and mosques (Fig. 6.4). Our longitudinal interviews with individuals with diabetes and hypertension suggest that they healer-shop between local biomedical facilities, pharmacies and chemical shops, herbalists and itinerant drug pedlars, and Christian and Muslim faith healers (de-Graft Aikins et al. 2014a, b). They also often bypass local facilities to access pluralistic health services in other parts of Accra: common locations include Dansoman, East Legon, Gbawe, Kaneshie and Osu (Fig. 6.3). Few individuals are on the National Health Insurance Scheme (NHIS) because they cannot afford the premium payments. However, the majority engage in dual use—the combination of prescribed pharmaceutical medicines with non-prescribed herbal medicines and supplements. This is expensive in terms of both finance and medical complications.

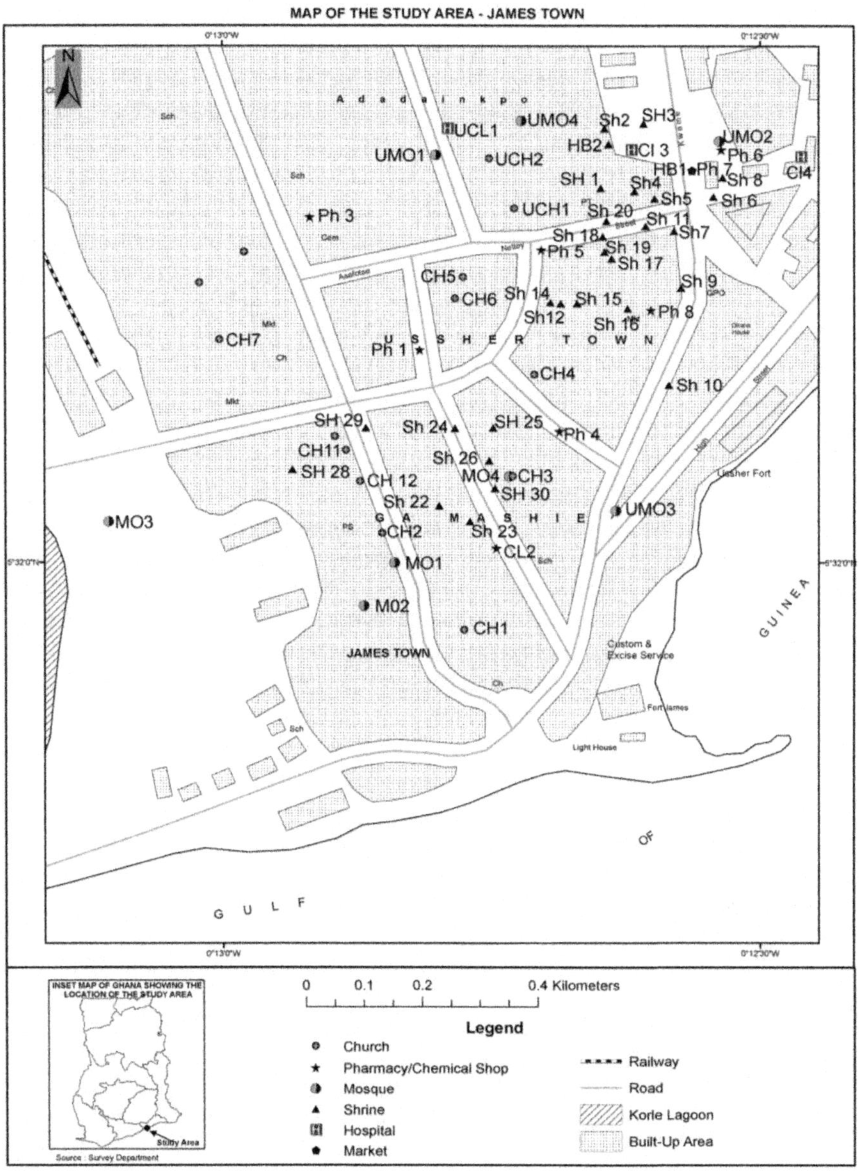

Fig. 6.4 Health facilities in Ga Mashie

Conclusions

Definitions of healthy cities incorporate the participatory power of city inhabitants. WHO, for example, suggests that healthy cities have 'a high degree of participation in and control by the citizens over the decisions affecting their lives, health and well-being'.[17] Current recommendations to address the complex public health challenges in low- and middle-income countries emphasize the need to empower communities to improve their health and prevent disease (Van Olmen et al. 2011). The challenge lies in facilitating a participatory approach to community health within a context, such as Ga Mashie, where development and change has been externally driven and instrumental and where, like other urban poor African communities, 'poverty presents a major barrier to health-promoting behaviours' (Greif et al. 2011).

The RIPS NCD Working Group is currently applying the 'social psychology of participation' approach (Campbell and Jovchelovitch 2000) to develop community-based CVD interventions in Ga Mashie. This approach has three elements: diagnosis, (psychosocial) intervention, and critical reflexivity. Diagnosis aims 'to understand, engage with and map out the local systems of knowledge of the community, the ways in which it conceives of itself and its mode of relation both inside and outside its boundaries' (Guareschi and Jovchelovitch 2004:316). Intervention aims to produce 'productive alliances' between different social actors with divergent knowledge, experiences, expertise, status and power. These productive alliances influence the structure, process and outcomes of interventions. Finally, the task of critical reflexivity is to recognize social reality as a 'field of open possibilities', to engage with this field and identify alternative possibilities that can transform reality. We have 'diagnosed' the context of CVD risk and morbidity in Ga Mashie using longitudinal multidisciplinary methods and developed preliminary 'productive alliances' with community members, community health workers, local NGOs and policy-makers. This collective group will inform the development of interventions to reduce CVD risk among the general population and CVD complications among individuals living with hypertension. Through this project we aim to identify the range of possibilities that exist to enhance community participation for sustainable health and wellbeing in Ga Mashie.

The Healthy Cities Programme of the City of El Khroub, Algeria[18]

Introduction and Context

Around the time that Algeria acquired independence (1962), El Khroub was a village with no more than 4000 inhabitants. Urban migration and the sedentarization of nomadic populations from the uplands created a first wave of new arrivals, who

[17] http://www.euro.who.int/en/health-topics/environment-and-health/urban-health/activities/healthy-cities/who-european-healthy-cities-network/what-is-a-healthy-city/healthy-city-checklist (accessed 9th November 2014).

[18] Abdelhamid Haroun, with the support of Jean Simos.

settled in a somewhat haphazard manner on land which had no formal acts of property, in some cases because they had formerly belonged to colonial owners. Proximity with the city of Constantine, only 15 km away, made the area a natural place for housing development, essentially apartment blocks.

During the 1970s an urban development plan was launched, this time in a structured manner. However, public infrastructure in the area continued to be insufficient for the needs of the population. Under pressure from local residents, the 1980s saw infrastructure such as secondary schools and post offices being put into place; in the 1990s a hospital extension and university lodgings were added. The architectural design is 90 % large buildings and towers (apartment blocks). At present the municipality of El Khroub contains three urban areas and four semi-urban areas. The population exceeded the 150,000 population mark at the 2008 census and continues to grow.

Passing from village status to city status, amid rapid population growth and in an impersonal concrete-based environment, has had an impact on the physical and mental health of the population: unhealthy behaviours such as violence, drug addiction, smoking and physical inactivity have emerged. A bird's eye view of the area gives the impression of a mineral world, with a serious lack of green spaces and sports grounds. Because of its rural origins, the lifestyle of the Algerian family has always deployed itself on a horizontal plane. The modern buildings in El Khroub are vertical; this difference has created difficult situations for the residents and for municipal services. Household waste management is one remarkable example: there is a completely different way to eliminate waste for a rural family living in an individual house with garden/fields and animals all around or for an urban one living in an apartment located in a multistorey building.

Setting up the Local Healthy Cities Programme

The Healthy Cities programme was introduced in 1996, on World Health's 'Healthy Cities for Better Life' day. The concept was to be adaptable to local cultural traditions. To ensure this, a participatory process was initiated. The local Healthy Cities committee fanned out into several thematic commissions responsible for assessing the current situation, identifying challenges and suggesting improvements. After 6 months, despite problems in securing data from the various administrations (there being no 'open data' principle), reports were published. The result of this brainstorming exercise was a four-pronged policy framework:

- Health across the whole city (while strengthening civil society initiatives)
- Health and environment (green spaces, waste management, water quality)
- Health and the city (transport, accessibility, refurbishment of buildings)
- Health at school (vaccinations, strategies against violence and against addictions)

A survey of local residents allowed the prioritization of the problems and their solutions. This methodology rested on a series of meetings aimed at informing and mobilizing residents in order to stimulate their engagement. This was not an easy task because at that time the whole country was living in a climate of violence.

Results

In 1997, the Healthy Cities programme became one of the mainstays of municipal policy and the local committee became a non-governmental organization.

The first result is that barriers between Municipal administration sectors have fallen, leading to a new form of mayoral governance. The programme is in a position to bring together all key stakeholders (several level authorities, economic and societal leaders, etc.) in the city.

Promoting Health Across the Whole City

A local centre for community health, which coordinates all Healthy Cities activities, was set up in 2001.

The mayoral office set up a citizen's advisory board to institutionalize the community engagement process initiated by the programme.

Health and Environment

After a long process with various stakeholders, a new master plan for household waste collection was set up, introducing new management processes. In 2013, this activity took up no less than 31 % of the municipal budget.

Starting in the year 2000, a programme entitled 'my city is green' was initiated. As a result the proportion of the municipality covered with green spaces increased from 1 to 7 %. Some patches of land being used for informal economic purposes such as small retail, occasionally exposed decision makers to the 'employment or environment' dilemma.

In the recent past, water-borne diseases such as hepatitis A, polio and especially typhoid fever were still important problems. A health map was drawn up showing the areas where new cases were likely to arise. The main cause was so-called 'cross-connections' between the drinking water supply and waste water pipes, attributable to the materials used and the technical knowledge of the construction companies involved. Other causes were non-treated private wells, healthy carriers handling food in restaurants, and insufficient hand washing. The local Healthy Cities association invested itself heavily in this area, pushing for prevention and putting an alert system into place. The annual number of cases of water-borne diseases has been zero for several years now.

Health and the City

In order to improve a general state of being described by local residents as 'sadness', a project to paint the outside walls was set up by local youths with two recognized artists, one from Algeria and the other from Mulhouse (France).

The city of El Khroub is in constant development and demographic flux, and must face mobility choices as it is spatially dispersed. Complex and difficult arbitration between conflicting interests has led to a new master plan for urban and inter-urban transport, the main objective to reduce motorized traffic in the centre. The new transport master plan uses new corridors which are wider and far from the centre. The preservation of equity in transport services was a vital factor in this plan; priority was given to public infrastructure and service access (such as to hospitals, day care centres, post offices, municipal offices). A new coach station was built outside the city, on the eastern side, to help reduce air pollution in the centre.

People with reduced mobility were also taken into account to a certain extent, by calling in a network of voluntary taxi operators. Annual contests take place to identify the best bus driver and the best taxi driver, in a bid to increase appropriation of the new policy.

Health at School

The school environment and the educational system are priority recipients of awareness-raising and communication regarding health. As well as this objective, improvement of school environments have taken place, such as rebuilding toilets and water reservoirs, and the widespread use of whiteboards in classrooms, which do not require chalk. At the beginning of 2014, of the 48 schools in the area 40 had benefitted from these improvements. A competition for the school with the most flowers and trees was been set up and officialized in the programme of every school. Vaccinations were organized for schoolchildren who had not been reached by previous campaigns.

The city of El Khroub has over 4500 students. Campaigns directed at this segment of the population include HIV prevention, women's health and food safety in catering facilities.

An alliance between the municipality, the association against cancer and the Healthy Cities association was created for the struggle against smoking. Its action plan will extend across several years, with awareness-raising, video-assisted lessons in schools, and a counselling service for those seeking to stop smoking.

In 2007, a specific study of physical violence and smoking led to an awareness of the seriousness of these phenomena. Defective town and country planning, such as a major trunk road cutting through the city, exposed inhabitants who needed to cross the city from east to west to danger: there have been high numbers of road traffic accidents, many of them with lethal consequences. Neighbourhood committees and the Healthy Cities association pointed out this intractable problem and have forward suggestions, but administrative red tape and the incompetence of certain officials have been serious obstacles to any solution.

Obstacles and Challenges

The environment within which the programme has evolved has been challenging for several reasons. Local decision makers have not always been able or willing to pursue this initiative, which began in 1996. Political circumstances have led to interruptions as each newly elected team implemented its own set of policies; the reluctance to complete previously established projects has become a serious obstacle to implementing projects in the long term. Starting again from scratch involves loss of valuable time, with social and economic stagnation the result, not to mention the waste of public money. Loss of access to data for monitoring and evaluation is a direct consequence of these changes. The arrival of each municipal team implies a latency period during which data is provided or they start again from scratch.

Conclusion and Perspectives

The city of El Khroub has seen its status evolve from a small village to a sprawling, polycentric city over an extremely short time-scale. Rampant urbanization has led to an excessive mineralization of public space, and the sharp population increase has led to challenges in identifying, prioritizing and providing the resources and needs of inhabitants. Many of the people arriving in the past few years have come from impoverished areas of neighbouring Constantine, making the national urban policy of 'new cities' look more like a social housing dumping process, from an aesthetic and sociological point of view. The redistribution of has inhabitants occurred in an efficient manner simply by moving people from their old horizontal sprawl into a new vertical sprawl. Lifestyles and behaviours have not changed, nor have they evolved. These facts have led to new trends, such as groups who used to live close to each other before now joining to appropriate odd patches of land. Life in these areas is stressful and without a true feeling of belonging to a community. Social cohesion, the necessary condition for any successful urban policy, is unable to emerge. Living together and adopting an urban culture are not yet reconciled. Territorial conflicts have led to the creation of gangs, with consequences which are known and have been described in many studies across the world. Violence, addiction and antisocial behaviour are the outcomes of a defective process of urban creation.

The Healthy Cities programme has brought positive solutions to at least some of these challenges, which is encouraging. The approach of promoting healthy activities through the schools makes sense in a country like Algeria, as it will produce future citizens already educated about and convinced of the importance of healthy practices.

It is time for new policies to emerge, for example, by introducing social diversity into newly built areas, upholding civil society and implicating inhabitants more directly, involving them more directly in changing their physical and cultural and social environment. Social cohesion is a work in progress, addressed in an array of projects. Mobilizing local residents around health must become a long-term educational programme if a healthy urban culture is to be achieved. Furthermore, local governance needs to be improved in order to set up a broad agreement to move forward, with an alternative policy able to kick-start an evolution towards participatory democracy. Neighbourhood committees and NGOs have a crucial role in making this happen.

Bibliography

Accra Metropolitan Assembly-UN Habitat (AMA-UN Habitat). (2011). *Participatory slum upgrading and prevention: Millennium city of Accra, Ghana*. Accra: AMA.

Adaptation au changement climatique en Afrique: plan d'action pour le secteur santé 2012–2016.

Afrifa-Anane, E., Codjoe, S. N. A., Agyemang, C., Ogedegbe, G., de-Graft Aikins, A. (2014, May). *The association of physical activity, body mass index and the blood pressure levels among urban poor youth in Accra, Ghana*. In Paper presented at Population Association of America (PAA) Conference, Boston.

Awuah, R. B., Anarfi, J. K., Agyemang, C., Ogedegbe, G., & de-Graft Aikins, A. (2014). Prevalence, awareness, treatment and control of hypertension in urban poor communities in Accra, Ghana. *Journal of Hypertension, 32*(6), 1203–1210.

Bart, F., & Bonvallot, J. (2002). L'invention de la ville en Afrique subsaharienne. *Regards sur l'Afrique Historiens et Géographes, 2002*(379), 59–65.

Campbell, C., & Jovchelovitch, S. (2000). Health, community and development: Towards a social psychology of participation. *Journal of Community & Applied Social Psychology, 10*, 255–270.

Chenal, J. (2013). *La ville ouest-africaine. Modèles de planification de l'espace urbain*. Geneva: Métispresses.

de-Graft Aikins, A., Awuah, R. B., Pera, T., Mendez, M., & Ogedegbe, G. (2014a). Explanatory models of diabetes in poor urban Ghanaian communities. *Ethnicity and Health, 20*(4), 391–408. doi:10.1080/13557858.2014.921896.

de-Graft Aikins, A., Kushitor, M., Koram, K., Gyamfi, S., & Ogedegbe, G. (2014b). Chronic non-communicable diseases and the challenge of universal health coverage. Insights from community-based cardiovascular disease research in urban poor communities in Accra, Ghana. *BMC Public Health, 14*(Suppl 2), S3. doi:10.1186/1471-2458-14-S2-S3.

de-Graft Aikins, A., Unwin, N., Agyemang, C., Allotey, P., Campbell, C., & Arhinful, D. K. (2010). Tackling Africa's chronic disease burden: From the local to the global. *Globalization and Health, 6*, 5.

Deuxième Stratégie de croissance et de réduction de la pauvreté (SNRP2) couvrant la période 2008–2011.

Document de *Politique d'aménagement urbain et Habitat*.

Rapport de l'Atelier Marché-Santé au Mali (Ministère de la santé du Mali, Direction régionale de la santé de Bamako 2009). [Report of the Healthy Markets workshop in Mali (Mali Health Ministry, Bamako regional directorate for health, 2009)].

Plan stratégique d'éducation à l'hygiène en milieu scolaire (Ministère de la santé du Mali, Division Hygiène Publique et Salubrité, 2008). [Strategic education plan for hygiene in schools (Mali Health Ministry, Division for Public Hygiene, 2008)].

Plan stratégique des pratiques d'hygiène à grand impact (Ministère de la santé du Mali, Division Hygiène Publique et Salubrité, 2010). [Strategic plan for high impact hygiene practices (Mali Health Ministry, Division for public hygiene, 2010)].

Durand-Lasserve, A. (2005). Land for housing the poor in African cities: Are neo-customary processes an effective alternative to formal systems? In N. Hamdi (Ed.), *Urban futures: Economic development and poverty reduction* (pp. 160–174). London: ITDG Publishing.

Durand-Lasserve, A., & Selod, H. (2009). The formalisation of urban land tenure in developing countries. In S. V. Lall, M. Freire, B. Yuen, R. Rajack, & J.-J. Helluin (Eds.), *Urban land markets: Improving land management for successful urbanization* (pp. 101–132). Washington, DC: Springer & World Bank.

Étude Dividende démographique, N'Djamena 2012.

Greif, M. J., Dodoo, F. N., & Jayaraman, A. (2011). Urbanisation, poverty and sexual behaviour: The tale of five African cities. *Urban Studies, 48*(5), 947–957.

Guareschi, P. A., & Jovchelovitch, S. (2004). Participation, health and the development of community resources in southern Brazil. *Journal of Health Psychology, 9*(2), 311–322.

Healthy cities: Guidelines for the development of Healthy cities: Projects and Activities. WHO, 1997.

Initiative 'Ville-Santé': Approche et expérience dans la région africaine de l'OMS (OMS 2002) [Healthy Cities initiative: Approach and experiences in the WHO Africa region (WHO 2002)].

Marchés-santé: Guide pour le respect des conditions d'hygiène sur les marchés alimentaires: (OMS 2007) [Healthy Markets: Guide for respecting hygiene conditions in food markets (WHO 2007).]. Retrieved from apps.who.int/iris/bitstream/10665/43616/1/9789242593938_fre.pdf.

Ministère de l'Economie, du Plan et de la Coopération Internationale (avril 2013) *Plan National de Développement 2013–2015*.

Ministère de la Santé Publique. (2008, octobre). *Plan national de développement sanitaire du Tchad 2009–2012*.

Myers, G. A. (2011). *African cities: Alternative visions of urban theory and practice*. London: Zed Books.

New African. (2013, May). *A future made in Africa. Interview David Adjaye*. No. 528, p. 70.

OMS/PNUE: Déclaration de Libreville sur la Santé et l'Environnement en Afrique. Libreville, août 2008.

Peterson, M. B., Dodoo, F. N., & de-Graft Aikins, A. (2014, May). *Poverty and mental health in three inner city communities in Accra, Ghana*. In Paper presented at the Population Association of America (PAA) Conference, Boston.

Première Stratégie Nationale de Réduction de la pauvreté (SNRP.1) de 2003 à 2006.

Quayson, A. (2014). *Oxford Street, Accra. City Life and the itineraries of Transnationalism*. Durham: Duke University Press.

Troisième Enquête sur la Consommation et le Secteur Informel au Tchad (ECOSIT3); N'Djaména, 2011.

UN-Habitat. (2014a). *The state of planning in Africa*. Nairobi: UN-Habitat.

UN-Habitat. (2014b). *L'état des villes africaines. Réinventer la transition urbaine*. Nairobi: UN-Habitat.

United Nations. *World urbanization prospects: 2011*. Retrieved from http://www.un.org/en/development/desa/publications/world-urbanization-prospects-the-2011-revision.html.

USAID. (2010). *USAID Country Profile—Niger. Property rights and resource governance—an overview*. Washington, DC: United States Agency for International Development. 17p.

Van Olmen, J., Ku, G., Bermejo, R., Kegels, G., Hermann, K., & van Damme, W. (2011). The growing caseload of chronic life-long conditions calls for a move towards full self-management in low-income countries. *Globalization and Health, 7*, 38.

WHO. (2007). Training manual for the healthy city programme.

Chapter 7
Healthy Cities in the Eastern Mediterranean: From Sanitation to Comprehensive Policy

Nastaran Keshavarz Mohammadi, Mohammad Assai, and Samar ElFeky

Background to the Eastern Mediterranean Region

The Eastern Mediterranean Region (EMR) of the World Health Organization comprises 22 member states with a population of nearly 583 million people (WHO/EMRO 2013a). Most of the countries are Arab except for Afghanistan, Djibouti, Iran, Pakistan and Somalia. The main language spoken in the region is Arabic (by about half the population) and other languages include Farsi in Iran (~76 million) and Afghanistan (~26.5 million), Urdu in Pakistan (177 million) and Somali in Somalia (8.5 million) (WHO/EMRO 2013a). In some countries, English and French are also spoken. This region is homeland of three major religions, Judaism, Christianity and Islam; the latter is the dominant religion in this region.

The area can be considered very diverse, not only in language but also in social, political, ecological and economic conditions. Some of the richest and poorest countries in the world are here. Social inequalities have resulted in health inequities within and between countries (WHO/EMRO 2012a). Several countries suffer from man-made emergencies, war and conflict which have forced many to leave their homes and live in neighbouring countries as legal or illegal refugees. When added to poverty and other local social factors, political factors negatively affect sustainable development and the health status of individuals and ecosystems, among both refugees and their host populations. This means EMR can be considered as a complex region for health and development programmes and efforts. Table 7.1

N.K. Mohammadi (✉)
School of Public Health, Shahid Beheshti University of Medical Sciences, Tehran, Iran
e-mail: n_keshavars@yahoo.com

M. Assai
Integrated Service Delivery (ISD), WHO-EMRO, Cairo, Egypt

S. ElFeky
WHO-EMRO, Cairo, Egypt

E. de Leeuw, J. Simos (eds.), *Healthy Cities*,
DOI 10.1007/978-1-4939-6694-3_7

Table 7.1 Some demographics and health indicators for the countries of East Mediterranean region

	% of Urban Population	GDP/capita (Average US$ exchange rate)	Adult literacy rate 15+ years	Out of pocket expenditure as % of total Health expenditure	Population with access to local health services			Life Expectancy at birth			ANC	Birth assisted by skilled health personnel	IMR	MMR
					T	U	R	M	F	T				
Afghanistan	23	497	27	83	57	79	54	64.0	62.0	–	60	34	77	460
Bahrain	100	17,379	94	14	100	100	–	75.0	78.3	76.4	100	100	7.2	20
Djibouti	84	1266	–	34	95	100	85	51.8	54.1	52.9	79	56	67	200
Egypt	42.9	2646	70	61	100	100	100	70.9	75.5	73.2	74	92	16.5	66
Iran	72.5	5655	83	58	98	100	95	71.1	73.1	72.1	–	–	18	21
Iraq	66	2932	–	19	83	86	80	70.9	74.6	72.7	66	87	21	63
Jordan	82.6	4445	93	25	99	–	–	71,6	74.4	73	99	99	23	63
Kuwait	100	46,537	95	18	100	100	–	77.5	78.6	77.7	100	100	10.7	14
Lebanon	85	9262	91	45	–	–	–	79.6	83.2	81.5	–	–	9	25
Libya	86	12,461	89	31	100	100	100	70.2	74.9	72.3	93	100	11	58
Morocco	58.3	2848	56	54	–	–	–	73.9	75.6	74.8	77	74	28.8	100
Oman	72	20,764	86	12	98	100	95	70.0	75.7	72.7	99	99	9.6	32
Pakistan	34.7	992	55	50	97	100	94	64.2	67.9	66.0	61	87	65.1	260
Palestine	73.8	1697	96	37	100	100	100	71.0	73.9	72.5	100	100	18.8	64
Qatar	100	82,248	96	16	100	100	–	78.0	78.7	78.2	100	100	7.4	7
Saudi Arabia	85	15,836	88	19	100	–	–	72.7	75.1	73.8	98	98	16.5	24
Somalia	37	284	25	–	–	–	–	–	–	50.0	26	33	109	–
South Sudan	10	1285	37	–	44	–	–	–	–	42.0	17	19	75	1000
Sudan	32.9	1328	56	67	71	–	–	–	–	59.8	74	73	57	730
Syria	53.5	2835	86	54	95	100	90	71.6	74.7	73.1	88	96	17.9	70
Tunisia	66	3831	78	40	95	–	–	72.7	76.6	74.7	96	95	16.8	56
United Arab of Emirates	81	39,619	92	19	100	100	100	75.1	80.2	77.4	100	100	7.1	12
Yemen	29.2	1219	34	75	68	–	–	61.1	62.9	62.0	45	36	68.5	200

Source: Demographic and health indicators for the countries of EMR, WHO/EMRO 2012a
GDP Gross Domestic Product, *ANC* Antenatal Care, *IMR* Infant Mortality Rate, *MMR* Maternal Mortality ratio

summarizes some of most important characteristics of countries in this region (WHO/EMRO 2012a).

Improving City Health

Like the rest of the world (Blas and Kurup 2010; UN-HABITAT/WHO 2010), EMR shares special health challenges in cities caused by urbanization and its consequences. The region's cities are faced with environmental and transportation pollution, communicable and non-communicable diseases, drug abuse, unemployment, homeless people, refugees, mental illness, physical disabilities, urban violence, sedentary lifestyle, crowding, insufficient access to safe water or sanitation and insecurity; these are all aggravated by armed conflicts and political instability.

Health projects in cities usually focus on different populations' needs and sometimes are environmentally oriented. There are two main frameworks for health initiatives at city level: Urban Health and Healthy Cities. Three interrelated approaches have been implemented in the EMR region; they include community-based initiatives (CBI), Healthy Cities, and Urban Heart. In implementing Healthy Cities to promote urban health, community-based programmes have been the central focus, and equity in health and participation has been adopted as central values.

Community-Based Initiatives (CBI)

Healthy City and healthy Village programmes in the region have been implemented predominantly as part of community-based initiatives (CBI) (WHO/EMRO 2009a). In 1988, the Eastern Mediterranean Regional Office of WHO (EMRO) adopted community-based initiatives as a strategy to promote community health by tackling social determinants of health. CBI approaches encompass four main programmes: basic development needs, Healthy Cities, Healthy Villages, and gender in health and development (WHO/EMRO 2009a). The healthy city programme was initially implemented in the 'Kooye 13 Aban' area of Tehran in 1991 as a component of CBI programmes, then expanded to 17 member states that implement at least one of these programmes, now covering a population of nearly 18 million (WHO/EMRO 2009a, b). The CBI programmes were found to work better in smaller towns and villages than in dense urban environments, and in rural areas rather than urban areas. CBI programmes promote community organization, empowerment, ownership and leadership of developmental processes, with each community identifying its own needs and mobilizing available resources. These initiatives have proven that community engagement, decentralization and intersectoral collaboration lead to effective and sustainable community development and health improvement (WHO/EMRO 2009b) and have recorded two decades of success in the region (WHO/EMRO 2009a). Related activities focus on improving literacy, water and sanitation, increasing immunization and establishing micro-credit schemes, in addition to many other areas; engaged communities have experienced improvements in health and quality of life.

Based on previous experiences in the region, several actions are suggested for the future:

- Institutionalization of community-based initiatives as part of a national health policy
- Building links between all health and developmental programmes at a national level
- Fostering community leadership for health and development
- Strengthening intersectoral partnerships and resourcing initiatives for health and development.

Healthy Cities in EMRO: History and Progress

The Healthy Cities initiative has been in practice in the Eastern Mediterranean Region since 1990, as one of the four components of CBI. It was launched in the Islamic Republic of Iran and expanded to Afghanistan, Bahrain, Iraq, Jordan, Oman, Pakistan, Saudi Arabia, Sudan and United Arab Emirates, adapting to different cities' particular needs and interests (WHO/EMRO 2010). In 2007, a training manual for use by trainers and members of Healthy Cities coordinating committees was developed to provide guidance for city planners and community members through a training course on the Healthy Cities programme (WHO/EMRO 2010) (Fig. 7.1).

World Health Day 2010 provided an opportunity to advance the urbanization and health agenda by securing the political commitment of cities in the region: 209 cities registered for the *1000 cities—1000 lives* global movement on this occasion and committed to implement healthy city Programmes. The excellent responses of mayors, governors and city planners encouraged the regional office to develop a clear, concise, simple and practical guide to help countries standardize their methods, mechanisms and processes of implementation for Healthy Cities, and to establish a list of tasks and criteria for cities wishing to implement the healthy city programme (WHO/EMRO 2010). CIP aimed to register a further 200 cities by the end of 2014.

The regional healthy city network was created in January 2012. Mayors of cities in Afghanistan, Iran, Morocco and United Arab Emirates joined the network by signing a formal commitment, contingent upon meeting certain criteria. First a team from the regional CIP conducted a planning workshop, and this was followed by each city coordinating committee developing a work plan to address some of the challenges facing the city and to improve the health and social status of its citizens (WHO/EMRO 2010).

The network now hosts a dynamic web site (WHO/EMRO 2013b) which includes an interactive map, tools to facilitate the exchange of national, regional and global experiences, and the sharing of success stories, photos and coming events. The site

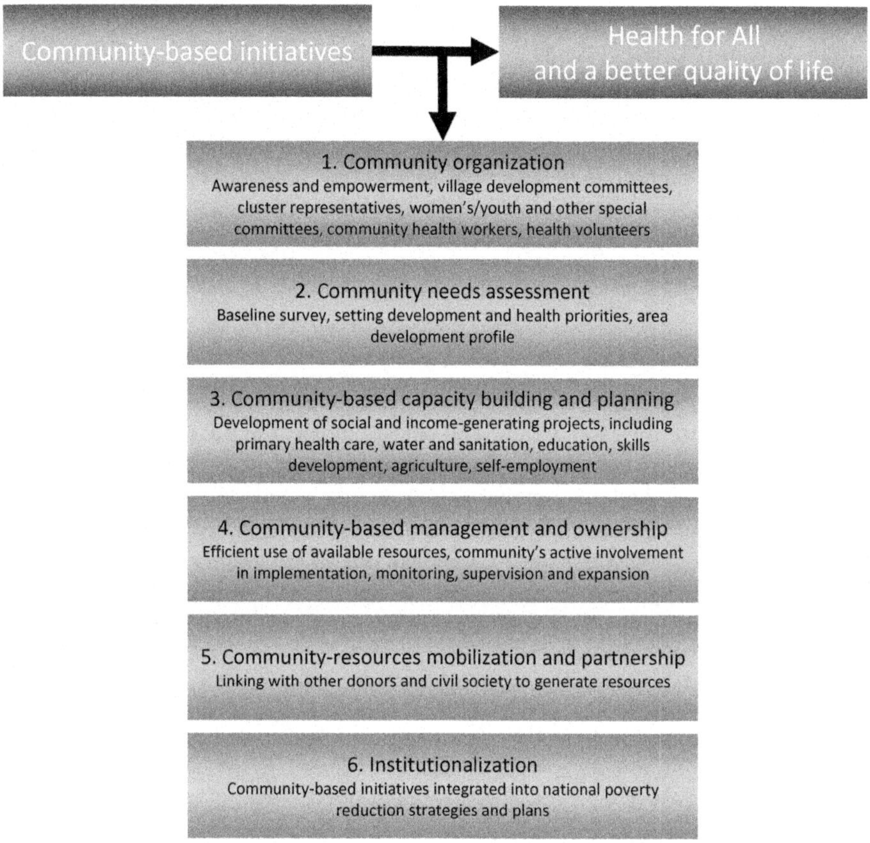

Fig. 7.1 WHO/EMRO (2009a). Major principles of community-based initiatives

provides links to guidelines, tools and information that can be used to provide technical support to countries to plan, implement, monitor and evaluate national healthy city programmes.

Any city can join the HCP by completing three steps:

1. Joining the regional movement on health and urbanization
2. Joining the regional healthy city network
3. Requesting assessment to qualify as one of the global healthy cities and receive a certificate from the regional office

The cities agree on a set of activities that will improve the health and social status of their residents and promote health equity, particularly in urban slums. They ensure that all monitoring indicators (80 measures that are integral to the guidelines) are in place and ready to be assessed by the joint regional and country evaluation team.

Any city, regardless of its current health and social status, can ask to join the regional network (WHO/EMRO 2010). The most important consideration is whether or not it has the political will and commitment to improve the health and social status of its residents and is willing to redirect its resources and adopt the policies, organizational structures and processes required for achieving healthy city status.

The regional office provides technical support and facilitates meetings, and offers technical support to reduce risk factors for non-communicable diseases through community ownership, sustained intersectoral collaboration and introducing the Healthy Cities programme (WHO/EMRO 2013b).

Urban Heart in EMR

WHO Kobe Centre for Health Development, established in 1995 with a focus on urbanization, developed the Urban Health Equity Assessment and Response Tool (Urban HEART) to synthesize global knowledge on social determinants in towns, cities and informal settlements (WHO Kobe Centre for Health Development 2012). Urban HEART has a 'health equity' component that assists ministries of health and cities to conduct a systematic assessment of unfair health conditions in the urban setting. For ease of analysis, the tool suggests reviewing evidence within four major policy domains (WHO/EMRO 2010):

1. Physical environment and infrastructure
2. Social and human development
3. Economics
4. Governance

After pre-testing the tool in Malaysia it was piloted in a number of low- and middle-income countries in different regions, including Brazil, Iran, India, Philippines and Zambia, in 2008–2009. Since then Urban HEART projects have been implemented in some countries of the region, including Egypt, Morocco and Tunisia.

The methodology of Urban HEART was adapted and used to meet local needs. Its approach enables decision makers to recognize equity challenges that face communities according to the priorities each community assigns them according to a red/yellow/green report matrix (WHO/EMRO 2010). The methods used to conduct assessments include a desk review of data produced by different sectors and surveys of selected samples of households.

The main purpose for conducting Urban HEART studies in the selected model areas was to identify the social determinants of health inequities across five domains:

- Health outcomes
- Physical environment and infrastructure
- Social and human development
- Governance
- Economics

Urban HEART combines research, partners' organizational data and community knowledge to assess urban equity in relation to these domains, and seeks to provide policy and assist decision-makers at national and local levels to identify the relationship between the health determinants and well-being of the population by identifying equity gaps in various parts of the city in comparison with benchmark values at national level. This enables the determination of appropriate, feasible, acceptable and cost-effective strategies, interventions and actions that will have a positive impact on reducing equity gaps through intersectoral action and community engagement.

This section synthesizes the experience of three cities in the REGION, showing how local city planners identified challenges and opportunities in addressing health inequities through the use of Urban HEART.

Iran

Iran is the third most populated country in the region, with a population of about 76 million in 1,648,000 km^2. It is bordered by the Persian Gulf to the south and the Caspian Sea to the north, Iraq and Turkey to the west and Pakistan and Afghanistan in the east. About 70 % of the population live in urban areas. The average literacy rate is 85 % (Global Health Observatory Data Repository 2011). Iran is classified as an upper middle income country (World Bank 2015). Iran is divided into 31 states, all working under central government administration (Iran Ministry of Interior 2011). The structure and function of city and village councils have provided opportunities for the participation of people in planning and implementing different development programmes (Iran Ministry of Interior 2011; Islamic republic of Iran Constitution 1997).

In last three decades several significant events have affected peoples' lives. They include the 1979 Revolution; armed conflict with neighbouring states, UN sanctions and—as in many other countries—changes in lifestyle. Armed conflicts in two neighbouring countries, Iraq and Afghanistan, have led to increased legal and illegal immigration of refugees, creating social as well as health issues in the nation. Despite these problems, after the revolution Iran made significant progress in providing public education and access to basic needs such as water, electricity, roadways and health services, especially in disadvantaged areas.

Health of Cities in Iran

Currently two frameworks by two authorities are working to improving health in cities. The older is Healthy Cities, run by the Ministry of Health in collaboration with other sectors. The second and newer one is Urban HEART, run by municipalities in collaboration with other sectors; both receive technical support from WHO.

Healthy Cities in Iran

Iran has the longest history in Healthy Cities in the region, with the pilot study of Aban 13 District in Tehran 1991 as a component of CBI (WHO/EMRO 2014). CBI occupies an office in the Ministry of Health, in the department of occupational and environmental health. Key partners in the community-based initiatives programme are the Ministry of the Interior, which addresses environmental issues such as solid waste management, and district governors' offices, which support needs-based projects. From 1991 to 2005 more than 83 CBIs with a range of small to large-scale programmes were implemented in 16 states, the majority to do with Healthy Cities or Healthy Villages (Keshavarz Mohammadi et al. 2005), with a particular focus on community health and development in disadvantaged communities.

Urban HEART in Iran

Urban HEART was officially launched in Tehran in April 2008, in collaboration with the Tehran municipality and WHO (WHO/EMRO 2014). Since then, Urban HEART has been adopted by the municipality as an approach for overcoming health equity gaps and improving the health of city dwellers. Improving and promoting urban health along with the development of the Healthy Cities approach is one of four objectives of the Department of Health of Tehran municipality (Asadi-Lari et al. 2010).

The municipality made some modifications in the domains and indicators: for example, nutrition and health were added to the four domains of Urban HEART, and a questionnaire in 12 sections including 65 indicators was developed (Asadi-Lari et al. 2010). The questionnaire was piloted in five selected districts before a door-to-door survey was run in 22 districts, polling 22,300 households. The municipality assessed urban health in two rounds, in 2009 and again in 2012 (Asadi-Lari et al. 2013); Table 7.2 compares the indicators in the two rounds of assessment.

GIS maps and matrices using the three colours indicating each community's assessment of priority issues were developed for each indicator, which made it easy to identify the health and equity priorities of each district and also comparison between districts. The data were provided to the mayors of each district.

In each district a briefing meeting presented the findings, focusing on disparities in neighbourhoods and discussing issues and priorities and potential community-based and intersectoral interventions (Asadi-Lari et al. 2013). Local districts mayors became responsible for taking action, including developing community-based programs and arranging the appointment of committees to meet the identified local needs. A range of activities implemented during and after the assessment included establishing a comprehensive urban health information system, developing public health education, facilitating the development of coordinated mechanisms for urban health management, initiating community health houses, providing the necessary

Table 7.2 Urban HEART indicators—comparing indicators in two rounds

Domain	Round 1 indicators (2009)	Round 2 indicators (2012)
Physical and infrastructure	1. Healthy water	1. Healthy water
	2. Accidents and injuries (5 indicators)	2. Traffic and non-traffic (domestic) accidents
	3. Air pollution	3. Occupational injuries
	4. Noise nuisance	4. Air pollution
	5. Access to public transport	5. Noise nuisance
	6. Solid waste management	6. Access to public transport
	7. Health centre utilization	7. Solid waste management
		8. Sanitary toilet
		9. Access to wastewater system
		10. Green area per capita
		11. Sport area per capita
		12. Children's playground
		13. Disabled-friendly areas
		14. Urban decay (to strengthen against earthquakes)
Human and social development	1. Education: NER/ GER/ primary school completion/ Higher education (4 indices)	1. Net enrolment rate
	2. Violence: domestic, street	2. Primary school completion rate
	3. Smoking/addiction	3. Higher education rate
	4. Smoke-free places	4. Domestic violence
	5. Mental health	5. Social capital
	6. Social capital	6. Woman-headed families covered by social security
Health	1. Safe delivery	1. Safe delivery
	2. Vaccination	2. In-time vaccination
	3. Teenage pregnancy	3. Teenage pregnancy
	4. Breastfeeding	4. Breastfeeding (exclusive BF and duration)
	5. Infant (IMR)/Under five (U5MR)/and maternal mortality rate (MMR)	5. Neonatal (NMR)/Infant (IMR)/ and Under five mortality rate (U5MR)
	6. Health related quality of life (HRQL)	6. HRQL
	7. Disability	7. Disability rate and access to special healthcare services
		8. Crude mortality rate and life expectancy (LE)
		9. Prevalence and incidence rate of myocardial infarction, stroke, hypertension, cancer, diabetes, asthma, and osteoporosis
		10. Health centre utilization
		11. Basic and complementary health insurance rate

(continued)

Table 7.2 (continued)

Domain	Round 1 indicators (2009)	Round 2 indicators (2012)
		12. Oral health and missed teeth
		13. Physical activity
		14. Pain prevalence and incidence
		15. Mental health
		16. Domestic violence
		17. Smoking and addiction
		18. Utilization of quit addiction services
		19. Suicide (and committing suicide) rate
		20. Wasting and stunting among under 5 years children
		21. Body mass index
		22. Nutrition habits
		23. Food insecurity
Economic development	1. Employment	1. Employment
	2. Residency in normal	2. Room and area per capita
	3. Fair financial contribution index (FFCI)	3. Household costs
	4. Household costs	4. Fair financial contribution index (FFCI)
	5. Absolute/ relative poverty	5. Catastrophic payment rate
	6. Social welfare index	6. Absolute poverty, susceptibility of households to poverty
	7. Human development index	7. Multidimensional poverty
		8. Polarization rate
		9. Unequal opportunities
		10. Human Development Index
Governance (municipality)	1. Annual reports	1. Satisfaction with municipality services
	2. Contracts transparency	2. Responsiveness of the municipality (hotlines)
	3. Satisfaction	3. Community participation (local elections)
	4. Responsiveness (Hotlines)	4. Population covered by health volunteers (in health centres) and health auxiliaries affiliated to the municipality
	5. Community participation (local elections)	
	6. Lawfulness	
	7. Standard activities	

(continued)

Table 7.2 (continued)

Domain	Round 1 indicators (2009)	Round 2 indicators (2012)
Nutrition	1. Calorie poverty	1. Integrated with 'health' domain
	2. Wasting	
	3. Stunting	
	4. LBW (IUGR/ NMR)	
	5. BMI: obesity	
	6. Food diary	
	7. Food costs	
	8. Cereal costs	
Cultural		1. Respect to values: truth, bailment, forgiveness, fairness, honesty, frankness, lawfulness, caring others, respect to elderly

physical structure and environment for participation of country members in community health, funding special health programs, and institutionalizing health inequity assessment and response (Asadi-Lari et al. 2013).

After conducting the first stage of Urban HEART, data was collected to measure urban health inequity, to enable appropriate responses to the identified problems (Asadi-Lari et al. 2013):

1. Evidence-based policymaking
2. Evidence-based practice
3. Intersectoral collaboration
4. Community-based initiatives (CBI)
5. Health systems response

Actions Taken

1. Endorsement of a 5-year development plan (2009) including annual assessments of inequalities in health and social determinants of health
2. Redistribution of budgets to give priorities to more deprived districts of the city
3. Integration of health inequity indicators into routine primary health care (PHC) information systems
4. Dissemination of information to mayors, neighbourhood councils, local authorities and organizations to obtain support for action
5. Funding of local community-based initiatives to tackle identified local priorities in 22 districts of Tehran
6. Establishment of the social determinants of a health-oriented urban PHC system in the district (reported as requiring the most urgent response to Urban HEART findings):

- Establishment of a strategic health board in Tehran municipality, chaired by an ex-minister of health
- Appointment of one "community health worker" and one "community social worker" for every 4000 people (almost 1000 households)
- Establishment of urban health houses, each covering 8000 people, and sponsored by the municipality

7. Establishment of an urban health management centre at district level, directed by a board of trustees, chaired by the district mayor and comprising all relevant sectors in the district
8. Establishment of a national health observatory committee for health inequity by the ministry of health
9. Modifying and adding indicators to develop a list of national indicators for the following round of urban HEART implementation
10. Expanding urban HEART to rural areas

The efforts and achievements of Urban HEART in Tehran are significantly observable, but some concerns need to be taken into account. One is the inadequate integration of interventions in current health systems in highly urbanized cities that are covered by Ministry of Health, with the potential to require costly parallel information gathering efforts or response interventions.

Egypt

Urban HEART in Gezerit El Warak, Giza, Egypt (WHO/EMRO 2012c). The Ministry of Health of Egypt has found that use of the Urban HEART tool by national and local governments can guide policymakers and key stakeholders to identify local social determinants of health and their impact on health equity. The tool provides evidence that can help local authorities prioritize urban problems and encourage city planners to allocate resources for interventions that will achieve health equity. Local health authorities acknowledge that Urban HEART can sensitize, mobilize and empower communities to attain health equity.

The study in Gezerit El Warak was planned during meetings at governorate level to identify the main criteria for site selection. The Ministry of Health, the governorate of Giza and WHO agreed on the selection of Gezerit El Warak because of its location as an unplanned residential slum area on the river Nile, with no access to motorized public transport, extremely narrow streets, and no means of public transport except the local ferry. Only two ferries (one private, the other one public) run daily services, but service finishes at 8:00 p.m.; this creates difficulties in emergencies when people need to get to a hospital quickly. There is no sewage disposal system in the area, and it lacks sanitary environmental facilities (Fig. 7.2).

On 30 June 2011, an orientation workshop was organized by the Ministry of Health and WHO to brief local stakeholders about the Urban HEART concept and methodology, and to explain how to proceed with the study.

Fig. 7.2 Mode of transportation for the citizens of Gezerit El Warak

The main objectives of the study were to:

- Assess the determinants that affect the health of the citizens in Gezerit El Warak
- Compare the social determinants of health of Gezerit El Warak with the standards of Giza governorates and at national level
- Determine the gap in health equity between Gezerit El Warak and the national level and share the results with policymakers and key local stakeholders.

In July 2011, a door-to-door household survey was conducted, supported by the local government and community, in partnership with WHO. Table 7.2 shows the indicators measured in this study.

Findings from Gezerit El Warak

Infant and neonatal mortality rate, fully immunized infants, female overweight and obesity, skilled birth attendance, unemployment and voter participation in elections were better than the national benchmarks. Among key risk factors, the prevalence of hypertension among males and females, and male overweight and obesity were higher than the national benchmark. Access to safe water and sanitation, literacy rate, and the prevalence of male tobacco smoking were far worse than the national benchmarks.

The key responses to the assessment included building safe roads, improving transportation, improving access to safe drinking water with the establishment of two stations for water purification, funded by community leaders and community volunteers, creating sanitary and waste disposal facilities, conducting tobacco control campaigns, screening people without ID, and providing equipment and medication for the local health centre (Table 7.3).

As a result of this study, the city planners decided to improve health-related behaviour such as smoking, poor nutrition and lack of physical exercise. Access to

Table 7.3 Measured indicators for Gezerit El Warak, Giza, Egypt

Factor	Local rate (%)	National rate (%)
The prevalence of female overweight	34.5	38.4
Connection to piped water	84	98
Connection to public drainage sewers	None	56
The illiteracy rate	36.4	25.2
Full vaccination of infants	95.1	91.7
Skilled birth attendance	72.2	71.7
Unemployment rate	7.9	11.9
Voter participation	52.7	41.0

quality primary health care services and maintaining the current coverage of the immunization programme were major concerns. Local authorities agreed to improve environmental health, including access to water and sanitation, and more local transportation and undertook to work with stakeholders to increase the literacy rate among adults. Future directions include ensuring community ownership at all stages of planning, implementation and monitoring; enhancing partnerships and mobilizing resources; and ensuring that collaboration is effective and sustainable (Table 7.4).

City planners shared the results of the Urban HEART study with local, national and international partners on 18 January 2012. City planners decided to pay attention to improve health-related behaviour, such as smoking, poor nutrition and lack of physical exercise. Access to quality primary health care services and maintaining the current coverage of the Expanded Programme on Immunization are the major focus in Gezerit El Warak. Local authorities will also improve the environmental health status, including access to water and sanitation, local transportation and will work with different stakeholders to increase the literacy rate among adults. Future directions include: ensuring community ownership at all stages of planning, implementation and monitoring; enhancing partnerships and mobilizing resources; and ensuring that collaboration is effective and sustainable.

Oman

Oman's health systems in all 11 regions and 61 wilayats (districts) recognize the importance of community participation in addressing the social determinants of health, so different approaches, including wilayat health committees, community support groups and community-based initiatives, have been adopted by health authorities to benefit from community participation for health.

The Ministry of Health in Oman, in collaboration with other sectors and communities, has succeeded in promoting health in the community, with the goal of improving quality of life rather than simply reducing morbidity and mortality. A total of 16 community-based initiatives have been launched by the ministry to strengthen the community focus of health services and programmes; examples

Table 7.4 Comparison of indicators from Gezerit El Warak with national averages

Health outcome indicators	Core indicator	Gezerit El Warak	National level
Mortality	Infant mortality rate per 1000 live births	***12.0***	25
	Neonatal mortality rate per 1000 live births	***6.0***	16
Disease-specific	Male hypertension >140/90 (%)	*12.8*	10.5
	Female hypertension >140/90 (%)	*14.5*	12.8
	Male overweight (%)	*36.7*	34.3
	Male obesity (%)	*21.1*	18.2
	Female overweight (%)	***34.5***	38.4
	Female obesity (%)	***38.0***	39.6
	History of diabetes mellitus (%)	***3.3***	3.1
Physical environment and infrastructure	Access to piped water (%)	*84.0*	98.0
	Water safety for drinking purposes (%)	**17**	100
	Water supply interrupted (%)	**100**	30.7
	Access of households to drainage and public sewers (%)	**0.0**	56.0
Social and human development	Illiteracy rate among adults (%)	**36.4**	25.2
	Fully immunized infants (%)	***95.1***	91.7
	Delivery assisted by skilled birth attendants (%)	***72.2***	71.7
	Prevalence of male tobacco smoking (%)	**55.9**	43.9
Economics	Unemployment (%)	***7.9***	11.9
Governance	Voter participation (last constitutional voting) (%)	***52.7***	41.0

Bold values indicates more than 20 % difference in performance compared to national level
Italic values indicates less than 20 % performance compared to national level
Bold italic values indicates good or similar performance compared to national level

include the Sur Healthy City project (2003) and Healthy City projects. Activities implemented include the provision of safe drinking water, the availability of healthy food choices in the market, community information centres, walking pathways, and promoting healthy active lifestyles in schools (WHO/EMRO 2012d).

Activities carried out by the districts' health committees have included

- Participating in national and international health days
- Participating in health exhibitions at district level
- Initiating different health programmes at district level, such as iron-deficiency anaemia, obesity in school children and tuberculosis social survey
- Conducting workshops on child health, smoking, cancer, HIV/AIDS and H1N1 influenza programmes
- Conducting health marathons

- Increasing health awareness by strengthening health education in the community
- Organizing health teams in the wilayat for cyclone and disaster planning.

The main focus of Healthy City interventions has been to promote healthy lifestyles by implementing health education programmes.

Bahrain

The Healthy City project was introduced in Bahrain at an orientation workshop on the concepts of healthy city with the assistance of the WHO regional office in May 2003. Following the workshop, it was decided that one of five governorates of Bahrain, Muharrag, should serve as the pilot location for implementing a Healthy City programme, and a multisectoral national team was established. In 2004, a national health survey was conducted in collaboration with WHO. The national committee agreed on the importance of community participation to maintain and enhance long-term credibility and the public image of the Healthy Cities programme. Several entry projects were implemented, and national campaigns were conducted to promote the programme and its components, including healthy schools, anti-drug addiction campaigns, awareness-raising about road traffic accidents, beach cleaning, graveyard cleaning, and seatbelts campaigns (WHO/EMRO 2005).

Conclusions

CBI and The Urban HEART studies, as dominant approaches to healthy cities in the region, brought different sectors and communities together under a common goal and shared responsibilities that could be used in future as a platform to sustain health development activities. Building national and international networks for Healthy City programmes healthy cities helped to facilitate the exchange of evidence and know-how between cities. The main obstacles to improving urban health seemed to be related to governance and the organization of civil society.

Regional experiences suggest that policymakers should be encouraged to generate and use the data to make better decisions for city planning. A supportive governance structure for implementing intersectoral action will assist in sustaining efforts and making efficient use of existing resources. Community participation and empowerment in the process of policymaking must be considered at all stages, from the initial assessment to evaluation of the intervention and monitoring of outcomes. Regular monitoring and evaluation of activities will identify strengths and weaknesses of interventions and maintain focus on the outcomes. Different experiences must be documented along with dissemination of best practices and lessons learnt.

References

Asadi-Lari, M., Vaez-Mahadavi, M., Faghihzadeh, S., Chergian, B., Esteghamati, A., Farshad, A. A., … Razzaghi, E. (2013). Response-oriented measuring inequalities in Tehran: Second round of Urban Health Equity Assessment and Response Tool (Urban HEART-2), concepts and framework. *Medical Journal of the Islamic Republic of Iran, 27*(4), 236–248.

Asadi-Lari, M., Vaez-Mahdavi, M. R., Faghihzadeh, S., Montazeri, A., Farshad, A. A., Kalantari, N., … Malek-Afzali, H. (2010). The application of urban health equity assessment and response tool (Urban HEART); in Tehran; concepts and framework. *Medical Journal of the Islamic Republic of Iran, 24*(3), 115–125.

Blas, E., & Kurup, A. S. (Eds.). (2010). *Equity, social determinants and public health programmes.* Geneva: World Health Organization.

Global Health Observatory Data Repository. (2011). *Iran (Islamic Republic of) statistics summary (2002–present).* Retrieved December 10, 2013, from http://apps.who.int/gho/data/view.country.10600.

Iran Ministry of Interior. (2011). Retrieved from http://www.moi.ir/Portal/Home/Default.aspx?CategoryID=5a7baacf-8edb-4d36-a913-247e6769f3be.

Islamic Republic of Iran Constitution. (1997). *Translation and publication office.* Tehran: Communication and Islamic relations Organisation.

Keshavarz Mohammadi, N., Bahreini, F., & Degpasand, S. (2005). *Situation analysis of community based health promotion programs in Iran.* Tehran: WHO/EMRO/Iran.

Tehran Municipality. (2008). *Launch of the urban health equity assessment and response tool (Urban HEART) 19–21 April 2008.* Tehran: Islamic Republic of Iran

UN-HABITAT/WHO. (2010) *Hidden cities: Unmasking and overcoming health inequalities in urban settings.* Switzerland: UN-HABITAT/WHO.

WHO Centre for Health Development. (2012). Urban *health equity assessment and response tool (Urban HEART).* Retrieved from http://www.who.int/kobe_centre/measuring/urbanheart/en/.

WHO/EMRO. (2005). *Report on the workshop of healthy cities programmes in Gulf cooperation council countries*, Cairo

WHO/EMRO. (2009a). *Community-based initiatives.* Retrieved from http://www.emro.who.int/entity/community-based-initiatives/index.html.

WHO/EMRO. (2009b). *Community-based initiative success stories 2.* WHO-EM/CBI/061/E.

WHO/EMRO. (2010). *A short guide to implementing the healthy city programme.* Retrieved from http://applications.emro.who.int/dsaf/dsa1088.pdf.

WHO/EMRO. (2012a). *Demographic, social and health indicators for countries of the Eastern Mediterranean.* Retrieved from http://applications.emro.who.int/dsaf/EMROPUB_2012_881.pdf.

WHO/EMRO. (2012b). *Urban health equity assessment and response tool project in Egypt.* Retrieved from http://applications.emro.who.int/hcn/Resources.aspx.

WHO/EMRO. (2012c). *Urban health equity assessment and response tool projects in Egypt, Morocco and Tunisia.* Retrieved from http://applications.emro.who.int/hcn/Resources.aspx.

WHO/EMRO. (2012d). Good practices in delivery of primary health care in urban settings. Community-Based Initiatives Series.

WHO/EMRO. (2013a). *About us.* Retrieved December 25, 2013, from http://www.emro.who.int/entity/about-us/index.html.

WHO/EMRO. (2013b). *Regional healthy city network.* Retrieved from http://applications.emro.who.int/hcn/.

WHO/EMRO. (2014). *Community-based initiatives: Islamic Republic of Iran.* Retrieved August 23, 2014, from http://www.emro.who.int/cbi/countries/islamic-republic-of-iran.html.

World Bank. (2015). *Country and lending groups.* Retrieved from http://data.worldbank.org/about/country-and-lending-groups#Low_income.

Chapter 8
Healthy Municipalities, Cities and Communities in Latin America: Strong Histories, Committed Futures

Marilyn Rice, Cris Franceschini, Nina Wallerstein, Raul Mercer, Karina Cimmino, Lucero Rodriguez, Ljubica Latinovic, Paola Pliego, Jose Osney Velandia Rodríguez, Gloria Libia Polanía Aguillón, Ana María Mahecha Groot, Teófilo Monteiro, Ana Rivière-Cinnamond, Ronice Franco de Sá, and Guillermo Hegel

Introduction[1]

The Latin American Region, represented by four subregions (South America, Central America, Spanish-speaking Caribbean and Mexico in North America), contains immense cultural, geographic, linguistic and political diversity. Within this diversity, dynamic and vibrant networks of healthy municipalities, cities and communities (HMC) have emerged.

In this chapter we provide a brief overview of the historical context of democratization and urbanization in the 1980s that supported the growth of health promotion, community development and the HMC movement throughout the region. We outline the major milestones over the last three-and-a-half decades and highlight case studies that showcase the range of initiatives, as well as their achievements and challenges. Over this period many tools have been developed to support countries in their efforts; a sampling of these has been summarized with links to the tools

[1] Marilyn Rice, Maria Cristina Franceschini and Nina Wallerstein.

M. Rice (✉)
CEO merci (Marilyn E Rice Consulting International)
e-mail: ricemarilyn2011@gmail.com

C. Franceschini
Consultant with merci and the Pan American Health Organization/World Health Organization (PAHO/WHO)

N. Wallerstein
Professor, Public Health Program, Department of Family and Community Medicine, and Director, Center for Participatory Research, University of New Mexico

R. Mercer
Coordinator of the Social Sciences and Health Programme at FLACSO (Latin American School of Social Sciences), researcher at CISAP (Centre for Research in Population Health) Durand Hospital, Buenos Aires, Argentina

E. de Leeuw, J. Simos (eds.), *Healthy Cities*,
DOI 10.1007/978-1-4939-6694-3_8

themselves. The chapter concludes with a brief analysis and some reflections on future prospects for the HMC movement in Latin America.

In 2008, for the first time in history, the United Nations reported that more than half the human population is living in urban areas, with this proportion expected to rise to 70 % by 2050. In 2012 the world's population surpassed seven billion, and it is projected to reach 9 billion by 2050. The region of the Americas is the most urbanized in the developing world, with 77 % of its population already living in cities. It is estimated that it will continue to urbanize in the coming two decades, when the urban proportion will rise as high as 85 % (UNFPA 2012).

Urbanization is more than a demographic issue: it is also a megatrend and a contradictory set of processes. Its origins are the perceived benefits of urban life, which are present in all countries but particularly in developing countries. While urbanization offers many opportunities, it also creates tremendous challenges for improving health and for health promotion to be effective.

Challenges

> The vast urban expansion in developing countries has global implications. What happens in the cities of the less developed world in the coming years will shape prospects for economic growth, poverty alleviation, population stabilization, environmental sustainability and, ultimately, the exercise of human rights worldwide. (UNFPA 2007)

K. Cimmino
Social Sciences and Health Programme at FLACSO (Latin American School of Social Sciences) Buenos Aires, Argentina

L. Rodriguez • L. Latinovic • P. Pliego
Mexico Ministry of Health

J.O.V. Rodríguez
Analysis in Health Equity project coordinator, Urban HEART

Hospital Pablo VI Bosa ESE I Nivel

G.L.P. Aguillón
Manager, Hospital Pablo VI Bosa ESE I Nivel

A.M.M. Groot
National consultant in environmental health and healthy settings, PAHO/WHO Colombia

T. Monteiro
International advisor in the area of sustainable development and environmental health, PAHO/WHO Colombia

A. Rivière-Cinnamond
Ecohealth Advisor, PAHO/WHO Peru

R.F. de Sá
Professor at the Federal University of Pernambuco

G. Hegel
PAHO/WHO Advisor in Environmental Health

Social inequalities are exacerbated in cities. In a sample of 37 selected cities in the world with an average Gini coefficient of 0.58,[2] nine Latin American cities were represented, and six of the nine were above the International Line of Warning (UN-Habitat 2009). Although urban poverty in Latin America fell from 31 % in 1990 to 29 % in 2007 (UN-Habitat 2010), the number of urban poor rose from 122 million to 127 million during these same years. Cities in Latin America often suffer from poor housing or informal human settlements with lack of basic services such as water, sanitation, electricity, roads, transportation and employment, and with lack of tenure, overcrowding and high density, unhealthy living conditions, hazardous locations, and social exclusion.

In Latin America, rapid urbanization has surpassed the countries' capabilities to offer adequate services. This is exacerbated by increased job instability in cities, leading to increased economic and social inequities. A recent report on major cities (UCLG 2009) highlights the strain put on them by the recent global economic crisis, resulting in unemployment and the loss of job opportunities, which in turn lead to deteriorated welfare and increases in social problems that inhibit the population's ability to meet basic needs such as housing, food and the acquisition of durable goods. The UCLG report also indicates that during economic crises, competition increases for scarce resources within the city government and between city and national governments. In 2010 we saw, from the most tragic human perspective in Haiti and Chile, the vital importance of planning and construction that assure that buildings and the urban infrastructure are safe in case of earthquake or other natural disasters (ISDR 2010).

Urban infrastructure affects the health status of the population by facilitating or hindering access to public services such as clean drinking water, sanitation, electricity and transportation, as well as to education, health services, communication and public spaces for recreation and sports. The complexity of the approach to these problems is increased by problematic social, family, community and gender relations. Urbanization also brings environmental change. Human health and well-being are highly dependent on ecosystems and their resources, such as food and drinkable water. However, over the past 50 years, the accumulation of people in urban areas, their consumption patterns, behaviours and activities, have caused a series of ecosystem changes that have had a major impact on the environment.

The economic, industrial and transportation dynamic of cities, and the consequent energy consumption, create a significant carbon footprint that has lasting implications for the health and well-being of the population in the cities and beyond. This translates into major inequities that are not consistently captured or detected

[2]The Gini coefficient is a statistical measurement commonly used to represent the inequality of income/wealth distribution among a nation's residents. It measures the inequality in the distribution of values (such as levels of income). A Gini coefficient of zero expresses perfect equality, where all values are the same (for example, where everyone has the same income). A Gini coefficient of one (or 100 %) expresses maximal inequality among values (for example, where a single person has all the income).

by health information systems. Managing the urban environment sustainably and equitably will be one of the greatest challenges in upcoming decades. While the region has been advancing in this regard, with fewer people living in poverty and increased literacy and life expectancy rates, the evidence tends to average out the large inequalities within countries and within cities (PAHO/WHO 2011a).

Unplanned and unwieldy urbanization is having an unprecedented impact on natural ecosystems, on the predisposition for natural disasters, as well as on consumption and waste discharge (UNEP 2002). An increasing number of urban poor are settling in potentially dangerous areas mostly outside the official and legal frameworks of building codes, land use regulations and land transitions, putting them at much higher risk (UNEP 2010).

In addition to insufficient infrastructures, major contributors to ill health in the region include violence, alcohol abuse, tobacco use, vector-borne diseases and chronic non-communicable diseases (CNCDs). These tend to be much more prevalent in the lower socio-economic populations, which again raises the issue of inequity and the role of the social determinants of health.

In order to address inequities, significant changes are needed in the way information is collected, analysed and used, and the development of appropriate tools and capabilities will be essential. Cities must develop healthy public policies focused on the social determinants of health, sustainable local development and health equity, and adopt planning and management mechanisms that include inter-sectoral collaboration and participatory decision-making processes (WHO 2008).

Three policy modalities can be identified from an analysis of intervention strategies on health determinants: (1) improving the health of the most disadvantaged population groups through programmes targeted by objective, (2) closing the gap between those in the poorest social circumstances and better-off groups and (3) addressing the entire health gradient, that is, the association between socio-economic position and health across the whole population (Solar and Irwin 2007). Such policies could be implemented in urban settings at a municipal as well as a national level. To bring these policies to the fore, strategies for healthy municipalities, cities and communities, grounded in the value base and framework of healthy urban governance and health promotion, become paramount.

The Need for Healthy Urban Governance

Healthy urban governance is key to developing and maintaining healthy cities. It requires defining and addressing the determinants of urban health from a multisectoral perspective. In an urban context, governance refers to the 'sum of many ways individuals and institutions, public and private, plan and manage the common affairs of the city' (UN-Habitat 2002).

Value-based urban planning can help ameliorate urban problems and provide opportunities for the populace to fulfil its physical, social and mental needs while actively participating in decisions that affect its quality of life. This is directly related to the opportunities for cities to build sustainable healthy environments. A more people-centred participatory approach can help ensure continuation of policies and programmes, even when there are major changes in government.

Governance models for urban health should include the active participation of all sectors of society (public, private and civil) in decision-making processes at the local level, as well as in mechanisms for planning, management, monitoring, and evaluation of urban policies and programmes. The active participation of all actors who are engaged at the local level, such as international and regional institutions, national and local NGOs, religious groups, social movements, the media, research, and the education sector, should be emphasized (Dodgson et al. 2002). However, many cities in the Americas exclude a large portion of their inhabitants, namely the urban poor, from the process of decision-making as elites or interest groups dominate the process; therefore, it is particularly important to foster the participation of excluded groups such as the urban poor or women.

Strengthening Local Health Systems and Supporting Community Participation

In the late 1970s and early 1980s, recognition began that a strict focus on medical care and vertical health issue-specific programmes were not having the desired impact on reducing illnesses and improving health in the populations of Latin America (Cueto 2004). The 1974, Canadian Lalonde Report of Chap. 3 (named after the minister of health) deemphasized the importance of the quantity of medical institutions and instead proposed four determinants of health: biology, health services, environment and lifestyle (Canadian Department of National Health and Welfare 1974). In 1976, Dr. Halfden Mahler, Director General of WHO, proposed that the World Health Assembly adopt the goal of 'health for all by the year 2000', which laid the groundwork for the historic meeting in Alma-Ata, Soviet Union, in 1978. The subsequent Primary Health Care Declaration of Alma-Ata urged that essential health care be made universally accessible to individuals and families through their full participation in health maintenance, and at a cost that the community and country could afford (WHO 1978).

PHC's basic principles included community participation and empowerment in all aspects of decision-making, medical and health services responsive to the real needs of the population with a focus more on health promotion and illness prevention, recognition of the interrelationship between health and development, and decentralization down to the local level of political structures. Implied in this approach were the training and mobilization of lay and community health workers working with traditional healers and birth attendants. It also recognized the need to improve the environment and the conditions that would facilitate and maintain the health of the population. The latter, by its very nature, implied improving living conditions and promoting multisectoral action.

Unfortunately, in Latin America (as in other parts of the world), even after the Primary Health Care Declaration, primary health care was interpreted as the most basic level for delivery of medical care services, and mainly for remote areas and the poorest populations, although even recently graduated physicians did not want to work in remote areas, or in the poorer areas of the cities. Vertical issue-specific or disease-specific initiatives were developed to address malnutrition, dehydration in infants and children, exclusive breastfeeding for the first six months of life and immunization campaigns. However, in the mid- to late 1980s, the Pan American Health Organization (PAHO), WHO's regional office for the Americas, began an initiative focused on the strengthening of local health systems, with an eye to supporting the more holistic approach promoted by the initial primary health care approach. A key part of the strategy promoted bolstering community participation in the identification of their health issues and then working hand in hand with the formal and informal health systems to develop appropriate and effective strategies to address them. Training of national and local leaders took place in all countries of Latin America, with follow-up to support participatory local planning and the implementation of health promotion and illness prevention programmes.

Paulo Freire, a great source of inspiration for many in Latin America, recognized that hunger and poverty created major obstacles to people's ability to learn, be productive, and become and stay healthy (Freire 2005). His approach to education as a means to empower and liberate the poorest populations began to take hold in Latin America in the early 1980s, especially through churches that were striving to enable populations to have more voice at a time when repressive military dictatorships were the predominant forms of government in many of the countries. Freire's philosophy helped to support the community-based orientation within PAHO that eventually led to the development of healthy settings approaches, such as Health Promoting Schools, Healthy Municipalities, Cities and Communities, Healthy Markets, Healthy Hospitals, and healthy approaches to urban planning and development.

Health Promotion

Many recent publications reassert the relevance of the Ottawa Charter for Health Promotion (WHO 1986) in the twenty-first century (HPI 2011; IUHPE 2007; Nutbeam 2008; Rice 2012; Wallerstein et al. 2011). Influencing sectors other than health is a key stewardship role of health promotion and of the health sector in urban settings; however, it is impossible for the health sector alone to manage a rapidly changing environment. The Ottawa Charter offered a vision for addressing social determinants such as shelter, education, a stable ecosystem, sustainable resources and equity, which can only be achieved by working across sectors.

Health promotion offers clear guidance for people living in cities, and the institutions serving them, to enhance their ability to improve unhealthy conditions and modify health and social determinants in order to improve their quality of life, health and well-being. Many countries have incorporated health promotion concepts and strategies into their policies, practices and programmes to promote and protect the health of their populations (WHO 2013), and global initiatives such as the Millennium Development Goals, the Sustainable Development Goals and Primary Health Care have recognized that health promotion is a central strategy for improving health and equity (PAHO 2011b).

National and city governments, together with civil society, grassroots organizations and private entities, and supported by academic institutions and international organizations, can design and contribute to the implementation of measures that can bring about significant change in the social, economic and environmental conditions of the majority of the Region's population (PAHO 2007a). Health challenges still loom large on the global agenda, including meeting the Millennium and Sustainable Development Goals, but establishing equitable trade practices and improving environmental ecosystems, health promotion strategies put forth by the Ottawa Charter, have led to a vibrant Healthy Municipality, Cities and Communities Movement in Latin America.

The Healthy Municipalities, Cities and Communities (HMC) Movement in Latin America

Working through healthy settings, such as the Healthy Municipalities, Cities and Communities (HMC) Strategy, has been one of the more successful strategies for putting health promotion into practice in the countries of Latin America (PAHO 2011b). Based on the notion of health as quality of life, the actions of the HMC Strategy focus more on the underlying determinants of health than on their consequences in terms of disease. The ultimate goal is to promote the processes that enable people to improve their living conditions.

The five main pillars of the HMC Strategy are building healthy public policies, promoting social participation from all sectors and walks of life, creating sustainable initiatives, incorporating collaboration from multiple sectors and partners, and creating systems changes through process and outcome modifications. The seven essential elements of the HMC Strategy are building public commitment to improve quality of life; ensuring and strengthening community participation during the planning, implementation and evaluation phases; developing a strategic plan to overcome obstacles and threats; building consensus and forming partnerships through various networks and projects; encouraging leadership and participation of all social sectors including the health sector; formulating healthy public policies at the local, regional and national levels; and conducting ongoing monitoring and evaluation to track and assess the progress of the initiatives (PAHO 2004). An HMC strives to achieve a social pact among civil society organizations, institutions from various sectors, and local political authorities in order to carry out health promotion actions with a view to providing the population with a good quality of life. The key to establishing an HMC is often to achieve a change of attitudes, concepts and models for promoting health in the broadest sense, through changes in policies, legislation, and services usually provided by the municipality. The Strategy promotes inter-sectoral action in order to address the underlying determinants of health, which require measures that are mostly beyond the mandate of the health sector.

In Latin America, the implementation of the HMC Strategy was developed to fit region—and country—appropriate contexts by incorporating characteristics that complemented the democratization and decentralization processes, with local government expansions of fiscal and political authority, that were taking place during the major societal upheavals of the 1980s in the Region. As military dictatorships fell, many countries reconvened citizen assemblies to rewrite constitutions to include language related to health and democracy. Because the HMC Strategy emphasizes community participation and local government control, these coinciding trends were very supportive of the HMC's subsequent adoption and rapid expansion in Latin America.

Another facilitating factor for Latin America's acceptance of the HMC Strategy was the Region's familiarity with, and commitment to, many of the movement's tenets. By the time the Strategy was being promoted, Latin America already had been involved in various initiatives that focused on local solutions to health problems, equity in health services and community priority setting.

Although almost every country of the Region has some HMC initiatives, the Strategy has been consolidated in Latin America by working through national and subregional networks. Included among these networks are the National Network of Healthy Municipalities of Argentina, the Brazil Network of Potentially Healthy Municipalities, the Costa Rica Network of Healthy and Ecological Cantons, the Cuba Network of Municipalities for Health, the Mexico Network of Municipalities for Health and the Healthy Municipalities and Communities Network of Peru. At the regional level, the HMC Network of the Americas facilitates the exchange of experiences among countries as well as the strengthening of the HMC Strategy.

A major difference between the HMC Network in Latin America and the one that was developed and briefly preceded it in Europe is that for the Americas, PAHO/WHO did not establish criteria for defining a healthy initiative, municipality, city or community as part of the network. PAHO/WHO outlined potential steps that could be followed and essential elements that could be included in establishing an HMC (PAHO 2004), and produced guidelines, materials and training activities to facilitate its development, monitoring and evaluation (OPS 1993; Pan American Health Organization 1992, 1994, 2004, 2006a, b, c, 2007b, 2011b, c, Kansas University). Some countries (Argentina, Cuba, Mexico, Paraguay and Peru) adopted their own criteria and procedures for certifying whether or not a site would be classified as HMC and what it needed to do to maintain that classification; however, the decision about whether or not officially to designate a site as a Healthy Municipality, City or Community was left to the discretion of each country and city. WHO/EURO, on the other hand, limited the number of sites that could be part of the network; the WHO/EURO office decided which sites would enter and remain in the network and membership was limited by criteria and numbers of cities per country. The European initiatives also tended to be almost exclusively focused on very urbanized settings (Healthy Cities), whereas in the Americas the Strategy focused on the different political and administrative authorities that were most effective in promoting democratic decision-making appropriate to geographic, social and cultural contexts, extending well beyond strict urban administrative borders. For this reason, the initiative in Latin America included municipalities, cities and communities.

The Importance of Local Governments

The Latin American trend toward the decentralization of social policies and the distribution and administration of resources underscores the key role played by local authorities, especially mayors. In the majority of countries in Latin America, political, administrative and legal power at the local level rests with the municipalities, most of whose authorities are chosen by direct election. These municipalities have jurisdiction not only over urban areas (where local government headquarters are generally located), but also over semi-urban and rural areas.

Given their leadership role, mayors and other local authorities are increasingly helping to define policies and execute programmes intended for the benefit of the communities under their jurisdiction; they therefore constitute a focal point for the coordination of multisectoral actions among state organizations, nongovernmental organizations and communities. The tendency is to give municipal authorities a greater role, reflecting the processes of decentralization, delegation of powers, and strengthening of democracy. The HMC Strategy supports such processes and changes at the local level with its emphasis on healthy public policy, inter-sectoral collaboration and social participation.

Local government has a key role to play in creating a healthy community or municipality, but it cannot play that role alone. Regardless of their dominant area of concern—whether the environment, health, social activities, education, safety, public works or any other—community members are also responsible for improving the living conditions, health and quality of life of the people living in their community, and are participants in that process (PAHO 2004).

The Current State of HMC Networks in the Americas

In 2007, through a general survey distributed through the HMC online network and through the PAHO country offices, municipalities reported on the planning, implementation and evaluation phases of the Strategy (Arteaga et al. 2007; PAHO 2008). Ninety-three per cent of the municipalities indicated they had signed a written agreement of political commitment to the HMC Strategy. Sixty per cent of the surveyed municipalities reported having a local health and well-being situational analysis, 70 % of those surveyed had an HMC Action Plan and 58 % had a follow-up and evaluation plan. It was concluded that the activities in which municipalities were engaged under the banner of the HMC Strategy varied greatly in the Americas, with differing degrees of success. Inter-sectorial collaboration and evaluation continued to be a challenge for the HMC initiatives.

The Pan American Health Organization established a Network of the Americas in HMC, which served as an ongoing resource and up-to-date reference for all the countries of the Western Hemisphere. This network had many dynamic features including:

- A dynamic web page launched in September 2005, with country and municipality reports, database information, newsletters and current events, connections, contact information, discussion groups, registration sheet and links with other networks (FLACMA, FEMICA, IULA, WHO, Municipal Secretaries of Health).
- Two list serves (English and Spanish).
- A quarterly newsletter (English and Spanish/Portuguese).
- A database with up-to-date information on countries that could be electronically searched.
- Case studies of specific country experiences and products.

In the country examples section of this chapter you will read about a few of the national networks.

In 2008, 12 countries[3] participated in a regional HMC survey carried out by the Pan American Health Organization and ISALUD University of Argentina (Meresman et al. 2010; PAHO 2008). In comparing the relative frequency with which different topics were addressed within HMCs, the regional survey showed lifestyle improvements as the highest priority: promotion of healthy eating habits, physical activity, and prevention of alcohol, tobacco, and drug abuse. Actions relating to the improvement of programmes and services came in second, and included vaccination campaigns, prenatal and perinatal care, prevention of teenage pregnancy, and prevention of sexually transmitted diseases. Environmental health actions in municipalities also appeared among those most frequently mentioned, including vector control, water and sanitation, waste management, and environmental quality (Table 8.1).

Since most countries demonstrated synergies between actions that have an impact on risk factors and lifestyles and others directed at social participation and having an impact on environmental and social factors, all of which are recognized as determinants of the main health problems, it seems clear that these two conceptually divergent trends are, in practice, complementary.

The results of the regional survey showed a need to strengthen the identity of HMCs and to update the conceptual framework and practical tools. The high degree of variability in the responses obtained relative to the 'most frequent spheres of implementation', 'topics addressed' and 'beneficiary population of the actions carried out' reflected the wide variety of versions of HMCs that have grown up during the last decade. This diversity in approaches may be seen as a strength, attributable to the need to adapt to new scenarios in the health environment and local government, and to changes in the demand for an offering of services, the transformation

[3] Argentina, Brazil, Canada, Colombia, Chile, Costa Rica, Cuba, Honduras Mexico, Paraguay, Peru and Uruguay.

Table 8.1 Comparison of the relative frequency of topics addressed by HMCs (Meresman et al. 2010)

Main topics addressed	Percentage with high and very high frequency (%)
Lifestyle improvement	
Promotion of healthy eating habits	73
Physical activity	70.5
Prevention of alcohol, tobacco and other drug abuse	70
Improvement of programmes and services	
Vaccination campaigns	75
Prevention of sexually transmitted diseases	61.8
Prenatal and perinatal care	53
Prevention of teenage pregnancy	51
Environmental health	
Vector control	67
Water and sanitation	60
Waste management	55.5
Environmental quality	51

of the configuration of players and sectors, and changes in institutional dynamics and management of public policies.

Most importantly, the identity of HMCs was not defined so much by their results as by their processes. The essential features of the methodologies and strategies applied (political commitment, community participation and inter-sectorial collaboration) remained constant, while the scenarios in which they operated, the topics addressed and populations with which they interacted were highly variable and diverse.

The survey responses identified key factors to strengthening HMC implementation processes: (a) institutionalizing the strategy with technical, political and legal tools, providing stability for the HMCs and their technical teams; (b) making available a legal–institutional framework that favoured multisectorial and multi-player convergence; (c) establishing multi-player and multi-institutional management councils; (d) establishing inter-institutional consultation instruments and abilities, and flexible, stable, integrated management and funding tools; (e) providing permanently assigned staff for collaborative work; (f) strengthening coordination and management competences and (g) making available integrated financing for collaborative projects. In these and other studies, common factors that contributed to success in social participation included increasing the use of participatory instruments (assessments, evaluation, participatory budgeting, etc.), establishing intersectoral committees and increasing the participation of civil society organizations in local projects and initiatives (Rice and Franceschini 2007; Franceschini et al. 2010). The use of virtual forums and networks was one of the most promising emerging strategies for channelling these participatory and collaborative processes (Meresman et al. 2010).

Some obstacles to effectively implementing the HMC strategies were political timing (elections and political transitions often led to delays or termination of initiatives); continual transfer and shifting of staff and financing; resistance by other sectors to collaboration with the health sector; lack of specific indicators to measure the HMC processes; excessive bureaucratic procedures; little technical capacity and awareness to use participatory methodologies; and the length of time needed to achieve measurable changes in promotion and prevention initiatives (PAHO 2011b; Rice and Franceschini 2007, 2008). Potential conflicts in the implementation process can be inferred from the need expressed for greater collaboration through networks, with some coordination at the central level, although with encouragement to operate independently. The challenge that arises from the tension between centralized coordination and decentralized management offers an opportunity to stimulate new technical interactions (PAHO 2008) and establish necessary competences at each level, making them complementary with one another. This favours both individualized processes, in keeping with local needs and available capacities, and strengthened national and regional HMC networks to improve opportunities for the exchange of learning experiences and resources.

Best practice recommendations from the survey supported the development of alliances at the local level, with strategic alliances at the central level as a requirement for long-term sustainability.

There was consensus regarding the progress and expansion of opportunities for social participation, with specific mention of valuable experiences such as participatory budgeting (PAHO 2008), the creation of regional advisory councils and the growing number of independent projects conducted by civic organizations. The integration of 'virtual' participatory forums and networks and e-government models were mentioned as emerging strategies that could be used to jump-start collaboration.

Finally, the larger Latin American HMC Networks were seen as critical for improving health and quality of life conditions based on equity. They have taken advantage of opportunities to integrate local authorities and sectors, and can position health promotion on the political agenda of its member countries to, in turn, contribute to local sustainable development and improvements in the social determinants of health, including urbanization.

Tools for Promoting Healthy Cities, Municipalities and Communities

- *Participatory Evaluation Guide for Healthy Municipalities, Cities and Communities* (PAHO 2006b): This guide, available in English and Spanish, provides guidance and tools for HMCs to evaluate their own efforts and contribute to the evidence base of the Strategy's effectiveness. It presents an evaluation framework that incorporates essential health promotion values and orients users through a step-by-step process to design and implement a participatory evaluation of their HMC initiatives. It has an extensive list of possible indicators for measuring and evaluating myriad aspects of the HMC Strategy. (http://www.bvsde.paho.org/bvsdemu/fulltext/guiaevaleng/guiaevaleng.html).

- *Healthy Municipalities, Cities and Communities: Evaluation Recommendations for Policymakers in the Americas* (PAHO 2005): This guide describes the importance of health promotion for political leadership and provides five recommendations for evaluation for policymakers and five for practitioners. For policymakers it suggests that they recognize that health promotion is a fundamental plan for pursuing equity and better health for all; that evaluation is part of all health promotion initiatives; that infrastructure must be developed to support health promotion evaluation; that collaboration among government sectors, private and public sectors, universities and non-governmental partners should be encouraged and supported; and that decision-making should take into consideration the results of health promotion evaluations. For practitioners it suggests that they adopt participatory approaches to health promotion evaluation; support the use of multiple methods for process and impact evaluation of health promotion initiatives; establish a fund for health promotion effectiveness evaluation using ten per cent of the total financial resources for health promotion initiatives; support adequate infrastructure for ongoing training and education in health promotion evaluation; and support applied research in health promotion evaluation. References to other tools available on the internet are provided.
- *Guide to Economic Evaluation of Health Promotion* (PAHO 2007b): This Guide, available in English and Spanish, provides a step-by-step guide on how to conduct an economic evaluation of health promotion interventions. It is written especially for health promotion practitioners and policymakers and assumes no prior knowledge of health economics. It aims at introducing non-economists to some of the language, tools and insights required to ensure the fidelity of the evaluation. The guide is also useful to those in charge of financing, implementing and evaluating health promotion interventions or to people interested in advocating health promotion strategies, by giving them access to the empirical literature on economic evaluation (http://www.bvsde.paho.org/bvsacd/cd65/finalecoeva.pdf).
- *Urban HEART and Health Impact Assessment* (WHO 2010; OMS 2010): The Urban Health Equity Assessment and Response Tool (Urban HEART) is a user-friendly guide for policymakers at local and national levels to address health inequities in cities. It consists of two key components:
  - Assessment, which analyses health outcomes and health determinants, grouped into four policy domains: physical environment and infrastructure, social and human development, economics and governance

  - Response, which identifies interventions and strategies for action from a list of best-practice interventions. While interventions may be modified to address the specifics of local contexts, the tool provides the basis for prioritizing appropriate interventions.

- The Health Impact Assessment (HIA) tool supports assessment of the health impacts of policies, plans and projects in diverse economic sectors using quantitative, qualitative and participatory techniques. HIA helps decision makers make choices about alternatives and improvements to prevent disease/injury and to actively promote health (see Chap. 18).
- In May 2011, a workshop was held in Bogota, Colombia, for 21 countries and 23 cities from Latin America and the Caribbean to train eighty participants in Urban HEART and Health Impact Assessment/Health Equity Impact Assessment (HIA/HEIA).[4]
- *Guide for Documenting Health Promotion Initiatives*: This Guide, available in English and Spanish, proposes a simple, standard format to support the documentation of health promotion initiatives. It is composed of a seven-part questionnaire and is accompanied by detailed guidelines on how to complete it. An adapted format of the guide was used as a basis for the first Hispano-American and inter-American contest of best practices in urbanism and health (http://www1.paho.org/hq/dmdocuments/2010/Bases_del_concurso_eng.doc). The winners of this contest are highlighted in the box in the case studies section below. A PAHO review of 15 health promotion case studies identified the mechanisms, processes and tools used to support health promotion activities in the region (PAHO/WHO 2011c). The key characteristics that supported health promotion included committed political leadership in key levels of government; initiatives and frameworks focused on health promotion and related subjects; strong political and legal frameworks, particularly at the national level; a high level of community organization and participation; attention to cultural and social norms; and the presence of local authorities and those responsible for policies, fully integrated into the process. The review also emphasized the central role of inter-sectoral collaboration, community participation, and decision-making power for planning, implementation and sustainability of initiatives.
- *Social Participation and Empowerment: Methodologies for Health Promotion* (Wallerstein et al. 2008). This train-the-trainers 40-h curriculum provides skill-building, team activities, case studies and strategic planning exercises for strengthening empowerment and community capacity-building within the health promotion and Healthy City movements. Modules include introduction and team-building; the social-political context of health promotion; health promotion

[4] Participants came from the following countries and cities: Argentina, Belize, Bolivia, Brazil, Chile, Colombia (Bogota, Cali and Medellin), Costa Rica, Cuba, Dominican Republic, El Salvador, Guatemala, Honduras, Guyana, Jamaica, Mexico, Nicaragua, Paraguay, Peru, Surinam, Trinidad and Tobago, and Uruguay. See the case study of Bosa, City of Bogota, Colombia, later in this chapter.

strategies; social participation timeline and community diagnosis; strategies for empowerment (including the methodology of Paulo Freire, theories of power and empowerment, working within diverse communities, negotiation and alliance-building, conflict-resolution, and health communication); and participatory research, participatory evaluation of healthy communities, and measuring effectiveness of participation for improving health. The curriculum is available in Spanish, Portuguese and English and is appropriate for university courses (with suggested literature) and training for health promotion professionals or community staff within Ministries or secretariats of health and non-governmental organizations (NGOs), and adaptable for community training(http://fcm.unm.edu/cpr/empowerment-curriculum.html).
- *Herramientas de comunicación para el desarrollo de entornos saludables* (Communication Tools for Developing Healthy Settings) (OPS 2006): This book, available only in Spanish, starts with a conceptual framework for putting health communications within the context of health promotion and the use of the settings approach (municipalities, schools, work places, etc.). It offers communication strategies to be used in this context, supplemented with communication methods and materials. It includes a section on how to evaluate communication as a means to develop and enrich healthy settings, accompanied by the theoretical models most used to promote changes in health behaviour.

These PAHO-supported tools have been used throughout the region to support mayoral, inter-sectoral and community capacity-building in the creation of healthy municipality, city and community initiatives that address the urbanization challenges facing Latin America. Tools that provide measurement and evaluation indicators in particular offer local processes and data for strategic planning and development of programmes, interventions, or policies to address specific health issues or inequities.

While evaluation of health promotion and HMC effectiveness in Latin American urban settings is beyond the scope of this chapter, it is worth mentioning that the interest in 'evidence' has been increasing, in books (Franceschini et al. 2010; McQueen and Jones 2007; Round et al. 2005; Tones and Tilford 2001; WHO 2001), conferences (Westphall and Wallerstein 2007), articles, and the creation by WHO of the Global Programme on Health Promotion Effectiveness (http://www.who.int/healthpromotion/areas/gphpe/en/). Evaluation of empowerment as a core component of community participation has also been growing (Kansas University; Kasmel and Tanggaard 2011; Laverack 2006; Peterson et al. 2006; Wallerstein 2006). The challenges are multiple, as cities and communities are complex dynamic systems and multi-level and multi-component interventions are needed to address major social determinants.

Determining the effectiveness of interventions remains critical, and the importance of uncovering lessons learned can build a literature and learning community of best practices. The next section provides some examples of case studies from the field, to showcase challenges to and facilitators for implementing HMC strategies and value-based urban planning.

Examples of Successful Initiatives in Latin America

Regional Programme of Action and Demonstration of Sustainable Alternatives to DDT for Malaria Vector Control in Mexico and Central America: municipal governments as strategic health assets (PAHO 2011b)

From 2003 to 2008, the DDT-UNEP-GEF[5] Project, carried out in Mexico and Central America, consisted of developing pilot programmes using an alternative model of comprehensive control of the malaria vector without the use of DDT, based on a successful Mexican experience and adapted to specific ecosystems. In 2002, a 63 % reduction in malaria incidence was seen in demonstration sites with high rates of malaria transmission, accompanied by an 86.2 % reduction in *Plasmodium falciparum* cases. In addition, institutional capacities were developed in these countries for control of the disease without the use of any type of insecticide, through a sustained programme of epidemiological and entomological surveillance, social participation, participatory planning and environmental management, primarily using an environmental care approach.

The key strategy was to enable participation by the community and municipal government. This was the first time in the subregion that municipal governments were successfully integrated into local activities to combat malaria, a responsibility traditionally considered to be under the purview of the Ministry of Health. The participating governments contributed by financing important infrastructure such as bridges, basic sanitation systems and reconstructing river banks; they provided materials, supplies and personnel to assist in community cleaning brigades, created permanent committees or staff positions (with adequate resources) to address issues related to malaria, established and enforced policies aimed at improving environmental management (such as proper solid waste disposal), and endorsed and promoted this model at national and international levels.

Another great achievement of this project was the identification and training of community leaders to serve as links between the community and the project's technical personnel. Community leaders assisted in the coordination of local activities, which in turn resulted in an increase of up to 63 % of community health agents in the pilot communities.

In this project municipal governments demonstrated their capacity to act as agents of change. They achieved this through the development and implementation of public policies and innovative management mechanisms that produced sustainable changes in the social, cultural and physical structures of their communities in order to prevent and control malaria. They successfully mobilized actors, sectors and resources, which resulted in better coordination of activities and more rational use of resources.

The project resulted in increased knowledge and improved skills among the population relating to the malaria-vector lifecycle and its control. The population

[5] DDT: an insecticide called Dichlorodiphenyltrichloroethane; UNEP: United Nations Environmental Program; GEF: German Aid for International Development.

demonstrated improved community environmental management (e.g. proper waste disposal), changes in attitudes and behaviours (improved personal hygiene), greater sense of responsibility about their and their families' health (e.g. keeping their properties clean), and less dependence on the public sector for the implementation of malaria control strategies (e.g. organizing and participating in cleaning brigades independent of health department technical teams).

This experience demonstrated that municipal governments can play a key role in the implementation of health promotion strategies. They are in a privileged position to act upon a variety of factors and levels, and to create the appropriate settings for the successful implementation and sustainability of such initiatives. They are also able to place health and health promotion on the local political agenda and to generate momentum for the discussion and resolution of community issues and problems without the creation of new or parallel structures. This indicates that municipal governments can be an important health asset, and that their incorporation into health promotion initiatives can be an effective and sustainable strategy.

Argentina, Territoriality and Health Promotion the Experience of the Healthy Municipalities Network[6]

Introduction

The process of implementing the Healthy Municipalities Strategy and, more recently, the Urban Health Strategy (PAHO/WHO 2010) is understood as part of the development of health promotion (HP) and health policies in Argentina (Mercer and Fernández 2010). An analysis of the development of health promotion in the Southern Cone countries (Argentina, Chile and Uruguay) shows that different paths were followed in accordance with each country's institutional logic, the priorities assigned by local decision makers, and the entry points for the formulation of health policies. In Argentina health promotion has achieved increased prominence in the health discourse as a result of the acknowledgement of its contribution to improving the health of the population.

The five principles outlined in the Ottawa Charter have been incorporated asymmetrically in Argentina. Initially, efforts aimed at reorienting health services and systems and promoting healthy lifestyles and settings (municipalities, schools and workplaces) gained more support in communication activities and within proposals as they were developed. Advances were also seen in the development of policies that regulated activities that affected health, in both legislative (enforcement) and normative spheres. Community participation and inter-sectoral collaboration did not appear as tangible goals at that time; possible explanations include the political context, in which successive interruptions to democratic processes by military dictatorships crushed any attempts at participatory and social mobilization.

[6] Raul Mercer and Karina Cimmino.

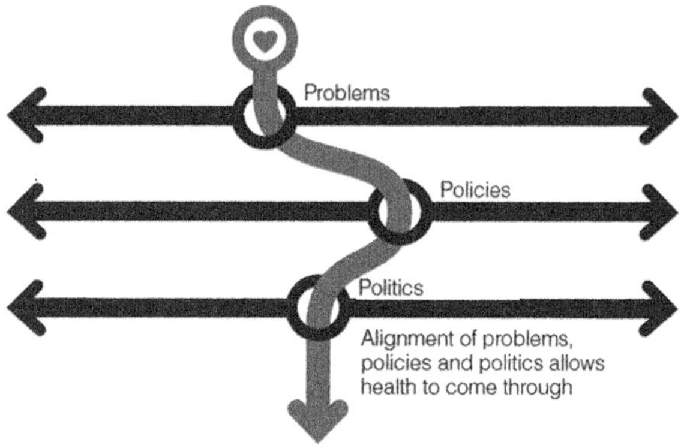

Fig. 8.1 Framework of non-linear processes in policy formulation (Kingdon)

The goal of this case study is to analyse the current development of the Argentina Healthy Municipalities and Communities Network (PNMCS, its acronym in Spanish), along with its historical development, the logic used in planning activities, the institutional support provided, and the sustainability mechanisms put into place. The case will close with a series of brief reflections on how some aspects of health promotion provide an opportunity for future-oriented strategic thinking and an opportunity to explore various ways of promoting health within a territory, along with thoughts about the scope and effectiveness of the Healthy Municipalities and Communities Network in addressing local issues related to implementation processes.

Entry Points for Healthy Municipalities Programmes

The development of the PNMCS in Argentina provides a clear example of how health promotion consolidated its identity based on a proposal centred on the notion of territories: the municipality as a space for transformation.

Kingdon's (2011) approach helps to better clarify the processes related to the development of health policies based on territories. According to the author, three main currents merge during policy formulation: problems, what should be done (through public policies) and what happens in reality (politics). These elements play a role in opening windows of opportunity for decision-making (Fig. 8.1).

First the 'problem' needs to be acknowledged by politicians, policymakers and the community in order to become part of the political agenda. The catalyst can be some kind of crisis or an issue raised by the media; or it might be the contribution of an agency such as WHO, as in the case of chronic non-communicable diseases. Second, proposals for the solution are required (public policies). These policies should be technically feasible as well as culturally, ethically and financially reasonable. Third, the political climate must be adequate. The search for these opportunities to insert an issue can be crucial to making a proposal viable (for example, pre-electoral periods, parliamentary debates and crisis).

Background

Historical landmarks help to explain the roots of the current HMC programme as well as to comprehend the trajectory and complexity of its processes.

According to a 2010 PAHO report aimed at evaluating health promotion capacities in Argentina (Mercer and Fernández 2010), the development of health promotion in the country is recent. Historically the area of care has had greater visibility, while prevention and promotion were neglected. In the last 10 years, a push to include health promotion took place, thanks to a rise in interest in territory and health (Healthy Municipalities), the development of policies, and the promotion of healthy lifestyles (Mercer and Fernández 2010). The most significant landmarks include:

- **1992**. The Ministry of Health, with the support of PAHO/WHO, adopts the Structural Programme of Environmental Protection, beginning with the Healthy Municipalities and Schools Programmes.
- **1997**. The Boca del Río Agreement is signed, creating the Latin American Association of National Healthy Municipalities and Communities Networks.
- **1998**. The SIMBEL Project (Municipal Holistic Health for Berisso, Ensenada and La Plata) is launched. This is the first proposal for the development of Healthy Municipalities from the perspective of the municipalities themselves.
- **1999**. The Medellin Declaration for better quality of life for the inhabitants of the Americas is signed.
- **2000**. The municipality of Gualeguaychú (Entre Ríos Province) requests support from PAHO/WHO to launch a Healthy Municipality initiative. The Ministry of Health spearheads the first meeting of Healthy Municipalities.
- **2001**. Guidelines for the National Plan for Health Promotion and the Prevention of Risks and Diseases are developed. The Tucumán Charter is signed as part of the Second National Health Promotion Forum. From this date the Healthy Municipalities and Communities strategy starts to take hold among local governments as the national government strengthens commitment to the development of healthy public policies through the implementation of local health promotion projects. This strategy expands through the Municipalities' Network, strengthening the leadership role of the national level in the process.
- **2002**. The fourth meeting of the Municipal Health Secretaries of the Americas is held in Rosario, Santa Fe Province.
- **2003**. The National Healthy Municipalities Meeting is convened on the topic of the challenges of local participatory management.
- **2004**. Argentina, through its Ministry of Health, joins the HMC initiative spearheaded by WHO and PAHO. The coordination unit of HMC was created by a ministerial resolution (PAHO/WHO and Ministry of Health of Argentina 2004).

This decision resulted in significant progress in the implementation of the HMC strategy by generating various initiatives aimed particularly at the promotion of healthy lifestyles. By December 2007, 387 municipalities had joined the network. By 2015, membership had increased to 1052 municipalities. From September 2013 to December 2015, 52 municipalities have been accredited through an external evaluation.[7]

This resulted in an important territorial extension of the network and a consolidation of the Healthy Municipalities and Communities identity. In the words of the coordinator, Laura Waynstok: 'We had been working on the Coordination Unit of the network, which was comprised of municipalities that represented their provinces or regions. The work of the Coordination Unit, at the time, had become very bureaucratic and focused on approving and accepting the documentation of the municipalities that applied for membership. At this moment, as the municipalities themselves started to question who would decide which municipalities would be part of the network, we geared more towards the issue of evaluation, indicators, and what to measure, and that took a long time to be defined'.

In line with more recent theoretical and political developments at the time, a new department was created within the Ministry of Health: the Secretariat of Health Determinants and Health Relations. Acknowledging the strategic nature of the local level in the design and management of public policies that bring together sectors, programmes and jurisdictions, and that are based on the needs and demands of each territory, the National Healthy Municipalities and Communities Programme was created (website of the Ministry of Health of Argentina).

- **2008**. Creation of the National Healthy Municipalities and Communities Programme

The programme included the network and helped to catalyse its functioning and the achievement of results. The words of the PNMCS Coordinator illustrate this statement: 'the municipalities themselves started to promote the Programme and to discuss the logic of the Network. As a result, the Program was created within the Ministry. The Program includes the network, which continues to meet annually and to carry out its own work, and the Program functions through the format of a network'.

Electronic newsletters reporting on local experiences are launched. As of 2013, 150 newsletters have been published (website of Ministry of Health of Argentina).

- **2009**. The World Fair of Municipalities and Health: Rights, Citizenship and Local Integrated Management for Development is held. This is the instigator of the virtual classroom as a management tool.
- **2010**. The Ninth National Meeting of Healthy Municipalities is held in Costa Salguero Centre, Buenos Aires.
- **2013**. The Twelfth National Meeting of Healthy Municipalities is held in Tecnópolis, Buenos Aires.

Six municipalities are accredited as Responsible Municipalities (Ministry of Health of Argentina 2013).

– Following its designation as a national programme, the Strategy has increased its resources and improved local management tools such as local health situational

[7]PAHO Argentina Web Page, January 2016: http://www.paho.org/arg/index.php?option=com_content&view=article&id=9961:una-veintena-de-localidades-acreditaron-en-2015-como-municipios-responsables-en-salud-&catid=333:arg03-salud-familiar-y-comunitaria&Itemid=512.

analyses, situation rooms and self-evaluations, leading to a strengthening of quantitative and qualitative information about the Strategy throughout the country.

– According to the PNMCS Coordinator, one of the conditions that facilitated the growth of the programme was the presence of a clear political decision:

The political decision that the work at the municipal level is fundamental to addressing health at the level of the territory has resulted in more availability of human resources to finance projects in the municipalities and to advance the strategies.

Strategic Orientations

The National Healthy Municipalities and Communities Programme defined the following strategic orientations:

(a) Holistic approach to the health determinants and conditioning factors (Fig. 8.2)

- Through this approach the programme seeks to identify inequity gaps, understand the incidence of the various factors in each of the dimensions described in Fig. 8.2, design public policies and promote the inter-sectoral collaborations needed to address and reduce inequities.

Source: Directrices del Programa Nacional Municipios y Comunidades Saludables Ministerio de Salud (Argentina), Ministerio de salud de la Nación Argentina (http://www.msal.gob.ar/municipios/images/stories/4-recursos/pdf/2013-09_institucional.pdf)

(b) Inter-sectoral development and strengthening

Within the framework of the strategy adopted by the national government, inter-sectoral and inter-jurisdiction actions are considered key. The complex configuration of the public health subsector in the various provinces, and the diversity among municipalities, pose challenges to national public policy in general and to the PNMCS in particular as they relate to the development of tools that, while being homogeneous to all jurisdictions, can also be adapted to the specificity of each territory.

(c) Development of institutional capacities

The decision to intervene, based on a complex and holistic approach to health determinants and their conditioning factors, requires the presence of institutions capable of addressing problems that arise at the local level. This approach requires a continuous and systematic process of validation and subsequent accreditation based on the improvements achieved.

Lines of Action

In order to orient and strengthen the approach of health determinants and their conditioning factors, the Ministry of Health defined various lines of work (Fig. 8.3).

Fig. 8.2 Social determinants of health

Life Cycle of the Process: From Membership to Accreditation

The development of the programme comprises four phases, defined as a process of inter-jurisdictional and inter-sectoral management through which the capacities of the municipalities to design and manage public policies are strengthened, resulting in improvements in the population's health (Fig. 8.4).

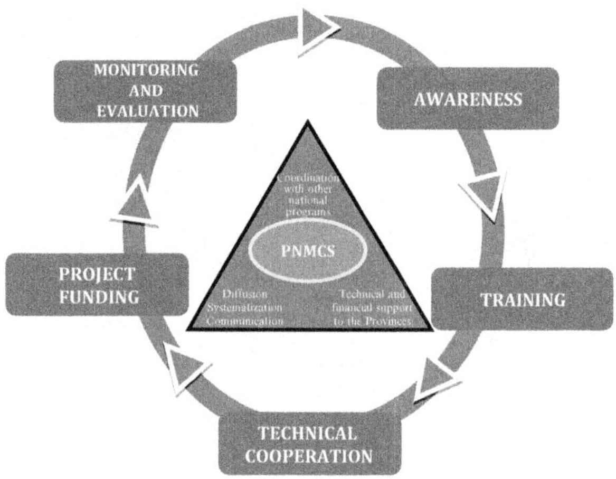

Fig. 8.3 Lines of action of the programme. Source: Programa Nacional Municipios y Comunidades Saludables Ministerio de Salud (Argentina)

Municipality **Associated**	Municipality **Full member**	Municipality **Responsible**	Municipality **Healthy**
Requirements	**Requirements**	**Requirements**	**Requirements**
•Submission of notice of intent to the Mayor • Presentation of a project by social organizations, public or private	• Training of technical teams • Designation of a municipal counterpart • Establishment of an intersectoral table	• Implementation of projects • Fulfillment of process goals • Positive external evaluation	• Fulfillment of result goals • Positive external evaluation
Entry	**Consolidation**	**Accreditation 1**	**Accreditation 2**

Fig. 8.4 Lines of action of the programme. Source: Programa Nacional Municipios y Comunidades Saludables Ministerio de Salud (Argentina)

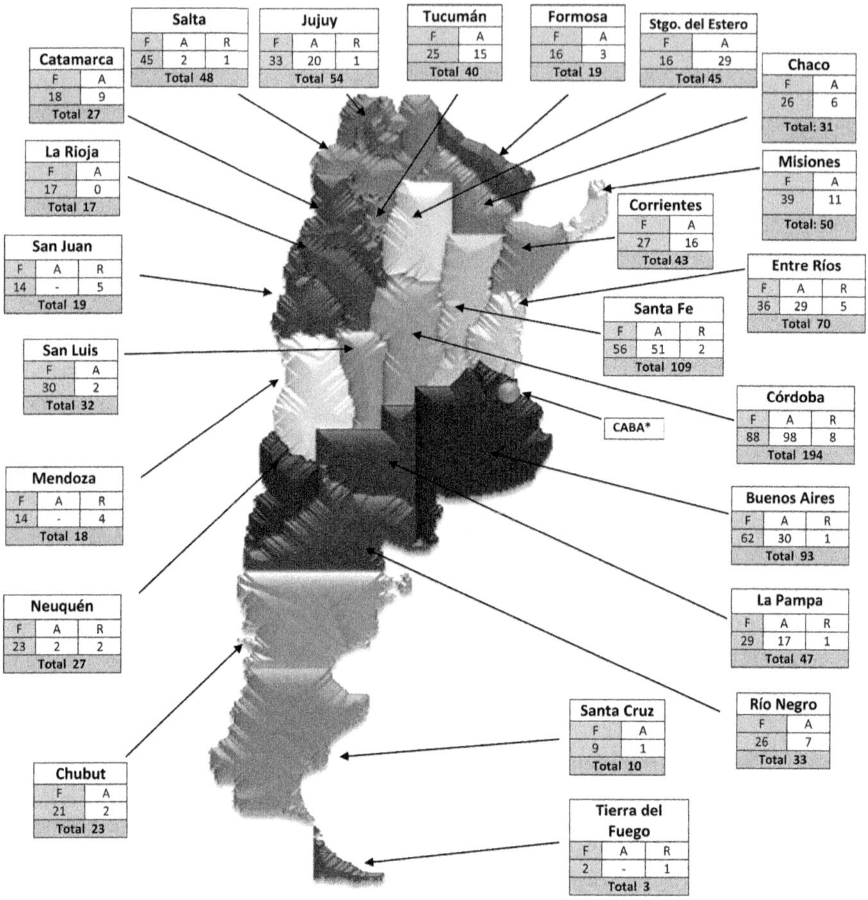

Fig. 8.5 Network of Healthy Municipalities in Argentina (2015) ($n = 1052$). *References*: F (Full Member), A (Associated), R (Responsible). *Source*: Programa Nacional Municipios y Comunidades Saludables Ministerio de Salud (Argentina). *Notes*: CABA (Buenos Aires City): although it does not appear in the list of the MOH, this district is developing experiences related to urban health. *Islas Malvinas*: no present experiences known as this territory is under the occupation of the UK

1. Phases. Following the path from an 'associated municipality' to a 'healthy municipality' implies advancing in the quality of the processes and development of institutional capacities at local level, as well as in the sustainability of holistic and equitable public health policies. The framework proposes a developing system of 'membership' in the programme which promotes the sustainability of the advances achieved.
2. Categories of Municipalities (Fig. 8.5) (Ministry of Health of Argentina 2013). The programme's activities seek to increase public action through an integral approach to the determinants of health, improving sustainability and inter-jurisdictional relations. This process is formalized through the phases of membership, consolidation and accreditation.

Associated municipality: A municipality that formally manifests its interest in starting to actively work from the perspective of the determinants of health in order to improve the health of its community.

Full-member municipality: A municipality that designates a counterpart, establishes an inter-sectoral committee, and has received training in the basic contents of the programme. Once this stage has been reached, the municipality is ready to sign the Territorial Agreement for Collective Health and other specific protocols.

Responsible municipality: A municipality which, based on systematically and agreed-upon knowledge about health problems as they relate to the four health determinants and conditioning factors, has started to develop projects and demonstrate positive changes in the process indicators established by the programme.

Healthy municipality: A municipality that has achieved results in the modification of factors that negatively determine and condition the collective health status in each of the four variables, and that has reduced inequity gaps according to the results indicators established by the programme.

Moving from the first category (associated municipality) to the second (full-member municipality) is often easier and faster than moving to the third category (responsible municipality), because moving up requires greater efforts, provision of resources (human and economic), and political and institutional commitment from the municipality. The factor that assists municipalities to advance is strong political and institutional support (political will from authorities and commitment to the programme's approach) that is continuous, transcending political and party administrations/periods. As such, political will should be translated into the availability of human and material resources. The main challenge to advancing up the ladder is weak or non-continuous political and institutional support.

Figure 8.5 maps the number of municipalities assigned to the various entry-level categories (Associated [A], Full-members [T]) and Responsible [R], according to provinces in Argentina (2015).

Virtual Classroom[8]

The virtual classroom is a local management tool created in 2009 by the PNMCS in conjunction with the National Institute of Public Administration (INAP, for its acronym in Spanish), using the Moodle platform. The use of ICT (Information and

[8]Ministerio de Salud. Programa Municipios y Comunidades Saludables. Aula Virtual. http://www.msal.gov.ar/municipios/index.php?option=com_contentandview=articleandid=349:x-jornadas-nacionales-de-municipios-y-comunidadesandcatid=7:destacados-separados349.

Communications Technologies) strengthens specialized communication channels and helps to promote the work of the inter-sectoral committees by providing a permanent space for capacity-building and technical assistance. It strengthens the work in other settings through the provision of information and exchange of ideas based upon field experience, and it promotes horizontal exchange among municipalities within the same province and with other provinces, with the intention of promoting a community of practice.

According to the PNMCS coordinator, the programme appeared at a time in which the use of technology was peaking, and was able to take advantage of this to maximize its results:

> The issue of social networks has greatly evolved. Technology supports the functioning of the network and this promotes exchanges, horizontal collaboration, online capacity building, etc. (Laura Waynstok, PNMCS coordinator).

The establishment of municipal teams in virtual classrooms allows them to receive permanent technical support from the national PNMCS team to advance through the various phases of the programme, and facilitates their continuous capacity-building. Through the virtual classrooms, specialized technical support is provided for the development of local situational health analyses, situation rooms and self-evaluations, which are all requisites for accreditation as a responsible municipality.

Knowledge Generation: Research

The research component is one of the most significant supports of health policies. The Ministry of Health has developed a policy to promote public health research through the National Commission for Health Research[9]. Through this initiative, in 2002, the Ministry of Health established research priorities subsidies and scholarships for individual and collaborative projects. Although the majority of resources were assigned to problems related to health care or health research, some projects were oriented towards health promotion and the social determinants of health (Salgado et al. 2011).

Reflections on the PNMCS

Health Promotion as an Opportunity for Dialogue

Health promotion is a paradoxical field when considering the importance assigned to it by international cooperation agencies like WHO and PAHO and the demands of their member states. While there is a decreasing tendency to provide support for

[9] Ministerio de Salud. Comisión Nacional Salud Investiga. http://www.saludinvestiga.org.ar/.

institution-building in the form of financial resources, technical personnel and cooperation resources on the part of the agencies, there is growing interest and mobilization on the part of countries to improve and update their health promotion proposals. There have been great advances since the Alma-Ata Declaration ('Health for All', WHO 1978), the Ottawa Charter ('Inter-sectorality and Health', WHO 1986), and HiAP by the EU presidency under Finland (Leppo et al. 2013); and the contributions of an HiAP focus explicitly on the reduction of inequities within the framework of inter-sectoral collaboration.

Regarding PAHO/WHO cooperation in Argentina, the acting representative in Argentina in 2013, Luis Codina, affirmed that

> The role of PAHO–Argentina in relation to the PNMCS is to give continuity to a process that dates back many years. As with many processes, it has its ups and its downs, moments of great advances and others of decline. It depends on what types of priorities are assigned, not only technically but politically, since in reality, it is a political movement, related to health policies and that incorporates other sectors. PAHO has supported this interest in highlighting the issue of municipalities in health policies. Among the health priorities of the Ministry this is not clearly articulated, but I believe it has created an interesting process in the municipalities, making it possible to understand health from an integrated perspective, more holistic and with greater participation.

Approaches to Addressing Health Within a Territory (Fig. 8.6)

The PNMCS is a strategy that has gained relevance in the country and which has continued to expand and consolidate. The programme supports important local management elements and helps to incorporate health into all municipal policies. The proposal to conduct local situational assessments, using indicators for all the determinants of health, highlights and generates relevant data at the municipal level. At the same time, the establishment of situation rooms[10] in municipalities helps to assert the relevance of the tools developed by the programme to improve local management. Currently in those municipalities with a situation room, data is becoming valued as a tool for decision-making and the development of policies and pro-

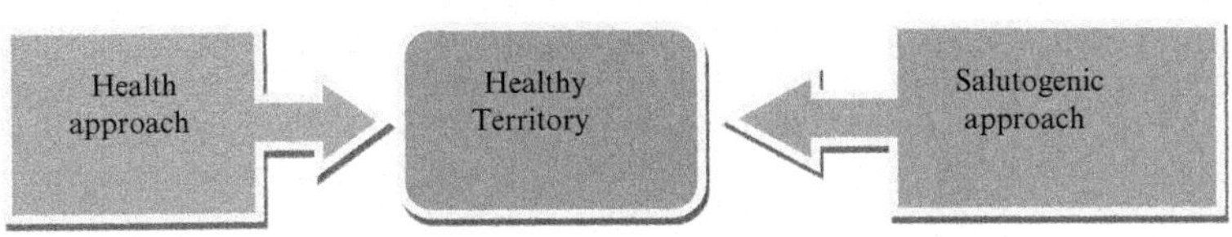

Fig. 8.6 Approaches to health within a territory

[10] The Local Health Situation Room is a physical and virtual space where the inter-sectoral committee analyses information to characterize the population's health situation and to determine action. The information, presented and disseminated through tables, graphs, maps, technical documents and strategic reports, supports evidence-based decision-making and promote democratization; as such, the room works as a tool for institutional management.

grammes. The accreditation processes promoted by the programme since 2012, which incorporate self- and external evaluations, have encouraged municipalities to improve documentation of their actions, to orient their actions, and to develop a baseline from which to determine the impact of programmes aimed at addressing health inequities within their territories.

A revision of published HMC experiences in Argentina indicates that in some cases the approach is still biased towards risk prevention more than health promotion. Similarly, a focus on health services still persists. Access to health services are connected to the environment (settings), lifestyles (nutrition, road safety, physical activity and tobacco) and health services. An analysis conducted in ten countries of the region (Meresman et al. 2010) demonstrated that the topics most frequently addressed under the framework of Healthy Municipalities and Communities were related to the population's lifestyle (healthy eating habits; physical activity; prevention of alcohol, tobacco and illicit drug consumption). In second place were programmes and services related to vaccination campaigns, perinatal health and the prevention of sexually transmitted diseases. Environmental issues were related to vector control, environmental health (water and solid waste management) and environmental quality.

It may be said that approaches based on social assets and which adopt a salutogenic approach (Lindström and Eriksson 2010) are yet to be considered within territorial contexts. Issues such as social capital, rights, social assets and community resilience are not perceived as tangible perhaps a carry-over from the dominant disease prevention model and its focus on risk reduction; however, they may also be interpreted through the growing visibility of salutogenic approaches in public health and the need to better understand the implications of this paradigm in addressing the production of health as a social good.[11]

Results

Some of the main results of the Programme, according to the key stakeholders interviewed, include:

> the ownership of the health issue by the municipalities … observed in the discourse. It is observed in the workshops how the discourse changes among the participants and mayors. It is difficult, but we have seen a growing participation of mayors in the meetings. We speak a common language in terms of intersectoral collaboration, situation rooms, etc. Another aspect is the vision of health in the municipality. We can tell when we speak with people during the workshops, we see that the mayor's discourse changes. Health is incorporated into the municipalities' discourse through the mayors. Another result is the linkage with other national and provincial programmes. The Programme also strengthens capacities for municipal health management. Along those lines, we generate many tools to facilitate decision-making and they are starting to be used by the municipalities. (Laura Waynstok, PNMCS coordinator)

[11] FLACSO. Curso internacional de Promoción de la Salud (CIPS). Producción Social de la Salud junto a Escuelas y Comunidades CIPS. http://flacso.org.ar/formacion-academica/cips/.

> The Program succeeded in incorporating an important number of municipalities that implemented different community projects, usually financed partially or totally by the national government. Some policies, such as those to control tobacco, greatly contributed to establishing tobacco-free public spaces in the majority of the country. Healthy physical settings, such as parks, squares, and walking paths have received sustained support in the municipalities, which in turn has contributed to increasing the population groups that engage in outdoors physical activity. The control of solid waste has also been commonly addressed within the Program, improving the situation in numerous municipalities. (Dr Raul Pitarque)

Since 2013 external evaluations for accreditation as responsible municipalities have provided a way to measure the programme's results in the territories. These evaluations have helped to identify which local actors (municipal political authorities, health sector personnel or others responsible for the health sector, and members of various municipal areas) have adopted the holistic approach to health as proposed by the Programme, and how this is reflected in concrete activities and actions at the level of the territory, as well as in its identity and sense of belonging.[12]

Challenges

Given the complex characteristics of the programme, it is anticipated that at the level of the territories the following challenges will continue to be faced:

- Inter-jurisdictional relations (nation, province, municipality).
- Consolidation of inter-sectoral and intra-sectoral work.
- Continuity of political support for the programme, independent of party issues. Institutionalization of the approach and management style in the municipalities' policies.
- Holistic approach to health and health promotion, questioning the disease prevention and health services paradigm.
- Technical support and monitoring of the municipalities to implement the proposed tools.
- Systematic documentation and evaluation of actions and programmes and the measurement of impact.

Other challenges were highlighted by the key stakeholders interviewed:

> I think the Program lacks an interface, with more content and institutionalization of the processes; these are processes that are sometimes external to the municipalities' own socio-political process, not in all cases, but in some. Therefore, it is necessary to institutionalize these processes in the municipality so they will stay in the municipality.

---

[12] These results came from six external evaluation reports conducted by PAHO consultants during 2013.

> It is important not to interfere with the local participatory processes since this may derail the essence of local empowerment and sustainability as well as affect the content of the work and how it gets incorporated into the local policies in a way that it becomes institutionalized at the municipality level. In my experience, here and in other places, we sometimes put pressure on some local movements and end up with something that does not fit the local context. (Dr Luis Codina, PAHO/Argentina Acting Representative)

Finally, and based on Kingdon's model, having greater baseline technical–political information about the territories and settings helps in describing scenarios that facilitate the recognition and visibility of a problem. Even if it is difficult to establish a causal relationship, it is a fact that the existence of opportunities for technical cooperation and the use of virtual technologies have generated mechanisms for sustainability and a greater allocation of budgetary resources, which are both a reflection of political will to promote the proposal.

For more information about the experience of Argentina, consult the PNMCS website: http://www.msal.gov.ar/municipios/.

Mexico: Healthy Communities and Environment Programme[13]

Introduction

The action programme Healthy Communities was implemented as a result of a 1999 constitutional reform of Article 115 and corresponding modifications to municipal responsibilities, within the National Health Programme of 2001–2006. It aimed to connect health with social and economic development, promoting the idea that every development project should take the determinants of the population's health into consideration. It also promoted the implementation of specific local-level actions such as training, education and communication for health and social participation through the development and implementation of municipal health promotion programmes.

More than 17 years ago health promotion work started at the municipal level through awareness-raising among local authorities, with the idea of creating new working structures with different sectors to determine priorities and plan corresponding actions. In this context municipal health committees and workshops were created. Healthy Municipalities started in 1993, adopting the strategy of systematically documenting actions related to health at the municipal level. This was the precursor upon which the Programme of Healthy Communities was established within the National Health Programme of 2001–2006. This action programme sought to fight problems that threatened the health of all people (families and communities), by reinforcing healthy behaviours that benefited the population, through the promotion of public policies that encouraged the creation of favourable environments for health, the reinforcement of the community's power over determinants of health, and the involvement of municipal governments and their populations.

[13] Lucero Rodriguez Cabrera, Ljubica Latinovic and Paola Pliego Fernandez.

Structure of the Programme

The programme continues today, and impacts significantly on 14 priorities in public health that seek to reduce morbidity and mortality from emerging diseases, with emphasis on chronic diseases. It also contributes to addressing the key elements that control those diseases, such as

- Circumstances and agents that determine a public health problem, often not dependent directly on the health sector but on other sectors, such as the management of solid waste, water and conduits.
- Modification of environments where people live, study, work and play.
- Capabilities of the population and their organizations for changing or acting on these environments.

This programme also focuses on and gives priority to the populations that live in areas that score low on the human development index.

The overall mission of the programme is to reduce the morbidity and mortality of priority diseases in public health by creating favourable environments for health, orientating municipal action in health promotion, and empowering the population to control its health determinants. Its main objectives are to improve the environments where people live, study, work and play by creating healthy public policies, addressing the determinants of health through shared municipal, social and public responsibilities, building and maintaining local, municipal and national networks, and strengthening the competencies of health personnel. The programme is based on establishing and strengthening partnerships among local authorities and members of the community and other sectors to improve collective health. There are four key concepts:

1. *Determinants of Health*. The effectiveness of health promotion interventions should be measured in terms of how it contributes to modifying health determinants. Health reflects the quality of life, well-being and functional capacity of individuals, which can be defined as positive or negative. The social, environmental, economic, cultural, political and institutional conditions determine the health conditions of individuals and populations, and interventions aimed at decreasing morbidity, mortality and disability.
2. *Healthy Environments*. This component is based on work among different groups and sectors; it encourages the exchange of national and international ideas and experiences that help to create and maintain clean and safe social spaces where people live, study, work and play. Positive environments for health can be achieved by coordinating actions among different sectors (local authorities, civil institutions and the population) to participate in the identification of resources as well as the development of programmes and specific plans for improving health.

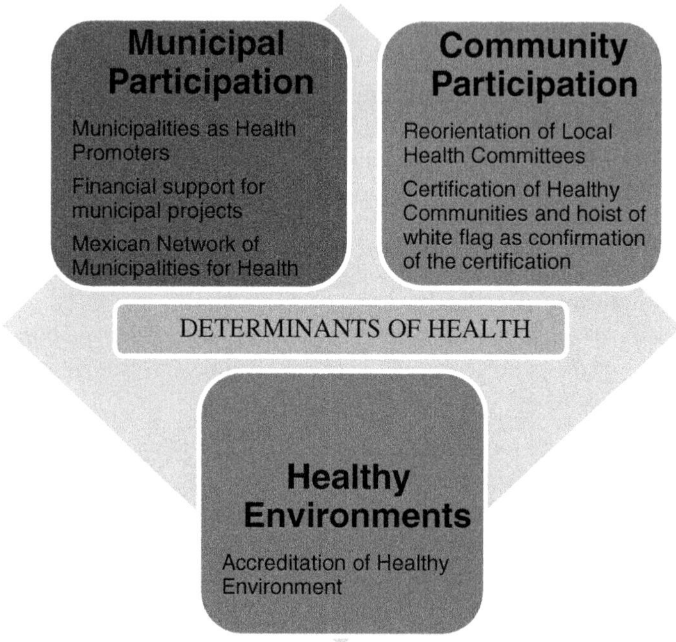

Fig. 8.7 The structure of the Healthy Communities and Environment Programme 2007–2012

3. *Municipal Participation*. Global trends towards the decentralization of social policies highlight the prevailing role that local authorities must play in the distribution and administration of resources, and this pertains to mayors in particular. It is worth mentioning that in most countries in the Americas, local political, administrative and legal power falls within the purview of municipalities, and mayors are generally elected by direct vote. The tendency is to strengthen the leadership of municipal authorities. Municipalities have under their jurisdiction urban, semi-urban and rural areas.
4. The Mexican municipality constitutes the framework for the creation of Healthy Environments and Communities, as it is the basis of the political and administrative, economic and social structure of the country.
5. *Community Participation*. Community participation in health is the specific and natural function of health promotion, which seeks to develop a social basis for the sustainability of the health services. Social participation implies the committed and active involvement in health of an organized society. Community networks in which the community commissioners periodically gather are forums to discuss health issues and to encourage the exchange of experiences and strengthening of capabilities (Fig. 8.7).

Strategic Lines of the Programme

The programme aims to encourage and empower the municipal health promoters in order to

- Contribute to the development of healthy sustainable policies which promote the participation of local authorities and society, as well as supporting all sectors in the definition of priorities that influence the determinants of health, and developing favourable environments.
- Encourage the participation of municipal health promoters in the development of projects correlated to public health.
- Support municipal projects related to public health by
  - Providing technical and financial support to projects which, based upon municipal health assessments, will impact on the health of the communities and be sustainable in the medium to long term.
  - Endorsing initiatives that target high-risk populations from the lower socio-economic groups.

In the strategic lines of the programme, different processes of great relevance are found:

- *Links with Action Programmes*: Linking is a way of working with others by assigning responsibilities and establishing commitments, deadlines and resources. For this to be successful, governing and functional tools must be provided, such as guidelines, technical and operational manuals, technical norms and counselling. In the federal states this relationship is achieved though the Healthy Community Committee of the state. The key to success is being able to solve the problems and meet the needs of the community, optimizing resources and avoiding duplication of effort.
- *Create alliances with different institutions of the public, social and private sector*: To achieve favourable environments for health it is important to establish alliances with different sectors that link to every single environment of importance to the programme.
- *Build social norms*: In order to achieve financial and social sustainability for the health services, it is fundamental to construct a social norm that recognizes health committees as agents that are jointly responsible and have the capability to make decisions.
- *Structures and levels of responsibility*: Because of the programme's national character, its leadership is the responsibility of the Health Secretariat, which defines the operating regulations to be observed by the states. Joint responsibility with the states is achieved through the participation of the respective technical areas in charge; the participation of the municipal government is fundamental. The Health Secretariat is responsible for coordinating the programme through the Health Promotion Directorate (attached to the Undersecretary of Prevention and Health Promotion).

- *States* participate in the coordination and evaluation of specific projects, and are accountable for the use of resources. Each state has a health promotion department that supports the development of actions and norms.
- *The health jurisdiction* is the technical administration unit. It reports to the state government and serves as a link to activities through coordination, planning, supervision and evaluation.
- *The National Committee of Healthy Environments and Communities* establishes the general rules for the operation of the programme and the criteria for its certification. It comprises the people in charge of the health promotion and prevention programmes.
- *Responsible entity for state health promotion*: The main purpose of this entity is to promote coordination with the people in charge of planning and implementing health promotion and illness prevention programmes in both general and local contexts of the state. This facilitates and assures supervision of programme management by the people in charge of the state health services.
- '*Healthy Environments and Communities "committees"*': These groups are created through the participation of the people in charge of health promotion and illness prevention projects in the states; they certify as 'healthy' those environments, municipalities, markets and schools which demonstrate improvements in the health conditions of the population.
- *The Health Promotion Jurisdictional Coordination Committee* is responsible for the organization, coordination and management of the programme at the municipal and local levels.
- *Primary Care Units* are responsible for providing integrated health services to individuals, families and communities through promotion, prevention, healing and rehabilitation activities. They apply the norms and procedures and participate in the programming, implementation and evaluation of the activities, involving the community and local government where possible.
- *Health Promoters* are attached to the primary care units. These important workers periodically visit their respective geographic units.
- *Local health committees* comprise people belonging to a community and its respective health personnel. They organize themselves in ways that improve local environmental and sanitation conditions.
- *Community health personnel* participate voluntarily in health promotion activities.

Description of the Experience

Positive results up to now, particularly in terms of convincing local and state authorities to work together with other sectors for the benefit of the population's health, have helped implement community organization schemes.

The first step in implementation is for the municipal president to convene a meeting of the people and lead a participatory process of decision-making by prioritizing, planning, implementing, evaluating and managing actions. Then the municipality participates in the institutional work of the state's health services, providing strategic support to high priority programmes such as 'Fair start in life' (*Arranque parejo de salud*).[14] Other priority programmes that have been developed with the participation of the municipalities and state's health services are Tuberculosis, Vectors, Accidents and Addictions, along with the National Health Card System (*Cartillas Nacionales de Salud*), Protection against Health Risks and the System for Social Protection in Health.

Once this stage of the process is concluded, the next step is to work on modifying the concept of public health in the country. This is not an easy task, and it involves the commitment and participation of many sectors. The participation of the municipality in health actions is necessary to be able to create development plans, partnerships, healthy public policies and the participation of a focused, convinced and committed community.

Example of a Successful Experience

By adopting the Healthy Communities Strategy, municipalities have been made aware and become empowered to address health as a priority in their government plans, and to involve other sectors in their projects. Nuevo Leon is an example of one such successful experience. In 1993, the First State Meeting of Municipalities for Health took place in Nuevo Leon; 11 municipalities participated. Since then Nuevo Leon has built the Nuevo Leon Network of Municipalities for Health (*Red Neoleonesa de Municipios por la Salud*), including the following important achievements:

- The state has had the privilege of serving as president of the Mexican Network of Municipalities for Health on several occasions and has held other high positions on the national board.
- During the period 2004–2010, the state submitted 176 project proposals for funding by the national government, of which 42 were approved.
- More than 120 communities have met the necessary health standards for certification, benefiting more than 200,000 people.
- The municipalities have participated in national forums which have facilitated exchange of experiences with other municipalities.
- Many municipalities have designed innovative strategies, some of which have been replicated both state-wide and on a national level. In some cases, there have been international acknowledgements of their value.

[14] To address maternal morbidity and mortality and closely related perinatal health challenges, the special 'Fair Start in Life' initiative ('Arranque Parejo en la Vida' in Spanish) was launched in 2001. The main purposes of the initiative were to address the health problems of newborns and children under 5 and the reduction of maternal mortality. The name of programme was meant to underscore the fundamental value of equality of opportunity.

Fig. 8.8 Accomplishments in funding municipal projects 2002–2013

	2002	2003	2004	2005	2006	2007	2008	2009	2010	2011	2012	2013
Presented	525	428	460	460	405	598	791	734	590	555	379	461
Supported	371	176	170	163	136	261	128	123	125	100	132	137

Results

There has been a large number of registered municipal experiences throughout the evolution of the movement of Healthy Municipalities and Communities in Mexico. It is worthy to note the various ways in which municipal governments, in coordination with the members of their communities, have faced problems affecting their community and developed projects that led to their solution: in ten years the programme has supported the development of 2022 municipal projects. It is worth mentioning the active participation of the Puebla, Oaxaca and Jalisco states, among others. All federal states have participated in the annual Health Communities and Environment contest and almost 70 % of the municipalities have become completely familiarized with the healthy community concept, developing different projects in response to their specific needs.

Accomplishments from 2002–2012

- 6386 municipal projects were conducted; 31.7 % obtained financial and technical support.
- 22 different public heath priority topics were addressed.
- Approximately 71 % of the supported projects have been evaluated.
- There was active community and municipal participation in support of projects based on high priority public programmes focusing on reproductive health and maternal mortality, dengue fever, rabies, diabetes, addictions, accidents and basic environmental improvement.
- The Mexican Network of Municipalities for Health actively participated in the National Health Council and the National Health Security Council.
- The Mexican Network of Municipalities for Health was a founding member of the Network of Healthy Municipalities, Cities and Communities of the Americas.
- Until 2012, 19 national meetings of The Mexican Network of Municipalities for Health were held.
- The number of maternal deaths decreased from 214 to 148 per 1000 in the poorest municipalities (2000–2006).
- There is representation from all regions of the country in implementing the Strategy (Fig. 8.8).

Challenges

- Incorporating the component of healthy environments as an element that emphasizes the creation of healthy environments and facilitates the modification of individual and collective healthy lifestyles.
- Building a social structure that empowers the population in the modification of the determinants of health.
- Achieving an effective exchange among the 14 most important public health programmes.

Promoting the National Network of Municipalities for Health and the regional and state networks so that they become true instruments of support for defining national policies that contribute to sustainable development on federal, state, municipal and community levels.

Analysis/Discussion

This programme is based on a dynamic concept, adapting in response to the diversity of each region, and its strategic vision adjusted continuously in order to support the process and achieve results. It has been demonstrated that the main strength of the programme is its ability to adapt local politics, economy and society to particular problems. During democratic transitions, the programme has become a basic instrument in the search for sustainable development.

The programme encourages municipal health promoters to change their attitudes in their understanding of health and the methods used to promote it, involving the participation of the population in decision-making through the establishment of healthy public policies, legislation, services and programmes that are usually provided by the municipality.

It is necessary for a community to be well organized and fully involved in health promotion actions if these are to succeed. The programme promotes the organized (informed, conscious and committed) participation of the community to modify the determinants of health and build environments favourable to improving the health conditions of the population.

The determinants of infant mortality in Mexico correspond to hygienic conditions. Factors such as drinking water and wood smoke have an effect on diarrhoea and acute respiratory infections. If municipal health promoters and communities are well organized, they may have a positive impact on the environment that in turn helps people achieve healthier lifestyles. They may also advocate the provision of such services such as water and trash collection. It is impossible to reduce priority public health diseases without modifying its environment and disease determinants; therefore, it is important to guide the authorities and to empower the population to act on both these issues.

Conclusion

Without a doubt the success of the programme is directly related to the creation, promotion and strengthening of networks. Currently, with federal resources allocated to support municipal projects and conduct national conventions, the Mexican Network of Municipalities for Health has achieved a high level of success, not only in the number of participants but also in the commitment demonstrated by these participants to the projects that have been designed, presented and implemented. In addition to showing the value of sharing knowledge, resources, information and experience, the programme demonstrates the reinforcement of relationships, solidarity between regions and, fundamentally, the conceptual evolution of projects as strategic actions of high impact. In the medium term these actions will help the Mexican Network of Municipalities for Health achieve the self-sufficiency and sustainability necessary to guarantee its permanence.

There are state networks in practically all the federal states. Some states have relied on their health jurisdictions to form regional networks to strengthen administrative transition schemes that will support the continuity of actions. These schemes have helped to support the consolidation of the Network of the Americas as well as the virtual space known as the World Network of Healthy Municipalities, Cities and Communities that facilitates the sharing of experiences and the highlighting of efforts in an international context. The programme demonstrates the fundamental principle that by modifying their environment, the health conditions of the people can be improved.

Colombia: Implementation of the Urban HEART (Health Equity Assessment and Response Tool). A Case Study[15]

Introduction

The Colombian cities of Bogota, Medellin and Cali participated in the Regional Workshop for Latin America and the Caribbean on Urban HEART that took place in Colombia in May 2011. They made a commitment to work with Urban HEART to assess urban inequities. In the city of Bogotá, and mainly in the area of Bosa, the Pablo VI Hospital has been utilizing the Urban HEART methodology as a tool to supplement its situational health analysis, making visible the inequities in local social territories and demonstrating their disadvantages related to health and well-being (Velandia 2012).

The experience presented in this case study not only offers lessons about the application of the tool, but provides evidence about how, through this instrument, one can begin to build healthy spaces based on informed decision-making and with evidence from a small geographic area. This policy tool establishes the relationship

[15] Jose Osney Velandia Rodriguez, Gloria Libia Polania Aguillon, Ana Maria Mahecha Groot and Teofilo Monteiro.

between health and its social determinants. It encourages an inter-sectoral dialogue based on local information, brought down to the micro-territorial level through primary health care (PHC). Hospital Pablo VI has implemented this strategy as a model for addressing health in the city. The threats to its implementation, along with accomplishments and effectiveness, are made evident through Urban HEART.

Description of the Experience

Bosa Township

The Capital District of Bogotá is divided into 20 townships, of which Bosa is the seventh. Bosa covers an area of 2.39 ha, which corresponds to 2.87 % of the total area of the district. It has a population of approximately 583,056 inhabitants,[16] of which 297.912 (51.1 %) are female and 285.144 (48.9 %) male. Bosa township is located in the southwest of the city. It can be reached by crossing the River Tunjuelito, which is a source of contamination and each year at risk of flooding, and its western border is marked by the Bogotá River, which is without a doubt one of the most contaminated rivers on the planet.[17] One of the greatest environmental risks to which the population is exposed is the result of Bosa's location between these two waterways, which causes flooding year after year due to deteriorating wetland areas that have constrained growth and where housing has been built based on social interests but, with obvious deleterious consequences to the population's health and quality of life.

It is estimated that a third of the population of the township is poor, and 5 % indigent. The township has the fifth highest incidence of poverty after Usme, Ciudad Bolívar, San Cristóbal and Rafael Uribe. In 2011, Bosa ranked third among the cities with low values on the index of living conditions, following Usme and Ciudad Bolivar: this translates into less access to services, low educational levels, lower quality of housing and larger families.[18]

Some populations are considered special because they face to certain risks: for instance, 16,313 people in Bosa are identified as having disabilities. In addition, a total of 28,000 people are displaced, and Bosa is estimated to be the second highest township within the Capital District in receiving victims of armed conflict. Additionally, according to the 2011 local health assessment of social participation, there are 625 families characterized as ethnic minorities: 85.76 % ($n=536$) of African descent, 7.05 % ($n=44$) Kichwas, 5.44 % ($n=34$) Pijaos, 1.44 % ($n=9$) Ingas, 0.32 %

[16](DANE, 2011—Departamento Administrativo Nacional de Estadística) (National Administrative Statistical Department).

[17]It is estimated that as the Bogota River flows into the Magdalena, it carries with it daily chemical and physical contaminants: 318 kg of chromium, 278 kg of lead, 140 t of iron, 1.11 t of detergents and 835 t of suspended solids, among others.

[18]The living conditions index measures the standard of living by combining the variables of access to and quality of services, education and human capital, household size and composition, and quality of housing; a higher rate reflects better living conditions.

($n=2$) Romani (gypsy). The urgent institutional services that these people need are insufficient, and inadequate governmental response in the face of the magnitude of the problem has served as the common denominator for the majority of activities and programmes intended to mitigate the effects of inequity and inequality.

The township comprises five units of zonal planning (UZP) distributed across 381 neighbourhoods legalized by the district planning secretariat (DPS). The UZP took the leadership for the application of the Urban HEART methodology. This geographical dimension (by UPZ) provides the basis for identifying the daily dynamics of populations in specific environments.

Development of the Experience

Hospital Pablo VI found that the Urban HEART methodology was a way to make visible the information gathered from years of experience in working with the community of Bosa, with the capital district and with the nation. To implement the methodology, sources of information were gathered from institutions and different sectors that had a relationship with the district level of the health sector. In 2012, the Urban HEART team of the hospital invited other sectors and institutions to participate by providing relevant and appropriate information. Among those who actively participated in the implementation of Urban HEART were:

- Hospital Pablo VI Bosa, ESE Level I
- District Health Secretariat
- Pan American Health Organization–Colombia
- Local mayor's office
- Water and Sewage Company of Bogotá
- National Police
- Local Social Integration Branch
- Colombian Institute of Family Well-being
- District Culture, Recreation and Sport Secretariat
- District Housing Secretariat
- Technical Support Unit–Local Board of Social Policy
- Board of Community Participation in Health
- District Planning Secretariat

The three fundamental elements of Urban HEART are: (1) the priority of solid evidence represented by indicators as having the greatest potential to influence health inequities, along with those that are of good quality and reliability; (2) intersectoral action and (3) community participation.

In addition to having an interdisciplinary and inter-institutional team to implement the methodology, consolidating an urban health network was considered necessary to gain the needed political support for a programme in health equity. This came together through advocacy with other sectors by promoting the importance and pertinence of Urban HEART. The technical support units–local boards of social policy approached the managers of different district sectors, the institutions, and the technical secretaries of committees, councils, boards and networks to promote the use of Urban HEART.

The design and measurement of the Urban HEART indicators took place at different points. In the first instance, locally available information was sought both from the health sector and in general, affording the opportunity to better understand the overall situation related to social dynamics as well as to fulfil the requirements for disaggregating data by units of zonal planning (UPZ), which in turn enabled comparison and analysis according to equity. Some information had to be left out when the territorial variable of the UPZ was not taken into consideration and only very local information was available. In order to resolve this deficiency some variables were used, such as nomenclatures, neighbourhoods and micro-territories, that allowed for the desired geographic coding. It was possible to clarify the sources for calculating each indicator, in some cases by using primary data, in others secondary data, and in yet others by referring to inter-sectoral sources. In the first instance these were data produced by information systems and hospital interventions reported by the district health secretariat (such as systems of epidemiological surveillance and population-based databases), and reports of the ministry of social protection (such as registry systems of live births and deaths). In the second instance, information was collected from sectors and institutions at both the district and local levels. In the third instance, inter-sectoral dialogues took place to obtain information that was essential but missing, such as numerators or denominators.

Once the information was collected with the needed characteristics, priorities were identified that were expressed in terms of indicators. In the first study, primary sources of both qualitative and quantitative information were found. As for arithmetic data, the health sector provided already established systems as well as sectoral sources that showed the magnitude of the variables, such as unemployment, food security and Primary Health Care on Line.[19] Some sources of qualitative data included territorial social agendas that were built upon integrated social management, as well as those for the social participation component developed by the hospital more than a decade earlier. The selected indicators corresponded in the first instance to a local health assessment, but they also followed the Urban HEART general guidelines. According to the policy domains recommended by Urban HEART, some modifications were made. Firstly, in this step of the process, the domains were reduced from five to four, eliminating the domain of governance, as other data with variables for UPZ relating largely to governance in the city were not available.

For the Urban HEART equity in health analysis, the hospital worked with 41 indicators, of which 11 corresponded to health results and 30 to general results of the social determinants according to available and dependable information. Of this group of 41 indicators, 7 were defined as basic indicators by Urban HEART, 17 of them were strongly recommended and the 17 remaining were part of the group of indicators from the Bogotá District, which were adopted by Urban HEART.

[19] Primary Health Care on Line is a digital application of the district health secretariat responsible for systematically consolidating information that characterizes populations. It was developed for the Primary Health Care Strategy in each locality in Bogotá. In the case of Bosa, Hospital Pablo VI is in charge of implementing this strategy and for disseminating information that is pertinent to this app. It currently has more than 100 variables registered.

Below is a summary of the validated instrument that was applied during 2010 to 2012 (Table 8.2).

The instrument allows for two dimensions of analysis. Looking horizontally, the situation relates to the particular indicator and its results in the UPZs of the city. Vertically, the status of the UPZ is shown in different political domains. Generating interrelationships among the health and social determinants results is basic for understanding the complexity of the social reality.

In line with this, the domains of human and social development, as well as those of economics and governance, show fewer expected results according to the defined benchmarks. Situations of alarm are evident in all the territories, with a focus on inequity in UPZ Tindal Sur and with different gradients in the other four territories.

The hospital found certain advantages in using the Urban HEART instrument, such as the Matrix of Health Equity, in analysing context, health and planning activities: it enabled comparisons among social territories of the city, defining benchmarks for identifying equity gradients in each of them; it not only showed health results, but included indicators of social determinants related to local priorities; it linked other sectors in participating in equity analyses; and it supported evidence-based decision-making.

Values Related to the Analysis of Health Equity in Urban HEART

Within the values generated or strengthened through the process of implementing the analysis of health equity in Urban HEART, several things were found. In health, inequity is understood to be the product of differences in three aspects: they are systematic, they have social origins (and are therefore changeable) and they are unjust. Since these differences are not due to chance, they may respond to similar pattern and targets. Additionally, these inequities, as the result of social processes, are embedded in the basis of society, acquiring dimensions that are structural, particular and individual. As unjust differences, inequities have an ethical aspect, generating different social conditions according to a scale of values, which leads to the discussion of rights.

A holistic approach to the concept of health is considered by understanding health from within an equity framework. This integrating concept was expanded beyond the social necessities to include aspects of physical, psychological and social well-being. In this way, the health–disease process was seen as a continuum of human development, along which opportunities could arise to carry out life projects based upon rights and capacities.

The concept of health as seen in light of social determinants is therefore understood as conditions of social interaction which cause either health or disease. From the perspective of social determinant, different situations affect the welfare of the population, ranging from its unique context, especially that which is shared among groups or communities, to its structural components, where public policies are implemented.

Table 8.2

RESULTADOS EN SALUD: SINTÉTICOS Y ESPECÍFICOS	49 APOGEO	84 OCCIDENTAL	85 CENTRAL	86 PORVENIR	87 TINTAL SUR	ALARMA (ROJO)	RESULTADOS ESPERADOS (VERDE)
Sintético: Tasa de Mortalidad en menores de 1 año x 1.000 Nacidos Vivos.	36,7	13,2	17,6	13,8	27,9	10,8	9,9
Sintético: Tasa de Mortalidad en menores de 5 años x 10.000 habitantes menores de 5 años.	37,9	13,6	18,6	20,1	31,9	23,5	18,5
Sintético: Razón de Mortalidad materna x 100.000 Nacidos Vivos.	0	0	97,8	0	174,6	40	34,6
Específicos: Tasa de Mortalidad por neumonía en menores de 5 años x 100.000 habitantes menores de 5 años.	0	5,4	4,7	0	0	13,4	11,5
Específicos: Tasa de Mortalidad por Enfermedad Diarreica Aguda (EDA) en menores de 5 años x 100.000	0	0	0	25	0	1,8	1
Específicos: Tasa de Incidencia de Tuberculosis x 100.000 habitantes.	2,7	7,5	6,6	6,3	7,4	14,3	10
Específicos: Tasa de Incidencia de Intentos de Suicidio x 100.000 habitantes.	43,6	38	64,2	110,6	33,6	71,7	27,3
Específicos: Tasa de Mortalidad en Accidente de Tránsito x 100.000 habitantes.	13,7	5,8	5,7	0	3,7	7,2	4,4
DETERMINANTES SOCIALES: INFRAESTRUCTURA Y ENTORNO FÍSICO							
Proporción de barrios con Cobertura de Alcantarillado por UPZ.	100	99,3	99,2	97	90	>99,1	<99,1
Proporción de barrios con Ocurrencia de Inundación.	0	0	1,4	9,1	10,7	5%	0%
Área de parques según metros cuadrados con relación al total de población.	4,1	0,7	2,1	1	3,7	2,03 M2	4,93 M2
Tasa de Equipamientos Generales (Recreación y Deporte, Cultura, Salud, Bienestar Social, Educación.	19	13	14	11	8	11	13
DETERMINANTES SOCIALES: DESARROLLO SOCIAL Y HUMANO							
Proporción de Bajo Peso al Nacer en Nacidos Vivos con peso inferior a 2.500 gr.	8,4	8,9	22	6,5	10,6	13,3%	10%
Proporción de Gestantes adolescentes de 10 a 14 años x Nacidos Vivos.	0	0,87	0,53	0,84	1,2	0,5%	0%
Tasa de Violencia Sexual según casos notificados x 100.000 habitantes.	62,7	91,3	116	132	155	75,8	34,7
Proporción de personas caracterizadas con Terminación de la Enseñanza Primaria	61,7	60,5	58,7	60,1	55,9	60%	84%
Proporción de personas caracterizadas que refieren no saber leer y escribir.	3,9	4,6	3,2	3,4	3,1	4,2%	2,5%
DETERMINANTES SOCIALES: ECONOMÍA Y GOBERNANZA							
Porcentaje de la Población Económicamente Activa Desempleada caracterizada.	39,6	40,4	35,5	39,3	27,6	30%	25%
Proporción de Familias caracterizadas en la que Alguno de sus Miembros dejó de Recibir Alguna Comida al día por Falta de Dinero.	9,2	9,6	7,5	11,5	44,1	10,8%	6,5%
Porcentaje de la Población Estrato Socioeconómico Bajo (dos).	53	95	99	46	94	87,4%	39,4%
Porcentaje de la Población Estrato Socioeconómico Medio- Bajo (tres).	45,7	0	0	0	0	2,9%	35,7%
Porcentaje de Cobertura de la población a través de la Atención Primaria en Salud.	23,4	36,4	51,4	30,7	14	40,4%	23,0%

Source: Hospital Pablo VI Bosa ESE I Navel, Matrix cf Health Equity 2010–2012

Within this framework the health sector interacts with other sectors, seeking strategic partnerships in favour of the health and quality of life of the population. Urban HEART thereby is converted into a tool for trans-sectoral dialogue to the extent that its impact on inequities is not limited only to actions by the health sector but rather establishes horizontal relationships with social beings in the city to create opportune responses that cause transformation of social inequalities.

Catalysts of Health Equity Analysis from Urban HEART

Speaking of catalysts, reference is made to the causes of or motives for transformation or the achievement of proposed objectives. Urban HEART helps to consolidate key situations in terms of making the indicators visible and the results manageable. The analysis of equity in health was in this way converted into a special project financed by the district health secretariat and carried out by one hospital in Bosa. Although the project is subject to a short lifetime because of the limited availability of resources, it has been an ongoing effort and a synthesis tool for positioning the health situation within the health sector.

Another catalyst emanating from local government is the extent to which political will promotes inter-sectoral strategies in the city. At this time, Urban HEART shows possibilities for analysing equity in depth on a small scale within the city, while serving as a guide for use by government entities to identify gaps and inform interventions.

One of the main reasons why it has been possible to launch Urban HEART in the city is the strong citizenship participation through social mobilization. Having a system of participation that responds to the different circumstances around which the community is organized is essential to the exercise of promoting health equity.

Even when community actors found it difficult at first to own the technical language of the indicators (since it is customary to refer to felt need and social perception in qualitative narratives), over time the dynamic strengthening of knowledge with reliable information generated satisfaction for those involved in community empowerment, dissemination and the ownership of the results of each activity. Such decisive participation guided a more equitable distribution of resources by taking into account territorial and population priorities.

Analysis/Discussion

Until now Urban HEART has served as a guide for prioritizing situations and inequities in health. Progress in integrating information systems for the public sector is in the early stages, but strengthening it may create links with other sectors and produce a broad response in health and social determinants.

Since this was a special project of the district, the process was constantly subject to contractual issues for financing an interdisciplinary team. Having profiles from different areas enriched the analysis and management that resulted from using Urban HEART as a methodology to incorporate an equity approach into a health situation analysis in a city like Bogotá.

The benefits of utilizing this tool through an institution like Hospital Pablo VI lay in the exchange of information with other actors in the city as well as in the joint systems for characterizing epidemiological data relating to the physical environment and the population. However, splitting the city into increasingly small pieces presented difficulties in standardizing benchmarks or goals consistent with inequalities; hence, a management response emerged that supported cooperation and complementarity with key stakeholders to foster health equity in the city.

The main findings evaluated the accuracy of geocoding at the desired levels within the city, testing the quality of the information reported by the official systems. To overcome barriers, one must be open to constant review of the data and ignore data that does not demonstrate reliability.

Conclusion

That ownership of the management of the instrument continues within the city is encouraging, since key stakeholders think the instrument holds possibility for designing a city that combats social segregation while integrating responses from multiple sectors aimed at improving quality of life and guaranteeing equal rights. During the nearly 3 years of local implementation of Urban HEART, several conclusions can be made regarding the analysis and response related to health equity.

The Analysis Component

- Difficulties were found in the information systems, as well as in the databases, ranging from the institutional level to district and up to national levels, especially relating to the required geographic variable and causing a considerable loss of information in some cases. In this sense, the results should be recognized as a representative sample of the city's reality. Although the land use UPZ category allows for trans-sectoral dialogue, for the hospital it is important to generate a similar instrument on the model of care for healthy micro-cities and cities.
- It was decided to adopt a social determinants approach for clustering indicators for the disaggregating information based upon geographic location. The scale used to test for social determinants was based upon health outcomes. The parameters used to differentiate between a critical or alarm situation were compared with the expected result in each specific case.
- The analysis was made from the perspective of the universality of human rights, so that a person's life conditions were identified within a particular geographic

area. It is anticipated that the tool will become better in its focus on population groups through future iterations. Other situations were prioritized according to local situation analyses and assessments; however, information sources were not found with the required characteristics.

In this regard the indicator is seen from a policy domain and has double value. In the first instance it can be understood as a statement: in other words, as the status of a given situation that affects the relationship between health and its social determinants. In the second instance the indicator represents an end in itself.

The Response Component

- Urban HEART can be translated into the 'urban heart of the town' since its fibres form the tissue that enables recognition as to where to find the most obvious gaps and how far they are from meeting the goal of building more just cities.
- Social participation is reflected in reality found by the instrument's institutional analysis, and it represents a challenge that needs to be faced by local government when considering the use of response strategies from the trans-sectorial perspective.
- Using a single instrument for orienting sectoral responses represents one of the main challenges of Urban HEART, even if the information is the same for each sector, since in the city each sector develops isolated analysis and prioritization exercises which can be at odds with each other. In this sense, the support of the Pan American Health Organization or World Health Organization in the country can be crucial for positioning Urban HEART within the district government.

The prioritization of issues such as the health of infants and young children, primary health care coverage, or teen pregnancy, viewed in the light of the Urban HEART equity matrix, generated responsive programmatic decisions such as the strategy for comprehensive early childhood care, the proposed pilot model for the implementation of primary health care, and research on preventive approaches to youth sexual and reproductive health.

Recife, Brazil[20]

The Healthy Municipalities Network of Pernambuco's research, capacity-building and inter-sectoral action built upon assets and local culture in promoting empowerment and equity.

It is no coincidence that the people in the North-eastern region of Brazil live with the reality of poverty, social inequity and resistance to change. The region is

[20] Ronice Maria Perreira Franco de Sa.

characterized by its strong attachment to tradition and by hierarchical relationships that have sustained iniquities over time (Freyre 1933). Promoting equity in the region is a Herculean task, since the patriarchal tradition of colonels who are above good and evil is maintained by current politicians, particularly those who represent geographic areas that are distant from metropolitan centres.

The Healthy Cities Network of Pernambuco (HCNP) was created in 2006 as part of an international cooperation project between Brazil and Japan, spearheaded by the Federal University of Pernambuco and in partnership with the Pernambuco State Government. The project began with an evaluation of the social capital of the five participating municipalities[21] that highlighted the harsh reality of the interior region of Pernambuco. The interdisciplinary team that participated in the survey identified a strong sense of disempowerment in civil society in these places; yet while historically, economically, geographically and socially disempowered, the participating municipalities expressed pride and confidence in the local culture: indicators of a people that cherishes traditional values. The projects favoured by local managers were, overwhelmingly, replicated from the federal government, which provided the main source of income for municipal employees or retirees. Youth expressed the desire to escape at the first opportunity, because of the lack of confidence in opportunities for jobs or social mobility in their home counties. Few experiences in social mobilization and collective achievements were presented.

The technical team responsible for the project initiated a mobilization process aimed at promoting reflection, increasing the sense of belonging and highlighting potential. Since the deficiencies were not very well known, maintaining the status quo seemed justified. After an initial period of suspicion, some of the most important local leaders joined the proposal and became the mainstay of local activities.

The project started with what was considered to be the dearest and most prominent community aspect: its culture and local art. This was used as a channel to mobilize different generations and social classes. A method for mobilizing, planning and prioritizing activities was developed in collaboration with local leaders and named the bamboo method.[22] This method is characterized by its focus on individual and collective strengths and opportunities and its capacity to spearhead actions to transform the local reality.

After the method was developed, the next step involved the interdisciplinary training of local volunteers from civil society and local management through a course, 'Health Promotion, Healthy Cities and Local Development'. The increase of individual and collective self-esteem was instrumental in the process of local change. The course, developed with the framework of the project, provided tools to be applied to the work and to promote reflection, as well as certificates and diplomas

[21] The participating municipalities were Barra de Guabiraba, Bonito, Camocim de São Félix, Sairé and São Joaquim do Monte.

[22] The bamboo method (de Sá et al. 2007) aims at mobilization for the planning and monitoring of priority collective actions. It is characterized by a focus on individual and collective strengths and governance for actions that promote alliances and a critical and reflective perspective on conflicts, controversies, and convergences among different actors.

accredited by the Federal University of Pernambuco. The university's contribution, by itself, triggered an important process of change and self-confidence. Actions were and still are prioritized and inter-sectoral teams from the university and state government support their implementation and monitoring.

The strict implementation of the work, with three interconnecting fronts—research, training and inter-sectoral action—promote local action and adherence by national and international partners, in spite of the various political–administrative inconsistencies that occur at the local, state and country levels. The continuous action of the Federal University of Pernambuco in monitoring, researching, promoting training and articulating partnerships (among the various disciplines within the university as well as with the rest of Brazil and other countries) has proven to be a factor contributing to the success observed in the particular case of these 'forgotten cities' in North-eastern Brazil.

Several evaluations have been conducted since the first one. Examples include research on social capital (Melo Filho et al. 2006) and participatory research of principles and values based on the participatory evaluation guide of the Pan American Health Organization (Akerman and Mendes 2006; PAHO/WHO 2006b). Other activities included collaborative work with international partners such as the Canadian Public Health Association (via the Brazilian Association of Collective Health) and a research project with the chair of Community Approaches in Health Inequity of the University of Montreal. All studies pointed to the importance of maintaining the evaluation research project which, in this case, functioned as an empowerment and capacity development strategy (de Sá et al. 2009).

The training developed by the technical staff of the Federal University of Pernambuco, supported initially by the Japanese International Cooperation Agency, was responsible for the sustainability of local action in the 2 years that comprised the first phase of the intervention. At the heart of the training was the assumption that the promoters of healthy municipalities, when enabled with broader health promotion concepts and equipped with tools for popular mobilization and prioritization of strategies and actions, would act as local focal points and supporters of the proposal. In fact, despite the large number of people who migrated from their initial sites and abandoned the initial action, it can be said that the vast majority continue to work as promoters of healthy municipalities, even if their actions are of smaller scope.

Research funded by the Foundation for Science and Technology of the State of Pernambuco showed that the graduates of this training referred to themselves as agents of local inter-sectorality. It is noteworthy that the profile of this social actor is quite heterogeneous as it encompasses volunteer representatives from civil society and local management, social action workers, professionals and health counsellors, among others. In this survey, six focus groups were conducted with 72 participants in total. It included a focus group in each municipality along with representatives of the state government. Both groups, promoters of healthy municipalities as well as the state managers, pointed out:

- Important elements that led to the success of the experience included hands-on action in the implementation of inter-sectoral local action, which was the bridge that connected people at the micro level, directly influenced the transformation processes, promoted the focus on otherness, and acted as agents of awareness and reflection.

- The experience was characterized by reflection in action and taking action based upon common desires and values, integrity, committed leadership, commitment to the empowerment of the population and of the people, commitment to local development, and citizen action beyond local partisan political issues.
- According to the respondents, and confirmed by monitoring reports, the promoters of healthy municipalities took several key actions: they organized people to clean the rivers and participate in ecological marches; held lectures on sustainability and the proper disposal of garbage; taught recycling; increased awareness of and vigilance in the use of pesticides (mainly for planting tomatoes in the cities studied); increased awareness, surveillance and reporting of domestic and school violence; offered courses related to and incentives for healthy eating; encouraged regular physical activity; conducted courses for income generation (support for local and traditional handicrafts); promoted local joint actions between managers and the population; and built and strengthened healthy public policies. There is still an ongoing effort to increase the population's awareness of its role as an informed and dynamic citizen. In this matter, the promoter of healthy municipalities has been a key player.
- Challenges of the experience were to increase visibility, to enhance communication, to document action systematically, to receive constant motivation and support from the supporting institutions (university and state government), to find space for meetings, and to resolve controversy over the institutionalization of actions.

Based on these results, training activities were intensified and expanded to Portuguese- and Spanish-speaking countries of Latin America and Africa as part of the Programme for Third Country Training within the Brazil–Japan Partnership Program. Since 2009, the Centre for Public Health and Social Development of the Federal University of Pernambuco has trained 75 official national representatives[23] to receive the same training offered in Pernambuco for the HCNP. An assessment conducted by the Japanese government using course graduates identified projects in their countries of origin that applied the concepts and tools learned in the training. The bamboo method was considered by almost 90 % of the graduates as adaptable, appropriate and easy to use when applied in their home countries.

The HCNP is currently conducting a mapping activity aimed at the development of a violence profile of the 23 municipalities that participated in the Healthy Cities Network of Pernambuco, with financial support from the Brazilian Ministry of Health. The workshops follow the same principles of participation, respect for local cultural traditions, working with promoters of healthy municipalities and

[23] Angola, Argentina, Bolivia, Cape Verde, Colombia, Dominican Republic, Guatemala, Honduras, Mexico, Mozambique, Paraguay, Sao Tome and Principe and Uruguay.

training new actors, focusing on capabilities, desires, possibilities and the capacity for governance of local actions. During 2013 visits to each municipality confirmed the need for continued commitment by the university, since its presence mobilizes local managers, allows monitoring and streamlines local health promotion actions.

Addressing inequities and the traditional hierarchical power structure in the inner cities of North-eastern Brazil is a task that cannot be accomplished in a few years. It requires persistence, sustainable actions and the establishment of trust within the population. The path to the changes that are necessary requires maintaining the triad of research, training and inter-sectoral action, focusing on personal and local potentialities, stimulating improved self-esteem, supporting the development of skills, promoting constant monitoring and positive inputs for the population, valuing engaged leaders through participation in regional and national events, supporting the empowerment of local society, and promoting equity and social justice as much as possible while considering the local political, economic and social context.

Peru: The Experience with the National Healthy Municipalities Network[24]

The creation of Healthy Municipalities in Peru started in the mid-1990s with the implementation of the national Healthy Municipalities Strategies (HME), developed within the framework of projects that were based on health promotion values and strategies. A Healthy Municipality is defined as one where 'citizens, institutions and organizations work together to achieve health, well-being and quality of life for its inhabitants'.[25] The initiative is coordinated by the Ministry of Health through the general Health Promotion Directorate. The concept of the Healthy Municipality is based on a holistic, multisectoral and transdisciplinary approach which promotes the idea that all sectors, including the private sector, assume co-responsibility for health. The HME includes municipalities, communities, educational institutions, households and families.

The HME vision is to 'strengthen local organizations and mobilize them to improve health, bringing together not only the adult population but also children, adolescents and youth, improving their capacities to identify priorities and implement intervention strategies and promoting healthy behaviours and lifestyles'.[26] It contributes to the implementation of the HiAP concept at the local level, promoting democracy and social co-responsibility.

[24] Ana Rivière-Cinnamond.

[25] http://promocionperu.blogspot.mx/2009/02/municipios-saludables-y-determinantes.html.

[26] http://www.munisanborja.gob.pe/index.php/red-de-municipios.

The Origins of the Network

The Healthy Municipalities Network was established in 1996 with the participation of the municipalities of metropolitan Lima, Barranco, Surco, Magdalena del Mar, La Molina, Ventanilla and Villa El Salvador, and the community organization Federation of Women Organized in Self-Managed and Allied Popular Eating Centres of Lima and Callao. Other communities later joined the effort, such as Jesús María, San Juan de Lurigancho and San Juan de Miraflores. By 2006, the network had expanded to the regions of Huánuco, Ucayali, Amazonas, Madre de Dios, Ancash, Piura, Puno, Ayacucho, Junín, San Martín, La Libertad, Lambayeque, Huancavelica, Cajamarca, Arequipa, Apurímac, Loreto, Pasco and Cusco; currently there are 22 regional networks to which a total of 753 municipalities are subscribed.

The Objectives of the Network are to promote consultations among municipalities of Peru and the rest of the world, promote efforts to improve the living conditions of individuals, families and communities, and contribute to participatory assessment processes and the implementation of health plans developed with citizen participation. Based upon these objectives, the network aims to promote and develop healthy public policies at the local level, promote citizen participation in the municipalities, encourage solidarity among municipalities, share experiences of municipal projects related to the development and well-being of the population, promote multisectoral consensus-building, train people in health promotion and expand the Healthy Municipalities and Communities Initiative.[27]

Activities carried out through the regional networks include the reduction of chronic infant malnutrition and maternal and neonatal mortality, the protection of environmental health, addressing climate change, road safety and alcoholism, land use and sustainable development, the treatment and reuse of sewage, recycling and the disposal of solid waste. The wide range of these activities demonstrates the breadth of coordination among multiple ministries at the local level, and especially among the health, education, environment, housing and agriculture sectors.

Currently the network continues to function; it coordinates its activities through a board in which mayors from different places participate. The operating mechanism for coordinating activities in different and distant places and municipalities is through periodic meetings. For example, the second meeting of the network, held on 12 July 2013, included representatives from the Ministry of Health, the General Directorate of Health Promotion, 50 local governments, the Board of the Healthy Municipalities and Communities Network, the Association of Municipalities of

[27] Red de Municipalidades Urbanas y Rurales del Perú: Network of Urban and Rural Municipalities of Peru.

Peru, the Network of Rural Municipalities of Peru [28] DISAS[29] and DIRESAS[30] of Lima and Callao, health directorates from 15 regions, PAR SALUD II,[31] the Ministry of Economy and Finance, PAHO/WHO of Peru and others. As an outcome of the meeting, a letter of commitment was signed in which stakeholders agreed to monitor and measure accomplishments and discuss progress during a third meeting of the Healthy Municipalities Network in November 2014.

At the local level, municipal work is accomplished through a string of separate but connected phases:

Phase 1, Awareness-raising and advocacy. In this phase, efforts are made to raise the awareness of local communities through meetings involving people as well as institutions in which ideas and projects are discussed.

Phase 2, Political commitment. A commitment document is signed that includes a resolution to form the council and board, and affirming the commitments, principles and basis for the work at municipal level with the actors present.

Phase 3, Organization. This includes the organization by the local authority of an inter-sectoral committee that has the objective of developing a generic situational assessment in the municipalities and communities identified.

Phase 4, Planning. During this phase a concrete proposal of activities and tasks is developed, accompanied by a definition of specific objectives and how coordination will take place in order to enrich the results through participatory processes.

Phase 5, Action. The planned activities are carried out at the community level, as defined in the work plan and according to the priorities of the participating individuals.

Phase 6, Social participation. Strengthening community participation, as well as promoting inter-sectoral action, are the pillars of the networks at the operational level. This is essential in order to promote healthy public policies.

Phase 7, Sustainability. Hand-in-hand with social participation is the identification of human and financial resources to support the priorities identified in Phase 3. This facilitates the achievement of programmed activities at the local level and ensures the quality of the health promotion and healthy municipalities processes.

Phase 8, Monitoring and evaluation. Finally, monitoring and evaluation of activities carried out is essential to assess project progress. This phase, which should be carried out on a continuous basis throughout the implementation of the project, helps contribute to its sustainability and continuity.

Through these phases, the Healthy Municipalities Network works to facilitate the implementation of projects at the local level with full community participation, to ensure that its values and principles are incorporated, and thereby to ensure the sustainability of activities over time.

---

[28] Dirección de Salud: Health Directorate.

[29] Dirección Regional de Salud: Regional Health Directorate.

[30] Micro Red de Campo: Micro Field Network.

[31] Mahier, Halfdan (1981) "The Meaning of Health for all by the year 2000", World Health Forom, Vol.2, No.1, p.5–45.

Concrete Examples from Peru

Through these holistic initiatives carried out at the municipal level, the social vision of health from the Primary Health Care concept of 1978 is renewed. The experiences and reflections of the last four decades have provided additional criteria, and explicit approaches in the healthy public policies framework have been added that have influenced the implementation of the concept of HiAP at the micro or local level.

Currently, many Healthy Municipalities have their own web pages, such as that of San Borjas,[32] and there are explicit references to the web pages of other municipalities (such as Lima's) that describe activities being carried out and keep their inhabitants informed.

Highlights of Other Experiences

In 2010, PAHO held the first Hispano-American Contest of Best Practices in Urbanism and Health as one of many activities in celebration of World Health Day. The objectives were to identify best practices in urban health, create a portfolio of best practices, and strengthen horizontal cooperation among countries by means of exchanging experiences and lessons learned. Over 130 submissions were received from 16 different countries.

The top five winners were

1. Reduction of social, environmental and security vulnerability of the inhabitants of the city of Medellin, Colombia, through a strategy of shared responsibility and community involvement aimed at managing risk.

    - Medellin has a total population of almost 2.5 million people, of which more than 30,000 families are located in unregulated high-risk sectors. This means that they are located in areas where geological and topographical land characteristics favour mass soil movements. These dwellings are located in areas where the channels of gorges are discharged, making them highly susceptible to flooding or torrential run-offs. There is also a high occurrence of fires, technological accidents, permanent mass events and the collapse of dwellings. In response, the Medellin Municipal Administration created a local emergency committee in charge of policies related to the prevention response to and recovery from natural disasters. Community links were established among 16 urban communes, and basic and free response services were provided based on ongoing research about local conditions. A network of volunteers was created to manage risks, consolidate neighbourhood emergency committees, upgrade evacuation plans, strengthen local management of public resources, and create

[32] http://www.munisanborja.gob.pe/index.php/fiscalizacion/itemlist/category/11-red-de-municipios-y-comunidades-saludables-per%C3%BA.html.

permanent monitoring of critical situations. Many active neighbourhood committees became involved in the project, including school-aged populations and representatives of the private sector. Community alarm systems and school safety programmes remain in permanent operation, and there has been a significant reduction in the number of city disasters and subsequent negative impact.

2. Promoting Healthy Work Environments in informal working units in Bogota, Colombia

   - Bogota has a population of approximately 7.6 million people, with a working-age population of approximately 5.7 million but an economically active population of only 3.8 million, of whom about 55 % are found in the informal economy. The informal work setting is characterized by people who are marginalized from industrial development and from access to formal employment. This means that they frequently work in precarious conditions without protection, with difficulty in mobility and exposure to daily violence, insecurity and health risks directly related to their work. The objective of this municipal programme was to prevent accidents and diseases related to work in this high-risk population. Some of the strategies used included training local leaders in health promotion and occupational problem prevention; participating on the district committee of occupational health to influence policies and advocate the inclusion of informal workers in the social security system of the country; developing a reporting system of health-related work events in the uninsured population; strengthening social capital through the organization of workers in the informal economy; managing risk control mechanisms devoted to improving working conditions in the manufacture of footwear, woodworking and preparation of furniture, and the manufacture of metallic parts; and guaranteeing comprehensive health care for workers in the informal economy.

3. Plan of sustainable mobility, City of Buenos Aires, Argentina

   - Policies were established for public transportation that included non-motorized pathways (pedestrian and cyclist) and non-polluting technologies (electric buses and hybrid autos). Road safety was the priority to be achieved through strengthening road networks and the reorganization of the transit system. Components of the plan included preferential roadways, a network of roads in Buenos Aires city that prioritized public and mass transportation; two-way circulation was established on the main avenues with special roads or lanes exclusively for public transportation. The Buenos Aires Bicycle Program established a network of cycle paths, with an integrated network of protected bike lanes and bike tracks connecting with principal paths reaching the margins of the city. Additional parking was developed for private bicycles, and an education programme was developed with information and communication materials to promote the use of bicycles by office employees. In road safety, laws were passed to impound vehicles that surpassed the maximum speeds or whose drivers did not pass the alcohol blood test, with temporary suspension of drivers' licenses. Courses were offered, including a children's traffic safety road, a mini-cinema and a mobile classroom.

4. Comprehensive tobacco programme of the City of Murcia, Spain
   - This programme was developed for the workers of the Municipal Council of Murcia with the aim of reducing the rate of tobacco addiction and improving the health status of non-smokers. A care unit developed for smokers was open during their working and free hours, accompanied by an information campaign about the risks of tobacco and rules regulating consumption. Sustainability of the programme was based upon inter-institutional collaboration, well-defined leadership, financial stability, controlled incorporation of new population groups, and feedback from relapsed smokers.
5. Environmental sustainability through the Millennium Institution: Trasmilenio in Bogotá, Colombia
   - The aim of the consolidated sustainable transportation system was to mitigate the effects of climate change and improve the quality of life of the citizens of Bogotá along with the environmental quality of the city. Programmes included monitoring and follow-up of the environmental parameters of comprehensive solid waste management, along with the efficient use of resources such as water, electricity, space and materials. Air quality was addressed through emission and noise control, water resources through efficient use and savings of water and the supervision of its quality, and energy resources through the quality of fuels and efficient use of energy. The project has a permanent monitoring scheme and an environmental observatory with standardized indicators to track progress and impacts.

For additional information consult: http://new.paho.org/hq/images/stories/AD/SDE/UH/informe_final_de_blanca_patricia1.pdf.

Guatemala[33]

The Healthy Municipalities Strategy in Guatemala was started in 1996 during the first Central American Meeting on Healthy Municipalities. In 2004, a ministerial agreement was signed which specifically established the health promotion and education department as responsible for the Strategy's implementation. About 45 mayors signed this agreement.

Since 2009, activities to strengthen the capacities of municipal governments to promote HiAP have been promoted and supported by the PAHO/WHO representative's office in Guatemala and by the Ministry of Public Health and Social Welfare,

[33] Guillermo Hegel.

along with various United Nations agencies and donors such as the Swiss Embassy. PAHO helped to establish an institutional coordinating board that brought together the ministries of health, education, agriculture, environment, food and nutrition security, along with the office of executive coordination of the president's office, German International Cooperation and Plan International. As a result of their work, an inter-institutional agreement on the Healthy Municipalities Strategy was established. Specific tools were offered to help with the organization and implementation of this Strategy, along with a Healthy Municipalities toolkit. The plan 'Zero Hunger' was formed whereby local governance was to focus on food and nutrition security. Through the implementation of the Strategy a network of Healthy Municipalities has been formed.

In collaboration with different agencies through a cooperative agreement, an Index of Municipal Management 2012 was established that includes the Index of Healthy Municipalities. This Index groups indicators and variables related to municipal work linked to prevention and actions that address the social determinants of health and healthy environments. It has been included in the strengthening of municipal policies and the publication of the National Ranking of Municipal Management, which includes information collected from 334 municipalities throughout the country.

Inter-institutional coordination still needs to be established at the municipal level, in order to promote health effectively at the local level. This will require dialogue and identification of best practices that are independent of political and governmental changes. Because of the epidemiological profile of the country, health promotion and education are key strategies to promoting good health, and the Healthy Municipalities Strategy will contribute to this at the local level.

The Future and a Way Forward

The trend of rapid urbanization will continue throughout Latin America, and medium and large cities will continue to grow exponentially. If cities and countries shy away from rational urban planning, there will be a corresponding increase in the development of informal and precarious settlements, without appropriate infrastructure and with limited access to services essential for the population to remain healthy and have good quality of life. We have seen in this chapter that some cities and countries in Latin America have recognized these trends and have begun to address them in a strategic, planned and integrated manner.

The traditional approach of creating alliances with people who have political clout and with institutions that hold formal power has been shown to limit participatory and collaborative dynamics, often by submitting community members and leaders to bureaucratic processes. In order for the strategy of multisectorial and multi-player partnerships (Overseas Development Institute 2007) to be successful, new relationships that draw from the most dynamic players and grassroots experiences similar to HMCs must be created. In some countries, alliances with institutions and players from cultural, youth, academic and private sector organizations (business and productive organizations) are still not being considered.

Progress still needs to be made in developing inter-sectoral and intra-sectoral activities (Rice and Franceschini 2007). The agenda of alliances and collaboration in initiatives across the most committed sectors (health, education, the environment) does not preclude the need for progress in actions that facilitate the mainstreaming of the HMC Strategy within the health care and health sector alone.

The jury is still out on the effectiveness and achievement of national networks. The degree of development of national HMC networks varies across the countries included in the PAHO surveys. Argentina displays significant achievements, and while Brazil has generated regional networks, some of which are extremely active and dynamic, it is only now consolidating a nation-wide network. The main challenge continues to be the availability of stable opportunities for exchange among networks, such as events requiring attendance, forums, joint publications and the like.

The need to promote in-service training should be based on practical applications of activities in the field of whichever competencies are needed. It is necessary to offer in-service training that includes critical reflection on personal practice as well as access to supervision, tutoring, study groups and forums that promote reflection and exchange among peers. This implies that the content of the courses should lead to frequent analysis of situations and the identification of the most successful strategies; and participation in HMC projects should provide physical space, time, and guidance for in-service training and learning from experiences. The continual cycle of actions, then reflections, then further actions, and further reflections, should be encouraged.

A broad initiative is required to establish alliances and networking with universities and resource centres, so that the continuous training of human resources will contribute to creating new interactions, putting theories into practice within academic institutions and communities, and strengthening technical resources. Evidence needs to be shown of the effectiveness of applying these approaches to local decision-making and to the implementation of local development strategies.

Among the priority areas that require training, the development of competencies in programme planning, management, and evaluation stands out, especially of those in associated management skills (communication, negotiation, problem solving and leadership). Lastly, the need to generate capacity in emerging topics, such as mental health promotion, violence prevention, and urban development, must be considered. Another priority need is the generation of strategic lines of research, through national councils that include renowned researchers from different disciplines and the mobilization of technical capacities within and across countries. A cross-cutting collaborative strategy is required to disseminate knowledge and strengthen capacities for the implementation of permanent research programmes.

It is necessary to encourage applied and participatory research models that will generate evidence of the results and the impact of the initiative, provide information to the communities that are involved in the research and evaluation, raise the awareness of the public, and provide an opportunity for technical training and education.

Looking to the future strengthening and sustainability of the HMC movement in Latin America, the following are some key suggested directions:

- *Strengthen the identity of the HMCs* by developing a new generation of concepts and tools that update the framework as well as the methodologies and approaches. This means expanding the production and availability of publications and resources, particularly those that address emerging topics and introducing elements for innovation.
- *Mainstream health promotion into HMC and urban health initiatives* by reviewing and revising public health programmes that provide essential and comprehensive services to populations in urban centres.
- *Scale up HMCs to become part of national policies and promote good governance models* to help overcome political and budgetary instabilities, while at the same time making them more suited to achieving the aims laid out in the Report of the WHO Commission on the Social Determinants of Health devoted to urban settlements and healthy urban governance.
- *Shift the currently predominant focus on lifestyles and* services to increase a focus on values-based urban planning/economic development/transportation/housing/mental health and prevention of violence. It will help to address the causes of ill health, as recommended by the WHO Commission on the Social Determinants of Health (2008), with all essential actors (housing, urban planning and development, finance, commerce, education, among others). This highlights the need to build bridges with more developed countries as well as the need to expand partnerships with sectors responsible for these areas of development.
- *Make health promotion services accessible to and effective for individuals, families and communities* through primary health care programmes, community action for health, and comprehensive integrated community development initiatives.
- *Develop public policies that position health promotion high on the political agenda* and that go hand-in-hand with strengthening regional and national networks that accompany the diffusion of participatory processes (scaling up) by encouraging cross-cutting, horizontal cooperation that values the uniqueness and diversity of the HMCs.
- *Develop greater integration of health actions with economically productivity activities*. HMCs need to be more connected with local production undertakings (generation of income), housing programmes and social inclusion initiatives.
- *Strengthen key factors that have been shown to improve the HMC implementation processes, such as* (a) institutionalizing the Strategy; (b) maintaining the stability of training and technical teams by reducing turnover and reassignments; (c) defining a clear legal–institutional framework, including criteria for different levels of Healthy Municipalities; (d) establishing multi-player and multi-institutional management councils; (e) integrating instruments for inter-institutional consultation and stable and flexible tools; (f) permanently assigning staff to collaborative work; (g) developing appropriate coordination and management competences and (h) making integrated financing available.
- *Collaborate with the new forms of social participation* (development of virtual networks and e-government tools) and social media by employing effective tools

that facilitate citizen decision-making, thereby improving governance processes through empowerment.

- *Focus future alliance strategies on the generation of new relationships*, engaging the more dynamic players and sectors such as cultural, youth, and academic centres, businesses, social movement organizations and productive organizations.
- *Make available nationally and internationally permanent training resource* for professionals involved in HMC initiatives, with emphasis on new topics (social determinants of health, participatory evaluation, empowerment, local development, values-based urban planning), providing specific tools that correspond with the level of responsibility and emphasizing coordination and associated management competencies.
- *Develop a permanent research agenda and mobilize existing technical capacities*, establishing national councils and urban health observatories in each country and with researchers from different disciplines.
- *Expand the knowledge and evidence base regarding the effective contribution of HMCs in improving the health and well-being of the population,* making use of existing evaluation and documentation guides and making visible the impact and processes of HMCs in terms of their complexity and multidimensionality. Expand upon currently available measurement tools, such as Urban HEART and Health Impact Assessments, and of information systems in order to inform policies and programmes appropriately.

Clearly, these are too many recommendations for any local, regional or national government to implement at once. However, based upon participatory assessments and ongoing priority-setting, these suggestions can help to reorient, reinforce or create Healthy Municipality, City and Community initiatives that will be supportive of value-based, inter-institutional and participatory urban planning and healthy urban development.

Acknowledgements Laura Waynstok (PNMCS coordinator, Ministry of Health); Raul Pitarque (secretary of health, Municipality of Olavarría, Buenos Aires, Argentina) (1995–2000, 2003–2007), social promotion manager, Project Vigia, World Bank–Ministry of Health, Argentina (2000–2003), president Argentine Association of Tobacco Sciences (2006–2008); and Codina (PAHO representative, Argentina)

References

Akerman, M., & Mendes, R. (Org.) (2006). *Avaliação participativa de municípios, comunidades e ambientes saudáveis: a trajetória brasileiramemória, reflexões e experiências*. São Paulo: Mídias Alternativas.

Arteaga, E., Rice, M., & Franceschini, M. C. (2007–2008). Documenting the current state of the healthy municipalities, cities and communities initiatives in the Americas. *International Quarterly of Community Health Education, 28*(4), 277–288.

Canadian Department of National Health and Welfare. (1974). A new perspective on the health of Canadians/Nouvelle perspective de la sante des Canadiens. Ottawa, Canada.

Cueto, M. (2004). The ORIGINS of primary health care and SELECTIVE primary health care. *American Journal of Public Health, 94*(11), 1864–1874.

Franco de Sá, R., Araújo, J. A., Freire, M. S. M., Senna Salles, R., Chuma, J., Royama, H., … Menezes Filho, A. (2007). *Manual do Método Bambu—Construindo Municípios Saudáveis*. Recife: Editora Universitária UFPE.

Franco de Sá, R., & Moysés, S. O. (2009). Processo avaliativo em Promoção da Saúde como estratégia de empoderamento e desenvolvimento de capacidades. B.Téc. Senac: a R. Educ. Prof., Rio de Janeiro, v.35, n.2, maio/ago.

Dodgson, R., Lee, K., & Drager, N. (2002). *Global health governance: A conceptual review*. Geneva: World Health Organization.

Franceschini, M. C., Rice, M., & Caballero, C. (2010). The application and evaluation of an assets-based model in health promotion in the Americas: The experience with the healthy settings approach. In A. Morgan, E. Ziglio, & M. Davies (Eds.), *International health and development: Investing in assets of individuals, communities and organizations*. New York: Springer.

Freire, P. (2005). *Pedagogy of the oppressed*. New York: Continuum International.

Freyre, G. (1933). *Casa Grande and Senzala: formação da família brasileira sob o regime de economia patriarcal*. Rio de Janeiro: Maia and Schmidt. 517p.

HPI. (2011). The Ottawa charter for health promotion: 25 years on. *Health Promotion International, 26*(Suppl 2), ii57–ii272.

International Strategy for Disaster Reduction (ISDR). (2010). *Making cities resilient. My city is getting ready: 2010–1022*. World Disaster Reduction Campaign. United Nations.

IUHPE. (2007). *The Ottawa charter for health promotion: A critical reflection*. Promotion and Education, Special Edition 2, 2007. Background to the 19th IUHPE World Conference on Health Promotion and Health Education, 'Health Promotion Comes of Age: Research, Policy and Practice for the 21st Century'.

Kansas University. *Community toolbox*. Retrieved from http://ctb.ku.edu.

Kasmel, A., & Tanggaard, P. (2011). Evaluation of changes in individual community-related empowerment in community health promotion interventions in Estonia. *International Journal of Environmental Research and Public Health, 8*, 1772–1791.

Kingdon, J. (2011). *Agendas, alternatives, and public policies*. Boston, MA: Longman.

Laverack, G. (2006). Improving health outcomes through community empowerment. A review of the literature. *Journal of Health Population Nutrition, 24*(1), 113–120.

Leppo, K., Ollila, E., Peña, S., Wismar, M., & Cook, S. (2013). *Health in all policies. Seizing opportunities, implementing policies*. Finland: Ministry of Social Affairs and Health. Retrieved from http://www.euro.who.int/en/about-us/partners/observatory/studies/health-in-all-policies-seizing-opportunities,-implementing-policies.

Lindström, B., & Eriksson, M. (2010). The Hitchhiker's guide to Salutogenesis. Salutogenic pathways to health promotion. IUHPE Research Report.

Marcer, R., & Fernandez, S. (2010). Mapeo de Capacidades de Promoción de la Salud en las Américas.

McQueen, D., & Jones, C. (2007). *Global perspectives on health promotion effectiveness. IUHPE*. New York: Springer.

Melo Filho, D. A., Franco de Sá, R., & Chuma, J. (2006). *Avaliação de capital social na área de abrangência do Projeto Municípios Saudáveis no Nordeste do Brasil*. Recife: Edições Bagaço.

Mercer, R., & Fernández, S. (2010). *Mapeo de Capacidades de Promoción de la Salud en las Américas*. Argentina: Organización Panamericana de la Salud (unpublished).

Meresman, S., Rice, M., Akerman, M., Vizzotti, C., Frassia, R., & Vizzotti, P. (2010). Contributions for repositioning a regional strategy for Healthy Municipalities and Communities (HMandC): Results of a Pan-American Survey. *Journal of Urban Health: Bulletin of the New York Academy of Medicine, 87*(5), 740–754.

Ministry of Health of Argentina. (2013). *Listado de Municipios Saludables. Cuarto trimestre*. Retrieved from http://www.msal.gov.ar/municipios/images/stories/3-info-municipios/pdf/2013-09_listado-municipios-miembros.pdf.

Ministry of Health of Argentina (website). *Programa Nacional Municipios y Comunidades Saludables*. Argentina. Retrieved from http://www.msal.gov.ar/municipios/.

Nutbeam, D. (2008). What would the Ottawa Charter look like if it were written today? *Critical Public Health, 18*(4), 435–442.

OMS. (2010). *URBAN HEART: Instrumentos de evaluación y respuesta en materia de equidad sanitaria en los medios urbanos*. Kobe: Centro OMS para el Desarrollo Sanitario.

Organización Panamericana de la Salud. (1993). Desarrollo y Fortalescimiento de los Sistemas Locales de Salud: la Participación Social en el Desarrollo de la Salud. Washington, DC.

Organización Panamericana de la Salud. (2006). Herramientas de comunicación para el desarrollo de entornos saludables (Communication tools for developing healthy settings). Washington, DC.

Organización Panamericana de la Salud, Argentina. Working document.

Overseas Development Institute y la Fundación para la Cooperación y el Desarrollo. (2007). *ODA multi-stakeholder partnerships*. Retrieved May 25, 2013, from http://www.globalknowledge.org/gkps_portal/index.cfm?menuid=178andparentid=179.

Pan American Health Organization. (1992). *Annotated bibliography of participation, health education and health promotion*. Washington, DC.

Pan American Health Organization. (1994). *Evaluation and support of on-going experiences with social participation in health development: Proposal for action research*. Washington, DC.

Pan American Health Organization. (2004). *Mayor's guide to quality of life*. Washington, DC.

Pan American Health Organization. (2005). *Healthy municipalities, cities and communities: Evaluation recommendations for policy makers in the Americas*. Washington, DC.

Pan American Health Organization. (2006a). *Empowerment and healthy municipalities, cities and communities (HMC) and health promoting schools (HPS)*. Washington, DC.

Pan American Health Organization. (2006b). *Guide for participatory evaluation of healthy municipalities and communities*. Washington, DC.

Pan American Health Organization. (2006c). *Healthy municipalities and development in the New Millennium*. Washington, DC.

Pan American Health Organization. (2007a). *Agenda de Salud para las Américas*. Washington, DC. 2008–2017. Retrieved from http://www.paho.org/Spanish/D/HAgenda_Spanish.pdf.

Pan American Health Organization. (2007b). *Guide to economic evaluation of health promotion*. Washington, DC.

Pan American Health Organization. (2008). *The 1st regional survey of healthy municipalities, cities and communities (HMC)*. Unpublished document. Washington, DC.

Pan American Health Organization. (2011a). *Second regional urban health forum and third regional health promotion forum*. Final Report. Washington, DC.

Pan American Health Organization. (2011b). *Trends and achievements in promoting health and equity in the Americas: Developments from 2003–2011*. Washington, DC.

Pan American Health Organization. (2011c). *Advancing health promotion in the Americas: Lessons from 15 case studies*. Washington, DC.

Pan American Health Organization Virtual Campus for Public Health. (2010). *Virtual course on health promotion*. Retrieved from http://cursos.campusvirtualsp.org/course/view.php?id=43.

Pan American Health Organization/World Health Organization (PAHO/WHO). (2010). Final Report: Second Urban Health and Third Health Promotion Forum, New York, November 2010. Washington, DC: PAHO/WHO

Pan American Health Organization/World Health Organization (PAHO/WHO) & Ministry of Health of Argentina. (2004). *Municipios y Comunidades Saludables. Lineamientos Generales*. Retrieved from http://publicaciones.ops.org.ar/publicaciones/cdsMCS/06/documentos%20institucionales/Lineamientos%20Generales.pdf.

Peterson, N. A., Lowe, J. B., Hughey, J., Reid, R. J., Zimmerman, M. A., & Speer, P. W. (2006). Measuring the intrapersonal component of psychological empowerment: Confirmatory factor analysis of the sociopolitical control scale. *American Journal of Community Psychology, 38*, 287–297.

Rice, M. (2012). Effective global health promotion achievements, tools, and strategies used in the Americas over the past decade. *Health Promotion Practice, 13*(3), 313–328.
Rice, M., & Franceschini, M. C. (2007). Lessons learned from the application of a participatory evaluation methodology to Healthy Municipalities, Cities and Communities initiatives in selected countries of the Americas. *Promotion & Education, 9*(2), 68–73. doi:10.1177/10253823070140021501.
Rice, M., & Franceschini, M. C. (2008). The participatory evaluation of Healthy Municipalities, Cities and Communities in the Americas. In L. Potvin & D. McQueen (Eds.), *Health promotion evaluation practices in the Americas* (pp. 219–223). New York: Springer.
Round, R., Marshall, B., & Horton, K. (2005, reprinted 2008). *Planning for effective health promotion evaluation*. Melbourne, Australia: Victorian Government Department of Human Services. Retrieved from http://www.one21seventy.org.au/system/files/resources/Planning%20for%20effective%20health%20promotion%20evaluation.pdf.
Salgado, N., Friel, S., Fotso, J., Khadr, Z., Meresman, S., Monge, P., & Patil-Deshmukh, A. (2011). Social conditions and urban health inequities: Realities, challenges and opportunities to transform the urban landscape through research and action. *Journal of Urban Health: Bulletin of the New York Academy of Medicine*. doi:10.1007/s11524-011-9609y.
Solar, O., & Irwin, A. (2007). Discussion paper for de Commission of Social Determinants of Health, Geneva.
Tones, K., & Tilford, S. (2001). *Health promotion: Effectiveness, efficiency and equity* (3rd ed.). Cheltenham, UK: Nelson Thornes.
UNEP. (2002). GEO: *Global Environment Outlook 3*. Retrieved July 27, 2010, from http://www.unep.org/geo/geo3/english/index.htm.
UNEP. (2010). Environmental Outlook of Latin America and Caribbean (GEO—Latin America and the Caribbean 3).
UN-Habitat. (2002). *The global campaign on urban governance: A concept paper* (2nd ed.). Nairobi, Kenya.
UN-Habitat. (2009). Global urban observatory.
UN-Habitat. (2010). *Estado de las Ciudades de América Latina y el Caribe*. Programa de las Naciones Unidas para los Asentamientos Humanos.
United Cities and Local Governments. (UCLG). Ciudades y Gobiernos Locales Unidos. (2009). El impacto de la crisis mundial sobre los gobiernos locales. Barcelona, España.
United Nations Population Fund (UNFPA). (2007). *State of the World Population 2007*. Retrieved September 14, 2010, from http://www.unfpa.org/swp/2007/english/introduction.html.
United Nations Population Funds (UNFPA). (2012). *State of the World Population 2012*. Retrieved May 25, 2013, from http://www.unfpa.org/webdav/site/global/shared/swp/2012/EN_SWOP2012_Report.pdf.
Velandia, J. (2012). Análisis de Equidad en Salud (URBAN HEART)-Localidad Séptima de Bosa. Bogotá: Hospital Pablo VI Bosa-ESE I Nivel.
Wallerstein, N. (2006). *What is the evidence on effectiveness of empowerment to improve health?* Health Evidence Network report. WHO Regional Office for Europe, Copenhagen.
Wallerstein, N., Gutierrez, M. L., & Gonzalez Sobera, M. (2008). Social participation and empowerment: Methodologies for health promotion.
Wallerstein, N., Mendes, R., Minkler, M., & Akerman, M. (2011). Reclaiming the social in community movements: Perspectives from the USA and Brazil/South America: 25 years after Ottawa. *Health Promotion International, 26*: ii226–ii236.
Westphall, M. F., & Wallerstein, N. (eds). (2007). Evaluation of health promotion effectiveness in Brazil. *Promotion and Education,* IUHPE, Supplement 1, 2007.
World Health Organization. (1978, September 6–12). Declaration of Alma-Ata. International Conference on Primary Health Care, Alma-Ata, USSR. Retrieved from http://www1.paho.org/English/DD/PIN/alma-ata_declaration.htm.
World Health Organization. (1986). *The Ottawa charter for health promotion*. Retrieved November 21, 1986.

World Health Organization. (2001). *Evaluation in health promotion: Principles and perspectives.* WHO Regional Publications, European Series, NO 92.

World Health Organization. (2008). *Closing the gap in a generation: Health equity through action on the social determinants of health.* Geneva, Switzerland.

World Health Organization. (2013). *Milestones in health promotion: Statements from global conferences.* Geneva, Switzerland.

World Health Organization, Centre for Health Development. (2010). *Urban health equity assessment and response tool (Urban HEART).* Kobe, Japan.

Chapter 9
Healthy Cities and Communities: The North American Experience

Trevor Hancock, Tyler Norris, Réal Lacombe, and Fran Perkins

The origins of the modern-day Healthy Cities movement can be traced back to the 'Beyond Health Care' conference held in Toronto in 1984 which included a day-long workshop on Healthy Toronto 2000, and to the inspired insight of Ilona Kickbusch in seeing in this the potential for a European Healthy Cities initiative; this history is dealt with in Chap. 4. But while an important part of the initial inspiration for the Healthy Cities movement came from Canada—and specifically from Toronto—Healthy Cities in North America from the outset took a somewhat different path, in particular by focusing more on healthy communities than on healthy cities. After first briefly discussing the state of urbanisation and some of the major health concerns in North America's cities, this chapter traces the development of the Healthy Cities and Communities (HCC) movement in North America, meaning Canada and the United States (while technically part of North America, Mexico is dealt with in the chapter on Latin America).

T. Hancock (✉)
School of Public Health and Social Policy, University of Victoria, Victoria, BC, Canada
e-mail: thancock@uvic.ca

T. Norris
Total Health Partnerships, Kaiser Permanente, Rockville, MD, USA

R. Lacombe
Former Directeur de la Santé Publique, RRSSS Abitibi-Temiscamingue, Rouyn-Noranda, QC, Canada

F. Perkins
Former Director, Healthy City Office, Toronto, ON, Canada

E. de Leeuw, J. Simos (eds.), *Healthy Cities*,
DOI 10.1007/978-1-4939-6694-3_9

Urbanisation and Health in North America

North America is highly urbanised, with 82 % of the population living in urban areas in 2011, up from 64 % in 1950 (UN Population Division 2012); by comparison, 73 % of the population in Europe is urban. A large proportion of the population lives in cities of millions; in Canada, for example, 'the three largest urban areas in Canada—Toronto, Vancouver, and Montréal—made up just over one third (35 %) of Canada's entire population' (Employment and Social Development Canada 2014). But what perhaps best characterises North American cities is their high degree of urban sprawl, which is harmful to both human health and the environment, and their high concentration of poverty, often in the urban core; in some cities, such as Detroit, the inner city has such high concentrations of poverty and has been so neglected by the city that it has run down to the point of collapse. Thus sustainability and equity are two of the common challenges that cities face, and that the Healthy City movement seeks to address.

Urban sprawl involves high levels of low-density suburban residential, commercial and industrial development, well separated by zoning ordinances and by-laws. This makes public transit inefficient and costly, and makes cars a necessity. The costs for other infrastructure (e.g. sewers, water and human services) to serve low-density development are high, adding to the inefficiency. This car-dependent urban form is both ecologically unsustainable and harmful to health, producing higher levels of air pollution and CO_2 emissions and low levels of physical activity, as well as high rates of motor vehicle crashes and pedestrian injuries; the long commute times contribute to mental health problems and loss of social capital, since commuting time is a loss of family and community time. Those unable to use a car—children, people with disabilities and some seniors—or who cannot afford a car are further harmed because of the restricted mobility they face, the social isolation that may result and their inability to access needed services and amenities.

The health consequences of urban sprawl were identified by Frumkin, Frank and Jackson (2004) as

- Poor air quality
- Physical inactivity
- Injuries and deaths from traffic
- Reduced water quantity and quality
- Mental health problems
- Loss of social capital
- Health concerns of special populations

To these may be added the health consequences of climate change (see Intergovernmental Panel on Climate Change 2014 for a recent authoritative discussion of these), since sprawling cities, with their high levels of car use, contribute disproportionately to climate change.

While suburban sprawl is a major problem facing North American cities it is not the only one. Another is poverty, and the associated problems of food insecurity, homelessness or poor-quality housing, high rates of unemployment or marginal employment, low levels of education and poor-quality environments. Associated with these conditions and the concentration of poverty we see the problems of violence, mental health problems and the problem of substance use, both in disadvantaged inner cities and in disadvantaged suburbs, and compounded by unhelpful federal or provincial/state social policies.

Other major concerns that relate to the issues of urban sprawl and poverty include a generally underdeveloped public transportation system and lack of green space, especially in low-income communities. These issues largely preoccupy those working to create Healthy Cities and Communities in North America.

North America: From Healthy Cities to Healthy Communities

The path taken in North America, with a focus on Healthy Communities rather than on Healthy Cities, can be traced to several factors. First, there was a concern that this approach should not be confined only to the large cities with populations in the millions, but that it should be relevant to all sizes of local government, including those with only a few thousand or even a few hundred people. The name 'Healthy Communities'—in Québec, *Villes et Villages en Santé* (towns and villages)—was widely adopted early on, building upon a strong history of local community organisation and development. Thus there are large numbers of small communities involved. Second, at least in Canada, local municipalities have not always been receptive to the idea of Healthy Cities, although in many cases public health has historically been part of their mandate. This seems to stem in part from the understanding that 'health' is about health care, and especially hospitals (which in Canada is a provincial responsibility), and perhaps a concern that this movement might be a surreptitious attempt by the provincial government to off-load health care and hospital costs on to municipalities. In addition, at least in Canada, the move to regionalised health care systems in most provinces over the past 20 years, with the integration of public health into these regional systems (Ontario remains the exception), has meant that public health has become separated from municipal government, to the detriment of both.

Perhaps for these reasons, while nominally engaged, the associations of municipalities in practice were not very interested or engaged, as is borne out in the case of the Ontario, British Colombian and short-lived national Healthy Communities initiatives, all of which tried explicitly to involve respectively their provincial and national associations of municipalities.[1] However, individual municipalities did engage, particularly in Ontario, where public health was, and remains, a part of local or regional government. (The experience in Quebec was different, and there the municipalities have played a larger role, as described later.)

[1] A finding of the more than 25 years of personal experience of the primary author of this chapter.

Third, in the United States in particular but also to some extent in Canada, there is a weaker role for governments at all levels than in Europe. Taxation levels tend to be lower, which means that resources are fewer, and it has often been difficult to secure sufficient funding from local, provincial or federal governments for long-term initiatives; in the United States—but not often in Canada—philanthropic funding has filled the gap. On the other hand, there is often a strong history of community organisation and mobilisation, particularly in the United States, where the Asset-based Community Development movement has its roots (see box below).

Asset-Based Community Development
Instead of focusing on needs and problems, John McKnight promoted an alternative path towards the development of policies and activities based on the capacities, skills and assets of lower income people and their neighbourhoods. By shifting to a capacity-oriented emphasis, communities take ownership of their issues. Kretzmann and McKnight (1993) note that community development takes place only when local people are committed to investing themselves and their resources into improvement. Communities are never built from the outside in or the top down, but from the inside out or the ground up. Outside assistance is often required, but this should be aimed at helping developing the communities' assets. 'Even the poorest neighbourhood is a place where individuals and organizations represent resources upon which to rebuild', the authors noted.

Fourth, there is a stronger tendency in North America—especially in the United States, but also very much in Canada—to focus on individualistic approaches to health promotion and to thus focus on 'lifestyle' and personal behaviour. This is in spite of the widespread recognition of the social, environmental and economic determinants of health and of a 'settings approach', of which Healthy Cities and Communities is the prime example.

This seems to be largely a political allergy in what are mainly conservative leaning provincial, state and federal governments to recognising the social, environmental and economic factors that underlie health: to acknowledging—never mind addressing—social and health inequity, and to accepting social and political responsibility for the determinants of health and health inequity. As a result, provincial or state-level Healthy Community initiatives sometimes pay lip service to the issues of equity and sustainability while in practice focusing on the usual suspects: tobacco use, physical inactivity and unhealthy eating.

An example can be found in the Canadian province of British Columbia, where the Ministry of Health's 'Healthy Families BC' programme, which describes itself as 'the most comprehensive health-promotion programme in Canada', illustrates precisely the pitfalls when a government is fixated on personal responsibility and 'healthy living'. The programme has 'partnered with employers, schools and local governments'—which is good—'to work on plans that will help British Columbians to live healthier lifestyles' (https://www.healthyfamiliesbc.ca/home/about-us). However, it does so 'by focusing on four key areas: proper nutrition, healthy lifestyles, resources for parents, and fostering healthy communities' and by supporting residents 'in adopting healthier lifestyles by providing access to physical activities, nutritious foods, tobacco-free environments and social connections that foster positive mental health'. Clearly, then, the 'Healthy Communities' component is quite narrow in its focus, with the areas of focus selected by the provincial government to meet their priorities, not by the communities, and thus far from the breadth and depth that should be encompassed by such a programme.[2]

Finally, there has been no WHO initiative in North America, because WHO is largely absent from engagement in this region. The Pan-American Health Organization (PAHO), WHO's regional office for the Americas, is headquartered in Washington DC but in effect is the Latin American and Caribbean Health Organization, and the predominant languages spoken are Spanish and Portuguese. No major WHO initiative in either Canada or the United States is known to have taken place, let alone a Healthy Cities initiative, and there has been no attempt to start one. This is not a criticism of WHO or PAHO, because the greatest needs are in Latin America and the Caribbean, where they have indeed developed and supported a Healthy Municipalities initiative for some 25 years, as described in Chap. 8; but it does mean that the Canadian and US Healthy Cities and Communities movements have been somewhat isolated from the activities in the rest of the Americas and around the world (and vice versa), although some partnerships have formed from time to time, and Québec's *Villes et Villages en Santé* initiative in particular has played an important role in the Francophonie, especially in West Africa.

Healthy Communities in Canada

The Healthy Communities movement in Canada is alive and well in some provinces, but almost absent in others. Only in the three largest provinces (Ontario, population 12.8 million; Québec, 7.9 million; and BC, 4.4 million) are there

[2] A similar criticism can be levelled at the Ontario Ministry's Healthy Communities Fund, which supports 'Provincial health promotion initiatives that address two or more priority areas—physical activity, injury prevention, healthy eating, mental health promotion, and preventing alcohol and substance misuse', as well as 'coordinated community mobilization activities to develop local healthy eating and physical activity policies that make it easier for Ontarians to be healthy' (http://www.mhp.gov.on.ca/en/healthy-communities/hcf/) and at the Nova Scotia Health Ministry's Healthy Communities programme which, while it focuses on 'creating supportive environments, healthy public policy and community action to promote health and well-being', does so primarily in 'lifestyle' areas (healthy eating, injury prevention and control, sexual health, tobacco control), although to its credit 'Reducing health disparities and inequities' is also an area of focus (http://novascotia.ca/dhw/healthy-communities/).

organised and reasonably well-funded provincial networks, although there is also a small but very active network in the Francophone part of the Atlantic province of New Brunswick (750,000 population); there is no national network although these four provincial movements have formed loose, but unfunded, connections.

There was for a short time a Canadian Healthy Communities Network (CHCN), established as a partnership between the Canadian Institute of Planners, the Federation of Canadian Municipalities and the Canadian Public Health Association, with funding from Health Canada, the federal government's health ministry. In a largely successful effort to prevent the Healthy Communities approach being seen as 'just a health initiative', as well as to re-establish historical links between public health and urban planning in Canada (see text box below), the CHCN was based at the Canadian Institute of Planners (CIP), a decision that over the long term has proven to be very beneficial, because the CIP has maintained an interest in Healthy Communities long after the CHCN has been de-funded (in 1991/1992, during widespread federal budget cuts), as discussed later.

The City Healthy: Canada, 1914–1919

'It is not so much the city beautiful as the city healthy that we want for Canada' (Commission on Conservation 1915).

The Commission on Conservation was established by the Canadian Government in 1909, modelled on the one established by President Roosevelt in the United States. It had a broader concept of conservation than we do today, expanding the purpose of conservation from the natural environment and physical resources to the social environment and human resources. It included a Public Health Committee because, as Dr. Charles Hodgetts, the Commission's Advisor on Public Health, said,

There are two important factors in the question of national conservation, the physical and the vital. The former relates to the protecting of our land, our forests, our minerals, our waters, our sunlight and our fresh air, and the latter to the prevention of diseases, to health and to the prolongation of life (Commission on Conservation 1912).

The Public Health Committee took it upon itself to address the issue of town planning because, as Hodgett put it, 'In housing and town planning we are dealing with most of the former [conservation of natural resources] and all of the latter [conservation of vital resources]'.

With its interest in health in towns, the committee recommended in 1913 that a national conference on housing and town planning be held and that Thomas Adams be invited. As described in Chap. 4, Adams was instrumental in establishing Town Planning as a profession in Canada and thus town planning in Canada can trace its roots directly to public health.

The information here is drawn in part from Armstrong, 1959.

Québec

Following the initial development of the concept in Toronto and then in WHO Europe, as described in Chap. 4, the first formal Healthy Community project in North America was established in 1987, not in Toronto (which did not have a formal initiative until 1989) but in the small Québec community of Rouyn-Noranda, under the aegis of Réal Lacombe, who went on to establish and to a large extent lead *Villes et Villages en Santé*, which has been going for more than 25 years and has more than 200 member municipalities (see Case Study 2).

Case Study 1: Rouyn-Noranda—A Healthy City for 26 Years[3]

At its creation in 1926, Rouyn-Noranda was two separate municipalities: Noranda, the 'company town', close to the mine and the smelter, and Rouyn, the 'sister' service town. They were amalgamated in 1986. It is now a city of 40,000 people, located in northwestern Québec at the heart of a mining region (gold and copper). For a long time it has been one of the most polluted communities in Canada, mainly due to the local copper smelter. Pollutants like sulphur dioxide gas, lead, arsenic, cadmium and mercury plague the locality and the region around it. In 1978 the area became the target of a major intervention of the Québec Government to address the environmental pollution problem as well as its impact on the population and workers' health. The Regional Public Health Department was deeply involved in that process.

In 1986, after the merger of Rouyn and Noranda, a new city council had to be elected. Local candidates were looking for fresh ideas for the electoral campaign. One of them, Gerald Marinovich, found it in the Healthy City idea, suggested by Réal Lacombe, the Regional Director of Public Health, who had just completed his MPH at the University of California Berkeley; there he had met Leonard Duhl, one of the two key conceptualisers of Healthy Cities, along with Trevor Hancock from Toronto.

Healthy City became the main focus of the electoral campaign, and Marinovich, a novice politician at the time, came top in the election. Only a few months after the election, the new city council officially adopted and announced its Healthy City project. It was the first in the Americas.

Implementation was quickly started. The first move was to hire a full-time coordinator, Denise Lavallee, a social worker with high community organisation skills. The second was to hold a widespread community consultation, using tools like the Kids' Place questionnaire (borrowed from Seattle) and vision workshops. Six thousand people participated, the majority being children. The third was a community forum that attracted 250 people, including representatives of more than 50 local organisations and institutions. The purpose was to establish priorities for the Healthy City project; not surprisingly, the environment became the first priority.

---

[3] Réal Lacombe.

Among the first sub-projects were the creation of a botanical garden around a small downtown lake which had been used as a dump for a long time, and a solution for the lead exposure problem that affected children of the Notre-Dame neighbourhood, close to the smelter. Both involved widespread community participation, and both were very successful. In the first 5 years the botanical garden, called 'Ā Fleur d'Eau' (a flower of water), benefited from direct contributions made by 3000 citizens and numerous private and public organisations. The lead project, started in 1990, involved all concerned actors including the Noranda Company (owner of the smelter), and was remarkably successful in reducing children's exposure to lead.

Those two projects are still very active, with continuing adjustments and improvements. A large number of additional initiatives have also been undertaken by Rouyn-Noranda Healthy City, many of them concerning poverty and other social issues. Over its 26 years of existence, the Healthy City concept and approach has profoundly changed the practices of the city administration and its partners, resulting in impressive secondary accomplishments. Rouyn-Noranda is very different from what it was before 1987 … and for the best! Indeed, as one key indicator of its success, Rouyn-Noranda Healthy City brilliantly survived five different Mayors!

Case Study 2: Villes et Villages en santé, Québec

- Source: The network's website http://www.rqvvs.qc.ca/

Villes et Villages en Santé (VVS), the Québec network of Healthy Cities and Villages, was established in 1988 and now has 208 municipalities as members, covering about 70 % of the population of Québec. No other network in North America is so closely linked with local governments, with whom it has a formal arrangement requiring municipalities to join through a resolution of their councils; moreover, a majority of board members must be municipal politicians. The network is a non-profit organisation but enjoys a privileged partnership with the National Public Health Institute of Québec and is largely funded by the Québec Government through the institute; this supports a staff of five people.

The mission and objectives of VVS are to 'Promote and support throughout Québec, the sustainable development of healthy living environments by focusing on the exchange and sharing between municipalities, the commitment of municipal decision-makers for the quality of life and their ability to mobilize their partners and citizens in action'.

Objectives of VVS

- Encourage decision-makers in the municipal sector to realise the importance of their decisions for the health of their citizens, both in their long-term strategic planning and in their everyday decisions

- Encourage them to work with their health partners and those of other sectors in the implementation of specific projects, support the health and quality of life of citizens and encourage the active participation of these groups and the community
- Promote and support the exchange of information and tools between the various local projects in Québec municipalities
- Defend the interests of local projects with relevant structures such as governments, the private sector and other associations

To join the network a municipality must meet several conditions of membership: In a formal statement, the municipality commits to its citizens to become a healthy municipality, undertakes to demonstrate a constant concern for health and quality of life in its decision-making and choice of actions within its own jurisdiction, agrees to share its knowledge and experience with other municipalities of the network and undertakes to allocate a minimum of human, material and financial resources to the implementation of projects selected by the city council. Member municipalities pay an annual fee to the network depending on the size of their population, from $35 CAD (<500 people) to $500 CAD (>200,000 people).

Ontario

The development of the Ontario Healthy Communities Coalition (OHCC) began at the same time as the Québec network but took longer to establish, in part because it is more community based and sought to engage with both community partners and key provincial organisations such as the Ontario Association of Landscape Architects, the Association of Municipalities of Ontario, the Ontario Professional Planners Institute, the Ontario Social Planning Council and the Ontario Environment Network, among others. There was a clear understanding that local communities rather than provincial organisations should set the agenda, so the coalition was structured to ensure that local communities had a majority on the board of directors. As a result, there is less engagement of the formal municipal government sector and more emphasis on local community organisations and informal communities.

The coalition was established in 1993. As of March 31, 2016, OHCC had 1,430 network members and subscribers from across Ontario. There are a 15-member board of directors drawn from across the province and many sectors, and a core staff of six people, half of whom are deployed across the province as community consultants. Most of its budget continues to come from the Ontario Ministry of Health and Long-Term Care, although it seeks and obtains funding from other sources for its many projects.

As of June 30, 2013, OHCC had 858 members (of whom 500 were organisational members), in 184 communities in Ontario. The community development approach is a central feature of the Ontario model, with a focus on process and some effort to weave together a variety of issues and take a holistic approach.

A good sense of the range of issues that the Coalition addresses and the funders it attracts can be seen from the following list of current and recent projects (http://www.ohcc-ccso.ca/en/about-us):

- *Healthy food for all: healthy and sustainable food systems in Ontario*, 2013–2015: a collaborative project of six organisations, funded by the Ontario Healthy Communities Fund
- *Accelerating rural transportation*, 2013–2014: a collaborative project with the Rural Ontario Institute, funded by the Ontario Trillium Foundation
- *Healthy Communities: an approach to action on health determinants in Canada*, 2009–2012: a 3-year project funded through the CLASP project of the Canadian Partnership against Cancer, in collaboration with three other provincial healthy communities networks
- *The future of the good food box*, 2009–2011: a collaborative project funded jointly by the Healthy Communities Fund and the Heart and Stroke Foundation's Spark Fund
- *FoodNet Ontario*, 2007–2010: a collaborative project funded by the Ontario Trillium Foundation
- *Healthy Communities and the built environment*, 2007–2008: a collaborative project funded by the Public Health Agency of Canada
- *Power down your power bills*, 2006–2007: in partnership with Rideau Environmental Action League, funded by the Ministry of Energy's Community Conservation Initiatives
- *Creating inclusive community organizations*, 2004–2006: funded by the Public Health Agency of Canada

The Healthy Community approach in Ontario spans everything from Canada's largest city to small municipalities and communities; the two case studies that follow describe the situation in Toronto (which no longer has an active Healthy City initiative for reasons made clear in the case study) and in the small township of Woolwich in the regional municipality of Waterloo.

Case Study 3: Healthy City Toronto 1989–1999—From Project to Office to Corporate Voice to Amalgamation and Dissolution[4]

Toronto, the capital of Ontario, was the birthplace of the modern Healthy Cities movement. It is the largest city in Canada, with a population of 2.5 million (2011 Census of Canada), and is part of an urbanised area, the Greater Toronto Area, of 6.3 million. It is one of the most ethno-racially diverse cities in the world, with 47 % of its population declaring themselves to be part of a visible minority in 2006; more than 140 languages and dialects are spoken by its citizens (http://www.toronto.ca/toronto_facts/diversity.htm).

[4] Fran Perkins.

At the beginning of the period covered by this case study, the City of Toronto was the old and reasonably progressive urban core, a municipality of 600,000 people within Metro Toronto, a generally very successful federated region of six municipalities, established in the 1950s and with a population in 1989 of about 2 million. The other five municipalities were the 'suburbs', although rapid suburban growth (sprawl, really) beyond the metropolitan boundary was under way. All this was to change in 1998, as described later.

In the 1980s the City of Toronto Public Health Department was at the leading edge of progressive public health. Following a groundbreaking 1978 report by the Board of Health, *Public Health in the 1980s* (Board of Health 1978), a sort of municipal version of the Lalonde report but more grounded and radical, it had reorganised to address the leading causes of death and disease, decentralised and regionalised, and had established innovative approaches to health promotion and advocacy. As part of this work, a new mission statement for the department declared its intention to make Toronto 'the healthiest city in North America' (Hancock 1986). This raised questions: What is a healthy city, how would we know it and how would we get one?

Some of these ideas were explored in the Beyond Health Care conference in 1984, organised by the Department of Public Health and the Board of Health in part to mark the centenary of the Board of Health (Beyond Health Care 1985). As described in Chap. 19, a forward-thinking day-long workshop, Healthy Toronto 2000, proved to be the seed that grew into the Healthy Cities movement in Europe. In Toronto, it led to the Board of Health establishing a Healthy Toronto 2000 Planning Committee in 1986.

In late 1988, after a 2-year process of research and consultation with the community and key stakeholders, the report 'Healthy Toronto 2000' was approved by the Board of Health (1988). The report laid out guidelines on how to create a healthier city, which included the establishment of the Healthy City Office (other recommendations included the creation of a Toronto Food Policy Council, which remains active to this day). The Healthy City Office (HCO) was established by a unanimous vote of Toronto City Council in 1989 with a two-pronged mandate: to work horizontally within the government corporation as a catalyst for change, and to work externally to build partnerships to help reach the goal of a Healthy City. Recognising the need to address the underlying conditions for health, the HCO was based within the city structure but outside public health.

From the very start the success and fluctuations in the role of the HCO were based on two key factors, political support and champions, and key people of influence within the civic structure. The project was guided by the core group of champions who were key decision-makers and acted as an advisory committee. The Office was stand-alone and reported directly to the Committee of Heads (of Departments), the highest decision-making body in the bureaucracy. It was new, innovative and exciting. It went through several incarnations, leaders and structures, but at its core was the WHO model for Healthy Cities, and it was grounded in health promotion principles.

The Toronto Healthy City Approach

Values	Triggers	Methodology

Visionary	Philosophical shift in public health	Developed a Healthy City Vision for Toronto
Integrative—the '3 Es' of environment, economy and equity	WHO conceptual leadership with local leaders at leading edge	Prepared a State of the City report
Address determinants of health	Political support	Strengthened citizen engagement and processes
Community participation connectedness and control valued	Champions within city bureaucracy	Established inter-sectoral coalitions and partnerships
Inclusive	Set budget and resources	Developed and guided the implementation of healthy public policies
Democratic decision-making	Adequate staffing	
Inter-sectoral action	Clear mandate	
Innovation and catalyst for change	Two major organisations within civic structure: staff and political	

Values and Process

The Toronto Healthy City model was a model for local governance, crossing all sectors and city departments. Residents, businesses, NGOs and the municipal government worked together to ensure a viable liveable city. It was clear that the goals belonged not just to public health or the Healthy City Office but the local government as a whole. Underlying all the strategies was the idea of democratisation: access to government processes and the importance of fostering the understanding of the Healthy City model.

Health City staff used three main 'process tools' to implement the strategies: visioning, analysis and action. Healthy City visioning is a method of identifying the hopes and dreams of the participants for the city. It is a positive approach. Residents wanted a clean, green, safe city where there was adequate employment and access to education and health. They wanted a healthy city in which to live, work and play. The Healthy City vision identified future trends, detailed hopes and dreams, and recognised past successes. The image used for the front cover of the State of the City report (below) is a child's vision of a Healthy City.

State of the City reporting was a way of organising facts for public and political use. Led by the HCO, different departments collected and analysed their pertinent data and helped prepare chapters for the report. This was to enable the city to research, analyse and benchmark data over time and identify areas for improvement and action. The chapters covered economy, environment, community health, education, housing, transportation and safety. Only one State of the City was produced because of cost and the availability of data. It was well used by many groups, and Toronto still prepares reports on individual issues (Fig. 9.1).

Action was based on the strategies of the Ottawa Charter for Health Promotion and the usual guidelines for Healthy City action, namely strengthening community action, creating inter-sectoral partnerships and developing healthy public policies.

Fig. 9.1 Toronto's first State of the City Report

Strengthening Community Action

Community capacity building and innovation were key to building skills and approaches with groups so that they could work on solving issues. They needed access to decision-makers and people with advocacy skills. Community grants were established, and groups were also recognised through The Healthy City Neighbourhood incentive awards (The Neighbourlies), which recognised groups that were helping make Toronto a better place to live.

In some instances the HCO incubated groups that were already formed but were having problems accessing key city decision-makers, both political and bureaucratic. One such group was called The Feet on the Street. The HCO offered them room in their office, so they would have an address, phone number and a place to meet in City Hall. The group was articulate and innovative, and eventually evolved to become one of the city's citizen committees, renamed the pedestrian committee.

From as far back as 1992 the HCO was part of the Green Tourism partnership, which brought together businesses, NGOs and individuals to establish a green tourism strategy. They produced an action report that survives to this day, and developed Green Tourism listings and guidelines. This initiative was part of a wider strategy called the Clean Air Strategy, led by the HCO; it was the first initiative given to the HCO by City Council, along with the Homeless initiative. By the time of the HCO's demise there were approximately 200 Healthy City initiatives in Toronto.

Inter-sectoral Action and Coalition Building

By establishing partnerships between citizens and city government, initiatives such as urban gardening, homeless persons' self-help and youth advocacy neighbourhoods were strengthened. An excellent example of success was the Clean Air Partnership (CAP), hosted by the HCO, an association of over 50 businesses, community groups and government agencies working on air quality issues for Toronto. It developed a programme called Air Share. It was complemented by the Toronto anti-smog strategy which the HCO coordinated across the city corporation, and which focused on greening the vehicle fleet and other initiatives to contribute to cleaning the air.

In 1996, following the civic election, a new mayor who had always supported the HCO undertook a radical reorganisation of City Hall. At this time the existing Toronto Mayor's and related Citizens Committees were moved into the HCO. These included the Mayor's Task Forces on drugs, on ageing and on community and race relations, as well as the Safe City Committee, and the Toronto Young People's Advisory Board, which had been a part of HCO since 1991. The HCO helped facilitate, coordinate and provide administrative support to these groups. This led to policy identification, public education and advocacy initiatives.

Healthy public policy was key to the success of the HCO. The positioning of the office enabled staff to work across city departments with internal and external partners, to recommend changes in policy. It also provided access to the city council through formal reports and recommendations that were debated. Many policy changes were implemented, such as the community safety strategy, clean air initiatives, youth employment initiative and homeless initiatives, to name a few.

Triggers

The Healthy City Office went through four clear phases.

1. *Innovation*: Toronto Health Department became famous during the 1980s for its new approach to public health, moving to a decentralised system of health promotion and initiating community development approaches to complement traditional public health methodology. The Healthy Cities movement grew out of this department and the strong connection with WHO. Initially there was much political commitment, and effective leaders in public health pushed the model forward. There was also a strong group of cross-department champions within city government (selected on the basis of being the rising stars within their respective departments) and a clear philosophy and model to follow. The city council provided adequate budget and staff and a clear mandate to follow. These were exciting times where many internal and external partnerships were formed.
2. *Choppy waters*: Political threats and fiscal cutbacks in 1993 suddenly created a political pushback to dissolve the HCO, and the council voted to disband it. After strong lobbying from the many community and inter-sectoral partners the HCO had developed—and who saw the benefits of the approach—the HCO was reinstated and the director changed.
3. *Stability*: In 1995 a new mayor was elected and the HCO budget and staff grew. Many policy initiatives were implemented and 3 years of stability and growth followed.
4. *Amalgamation, chaos and jockeying for power*: In 1998 a new, right-wing provincial government imposed amalgamation on the six municipalities of Metropolitan Toronto, in spite of the strong rejection of this approach by all six municipalities and by the citizenry in referenda in five of the six municipalities, by margins as high as 70 %. The old City of Toronto became amalgamated with the five cities and became a city of over two million. Politicians and bureaucrats with no concept of the Healthy City approach or innovative public health took key positions. There were no longer any political or internal champions for the Healthy City.

Downsizing and positioning for power led to large numbers of established staff and leadership leaving. Resources were reassigned from the Healthy City office. The chaos caused by this forced amalgamation still resonates over 14 years later. Many processes and committees did survive the amalgamation, but no structure was left with the mandate to work across departments and with internal and external partners.

Inevitably, the rise and fall of Toronto's Healthy City Office was, in the final analysis, political.

Case Study 4: Woolwich Healthy Communities

Source: http://www.healthywoolwich.org/

The Township of Woolwich is a rural community of 23,000 people in the region of Waterloo, about 75 miles west of Toronto, in rich farmland. Several attractive small towns, notably St. Jacob's with its large farmers' market and heavy presence of Mennonite and Amish people, have made it a major tourist destination.

Woolwich Healthy Communities (WHC) is a volunteer group that 'strives to promote and improve the health and well being of our community, economy and environment'. Formed in 1991, WHC began with a vision workshop; one of the most striking findings to emerge, which has been a central theme ever since, was that while the Grand River, which runs through St. Jacob's, was depicted in the images that people drew as being central to the life of the community—a place where people walked, swam, fished, played, boated, etc.—in reality that was not the case. This led to the creation of a Clean Waterways workgroup, which is still active:

The *Clean Waterways Group* improves water quality throughout Woolwich Township. Working with volunteers and landowners, we improve the health of Woolwich waterways by stabilising and rehabilitating creek banks. Schools and community members have helped the core volunteers to plant native trees and shrubs and install fencing along upper sections of the Canagagigue Creek. These activities reduce erosion and prevent pollutants from entering the waterway (*source:* http://www.healthywoolwich.org/Clean%20Waterways.html).

In 2008 the Clean Waterways group was honoured by the Grand River Conservation Authority with a Watershed Award for helping farmers protect and improve the quality of the creeks flowing through their land.

In addition to the Clean Waterways group, WHC in 2014 had four other working groups: trails, on-road cycling, environmental enhancement and a co-ordinating committee. Other groups, such as a well water quality group and a sustainable community group, have sprung up from time to time to address particular needs and interests.

- The environmental enhancement committee was formed in 2004 and in addition to developing an environmental enhancement plan for the township, it has developed and is managing a reforestation plan: Trees for Woolwich.
- The trails group promotes hiking and biking in the township, producing a map and guidebook to the 80 km of public trails, and organising hikes.
- The on-road cycling group promotes and encourages on-road cycling in the township and surrounding area, working with township staff and organising rides.

In 2007, the Ontario Healthy Communities Coalition awarded WHC a Community Recognition Award.

British Columbia

A Healthy Communities network was originally established in 1989 with provincial government funding, but lost its core funding in 1992 during the recession and subsequent cutbacks, and went out of existence. In 2005, with provincial Health Ministry funding, a new healthy communities initiative known as BC Healthy Communities was created; the Union of BC Municipalities (UBCM) acted as the administrative host. BC Healthy Communities (BCHC) became an independent NGO in 2012 and has worked to diversify its funding sources. As is the case in Ontario, membership is rather loosely defined and the focus is on community development and a holistic, integrated approach. Currently BCHC is working with close to 200 local governments and communities across the province.

As with the OHCC, its areas of focus are broad:

- Age-friendly: healthy, safe communities for everyone to live and thrive in
- Arts and culture: promoting arts and 'creative' activity for health and well-being
- Health and built environment: designing communities that support healthy choices
- Health promotion and chronic disease prevention: upstream approaches for improved health
- Healthy families: inclusive participation and involvement for all of community health
- Inclusion and diversity: fostering conditions for diversity and inclusion
- Sustainable communities: engagement and planning for healthy environments
- Youth: youth empowerment for health and well-being (http://bchealthycommunities.ca/focus_areas)

BCHC has a staff of nine people with a wide range of experience and expertise, some based in various regions in BC. They provide facilitation, consultation, leadership training, community planning and engagement, research and evaluation services that 'support community groups and local governments to assess, enhance and mobilize the capacities they need to continually create and improve community health and well-being'. Current programs include the following:

- *Plan H*, which supports local government engagement and partnerships across sectors for creating healthier communities
- *Municipal Alcohol Policy* (MAP), which helps support healthy communities by providing clear guidelines for alcohol use in municipal settings and contributing to a culture of moderation in the community
- Development of *Youth Councils* in Victoria, Surrey and other municipalities
- A provincial *Age Friendly Communities Recognition Award* designed 'to help local governments increase awareness and garner support and commitment to the age-friendly initiative from key partners in their communities'
- *Building Shared Leadership for Climate Action*, which 'supports multiple sectors to learn about, develop and implement effective approaches for engagement and collaboration on climate change'

- The Fraser Basin Council's *Smart Planning for Communities Program*, which provides 'opportunities to support and foster local government actions on community sustainability' (http://bchealthycommunities.ca/programs).

In addition, BCHC runs a number of projects within just one or two municipalities.

New Brunswick

The Acadians are a Francophone community, about 25,000 strong, inhabiting the Acadian region of New Brunswick and other parts of the Maritime provinces. Expelled to Louisiana and other places by the British in the eighteenth century, they became the Cajuns; but some later returned to their Canadian homelands. The Mouvement Acadien des Communautés en Santé (MACS-NB) is a Francophone, provincial network that was established in 1999 in New Brunswick and has some 80 members' groups, including 15 municipalities, a number of colleges, health centres and hospitals, and 35 Healthy Schools (http://www.macsnb.ca/index.php). Like other Canadian networks, its five staff members focus mainly on community development and capacity building, with special attention to linguistic and cultural inclusion.

These provincial networks take a broad approach that links environmental, social and economic factors, and all facilitate and support collaborative action within communities. The Québec initiative has a particularly strong focus on and link to municipal governments, while the Ontario initiative has a strong focus on community-based organisations and networks; the BC initiative, learning from the experience of its two senior partners, does both, being located at the Union of BC Municipalities but having a strong focus on community capacity building. All have somewhat similar funds and staff, and manage to reach, work with and support a large number of communities with what are quite modest budgets.

The Healthy Cities and Communities Movement in the United States[5]

> 'A healthy community is a garden to grow people in.'
> James Rouse, community developer and philanthropist 1914–1996.

---

[5]Tyler Norris. As coordinator of the Colorado Healthy Communities Initiative in the early 1990s, Tyler Norris, MDiv, served as an early leader in the US Healthy Communities movement, helping spread the growth and deepen the impact of the field to this day, since the late 1980s. This case study was in part adapted from two special issues of the National Civic Review, celebrating 25 years of Healthy Communities in the United States. See http://onlinelibrary.wiley.com/doi/10.1002/ncr.v102.4/issuetoc.

The Healthy Cities and Communities movement in the United States comprises thousands of independent multi-issue, place-based, cross-sector, trans-partisan, collaborative initiatives working to improve the well-being of people and places.[6] Signs of the movement can be seen everywhere across the country, in visionary planning, priority-setting and civic mobilisation efforts, and operating at neighbourhood, institutional, municipal, regional, state and national levels working towards resilient equitable community development. There is no one national organisation serving as a coordinating entity, no primary funding source and no common branding protocol. Many of the initiatives do not use the terms 'Healthy City' or 'Healthy Community' in their names, although they share common values and methods and seek to deliver similar types of impacts.

Absent a unifying identity separate from the patterns that are visible in the nature and spread of its work, the US movement is best viewed as a widely distributed *praxis* of democratic processes of civic engagement, collaboration and innovation, shaping practices, policies and investments in and across sectors and disciplines. The US movement is less a field, than a boundary-crossing field of fields, part of a broader movement that embraces themes such as sustainability, liveability, equity, safety and resilience. Distinct from a centrally organised enterprise, it has emerged as a kind of civic operating system, a DNA-like phenomenon expressing itself in a diversity of settings via common memes and approaches to inclusionary participation, whole-system design and strategies that can concurrently contribute to both population health and complementary factors ranging from economic opportunity for all to preservation of healthy ecosystems.

Inspired by the Healthy Cities project in Europe, and advanced by the early leadership of individuals such as Leonard Duhl, Trevor Hancock, Kathryn Johnson (Health Forum), Judith Kurland (Boston), Michael McGinnis (Healthy People), John Parr (National Civic League), Joan Twiss (California Healthy Cities), Peter Lee (South Carolina Department of Health, and later Massachusetts Healthy Communities Partnership), and Mary Pittman (Health Research and Educational Trust, and then later Public Health Institute) the US movement initially emerged on both coasts in the late 1980s. The National Civic League, the Health Forum, United Way Worldwide, the YMCA of the USA and other national partners in the early 1990s activated work in hundreds of communities across the nation that were inclusive of, but not limited to, the jurisdiction of cities.

There are many highlights from this early phase of development, several of which have been ongoing for decades:

- *The California Healthy Cities and Communities Program* is state-wide; established in 1988, it was the first and remains the largest of its kind in the United States. The initiative is part of the Centre for Civic Partnerships, in turn part of the Public Health Institute, and has sparked citywide projects across the state, endorsed by municipalities: 'For the last 25 years, the Program has engaged thousands of

[6] http://www.communitycommons.org provides links to stories from hundreds of US initiatives, as well as access to community tools and GIS mapping resources, and ongoing leadership dialogue concerning the future of the US movement for healthy people and places.

municipal and public health officials and community leaders from all walks of life to address the leading health threats through programs, policies, and environmental interventions' (California Healthy Cities and Communities Program, 2013). Over time more than 75 communities have received grants from the initiative, ranging in size from San Francisco to small desert communities (http://www.civicpartnerships.org/#!ca-healthy-cities-and-communities-progra/cjhg).

- *Healthy Boston*, a city-led process, started with two dozen grass-roots-style neighbourhood-level initiatives. It inspired a state programme (Healthy Communities Massachusetts)—there is still an Office of Healthy Communities in the Massachusetts Department of Public Health—which evolved into the Massachusetts Healthy Communities Partnership (http://masspartnership.org/). One of their key activities, from 2003 to 2010, was to bring together '68 cross-sector teams from communities across Massachusetts … on a monthly basis to learn and practice the application of the Healthy Communities principles'. This had a 'positive long-term impact on sustainable, cross-sector, and community-driven collaboration as a basis for generating constructive change' (Valovic et al. 2013).
- *The Colorado Healthy Communities Initiative* was a partnership between the Colorado Trust (itself the legacy of the sale of a public hospital to a private health care system, a unique funding model for a Healthy Communities initiative!) and the National Civic League, which is headquartered in Denver. They created a well-funded 8-year multi-community project with a built-in evaluation. The evaluation report noted that 29 communities received planning grants, and of these 27 received $100,000 implementation grants. The communities ranged from inner city neighbourhoods to fairly remote and rural regions.

Five years into the project, evaluators found engaged projects with diverse stakeholders and an average of six projects per community, focusing on a mix of specific issues (e.g. transportation) and broader community development and participation processes. The communities were mostly achieving important outcomes and benefits, and a few 'were beginning to show progress toward long-term goals such as community decision-making and governance' (Colorado Trust 2002). Ten years later, there was found to be a strong correlation between each community's success and its 'quality of process' rating. The three most important success factors were being in touch with community needs, having a high level of social capital (committed participants) and having a strong leader with good facilitation skills; the greatest challenges identified by participants were obtaining sufficient funds and attracting and engaging enough committed people (The Colorado Trust 2002). Nineteen communities were still active, and the researchers concluded:

What seems to sustain the projects now, 10 years since the planning phase began, are two things:

- A good process, worthy of the continued investment of time and energy
- Recognisable attainment of goals, concrete impact on root problems and visible signs of impact on community processes and practices.

Now, many of the early CHCI sites are leaders with the Live Well Colorado initiative, a state-wide campaign focused on healthy eating and active living (www.livewellcolorado.org).

- *The Health Forum* was a leading-edge California-based hospital/health care association (now incorporated in the American Hospital Association's Health Research and Educational Trust) which promoted the Healthy Communities approach in the health care sector and beyond. The Forum initiated an annual Healthier Communities Award and a year-long leadership training programme, the Creating Healthier Communities Fellowship, which ran for 12 years from 1990.

From these early beginnings the movement has blossomed. Its accomplishments and experiences can be found in a series of articles in two special editions of the *National Civic Review* (102(4), 2013 and 103(1), 2014) published to mark 25 years of Healthy Communities in the United States. The US movement is reflected in the leadership strategies, policies and investments of thousands of diverse community collaboratives involving hospitals and health systems; philanthropies; Chambers of Commerce and community development partnerships; United Way agencies; YMCAs; K-12 schools and institutions of higher learning; local, state, and national governments; faith-based partnerships; media; entities such as the Convergence Partnership and the Partnership for a Healthier America; and all manner of civic and social networks.

Arising as if a healthy immune response to the great challenges of the day, these initiatives are rooted in the finest traditions of American participatory democracy. The foci and impacts of the US movement include access to healthy, fresh, regional affordable foods and beverages, more active forms of transportation and daily physical activity, a strong focus on health equity and social inclusion for those groups with the highest disease burden, and meaningful access to life potentiation and providing a family wage via education, skill development and connection to opportunity both in the emerging knowledge economy and in traditional manufacturing. These local and regional efforts are characterised by compelling modes of civic engagement and innovation that build social capital such as trust and reciprocity, and are guided by leadership working skilfully across lines of politics, perspectives, sectors, issues, jurisdictions and generations. In rural areas they have sought to preserve local culture and ownership while boosting the prospects for economic prosperity. In metropolitan areas these approaches have helped fuel an urban renaissance that is the engine of smart growth and resilience, and which can be characterised as homeland security from the inside out.

More recently, Healthy Community efforts are being connected, fortified and invested in by forward-looking regional, state and national-level organisations and agencies, as well as by diverse funders and social impact investors. They are gaining access to robust data engines and social engagement platforms with GIS mapping capability and crowd-sourcing to help tell stories, target interventions, track outcomes, facilitate peer learning and build constituencies for action.

A dialogue-based, nationwide participatory research project that included hundreds of US communities and thousands of civic leaders conducted over a decade ago (Norris and Pittman 2000) was revalidated as a core operating system for the field by many leaders in the movement.[7] Across a diversity of initiative types, these

[7] The 'seven patterns' initially defined in 2000 by the Coalition for Healthy Cities and Communities were reaffirmed at an annual meeting of the Association for Community Health Improvement held in Jacksonville, Florida, in March 2010.

seven characteristics or patterns of a healthy community were gleaned from hundreds of dialogues engaging thousands of participants around the United States. The findings from this study suggest that a healthy community:

- Practises ongoing dialogue among residents to build relationships and the capacity to effect positive change
- Generates leadership everywhere within the community, and fosters a leadership style that emphasises facilitation and collaboration, encouraging coalitions and partnerships
- Shapes its future based on creating and implementing a shared vision of the community
- Embraces the diversity of its residents and the gifts and competencies they bring
- Gathers information about its assets and needs to inform decision-making and policy
- Connects people to community resources
- Creates a sense of belonging and shared responsibility among its residents

The graphic below illustrates the seven patterns of what brings greatest efficacy and results to health and quality of life through collaborative partnership.

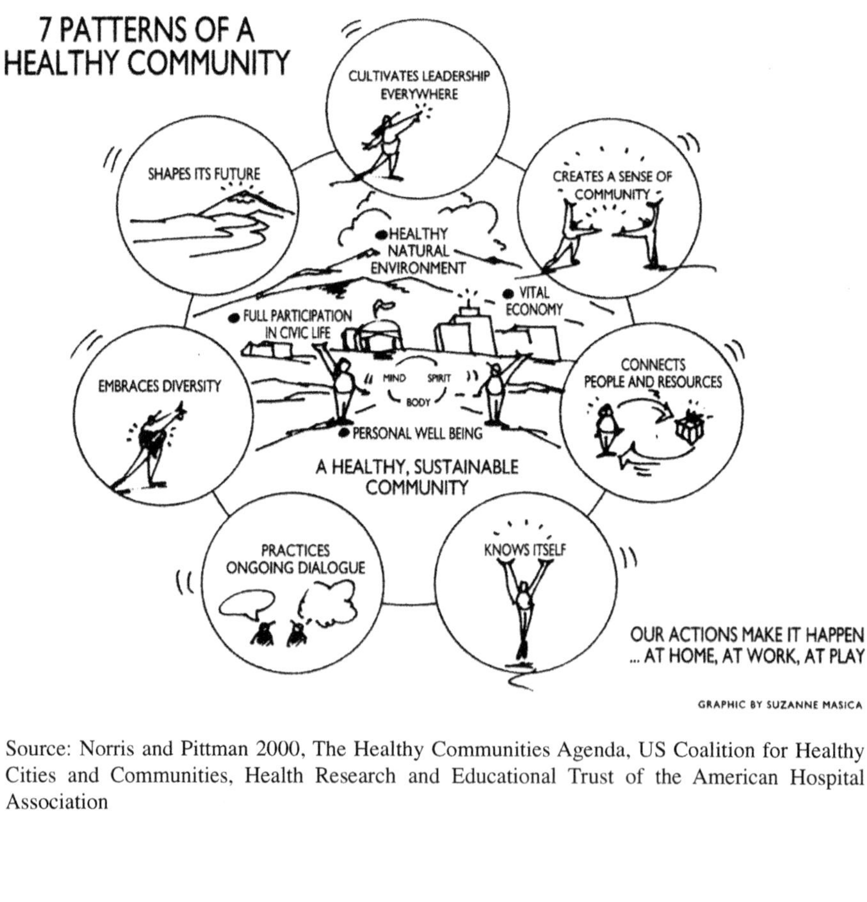

Source: Norris and Pittman 2000, The Healthy Communities Agenda, US Coalition for Healthy Cities and Communities, Health Research and Educational Trust of the American Hospital Association

In an era when American democracy and the prospects for equitable economic prosperity are challenged by toxic partisanship and the influence of powerful vested interests, resident-engaged local democracy rooted in civic life and asset-based result-producing creativity is a welcome occurrence. Three important themes to consider as the movement enters its second quarter century include the following:

- *Boundary-crossing leadership*: whatever divides us, that which connects us is greater still. The leaders of the healthy community movement emerge from diverse backgrounds and are found in every sector and setting.
- *Complementary benefits for impact at scale*: a good solution solves many problems. Those leaders and initiatives with the most widespread and sustainable support and impact over time practise *systems-thinking*. They work to coalesce partnerships with players who may have *divergent* interests and missions, but can benefit from *convergent* strategies, practices, policies and investments.
- *A blend of art and science*: assuring access to accurate and granular data on current realities, trends and outcomes over time and effective processes for community meaning-making, discernment and implementation, go hand in hand. The use of GIS data engines and mapping tools can power robust community assessments linked to an ever-expanding evidence base on 'what works': that is, developing the community will and accountability to act with 'dose-sufficient' approaches of reach (population), intensity (strength) and duration (time).

A quarter century of learning from the US Healthy Cities and Communities movement offers insight and constituency for wise investment and policy choices aimed at equitably and sustainably improving economic, ecological, social and population health. Steeped in the science of the determinants of health and the arts of civic engagement and community building, this diverse and widely distributed movement of public, private, non-profit and sector collaborations has matured into a formidable outcome-producing force for the sustained renewal of the nation and its communities.

New and Emerging Developments in North America

One of the reasons that the HCC movement has persisted all these years, in spite of a relative lack of government funding, is that it is very flexible and adaptable. Because what constitutes a healthy community is defined by the community itself, the issues that are important to that community will differ from those of neighbouring communities, and moreover will change over time. As can be seen in the examples given above, HCC initiatives cover a very wide range of environmental, social, economic and other issues. They intersect with a number of other city- or community-focused initiatives, including those trying to help cities and communities be sustainable, safe, green, liveable, child friendly, age friendly and so on. In fact, all of these 'movements' help to improve health and well-being or, more broadly, human development, and thus are complementary to, and may even become the main focus of, an HCC initiative.

A particularly important theme that has emerged in North America in the past decade has been the links between the health of a community and urban planning, which has helped to reunite the public health and urban planning professions. In Canada this has included an ongoing interest in Healthy Community concepts at the Canadian Institute of Planners (Christy et al 2012).[8] Their resources include the following:

- *Healthy Communities legislative comparison survey*: a high-level survey of planners across Canada to compare the legislative, administrative and policy frameworks that support the creation of healthy communities (December 2013)
- *Healthy Communities practice guide*: designed for planning practitioners to help them in their work towards healthier communities, and to increase the understanding of the supporting role that health practitioners can play in reaching their common goal (2012)

Another recent Canadian development has been Healthy Canada by Design, which was

> a partnership of national health, planning and transportation organizations, regional and local health authorities, non-governmental organizations and university researchers that are collaborating to bring health considerations into land use and transportation planning processes with the goal of creating healthy communities that support active transportation and physical activity. (http://hcbd-clasp.com/about-us/)

The partnership was established in 2009 and was funded through the Canadian Partnership Against Cancer's Coalitions Linking Action and Science for Prevention (CLASP) project; its secretariat was based at the Heart and Stroke Foundation of Canada. In its 5 years of operation it supported the creation of a wide range of resources, available though its members' websites, and it hosted regular webinars and an active blog. Although the funding has ceased, the resources are still available through the legacy website.

There has also been a strong interest in urban planning and the built environment in the United States, where in addition to the groundbreaking book *Urban Sprawl and Public Health* (Frumkin, Frank and Jackson, 2004), the CDC has a Healthy Community Design initiative (http://www.cdc.gov/healthyplaces/) which includes a toolkit and checklist, as well as resources for a wide range of issues including accessibility, aging, air quality, children's health, gentrification, healthy food, injury, mental health, physical activity, social capital, transportation and water quality. In addition there has been a four-part PBS TV series (http://designinghealthycommunities.org/about/) and book, *Designing Healthy Communities* (Jackson 2011).

---

[8] See http://www.cip-icu.ca/Topics-in-Planning/Healthy-Communities#.

Of particular note and importance is the growing importance of health impact assessments in the United States (National Research Council 2011). The Health Impact Project, funded by the Robert Wood Johnston Foundation and the Pew Charitable Trusts, is a broad-based partnership that has encouraged and supported such assessments in America. By 2013, more than 275 HIAs had been identified, many in communities (see interactive map at http://www.healthimpactproject.org/hia/us).

Reflections and Conclusion

The Healthy Cities approach has been quite widely, if somewhat inconsistently, adopted in North America and has remained alive and reasonably well for a quarter century. It has adapted to focus as much or more on smaller communities as on large cities and towns and to be based more in communities than in municipal governments; and it has been adopted by a wide variety of local, state and national partners from the public, non-profit, philanthropic and private sectors.

What has been missing, on the whole, has been its widespread adoption by provincial/state or federal governments as an instrument of public policy. Significant funding and support from these levels of government have been largely lacking, or at too low a level to ensure an effective 'dose', and too subject to short-term funding commitments. As a result provincial or state organisations are the exception rather than the rule, and national organisations are absent in both Canada and the United States.

Moreover, there has been a disturbing tendency in many governments to see Healthy Cities and Communities as simply a community-based strategy for supporting the government's priorities for individual behavioural change focused on a small set of standard health factors affecting lifestyle: smoking, physical inactivity, unhealthy eating and sexual behaviour, and unintentional injury stemming from 'risky behaviour'. On the positive side, a wide range of partnerships have been developed to address a broad set of issues. In particular there has been considerable interest in the health impacts of the built environment and transportation, and a re-forging of the historic links between public health and urban planning. Other commonly addressed issues include sustainability, food security, community safety, age- and child-friendly design and air quality.

The first quarter century of Healthy Cities and Communities in North America has laid a foundation for civic action across wide sectors of society and in communities of all sizes. However, there has not yet been much emphasis on some of the more profound issues that underlie the concept of HCC, in particular the issues of social equity and justice, ecological sustainability and reform of the democratic process. These are the next challenges for Healthy Cities and Communities in Canada and the United States.

References

Armstrong, A. (1959). Thomas Adams and the Commission on Conservation. *Plan Canada, 1*(1), 14–32.

Beyond Health Care. (1985). Proceedings of a working conference on healthy public policy. *Canadian Journal of Public Health, 76*(Suppl. 1), 1–104.

Board of Health. (1978). *Public health in the 1980s*. Toronto: Local Board of Health, City of Toronto.

Board of Health. (1988). *Healthy Toronto 2000*. Toronto: Local Board of Health, City of Toronto.

California Healthy Cities and Communities Program. (2013). *Twenty five years*. Oakland: The Public Health Institute.

Christy, H., Harrison, D., McKibbon, G., Miro, A., Pantelimon, O., & McKibbon, A. (2012). Our 21st century challenge: Healthier communities. *Plan Canada,* Spring: 53–55.

Colorado Trust. (2002). *Colorado healthy communities initiative: Ten years later*. Denver, CO: The Colorado Trust.

Commission on Conservation. (1912). *Annual Report* (p. 148) Ottawa: Commission on Conservation.

Commission on Conservation. (1915) *Annual Report* (p. 270H). Ottawa: Commission on Conservation.

Employment and Social Development Canada. (2014). *Canadians in Context*. Retrieved February 26, 2014, from http://www4.hrsdc.gc.ca/d.4m.1.3n@-eng.jsp?did=6

Frumkin, H., Frank, L., & Jackson, R. (2004). *Urban sprawl and public health: Designing, planning and building for healthy communities*. Washington, DC: Island Press.

Hancock, T. (1986). Public health planning in the City of Toronto. Part 1. Conceptual planning. *Canadian Journal of Public Health, 77*, 180–84.

Intergovernmental Panel on Climate Change. (2014). Climate change 2014: Impacts, adaptation, and vulnerability.

Jackson, R. (2011). *Designing healthy communities*. San Francisco, CA: Jossey-Bass.

Kretzmann, J., & McKnight, J. (1993). *Building communities from the inside out: A path toward finding and mobilizing a community's assets*. Evanston, IL: Center for Urban Affairs and Policy Research, Northwestern University.

National Research Council. (2011). *Improving health in the United States: The role of health impact assessment*. Washington, DC: The National Academies Press.

Norris, T., & Pittman, M. (2000). The healthy communities movement and the coalition for healthier cities and communities. *Public Health Reports, 115*(2–3), 118–124.

United Nations, Department of Economic and Social Affairs, Population Division. (2012). *World Urbanization Prospects: The 2011 Revision—Highlights*. Working Paper No. ESA/P/WP. 224.

Valovic, S., Lee, P., O'Connor, C., & Ayers, A. (2013). *Building healthier communities: An evidence-based training model to increase effective cross-sector collaboration*. Presentation and poster at the American Public Health Association Annual Meeting in Boston, MA. Retrieved from November 5, 2013, from http://masspartnership.org/news/18/99/MassForum-Impact-Presented-at-APHA.html.

Chapter 10
Healthy Cities in Europe: Structured, Unique, and Thoughtful

Helen Wilding, Ronald Gould, Julia Taylor, Annette Sabouraud, Patricia Saraux-Salaün, Dionysia Papathanasopoulou, Antonio de Blasio, Zsuzsanna Nagy, and Jean Simos

The World Health Organization's European Healthy Cities Network

Introduction

The European Regional Office of the World Health Organization (WHO/Euro) launched their Healthy Cities project in 1986 as a vehicle to advance the Health for All Strategy and put into action the words and principles of the Ottawa Charter for Health Promotion (Ashton et al. 1986). The project endeavoured

> to put health high on the agenda of political decision makers, key groups in the city and the population at large … to develop feasible strategies for reorienting public health endeavours at city level and to make prevention and health promotion a highly visible and community-supported enterprise. (Kickbusch 1989, p. 77)

In fulfilling these aims, the project anticipated improvements in the health and well-being of the people who live and work in Europe's cities (Tsouros 1990).

H. Wilding, B.Sc., M.B.A., M.Sc.
UK Healthy Cities Network, Newcastle upon Tyne, UK

R. Gould
Lord Mayor of the City of Liverpool, Liverpool, UK

J. Taylor
Liverpool Healthy Citie, Liverpool, UK

A. Sabouraud
Health of the City of Rennes, Rennes, France

P. Saraux-Salaün
Générale Déléguée à la Cohésion Sociale et aux Territoires, MSP, Nantes, France

D. Papathanasopoulou
Hellenic Healthy Cities Network, Amaroussion, Greece

E. de Leeuw, J. Simos (eds.), *Healthy Cities*,
DOI 10.1007/978-1-4939-6694-3_10

Working directly with local governments rather than national ministries was a new departure for WHO/Euro. In doing so, WHO recognised that cities could make a difference to public health as they are the lowest level of political mandate and offer good prospects for participation (Ashton et al. 1986; Hancock and Duhl 1986). The initial intent was to work with four to six cities for 5 years (Ashton 1992) but the 'healthy city' idea caught on and as a result a total of 34 cities were involved in that first phase (1987–1992) (Tsouros 2009). The extent of interest was such that national networks started to form within Europe and elsewhere around the globe—the project felt more like a social movement (Kickbusch 1989; Tsouros 1990). Nearly 30 years on and the 'long-term development project' (Tsouros 1990, p. 11), now referred to as the WHO European Healthy Cities Network (WHO-EHCN) has moved into its sixth 5-year phase (2014–2018) shaped by the new WHO European policy and strategy for health and well-being, Health 2020 (World Health Organization Regional Office for Europe 2012). The intention is to designate around 100 cities as members and reach an even greater number of towns and cities through 30 accredited national networks. The Strategic goals of WHO/EHCN (Box 10.1) still remain very close to the original project's aims.

As it was a completely new initiative taking place in a variety of national policy contexts, there was no prior evidence that the efforts of Healthy Cities would improve health. However, the underlying theory implies it should work (de Leeuw and Skovgaard 2005). The next few subsections describe four dimensions of the underlying theory drawn on by WHO/EHCN (ecological model; salutogenic perspective; community participation; and management and change) and provide insights into related practice. The section subsequently discusses the potential influences of a healthy city's national context and highlights the complexities involved in evaluating the work.

Ecological Model

Healthy Cities (and other healthy settings approaches) draw on an ecological model of health. This model recognises that health (or disease) status is the emergent property of the complex and dynamic interplay of a range of environmental, social, behavioural and personal factors that impact on an individual positively or negatively throughout

A. de Blasio
Hungarian National Network of Healthy Cities, Pécs, Hungary

Z. Nagy
Healthy City Foundation of Pécs, Pécs, Hungary

J. Simos (✉)
Institute of Global Health, University of Geneva, Chemin des Mines 9, Geneva, CH-1202 Switzerland
e-mail: jean.simos@unige.ch

Box 10.1: Strategic Goals of the WHO European Healthy Cities Network

To promote action to put health high on the social and political agenda of cities

To promote policies and action for health and sustainable development at the local level emphasising addressing the determinants of health, equity in health and the principles of the European policies Health for All and Health 2020

To promote intersectoral and participatory governance for health, health and equity in all local policies and integrated planning for health

To generate policy and practice expertise, good evidence, knowledge and methods that can be used to promote health in all cities in the European Region

To promote solidarity, cooperation and working links between European cities and networks of local authorities and partnerships with agencies concerned with urban issues

To increase the accessibility of the WHO European Network to all Member States in the European Region

(Source: World Health Organization Regional Office for Europe 2013)

their life (Dooris 2005). In the 1980s, Healthy Cities pioneers (e.g. Giroult 1989) drew from the field of human ecology which uses principles from natural ecology to study the interactions between humans and the human-designed and natural environments in which they live. Six elements of an urban environment can be distinguished (social; economic; psychological; physical; chemical and healthy biological) and it was proposed that inequalities in access to health improving environments may have a greater impact than inequalities in income (Giroult 1989).

In the years since WHO/EHCN launched, research has continued to reinforce this model, with greater knowledge of the role of social and cultural resources (Ziglio et al. 2000) and social determinants (Tsouros 2013). There is a greater understanding of the need to balance changes to settings with interventions that enable individuals to benefit from these changes (Shareck et al. 2013). At the same time, there is emerging and tentative evidence that the narrower approaches which focus on health education to encourage lifestyle changes may exacerbate health inequalities whilst focussing upstream holds more potential to reduce them (Lorenc et al. 2013). However, research does still seem to focus on identifying cause and effect and isolating single variables in a reductionist manner (de Leeuw 2011), rather than looking at the emergent effects of multiple interdependent factors.

Ecological public health is described as drawing from, and adding to, the sanitary-environmental; social-behavioural; bio-medical and techno-economic models of public health and therefore requiring the input of multiple professions with a sustainable development focus (Rayner and Lang 2012). Within this ecological model of health, it is plausible that appropriately working with, and through, other sectors to modify the determinants of health will improve public health. It is also plausible that if this is done in a way that leads to greater equity in the conditions in which people live then inequalities in health will be reduced. Therefore, a practical implication of adopting the ecological model is the importance of intersectoral action. Since its inception, WHO/EHCN has required Healthy Cities to establish intersectoral partnerships and cities have developed both formal and informal working arrangements which have developed over time (Green et al. 2009b; Lipp et al. 2013). Working with and through intersectoral partnerships is challenged by regular changes to structures and personnel (Lipp et al. 2013) with the implication that establishing intersectoral partnerships is not a one-off event but an ongoing process of continually striving to maintain and develop a way of working. However, in spite of the plausibility of the importance of working across sectors, research into general English public health partnerships concluded that there is no clear evidence that partnership working actually leads to health gains (Smith et al. 2009).

Salutogenic Perspective

In addition, Healthy Cities was informed by Antonovsky's introduction of the salutogenic perspective and his question 'what creates health' (Kickbusch 1996, 2003). It has been proposed that it is this salutogenic focus that marks out WHO/EHCN more than its rejection of the medical model for a social one (Kelly et al. 1993). Antonovsky (1996) critiqued the prevailing pathogenic focus—seeking to understand and act on the risks that leads to a disease—because in this perspective health equates to being low on risk. He proposed that it would also be productive to seek to understand and act on salutary factors (those that actively promote health). Drawing on Antonovsky's work, WHO advocated that healthy settings required a shift of 'focus from the deficit model of disease to the health potentials inherent in the social and institutional settings of everyday life' (Kickbusch 1996, p. 5). Unfortunately, in spite of the challenge offered by Antonovsky, the salutogenic model has not received much ongoing attention in health promotion research (Ziglio et al. 2000; Mittelmark and Bull 2013) and the pathogenic epistemology still has a strong presence in research on the social determinants of health (Clavier and de Leeuw 2013). It seems that the 'safety' of disease prevention approaches with their 'empirical knowledge, sophisticated techniques and methodologies, and evidence that many problems can be understood and managed' (Antonovsky 1996, p. 12) have proven too attractive to researchers and their funders.

The adoption of a salutogenic perspective has not been a specific focus of WHO/EHCN evaluation but it is interesting to note that the common Healthy Cities Indicator set (Webster and Sanderson 2013) includes mortality and cause of death data stemming from the pathogenic paradigm and no equivalent positive health measures. This suggests that the salutogenic perspective may not be embedded as part of practice. This continued use of disease-related measures is a more general area of concern and there are calls for public health to more proactively link to developments on measuring well-being (Kickbusch 2012).

Community Participation

The Ottawa Charter and therefore Healthy Cities also highlights the importance of community participation and empowerment drawing on the participation of some key players in women and gay rights movements (Kickbusch 2007). Duhl, one of the early adopters of the term 'healthy city', highlights how success requires conditions where people of a city can work together to deal with the determinants of health to create 'their healthy city' (Duhl 1986, p. 59). Getting participation and community ownership is seen as fundamental to the sustainability of health improving policies and practice as governments change (Boonekamp et al. 1999; Werna et al. 1999) and is particularly important for issues that require sustained activism (Nutbeam 1998). This part of WHO's approach was critiqued for being atheoretical (Dooris and Heritage 2013) and particularly challenged the dominant medical public health approach (French in Berridge et al. 2006). However, there is now a greater understanding of the potential health gains from increasing people's sense of control, for example, through research that links low control at work to poorer health (Marmot and Bell 2010).

Healthy Cities have used a range of approaches to inform, consult, empower and work in partnership with local people (Heritage and Dooris 2009; Dooris and Heritage 2013). Research that took place in Valencian Healthy Cities highlighted that community engagement had helped public sector become aware of communities' views, but there had been little participation in actual decision-making (Boonekamp et al. 1999). More generally within research, concern has been raised that whilst community empowerment can have positive health and social outcomes, practised badly it can harm health (Popay et al. 2010). A systematic review concluded that initiatives have demonstrated improvements on some determinants of health but not to population health or the quality of services (Milton et al. 2012).

Management and Change

Finally, Healthy Cities also highlighted that the ecological, salutogenic and participative focus invited different styles of management and development. Traditionally public bureaucracies were associated with prescriptive requirements, rational planning, hierarchy and control. However, a consequence of the ecological model of health with a focus on dynamic, interacting factors and the involvement of different sectors as well as the community is the need to accept that prediction, certainty and control are neither desirable nor feasible. Concerns like health, sustainability and equity require

> a new style of management in which power is wielded by influence rather than authority and is shared with people rather than held over them; involves negotiation rather than direction, requires collegial rather than hierarchical relations, enhances collaboration rather than competition, and takes a holistic rather than a sectoral approach (Hancock 1993, pp. 15/16).

Healthy Cities drew on management theorists such as Tom Peters who highlighted that working within complexity required an adaptive learning approach (Kickbusch 1989). In embarking on WHO/EHCN, WHO/Euro had to adopt a more facilitative approach (Ashton et al. 1986). It also emphasised networking as a new form of governance that unites independent members to a common purpose and constantly creates 'new connections and relationships' (Kickbusch 1999, p. 452). Networks also offer a strong platform to advocate for national or even global change (Adams 1989) so that the global and the local can influence each other (Kickbusch 1999). Framing Healthy Cities as a 'movement' brought into focus the importance of social change. This type of change is slow, unpredictable and opportunistic (Nutbeam 1998) and needs to draw on organizational and community development practices (Paton et al. 2005) rather than traditional change management.

No research has been undertaken as to the degree to which Healthy Cities have adopted the new managerial paradigm but the more 'traditional' approach does seem to re-surface in a positivist focus on measurement and indicators (Stevenson and Burke 1991) and rational approaches to city health planning (Green et al. 2009a; Green 2012). It has also been claimed for all its emphasis on participation and recognising a range of contributions, WHO still retains an expert-led model in its own practices (Seedhouse 2004) and in spite of using the language of social movements, the work of Healthy Cities are professionally dominated and aligned to bureaucracies (Stevenson and Burke 1991; Baum 1993).

European Healthy Cities in Their National Context

Whilst WHO/Euro chose to 'leapfrog' national governments to work directly with cities, the national policy context is not 'neutral' to the endeavours of a Healthy City and the health of those who live or work there.

It has been noted that the Health for All/Ottawa Charter principles and ideas have not inspired a Healthy Nations project (Lindstrom and Eriksson 2009) and that the 'political ideology of many European governments does not acknowledge the role played by socio-environmental conditions in determining health' (Goumans and Springett 1997, p. 314). National context will influence where cities allocate their resources, including staff time and energy, so Healthy Cities are likely to be more successful where there is a supportive national strategy (Goumans and Springett 1997).

Healthy Cities requires moving away from a medical model (Kickbusch 1989) and many of the ideas the movement emphasise population not individuals; health not illness/disease; holistic focus rather than reductionist; and emergent rather than linear models of causation (Dooris 2005). However, research in England, for example, identified a prevailing bio-medical paradigm (Smith 2008) which provides a key cultural barrier to the realisation of Health for All principles in day-to-day practice (Berkeley and Springett 2006). More generally in the UK, in spite of government commissioned reports which acknowledge the importance of action on the upstream determinants (Hunter 2007), there is still a tendency for policy and practice to revert to focussing on actions to modify individual lifestyle behaviours. This 'lifestyle drift' (Popay et al. 2010) is reinforced through the amenability of lifestyle interventions to be researched through experimental designs; through the interests of research funders; and through policy, departments and specialties that are structured by disease or lifestyle silos (Katikireddi et al. 2013).

An additional barrier to the success of Health for All initiatives, identified in research in England, is the new public management ethos (Berkeley and Springett 2006). New public management (NPM) has led to the use of a target-driven planning and performance management, which requires predictability, certainty and controllability. This impacts on those at a local level. For example, the English NHS performance regime influences not only the decisions and actions of those with a role in improving health inequalities but also the way in which they frame the issue in their discourse—actors focus 'downstream' on early detection and secondary prevention, rather than the broader upstream interventions where there is less certainty and longer time horizons (Blackman et al. 2010, 2012).

The national policy context can therefore create barriers for a healthy city to be true to the underlying evidence base in its day-to-day work.

Evaluating WHO/EHCN

Whilst WHO/Euro initially understood WHO/EHCN as a series of 'natural experiments' (Tsouros 1989) with cities as field laboratories (Tsouros 1995) and involved academics from the start (de Leeuw 2001), evaluation has always been problematic. At a fundamental level, questions have been raised about the purpose of Healthy Cities evaluation with a tension between research that supports practical learning within cities and across the network and that which produces valid, generalisable evidence (Boonekamp et al. 1999; de Leeuw and Skovgaard 2005; Dooris 2005).

There was never any intention to pin-down and operationalise the Healthy Cities concept and it was understood that it would take a different form in different cities (Hancock and Duhl 1986). The rapid early expansion of Healthy Cities and the variety of ways that cities understood and implemented the concept led to the conclusion that traditional, validated research paradigms would not be possible (de Leeuw 2001). Whilst 'old' public health favours the epidemiological framework and randomised control trials through which effectiveness can be established, 'new' public health requires an expansion to include 'new' generations of research methods, with a post-modern underpinning (Davies and Kelly 1993). It is important to analyse the achievements of healthy settings approaches by understanding social and political processes (Kickbusch 2003) and methods such as participatory action research have been suggested as a way of keeping with the intersectoral, participative ethos (Dooris et al. 2007). Case studies, informed by theory of society and systems theory, have been identified as a suitable evaluation strategy for understanding environment and policy interventions (Eriksson 2000). Theory-based evaluations, logic models and realistic evaluation have also been noted as holding potential (de Leeuw and Skovgaard 2005; Dooris 2005).

WHO/Euro commissioned evaluations of Phases I and II which resulted in internal reports. From Phase III onwards evaluations have also resulted in more 'scientific' output in the form of special issues of Health Promotion International (Volume 24, Supplement 1, 2009 and forthcoming) and Journal of Urban Health (Volume 90, Supplement 1, 2013). Resource limitations mean that these centralised evaluations are necessarily dependent on the self-reporting of Healthy Cities and may miss a wealth of relevant material being generated in the natural course of their work (de Leeuw 2013). More generally, funding for Healthy Cities research has been difficult to get (de Leeuw 2003) and as a result, 20 years into the project, it was still being noted that little evidence has been documented on the vast range of activities and their impact (de Leeuw and Skovgaard 2005; O'Dwyer et al. 2007).

Conclusion

This section has described the underlying theory drawn on by WHO/EHCN and provided insights into the degree to which these are realised in practice. However, there is no evidence that gives a definitive answer to the question—does all this activity improve health and address social inequalities in health? Answering this question may not be possible, not only because of methodological challenges but also because the nature of the national context within which a Healthy City is working may mean that their day-to-day practice is not true to the underlying theory.

Whilst WHO/Euro can not claim success in terms of health gain or an impact on social inequalities in health, it can however claim 'success' in its aim of keeping health high on the agenda of local governments and supporting Healthy Cities to continually strive to experiment, learn and adapt in their work to create better conditions for health even when the wider institutional and policy setting they themselves are working in will lead to tensions and constraints to success.

The next section of this chapter presents descriptions of the experiences of some emblematic European cities as they have participated in WHO/EHCN and in doing so illustrates these issues and the diversity of Healthy Cities concept implementation.

Theory, Policy, and Practice of Value-Based Urban Health Planning: Liverpool Healthy City[1]

'*Few cities in the healthy cities movement could match Liverpool's past and present public health credentials. Liverpool is one of the most committed and inspirational members of the European Healthy Cities*' (Dr Agis Tsouros, Director of the Division of Policy and Governance for Health and Well-being, WHO European Office).

The economic, social and political fortunes of the once second city of the former British Empire markedly declined after the 1960s.

[1] Ronald Gould and Julia Taylor.

Table 10.1 Inequalities in health in Liverpool 1986: comparison between local government employment rates and standardised mortality ratios

Unemployment rates		Standardised mortality ratios	
Most deprived wards	Least deprived wards	Most deprived wards	Least deprived wards
Over 33.6 %	00–20.1 %	00–113.5	Over 156.5

Source: Ashton and Seymour (1988)

By the start of the 1980s this had led to record levels of unemployment exceeding 30 % and almost double the national average. With the poverty of hope and opportunity, the quality of life in the city began a downward spiral.

Riots in the city in 1981 focused international attention on Liverpool and the unhealthiness of an environment in part characterised by squalor, poverty, unemployment and inadequate education. At the national level the attempt to suppress the launch of the Black Report 'Inequalities in Health' 1981 provoked an unexpected response and trigger for action.

Involvement and internships in the European Office of WHO by future Public Health Leaders initially at city, then regional and more recently national level illustrates that Liverpool was involved in the gestation, birth and ongoing development from the launch to the intention to continue in Healthy Cities Phase VI.

Liverpool was highly motivated in taking the WHO 'Declaration of Health for All' (Alma Ata; 1978) forward and together with ten other European cities began working as a network in 1986 to share good practice and to implement the healthy cities approach at the local level .

The way the healthy cities movement in Liverpool began to address issues and tackle problems would make it unique and a trigger for change by advocating a new way of working, bringing agencies and communities together working in partnership, sharing a common set of values about how and why inequalities were unjust and unavoidable and how this should be tackled (Table 10.1) (Buck and Taylor 2009).

One of the first actions was to launch 'The Liverpool Declaration on the Right to Health'. This was produced by the Healthy Cities Intersectoral Committee and launched at the first Healthy Cities International Conference held in Liverpool in 1988 (Fig. 10.1).

A key strength of the healthy cities approach is the requirement for political support at the highest level and the involvement of key decision-making partners in the governance of the city. It was crucial to get the city council involved in order to move health away from medical health targets and ensure that the outcome of council decisions was focused on the likely impact on the health and well-being of the citizens.

Liverpool Healthy Cities has continued over time to enjoy strong and loyal support from both the City Council and the health services. It has continuously sought and developed strategic links with national and local intersectoral partners to develop synergies with policies that strengthened its base.

It has strived to ensure that the values and principles of 'Health for All' are at the heart of policy implementation.

WORLD HEALTH ORGANISATION
HEALTHY CITIES PROJECT
THE LIVERPOOL DECLARATION
ON THE RIGHT TO HEALTH

Equity in Health
Community Participation
Partnerships for Health
Health Promotion
Primary Health Care
Research for Health
International co-operation

LIVERPOOL • HEALTHY • CITY
2000

Fig. 10.1 Cover page of the Liverpool Declaration on the Right to Health

Those involved in directing and coordinating local healthy cities action are required to be extremely skilful in managing change and political change in particular. Political governance in Liverpool is rarely the same as at the national level which provides both challenge and opportunity to implementing national policy.

A new political party was elected in 1998 and acknowledged the WHO requirement to designate a specific Cabinet Member to be the lead politician for Healthy Cities. Councillor Ron Gould states:

> *I was parachuted into the Healthy City programme by the senior members of my party because no one else was willing to take up the challenge! The co-coordinator became my tutor in all things and nurtured my understanding until it became second nature to live and breath all the Healthy Cities programmes. Working together our goal was to influence not only our health partners, the Primary Care Trust, but most importantly the whole political spectrum of the City Council—a must for any vision to succeed and for the values of the approach to act as a*

Table 10.2 Adult smoking prevalence

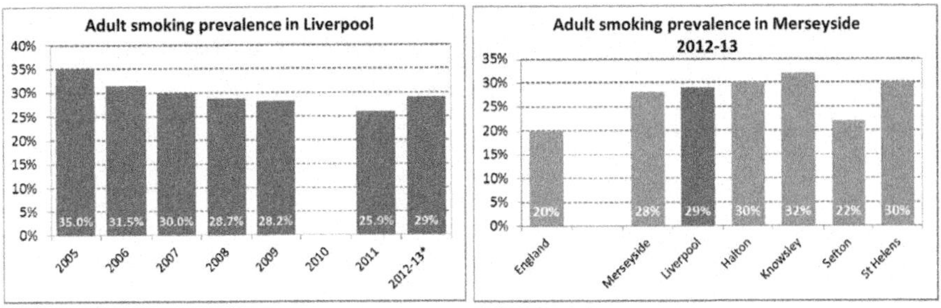

Source: adapted from Merseyside Lifestyle Survey 2012–2013 and Jon Dawson Smoking Surveys 2005–2011

catalyst for action. Greater awareness of the fact that smoking is the leading cause of preventable death and as a pharmacist prescribing for the many sufferers of smoking related diseases I acknowledged the problem and vowed to do something about it.

After months of deliberation I took the plunge and attended a meeting of the City Cabinet to declare I wanted to take Liverpool Smoke Free. I was told it was too difficult; it would never happen I was wasting my time, however with the support of our Health Partners, Liverpool Smoke Free became England Smoke Free and the day it was voted through Parliament many of us shed a tear or two.

Thousands of lives saved, health outcomes still being evaluated probably one of the biggest steps forward in Public Health for decades.

Political support is invaluable, however local politicians have many competing calls upon their time and are mostly on a steep learning curve when newly appointed to the Healthy Cities role (Table 10.2).

The Healthy Cities function in Liverpool has changed over time from project and programme development to become a source of strategic advice on the governance aspects of the public health function within the local authority, as well as ensuring that the values of Healthy Cities underpin the processes and priorities for sustainable health and well-being for the people of the city.

Weathering the storms of yet further organizational change as a result of national policy and the austerity measures placed on local government impacting particularly on cities like Liverpool necessitates renewed vigour.

The city is currently reviewing its Health and Wellbeing Strategy through changing the emphasis to ensuring that governance structures and processes are in place to fully support intersectoral action for health and well-being.

The intention is to make the understanding of and engagement with the principles of Health 2020 for both governance and action the basis on which the Joint Health and Wellbeing Board will achieve its responsibilities and be accountable to the people of Liverpool.

Rennes: A Healthy City Since 1986[2]

Working Toward the Health of the City and Its Inhabitants: A Tradition for Rennes, a Continuous Drive Energised by a Visionary Project

Box 10.2: The City of Rennes
Rennes is a city of 208,000 inhabitants, capital of the Brittany region, one of 18 regions that make up France. The Rennes metropolitan area is made up of 38 communes and 416,000 habitants. It is a dynamic city from the viewpoint of its economy, as well as from that of teaching and research, and leads the 'hit-parade 2012 of cities that are a pleasant place to live' drawn up by *L'Express* magazine.

Its population is young by European standards, and 50 % of Rennes' population—30 % of them students—are aged under 30.

Health is on the whole better than the national average. However, socio-economic and territorial disparities must be noted, as well as an avoidable premature mortality rate, greater than that for the national population, that affects under-65s. Among the causes of death, the situation is particularly unfavourable with regard to suicide and alcohol-related diseases. Mortality due to cancer and diseases of the circulatory system remain below the national average.

Origin and Context of Rennes Healthy City Programme

In March 1986, the European Office of WHO invited some 20 cities of the Europe region to become acquainted with the Healthy Cities project with a view to an experiment. Rennes, along with ten other cities, eagerly and quickly committed to the project, which was all the more attractive as it reflected directions in which the municipality wanted to move. Health has indeed been a constant focus at Rennes since the early 1980s, which was rare in France at the time and even nowadays is not very widespread.

In 1981, the mayor decided to draw up a health plan for the city by setting up a committee to 'identify the entire city's health needs, draw up a list of available resources, current and future, in the various healthcare-related domains (public, private, curative and preventive)'. This committee was composed of representatives from medicine, associations and unions, and interested citizens.

The prospects and guidelines of the plan were implemented early on in the next mandate in 1983, with particular focus on

- Creation of a health delegation within the city council and the appointment of a deputy mayor for health.

[2]Annette Sabouraud, with the support of Jean Simos.

- Reorganization of the Municipal Hygiene Office which would become the Hygiene and Health Department with the aim of guiding its activities towards a wider public health policy.
- Implementation of a coordination *santé-affaires-sociales-logement* (CSSL—health-social affairs-housing) group within the municipal administration.
- Creation of a health committee and a local health observatory.

The directions of the Healthy Cities project were in line with the general policy of the City of Rennes, which had set the following objectives:

- Reduce inequalities.
- Strengthen cohesion and social mix in all sectors.
- Encourage participation and consultation.
- Open up the city to other regional, national and international authorities.
- Start up a decentralised policy for the neighbourhoods.
- Improve the access of culture to everyone.

Other assets supported the implementation of Healthy Cities in Rennes.

- A tradition of dynamic and diverse associations as well as a strong attachment of the population to the area and local life

Attention was given to preserving the environment that encourages the 'socio-ecological health' approach advocated by the organisers of the WHO project. Demographic, economic and academic development, together with that of the health sector for the city and its metropolitan area (notably the existence of a major teaching and research hospital, the *Ecole Nationale de Santé Publique*, a vital stakeholder in public health), and a firmly established mutual health insurance sector were key partners in the project's development from the start.

- Average size of the city, which makes dialogue and cooperation easier
- Expectations of local officials with respect to the WHO project for cities:
  - Increased effectiveness owing to a rigorous and dynamic framework.
  - New opportunities for fruitful dialogue with cities in other countries and cultures.
  - Strong feelings of taking part in a 'common struggle' to be fought beyond local actions.

Evolution: Change and Continuity

Project Beginnings 1986–1987

The official announcement of the 'Towards health for all' charter and the setting up of a Rennes Healthy City committee created new drive. The committee rapidly gathered around 60 partners from highly diverse institutional or association organizations, as well as elected representatives and interested parties whose role was to develop a Healthy City plan.

Successive Plans

In 1989, the first Healthy City plan fell within the extension of the 1981 Health Plan, given the convergence that it showed with the recommendations of the WHO project and those of the Ottawa Charter. Drawing on these, it opened up innovative perspectives, energised new drive and was used to establish a wider, better structured and more visible programme. For instance, actions in progress such as 'Outward-looking schools' or 'neighbourhood policy', youth policy, were 'enriched' by the Healthy City spirit, making it possible to:

- Strengthen and develop healthcare actions at school that were not just implemented alongside those already conducted (culture and sport, for example) but genuinely integrated into them and encouraging a global approach.
- Incorporate a healthcare perspective into the 'neighbourhood policy' and closely develop the aspirations of the population and the needs identified, diverse health promotion actions (sport and health, food, dental hygiene, addictions, sexual behaviour, etc.) also with the perspective of identifying 'neighbourhood health contacts'.[3]

The Youth Health Programme joined with the activities of the *Centre Information Jeunesse* (Youth Information Centre), community centres and various youth organizations for health promotion and information operations that not only were well received but also enabled young people from very different backgrounds (students, apprentices, young employees) to meet and get to know each other.

1993: The new Healthy City plan aimed to strengthen the Rennes Healthy City policy and support new initiatives by proposing to:

- Consolidate the actions and projects capable of reducing health inequalities, with strong emphasis laid on citizen consultation and participation.
- Perfect a global health-environment-quality of life policy to be incorporated into all municipal decision-making levels.
- Intensify the city networks and expressions of solidarity (city-countryside, twinned cities, decentralised cooperation, etc.).

2005: The 'Acting together for the health of everyone' plan reflected both a lasting commitment to the Healthy City project and a growing determination to further unify efforts. It was part of a new phase in the WHO project, drawing on experience acquired and accounting for new data: more specific information on the health of the population, emergence of new needs given changes in the local situation and society, popularisation of the notion of sustainable development, and the like.

The following can be noted among other things: the significant development of the Health–Environment Advisory Committee and neighbourhood networks, implementation of new action programmes for different population categories, prevention of high-risk health behaviour, use of an environmental scoreboard, and a 'Zéro Phyto' programme to encourage the use of products containing no pesticides or chemical fertilisers.

2013: The local health plan continued to carry out a policy designed to promote health for all at every stage in life. Both ambitious and specific, it worked to strengthen and diversify intersectoriality and cooperation.

[3] Now represented by the health coordinators as part of the local health contract.

Using the priorities of the 2005 plan integrated into the health objectives of the current municipal programme, the 2013 plan was built from consultation with more than 20 services of the city that met in an 'interservice' steering committee. This cross-functional organization had outward-looking features and saw many intersectoral partnerships set-up. The city advisory committee, where partner consultation occurs, is composed of the following themed working groups:

Health environment and habitat

- Animals in the city
- Access to care
- Prevention of addictive behaviour
- Promotion of mental health
- Sexual and emotional life
- Nutrition and physical activity

An accessibility community committee was established and the creation in 2010 of a local mental health council once again showcased the unifying role of the city.

The local health contract (CLS) makes official the institutional partnerships with government departments, the hospital, social security, School of Public Health, and departmental and regional bodies, with a view to developing joint multi-annual programmes. This local contract encourages and strengthens the actions of each party and promotes the diffusion of a health promotion culture.

Main Characteristics of Rennes Healthy City Over the Years

Participative Approach that Blends in with the Local Situation and Enriches the Current Actions

From the start, the organisers of Rennes Healthy City were keen to integrate their projects not only in traditional health actions (vaccination, public hygiene, etc.) but also in actions conducted in other sectors, such as school and university, culture, habitat, urbanism, environment, association and neighbourhood life, and international relations.

The government, region, department, neighbourhoods and various institutions of different levels and many kinds (administrative, technical, NGOs) who were associated from the beginning were more formally involved with the local health contract.

Drive to Increase Health Issues Further and More Explicitly Throughout Local Policy

In 1993, a study based on a multiple criterion method, 'integration of health criteria in municipal decisions on urban planning', was conducted with a view to using a 'health grid'. Rennes was subsequently involved in the work of the European Healthy Cities network on health impact assessment. It was the first French city to use this method.

Recognition of the Links Between Urban Planning, Environment and Health

Evidence of this lies, among other things, in the 'urban project and health policy at Rennes' in Agenda 21 of Rennes Metropolitan area,[4] and in conducting studies on the health impacts of an urban planning project (Tollec et al. 2013).[5]

Continuous Action and Intersectoral Convergence

Convergence has only increased over the years between the city and its institutional, private or NGO partners.

National and International Receptiveness

Rennes Healthy City has been regularly involved in the activities of WHO's European Healthy Cities network. It was one of the first to be part of the Healthy City approach, and was recognised by WHO in 1987 as one of the 11 pilot cities of the project. Keen to be joined by other French cities and by cities from other French-speaking countries, it took the initiative to create two networks: in 1987 the French network of WHO Healthy Cities,[6] which now has 80 cities and four metropolitan areas, and in 1988 the French-speaking network of WHO Healthy Cities. Following the latter initiative, the city was designated the WHO collaborating centre for French-speaking Healthy Cities. Its scientific committee became a health and sustainable development association[7] (Santé et Développement Durable, S2D) in 2001 to meet the changing requirements for WHO collaborating centres.

Twinned healthy cities were established and undertook themed encounters (such as school and health, or disability). Some cities twinned with Rennes subsequently joined the Healthy Cities movement, either in the European network (Poznan) or in the German network (Erlangen); Sétif (Algeria) is part of the French-speaking network.

Rennes observes and is enhanced by experiments conducted elsewhere, as it is a partner of the National Health and Nutrition Plan and member of the Association of Local Elected Representatives for Promoting Public Health in Cities as well as the Healthy Cities of the West (of France).[8]

Rennes works with the health authorities of the Brittany region (Regional Conference for Health and Autonomy, Regional Conference and Public policy coor-

---

[4] http://metropole.rennes.fr/politiques-publiques/elus-institution-citoyennete/l-agenda-21/

[5] See also the follow up of this project in the partnership "*Réseau Bretagne Urbanisme Santé*" at http://rbus-eis.org

[6] http://www.villes-sante.com/

[7] http://www.s2d-ccvs.fr/

[8] Angers, Bordeaux, Brest, Brive-la-Gaillarde, Cherbourg-Octeville, La Rochelle, La Roche-sur-Yon, Lorient, La Rochelle, La Roche-sur-Yon, Lormont, Nantes, Quimper, Rennes, Saint-Brieuc.

dination committee), to promote the health of the cities and underline the importance of health promotion policies.

Many national and international conferences have been organised or jointly organised to bring together the cities of different networks or urban planning and environment professionals.[9]

Factors that have Cemented Adhesion to the Healthy City Programme and Its Development

This is mainly a strong and unwavering political drive, with highly motivated personnel and resources that have developed considerably in orientation as well as in organization.

The different names and missions of the health service clearly show this change: the Hygiene Office, created in 1907, became the Communal Hygiene and Health Service in 1983 and was known in 2014 as the Public Health and Disability Department.

Relevance of the Healthy City Project and Its Requirements

The remarkable continuity of the Rennes Healthy City programme reflects the relevance and the innovative, creative and genuinely visionary nature of the project presented to the city 28 years ago by WHO. The project, founded upon the notions of intersectorality, partnership and participation, has proven to be most fruitful, for the development of a rich and productive human and institutional relationship.

But the complexity inherent in such a policy, the cross-functionality that it imposes, brings many challenges. These may be administrative, but also include political reactions based sometimes on the fear of losing power, the feeling of a badly focused or utopian scope of action, movements not always immediately controllable, source of efforts dissipation. A certain maturity of the programme can also bring risks, such as growing institutionalisation that can lose sight of the specific nature of the approach, and particularly its intuitive and changing nature. Flexibility, perseverance and creativity are key factors that can overcome these obstacles.

The recommendation of the project founders to 'build while walking', sometimes with small steps but always focused on a global objective, remains relevant, provided that there are regular reviews to assess where the walking has led and to draw lessons from it, and identify new useful factors and new allies. All this requires resolute determination from the various stakeholders, conviction and expertise from professionals, the availability of partners, and above all the vigorous support of WHO.

---

[9] Examples are the international Santexpo Health-Environment in the city (1992), the National Health Forum for the Environment and Chemicals, responsibilities and responses of local authorities (2007), the annual business meeting of WHO European Healthy Cities Network (2007), the regional urban planning days, for environment and health, acting for health and sustainable development, tools for action (2011), etc.

Nantes Healthy City[10]

Introduction

Situated in the north-west of the country, close to the Atlantic Ocean and guarding the entrance to the Loire valley, Nantes is the sixth largest city in France with a population of nearly 300,000 and a metropolitan area (conurbation) of around half a million inhabitants. The city has been nominated vice-chair of the French network of WHO Healthy Cities and joined the European WHO Healthy Cities network in 2012. In 2013, Nantes was designated a 'green capital' and the 'European capital of engagement for environment and quality of life' (City of Nantes 2013a; Photo 10.1).

Since 2010 Nantes has had a health observatory, which enables health monitoring at city sub-sector level (based on the French Iris statistical system, with sub-sectors of around 2000 inhabitants) (ORSLA 2010). It has therefore proved possible to document social and health inequalities between territories, such as a significant increase in risk of early death in some marginalised areas. Much of this inequality appears at an early age: at age 6, the prevalence of overweight and tooth decay are two to three times higher in certain disenfranchised areas compared to the city average (City of Nantes 2012).

Photo 10.1 Nantes, green capital of Europe in 2013

[10] Patricia Saraux-Salaün, with the support of Jean Simos.

City authorities are aware that, while behaviour plays a role, individuals are not solely responsible for their health. Crises can and do appear in the course of a person's life: losing one's spouse, job or apartment can—especially in a context of economic difficulties—lead to a person's social downfall and through to ill health. Beyond personal behaviours, society plays a structural role by defining the socio-economic, cultural and environmental context which, by impacting living and working conditions, has an effect on the health of the population (CSDH 2008).

The Local Healthy Cities Programme: A 'Public Health Mission'

As an active member of the French and European networks of WHO Healthy Cities, Nantes has concentrated its public health mission (*Mission santé publique*, MSP) on three topic areas:

- Social and territorial health inequalities
- Urban health
- Community health with buy-in from local inhabitants

Concepts such as the determinants of health and health gradients underpin political action in the field. Public health defines its mandate as 'promoting all factors conducive to health, up to healthcare but without including healthcare' (City of Nantes 2013b). A Local Health Contract (*Contrat Local de Santé*, CLS) was set up in 2012 with the authorities of the Region. It expresses the intentions of local policy makers and supports the daily work of the public health mission, which structured itself into three poles in September 2013: territories, content and resources.

Concentrate Resources Where They Are Most Needed

Sub-dividing the territory into four sectors (North, East, West and South) has allowed public health to concentrate on areas defined as priority areas based on statistics from the health observatory. A little less than one-third of the population are targeted by the MSP activities, which include the municipal school health service, the public showers, health workshops in marginalised areas and health conferences (Focus Santé) organised with local inhabitants in marginalised areas, or also in the city centre, in the tradition of people's education conferences.

Community health is rolled out according to three priority populations. The first is children and their families, with the municipal school health service concentrating not only on schools but also on places where children go at various times outside school (community centres, child-minding NGOs). The second focus is on marginalised populations in priority areas, through the local health networks, which bring together health and social professionals with neighbourhood representatives, and severely marginalised people (often homeless or without heating or hot water) who are among the 70–110 people who visit the public showers every day. The public showers institution, open 7 days per week, has seen its patronage steadily increase and actually double in the space of 8 years, demonstrating its usefulness in times of economic uncertainty. They were renovated in 2013 to cope with increased demand and to enable the development of prevention and health promotion initiatives aiming at re-establishing access to health for the users.

Regarding access to health rights and to healthcare, Nantes' third focus is on helping people improve their health insurance coverage if their earnings are slightly too high for them to be eligible for social benefits. This system was initiated when a large-scale local survey conducted in 2011 showed that many people do not access healthcare when they need it.

In 2013, another survey, on 'barriers to health rights and to healthcare', included 480 inhabitants and health professionals in a marginalised area (Aurillon-Van Den Broucke 2013). Results showed that disposing of limited resources, having to pay for the consultation on the spot, fear of diagnosis, and having to book appointments ahead of time were the main barriers preventing access to healthcare. It follows that the local pharmacist plays a central role in health access for these marginalised populations. In the framework of the French urban regeneration programme and the CLS, three action points were developed based on the responses to the survey:

- To inform and educate urban planners about concepts such as the social determinants of health and urban health, and about the tools that can help planning projects take health into account, e.g. health impact assessment before a project is operationalised.
- To set up and disseminate tools to facilitate stakeholder participation for the inclusion of health in renovation projects: focus groups, walking diagnosis (a participatory citizen-centred local micro-problems tool).
- To develop a project aiming at creating an inter-disciplinary multi-site Health Centre (Maison de la santé) in conjunction with local partners.

Transversal Action

Health is also included in the transversal strategic plans, which span several topic areas such as school achievement initiatives, equality between women and men and between girls and boys, nutrition (combined with physical activity) and environmental health.

In order to evolve towards a common culture, the public health mission collaborated in 2013 with the Greater Nantes (Nantes Métropole) administration for public spaces, leading to the inclusion of a chapter in the 'Charter for public space planning: How to integrate health into urban planning' (with 12 key objectives). It is indeed necessary to make people understand that the health of populations is not only a function of the availability of healthcare, but above all an environmental question. MSP has developed expertise in urban health by adapting and sharing its environmental public health plan and its tools (diagnostics en marchant, focus

Photo 10.2 Traffic appeasement and cycling promotion measures down town

groups, training programmes, maps, etc.) with urban planners, city authorities in charge of parks and green spaces, local farmers, etc.

The Nantes Plan for Environmental Health

The plan for environmental health (acronym PLASE) launched by Nantes in January 2012 is the expression of a strong political commitment in favour of environmental health. So far it is one of the most detailed plans of its kind in France (Guyard 2016). Its objective is to improve the health of local inhabitants, since reports now show the many linkages between environmental factors and health (e.g. allergies, cancer, cardio-respiratory diseases). One important example is protecting the inside of buildings from excessive noise, or promoting walking and cycling (Photo 10.2). Another important part of the plan is improving the quality of indoor air. Indeed, measures intended to save energy can sometimes lead to decreased indoor air quality.

Radon is also an important problem for indoor air quality in some areas. A 'radon screening' campaign launched in 2008 first concentrated on buildings open to the public before being opened up to people living on the ground floor in an area known for high radon levels (*Sillon de Bretagne*) (OSCAR 2014).

Another important aspect of the plan for environmental health is the prevention initiative against Legionnaire's Disease, directed at municipal buildings. A new collaborator has been recruited to manage this initiative, which involves not only public health but also the municipal departments for Sport and of Public buildings.

Within the framework of the plan for environmental health and the second regional health-environment plan, a cartographic survey of environmental nuisances and threats is under way. A pilot project supported by *Nantes Métropole* and the French government, this survey aims at identifying the sectors and populations most exposed to environmental problems. The objective is to provide a tool which will facilitate the integration of health and environmental concerns into other policy sectors such as housing (so-called OPAH and PLH plans), urban planning (PDU plan) or transport (Chronobus).

A discussion has also initiated with other municipalities around Nantes on the topic of childhood lead poisoning (saturnism). The first step is to identify children potentially at risk and inform doctors in private practice whose catchment areas include old buildings (essentially, those built before 1948). Refurbishment of degraded housing stock is planned in the framework of the next round of programmed operations for housing improvement.

Focus on Nutrition

Municipalities have the mandate to develop programmes to help various local populations—young children, senior citizens, marginalised communities, persons with special needs—adopt behaviours and lifestyles conducive to better health. Examples are policies that help increase daily physical activity, that improve the nutritional value and taste of school meals, or that support NGOs so they can increase the quality of the food that they distribute to those in need. Nutrition, one of the major determinants of health, is as much an individual as a collective responsibility.

Nantes has been an active member of the national plan for nutrition and health (Plan National Nutrition Santé) ever since 2009 (PNNS 2011). MSP leads this programme locally, in collaboration with the municipal departments responsible for: children, education, sport, parks and green spaces, solidarity, nursing homes, etc. Many other local stakeholders are implicated in this programme, such as service-delivery NGOs which are well implanted in various public spaces and under-privileged areas, the idea being to use pre-existing networks in order to empower citizens with appropriate knowledge about nutrition, thus leading to more healthy behaviours.

MSP rolls out the national plan for nutrition and health using information and education for health from individual counselling to group interventions around nutrition and physical activity, participation in public gatherings and events, creation of environments conducive to health (shared urban gardens, picnic areas around fruit trees, open-air markets, etc.) (Gassiot 2015).

Three actions illustrate the transversal nature of nutrition policy in Nantes:

- The observation of health, especially health inequality, starting in childhood
- The creation, at the suggestion of a focus group of local residents, of an open-air market in the north (*Nantes Nord*) to diversify food sources with a view to increasing the intake of in-season vegetables and fruit, to reinforce access to the local area for all residents, to develop knowledge and knowledge sharing about

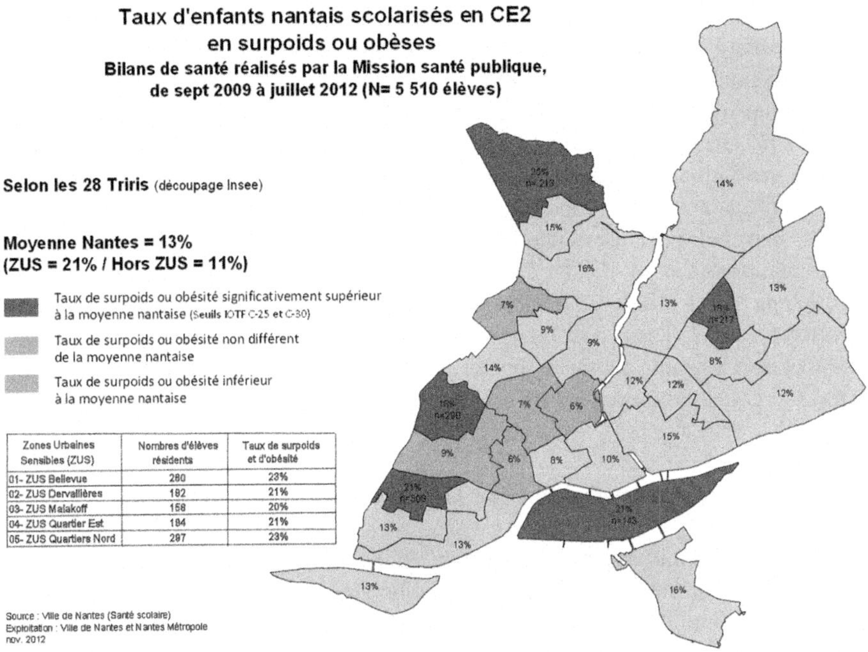

Fig. 10.2 Overweight and obesity prevalence among Class CE2 children (around 8 years old) in Nantes. Source: *service de santé scolaire municipal de Nantes*, 2012

nutrition (a health forum is held in the market every week), and to combine nutrition with physical activity thanks to a walking trail created by a local youth group.

- The improvement of workplace and schools canteens in municipal buildings, towards balanced diets, food that is in season and produced locally, accessible prices, improved decoration and atmosphere, and increased knowledge enabling people to forge a critical mind about nutrition (Fig. 10.2).

'Living Together': Mental Health and Risk Reduction

A network for mental health created in Nantes in 1994 subsequently evolved into a local council for mental health, in order to improve coordination and advocacy for mental health, structured according to four key areas (Saraux 2012):

- Enabling people with mental health problems to stay at home, thanks to interdisciplinary health staff in so-called 'sector points' (points de secteurs).
- Initial and continued education for healthcare and social services staff.
- Communication.
- Living in the city with a chronic psychiatric disease.

The health of adolescents and young adults is also a shared topic area, founded on risk reduction, a specific plan addressing alcohol consumption, and initiatives in favour of sexual and emotional health. Nantes is a university town. Faced with cases of acute intoxication with alcohol, the MSP and the Mission for Public Quiet (*Mission de Tranquillité Publique*) created the first alcohol plan back in 2008, associating prevention, education, monitoring, rule of law and repression. Seventy NGOs, groups and institutions backed the plan and now work together to reduce hazardous drinking and associated risky behaviour. A study on hospital admissions for drunkenness among the 10–24 age bracket found an average of one entry per day, among which 7 % were transferred in intensive care, with no associated death reported for 2012. These data indicate that there is continuity between the identification of at-risk young people by police and first-aid personnel (working together in pairs) and patient management on arrival at the emergency ward (OSCAR 2011).

Managing and Empowering Partner Organizations

The public health mission is also a gateway for health NGOs requiring information, authorisations for their own activities in public areas, knowledge sharing, or grant applications. Some 189 NGOs are known to the MSP, and every year around 50 of them apply to local government bodies for subsidies related to Healthy Cities.

Perspectives and Challenges

Favourable factors in Nantes evidently include recent local history and political will. The city succeeded in keeping the school health service and the municipal public showers (Samson 2015), and nominated public health personnel, while other cities in France were making rather different choices (public health is an entirely voluntary task at the municipal level in France, with the exception of hygiene, public sanitation and the vaccine register).

Three challenges still hamper the implementation of an ambitious public health policy in Nantes:

- Blockages due to the inability of professionals from various disciplines and sectors to work in a truly transversal and trans-sectorial way. The same words may not have the same meaning for a public health manager or for an urban planner.
- Insufficient advocacy for the improvement of living conditions coming from key people and institutions in the health sector, whereas such advocacy has been forthcoming from the environmental sector.
- The difficulty for municipalities to recruit new personnel with expertise in public health: due to budgetary constraints they have had to rely mostly on civil servants transferred from other positions within the municipalities.

Conclusion

Public health is transversal by its very nature. It aims at 'improving health up to but not including healthcare' in conjunction with all the departments of the municipality or conurbation, insofar that they all have an impact on social and/or environmental determinants of health. Other essential partners for public health are the NGOs (which may or may not be active in the health sector) and the inhabitants themselves. It is therefore essential that civil servants be allowed—and encouraged—to work with other sectors than their own, because all policy sectors have potentially positive or negative impacts on health.

Athens: Establishment of the Healthy City Programme[11]

Introduction

Athens is one of the most ancient cities in the world bearing the roots of its foundation in 3500 BC. Therefore no one can deny that as a city is sustainable, healthy, secure, and no matter how authorities and systems have changed people always found a place here that they could call home.

Even in ancient years, Athenians had a strong tradition in participatory governance. Greek philosophers had thoroughly analysed the way to structure and to rule a city so that the quality of life, mental and physical health could be promoted. The famous philosopher Aristotle, founder of the philosophy school Lyceum, describes in his books the politics and the scheduling that cities had created at the time. Their goal was to build a city in such a way so that people would be mentally healthy, stable, organised and happy. That, he called, 'not only for living but for living well'. And that became the motto of Athens Healthy City Project (Photo 10.3).

As we forwarded in today's world, societies in the western world made organizations, foundations that would help one another thrive through matters of economics, politics, environment, or health. In 1993, Athens joined the Healthy Cities programme.

The philosophy of the project is to bring health issues highly in the political agenda of the city, and to make them a fundamental matter in political decisions. The city's mayor in 1993, and all the mayors that followed him, found the methodology of the Healthy Cities programme very helpful.

From the start, city authorities had realised that this project had the necessary holistic approach and that its implementation could cover the vast majority of programmes and actions of the municipality. They realised that the implementation of the project could be used as a tool in order to bring together different people from

[11] Dionysia Papathanassopoulou, with the support of Popi Yannopoulou and Evagelia Tsapatsari.

Photo 10.3 The Acropolis of Athens and the Athens Healthy City logo

different fields and municipality departments and make them work together under the umbrella of WHO. All of course, for a better quality of life in Athens. It was then that it was decided to adopt the programme and use it as a 'growth engine' for developing plans and decisions through all municipal services.

The first step was to follow the instructions of WHO for Healthy Cities, and adapt them in our own political, administrative and legal system, and then to continue by giving the programme a structure.

Strong Political Support

A special discussion was organised in the city council and the head of the European Healthy Cities network, Dr Agis Tsouros, was invited to present the Healthy Cities programme as well as its philosophy, methodology and benefits. A long discussion took place and the city councillors from all political parties raised any question they had. The councillors liked the programme and the presidents of the political parties gave the permission to start and work fulfilling WHO's criteria for Healthy Cities with purpose to apply to be project city in the Europeans Healthy Cities network. After that, the council decided the creation of a special bureau for the Healthy Cities project with its own budget and a coordinator with special qualifications to run it.

The Healthy Cities office was then established in the Directorate of Social Welfare and Health. To ensure the existence of strong political support, the city council decided that the mayor should be the political principal, responsible for all matters concerning the programme and the Healthy Cities office.

The management group consisted of a representative of each political party, the deputy mayor for Health and Social Welfare, the deputy mayor for Environment and the city coordinator (Figs. 10.3 and 10.4). His head is the mayor.

Every 4 years, after local elections, one of the first decisions of the new city council is to decide the parties' representatives in its management group. This group embodied from the very start the vision, the viability and the strong political will to work as a Healthy City. That was not only essential in order to motivate all the political and administrative forces under a common purpose but also to adhere to the main cause of the project.

Intersectoral Cooperation

To ensure intersectoral cooperation, the city council appointed a steering committee (Figs. 10.3 and 10.4). Head of this committee is the mayor and its members are the deputy mayors or councillors, the directors of the main directorates or agencies of the municipality (that deal with matters of health, social welfare, environment—urban and physical, civilisation and athleticism) and lastly the city coordinator (Fig. 10.4). The main duty of this committee is to plan the course of action and make sure that it gets implemented, and to align the different departments.

For both, the management group and the steering committee, the purpose is, furthermore, to plan, as well as to ensure that the intersectoral bodies work harmoniously. They are assigned the task of spreading the philosophy and methodology of the Healthy Cities working group. A body of experts was needed from all the involved departments for this specific assignment, and each department appointed an expert to the working group. Leader of this group was the city's coordinator.

The working group meets once a month to discuss and plan its course of action. Its main mission is to establish cooperation between the departments, and the implementation of the programme. Members are continuously informed by the coordinator about all ongoing business in the European Healthy Cities network and share all data that they were given with their departments and associates. All of the above is a basic step in helping the different departments to organise their projects and later on propose them to their superior committees.

This course of action also guaranteed the use of the Healthy Cities methodology in their projects. The idea was that to promote health and quality of life and for that to be succeeded all the city's departments had to preserve this idea in their daily work.

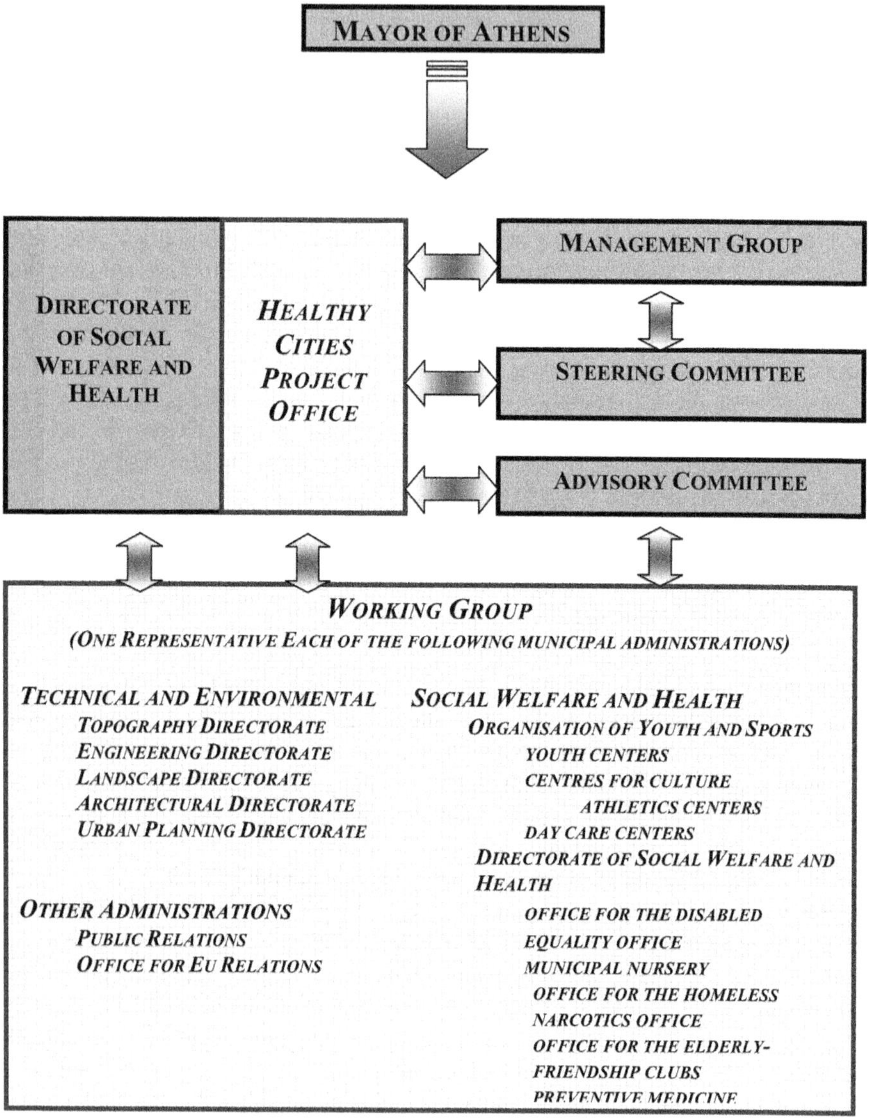

Fig. 10.3 Organization Chart of the Healthy Cities Project in Athens

It is important to underline that one of the working groups' most important assignments, was to prepare the profile and the plan of the city. That task required hard work and cooperation, but in the end it was a remarkable way to harness every source of the municipality accordingly and with harmony.

Management Group

- Mayor of Athens
- Deputy Mayor of Health
- Deputy Mayor of the Environment
- A City Councillor from Each Political Party
- Director of Social Welfare and Health
- Coordinator of the Healthy Cities Project in Athens

Steering Committee

- Mayor of Athens
- Deputy Mayor of Health
- Vice President of the Organization of Youth and Sports
- Vice President of the Municipal Nursery of Athens
- Vice President of the Cultural Organization.
- Deputy Mayor of Public Relations and International Cooperation
- Deputy Mayor of City Planning
- Deputy Mayor of Landscaping
- Members of the Management Group
- Directors (or Representatives) of Competent Municipal Service

Advisory Committee

- Members of the Management Group
- Directors of the Seven Municipal Districts
- Representative of the Union of Public Servants.
- Representative of the Labour Center of Athens
- President of the Chamber Of Commerce
- President of the Chamber of Hotels
- President of the Technical Chamber
- President of the Consumers Union
- The President of the National Council of Greek Women and Representative from the Women's Lobby
- President of Greek Industrial Association
- Representative of the Greek Archiepiscopal
- Representative of the Greek University from the Branch of Public Health
- President of the Medical Association
- Representative of the Dental Association
- Representative of the National School for Public Health
- President of the Federation for the Disabled
- Representative of the Polytechnic University of Athens
- Representative of the Technological Educational Institution of Athens from the School of Health Executives
- President of Social Workers' Association
- President of the Greek Sociologists' Association

Fig. 10.4 The Main Administrative Committees of the Healthy Cities Project in Athens

Citizens Involvement in Decision-Making Process

The Municipality of Athens is quite large in population. In 1993, Athens had a population of 785,085 citizens. It is also the centre of the greater Athens area including all other municipalities and suburbs. The population of this wider area is much greater, and the idea that we should involve citizens in the decision-making process, came out as a great challenge.

In order to make this work, firstly, an advisory committee was set up, consisted of representatives of different but important organizations of the city and led by the mayor.

This advisory committee's task was to inform different factors (governmental or not) of any new project and build with them the alliances needed, for a strong healthy and long-term cooperation. Secondly, contacts with organizations representing groups of citizens were made, for example, associations of women, of the disabled, and of workers. These new relationships gave the opportunity to decide the municipality's policy for these categories of citizens in cooperation with their representatives.

Lastly, contacts were made with people who were already assigned to special programmes, like the 'renovation of schoolyards programme' planned and implemented with each school community. The Healthy City team got to know children, parents, teachers, etc. As a result of this structure and policy, they built healthy and long-term cooperation with many organizations that involved different sectors of the municipality and widely different parts of the population as well. This opened their eyes to see a much greater variety of the problems people are facing every day, a new perspective on the job that they were doing for the people, and of course the necessary support from the people that they really work for: the people of Athens.

Lessons Learnt

As years went by, the Healthy Cities programme in Athens survived political and social changes and different local authorities, thanks to its structure, the strong and continuous political support, the involvement of different sectors and the involvement of citizens.

As one can easily imagine, many things can change in more than 20 years, some for good and some for worse. The Healthy City team stood through the difficulties, and came forward stronger. Key factors in this success and maintenance were the following:

- Strong political support.
- The fact that the mayor is also head of the project.
- Main involvement of different political parties.
- The common vision of all for a better city.

The holistic approach of the city that brought together different sectors and experts. All of them saw the programme as part of their own dream for a better city.

The involvement of so many different experts in their daily duties made the philosophy and methodology of the programme spread in all the municipality sectors as a way of working for the city and the citizens.

The author of these lines, a former Healthy City team leader, has to say that the keys to success were no other than the programme's great start, its sturdy foundations, and the fact that each and every one of those who worked with the Healthy City team found that through it there was an opportunity to help, to make their own city a better place in acts and not words, and the fact that they would do good for the purpose of a better change; these ideals automatically set aside those who would not have been helpful for the network or the city, and embraced all those who for years had ideas and hopes for the city of Athens and could finally make them real. It was a pleasure working with all of them.

Healthy City in Pécs[12]

Context and Background: The City of Pécs

The city of Pécs lies in the south-western part of Hungary near the Croatian border. Pécs—which is the fifth biggest city in Hungary—is an educational, cultural and administrative centre with a wide range of monuments from medieval times and the Turkish occupation (1541–1686). The University of Pécs, one of the biggest in Hungary, was also the first, founded in 1367.

The population of Pécs is about 156,000, with a significant number of national minorities like Germans and Croatians. The city was awarded the UNESCO 'Cities for Peace' prize in 1998 for its endorsement of the cultural heritage of its minorities, and for treatment of refugees of the Yugoslavian War (1991–1995).

Pécs was one of the founding members of the WHO European Healthy Cities project in 1987. From 1991 the Healthy City Foundation of Pécs (founded by the city) has performed as the coordinator of local Healthy Cities projects, based on the resolution of the city council. Strategic decisions and plans are made by the inter-sectoral steering committee. To ensure political commitment the chair of the steering committee is always the current mayor, and its members represent various professional areas and political parties of the local government.

Phase I (1987–1992)

The Healthy City Foundation of Pécs is also the centre of the Hungarian National Network, which was formally founded as the Hungarian Association of Healthy Cities in 1992 with ten cities after 4 years of informal networking. In 2014, the national network had 18 member cities from across the country (Fig. 10.5).

One of the benefits of being a member of European Healthy Cities is to learn new methods and tools of improving public health and city health. The first new tool was the City Health Profile, a public health report including information on and analysis

Fig. 10.5 Logo of the Association of Hungarian Speaking Healthy Cities

[12] Antonio Di Blasio and Zsuzsanna Nagy.

of health and the social determinants of health. During Phase I Pécs started to prepare its City Health profile, which was completed in Phase II. It has undergone several revisions since then, the latest in 2013.

Phase II (1993–1997)

In the second phase, following WHO priorities, the focus was on organising local actions and programmes. The most important project in this Phase was the anti-smoking campaign run between 1995 and 1997 and financed by the World Bank with a one million USD budget. Pécs worked together with other European and Hungarian partners (e.g. Horsens Healthy Cities and a Hungarian communications company) under the supervision of the Hungarian Ministry of Welfare. Children, youths and groups such as teachers, healthcare workers and politicians who might set the example for people were the main target groups of the programme.

The main elements of the local project were a survey about smoking habits, information leaflets about the harms of smoking and guidance for people who would like to quit, training for teachers and health visitors about prevention among children and youth, competitions for pupils and students, and an intensive media campaign. The survey and the media campaign were carried out at national level.

Awareness of the national campaign was 60.6 %. The proportion of the people reached by the media campaign was 76.7 %. Smokers found the programmes more useful than non-smokers. Unfortunately the Ministry of Welfare did not use the experience gained from the project, which could have provided a useful base for further anti-smoking measures and activities at that time.

Phase III (1998–2002)

The themes of the third phase were sustainability, equity and social improvement, and the focus was on integrated health development planning. The previous phases had demonstrated that differences in health status were primarily associated with social determinants of health, and that improvement required the cooperation of various sectors and disciplines. This led to a paradigm shift, from a health plan to a health development plan, moving the emphasis to improving the social determinants of health.

The new approach started from the fact that local government decisions are usually based on policy papers, development plans and action plans that define medium- and long-term activities for certain professional areas. Based on this idea, if the conceptual aims of every professional sector are defined supporting the health status as a priority then there is a much better chance that the activities of certain policy areas—operative decisions based on the concepts—will function in a more health-supportive way. At the same time, as public policy is based on various professional policy papers, if these are developed with a strong focus on health supportive functions, they will support the health component of public policy as well.

Principle defining strategies in connection with health promotion put special emphasis on factors such as multisectoriality, equity, and wide, multi-level partnerships for accountability and responsibility as criteria of successful programmes. These factors were considered from the earliest stages of conceptual planning. By the end of Phase III, Pécs had prepared its health development strategy from the new perspective of health development planning.

Phase IV (2003–2008)

Being a member of an international movement like Healthy Cities offered a great opportunity to work with other cities and participate in international projects facilitating the implementation of the core themes of the fourth phase, such as physical activity and active living or healthy ageing. Pécs was one of eight European cities that participated in the project 'Victims of Crime', with the purpose of learning and sharing experiences in the field of victim protection and victim support, particularly focusing on elderly people, women, and children (2005–2008).

In 'Shape Up Europe!' 26 cities participated from 25 EU members. 'Shape Up Europe!' was a school–community approach to healthy and balanced growing up, with the aim of decreasing childhood overweight and obesity (2006–2009). Pécs took part in the project with a kindergarten, a primary school, and a secondary school for Roma children.

Phase V (2009–2013)

Introducing the health impact assessment in the daily practice of Pécs City Council was part of the overarching goal of Phase V, which was to incorporate health and health equity in all local policies. The project aimed to fulfil this strategic goal and to fit in the process of health development planning started in Phase III. The Healthy City Foundation of Pécs initiated a 'health filter' to test every proposal of the city council.

In December 2011, the city council passed a resolution that all strategic documents of the city (every mid- and long-term conception, programme, strategy, etc.) discussed by the COUNCIL first had to be sent to the Healthy City Foundation for an assessment based on Healthy Cities principles. The aim of the resolution was to draw decision makers' attention to ways to consider health issues or how to emphasise them in strategies which in the long run provided frames to City Council decisions.

Three strategic documents have been examined so far:

- Integrated City Development Strategy.
- City Environmental Programme.
- Concept on Planning Social Services.

The city lacked physical activity programmes, especially for elderly people. The Healthy City Foundation considered it important to initiate such programmes, with deference to the Phase V priorities of the Healthy Cities project and the EU

Presidency for the year 2012 priorities as well, to provide regular affordable physical activity for pensioners. 'Fit Granny!' was a swimming programme started in 2011 for pensioners, initiated by the Healthy City Foundation with the cooperation of Pécs Swimming School and the financial support of Pécs Sport Non-profit Company. The programme provides aqua gymnastics and swimming supervised by a physiotherapist and a swimming instructor.

Connecting to the core priorities of healthy settings, active living and healthy food and diet, 'Health promotion at workplaces' was an outstanding project between October 2012 and September 2013. Workplaces have a strong impact on employees' physical, mental and social well-being, and indirectly may influence the health of their families and communities. In 2012, city companies had the opportunity to submit applications within the framework of the EU Social Renewal Operation Program. The planning, organization and implementation of health development projects in the successful companies were coordinated by the Healthy City Foundation. Main priorities were healthy nutrition, physical activity, mental health and anti-tobacco activities, and these translated into programme elements including screenings, presentations, group sessions, physical activity programmes for groups (swimming, hiking tour, bicycle tour) and health days.

Future

The Healthy Cities Project in Pécs celebrated its 25th anniversary in 2013, a triumph of sustainability. Phase VI (2014–2018) has already begun with many public health challenges for cities and towns.

Healthy Cities has to be able to be revise and apply new approaches to enhance health development at the local level. In the Carpathian basin many cities in Slovakia, Romania, Ukraine and Serbia, where Healthy Cities are not so extensive, have a high number of Hungarian inhabitants who are interested in joining the Hungarian Network of Healthy Cities.

In 2014, the national network, the Hungarian Association of Health Cities, amended its constitution to allow cross-border Hungarian-speaking cities to join the network, and to express this new function changed its name to the Association of Hungarian Speaking Healthy Cities.

Geneva Healthy City[13]

Introduction

In February 1991, the government of the canton of Geneva decided to make Health for All, as defined by WHO, the cornerstone of its health policy. It entrusted its newly created public health directorate with preparing a candidacy for the European

[13] Jean Simos.

network of Healthy Cities. This decision was confirmed in April 1994, when Geneva committed itself to Phase II of this network.

In Geneva, Healthy Cities activities were always tightly integrated with the other activities of the public health directorate, thereby benefiting from the same budget earmarked for prevention and health promotion. According to Healthy Cities principles, a steering committee was set up which included representatives from other sectors of the cantonal administration, from municipal authorities and civil society organizations. Beginning in Phase III, and in keeping with a core theme of that phase, the local Healthy Cities programme covered the social and health aspects of Geneva's Agenda 21 (first adopted by the cantonal parliament in March 2001, and renewed every 4 years since then). A local plan for environment and health was also set up, and was implemented within the Healthy Cities framework. International collaborations were actively pursued with other French-speaking cities, through the WHO Collaborating Centre in Rennes, France, which subsequently became the International association for health and sustainable development (S2D).

Many actions have been carried out over the short, medium and long terms under the Healthy Cities banner. Here two of them are presented as a testimony of the lasting influence of the Healthy Cities approach in Geneva.

*The Green Fork Label (*Fourchette verte*)*

Origin of the Project

Concern about unsanitary conditions in a handful of restaurants led to initial brainstorming about how to improve hygiene across all catering services in the Canton. Since legal constraints did not allow the local government to establish a blacklist, the opposite strategy was chosen: to establish a list of 'good' restaurants. The concept was expanded to cover not only basic hygiene but also other health-related aspects.

This brainstorming process occurred at an appropriate juncture, as the canton's public health authorities were preparing Geneva's candidacy for the European Healthy Cities network and were on the lookout for new projects. In 1993, a working group was set up in order to define the criteria which would enable restaurants to acquire the Green Fork label. The future local Healthy City coordinator led this group, which included experts from fields as diverse as communications, economics, prevention (alcohol and tobacco), police and security, nutrition and consumers' rights, as well as the chairman of the local restaurant association.

Project Roll-Out

At the end of 1993 restaurant managers interested in promoting the enjoyment of healthy eating could apply for the Green Fork label on a voluntary basis. The first target group was adults taking their lunchtime break outside the home, because statistics at that time showed that this covered around 40 % of the working population. The criteria that the restaurants had to meet were the following:

- A healthy and well-balanced dish of the day, different on each day of the week. A 'nutritional balance' concept (see Fig. 10.5) was applied to this dish: one protein-rich food product in limited quantity (with a precise maximum quantity defined for each type of food); a carbohydrate-rich food product whose quantity might vary according to the requirements of the customer (according to age, weight, hunger, etc.); and one or two fibre-rich products also containing vitamin C. Further non-binding recommendations were made, in particular to do with the use of fats and oils.
- Non-alcoholic beverages at a reasonable price: at least one mineral water, one milk-based drink and one fruit juice must be cheaper than the cheapest alcoholic drink (usually beer).
- A single, clearly marked non-smoking area covering at least one-quarter of all seats (these criteria were drawn up before the Canton of Geneva banned smoking in all closed public spaces, in 2009).
- Compliance with food safety regulations (Fig. 10.6).

The Green Fork initiative met with an enthusiastic response from the health sector and from many members of the public; however, most restaurant managers remained unconvinced, in keeping with a strong anti-establishment sentiment prevailing in that profession. Many were wary of any label attributed by the public sector, believed that keeping even a minority of their seats for non-smokers was economic suicide,[14] and were convinced that their food was nutritionally sound anyway.[15] Resistance was also present on the side of the administration, which was not accustomed to suggesting rules that could be followed on a voluntary basis, and who were concerned that restaurants with or without the label might end up being treated unequally.

The initial years of the Green Fork concept were therefore far from straightforward, and its supporters were at pains to prevent political authorities from calling for an early termination of the project.

Evolution of the Concept

One category of restaurants welcomed the Green Fork concept: work, school and office canteens. This sector was dominated by a few large, well-structured catering companies, whose workforce either already had high-level training or was in the

[14] Around 33 % of the adult population in Switzerland were smokers in 1992 (Source: Swiss Health Survey), but most restaurant managers believed that smokers formed a majority of the population and were unconvinced by officially certified data.

[15] This was not true for many restaurants in Geneva; according to an ad hoc study by the Geneva Health Department, nutritional knowledge among restaurant manager and cooks was as low as in the general population at that time.

Fig. 10.6 Logo of Pécs Healthy City

process of acquiring it on the job. There was an opportunity for these kitchens to use their know-how in quality control and process management to integrate Green Fork nutritional concepts. This had the advantage of coinciding with the initial objective of the project: using the lunch break to help change the behaviour of the working population in other life settings, and using the catering profession as a catalyser for best practices.

Over time the Green Fork label was adapted for cafés, restaurants, self-service restaurants, canteens (for children and for senior citizens), day-care centres, etc. The greatest success was in the school canteen sector, where the concept was able to influence eating behaviour at home for thousands of children and adolescents.

The criteria have evolved over time: for example, the requirement to have a non-smoking sector was scrapped when smoking in public places was banned—in 2009 in the canton of Geneva and in 2010 at the federal level. Other requirements have been added, such as environmentally friendly criteria for waste disposal (left-over food, oil, packaging, etc.).

The Green Fork concept now forms part of the general health promotion strategy within the Healthy Cities programme. The employees of the civil organization that runs Green Fork (despite occupying less than two full-time positions) have given many training courses and conferences and participate as exhibitors in many cultural and sporting events.

Institutional Aspects

Challenges linked to the almost atavistic distrust of certain restaurants towards the administration, as well as the administration's own lack of operational flexibility, led to the creation of a civil society organization (CSO) which has been managing the Green Fork project since 1995 within the framework of a government grant. This operational solution has been highly successful in reducing mistrust towards the project and simplifying procedures, and has facilitated the inclusion of consumers who test whether the criteria are being respected by the various caterers participating in the project.

From 1996 to 1997 the Green Fork label was adopted by the health promoters of Canton Vaud and Canton Ticino. These success stories outside Geneva led to the creation of an overarching Swiss Green Fork Federation, which brought together the various CSOs and other cantonal initiatives and rapidly grew to be in a position to accept funding from the Swiss Foundation for Health Promotion. From the onset the chairperson of this umbrella NGO was a politician belonging to the executive of one of the cantonal governments (Fig. 10.7).

Support for has extended beyond the borders of public health. Ever since the 2001–2002 school year, the Geneva Department of Education has supported the use of the label by all cafeterias serving schools in the public sector, investing over 0.5 million Swiss Francs (around 0.5 million USD) in the project every year a contribution higher than the amount committed by the Health Department. Green Fork was supported by almost all political parties in the Geneva cantonal parliament in 2001, and has been renewed every year to now (2015).

Quantitative Appraisal

At the end of 1995, after 2 years of direct administration of the project by local government and just after the creation of the local CSO, 50 restaurants in the Canton of Geneva carried the Green Fork label for a total of 6982 seats, of which around 43 %

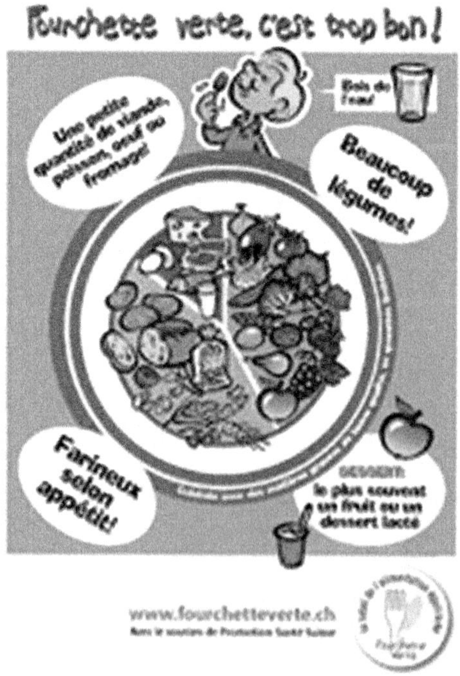

Fig. 10.7 Green Fork (Fourchette verte) poster demonstrating the balanced dish concept

(3019) were in non-smoking areas. By 31 December 2014, some 295 restaurants were in the project, for 38,189 seats, all of which were non-smoking due to the change in legislation. At national level (including Geneva), there were 1387 Green Fork restaurants for 113,964 seats, scattered over half of Switzerland (see Fig. 10.8).

This impressive progress is mainly due to systematic uptake by school canteens in the participating cantons (see Fig. 10.9). In Geneva only 10 % of the nearly 300 restaurants are independently run, against 25 % of office canteens and slightly over 40 % of school canteens.

Finally, brand awareness is a further mark of success. Around four out of five residents of Geneva have heard of Green Fork, and its public image is positive.

Discussion

The objective of supplying the population with a choice of healthy and balanced dishes in collective catering facilities has been met.

Green Fork has positioned itself as a flagship in the health promotion strategy of the Canton of Geneva, serving as a model for other projects and inspiring other cantons across the country. It is a good practice, sharing many characteristics with other successful health promotion interventions:

- Time is of the essence. The project requires sufficient time in order to prove its value. This means that the promoters of the project must themselves be convinced—and patient. Support within the administration and among local politicians is essential.
- A coalition must be built up—in this case with caterers, restaurant managers, other public health actors, etc.

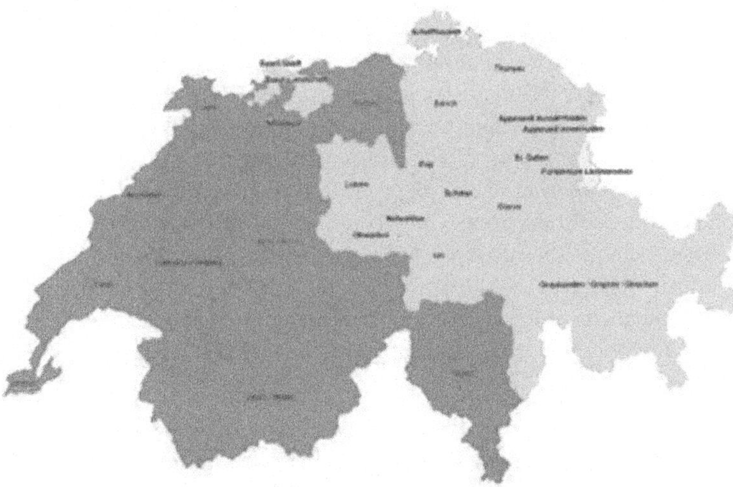

Fig. 10.8 Coverage (in *green/dark*) of the Swiss Green Fork Federation in October 2013

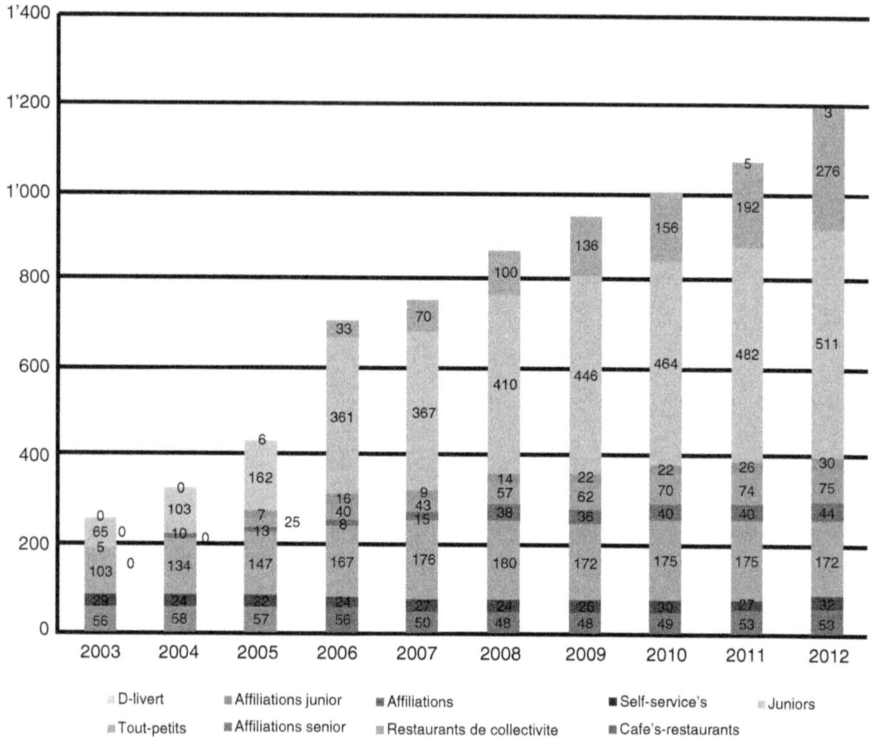

Fig. 10.9 Evolution of the number of Green Fork restaurants in Switzerland (colours correspond to different categories of restaurant). Source: www.fourchetteverte.ch, as at 3 August 2015

- Collaboration with other sectors—in this case, education—is a recipe for success.
- Flexibility is essential, especially in the initial phases of the project. Whereas downtown restaurants were initially expected to be the participants of choice, it rapidly became clear that canteens in offices, schools and other institutions were more receptive and more able to implement the project. This led to additional benefits, such as the ability of the project to have a positive influence on the development of children, in a society where they are more and more likely to have lunch outside their home. In schools and crèches, children discover and learn basic behaviours including healthy nutrition. The Green Fork approach enables the inclusion of the staff of these institutions and of the children's parents in the management of the project. By supplying an example of healthy behaviour, the project helps these institutions provide a healthy environment for the children.

Finally, Green Fork is in ongoing dialogue with the main networks and projects linked to nutrition and healthy body weight, as well as with policy makers, private companies, consumers and other stakeholders. It forms a highly active and significant part of health promotion strategy at the local level.

Participatory Approach Based on Future Scenario Workshops

During Phases IV and V of the European WHO Healthy Cities programme in the 2000s, Geneva launched a project entitled 'Community health promotion and quality of life in the canton of Geneva—Make your voice heard! Quality of life and health in your neighbourhood'. This project mainly took place in two neighbourhoods, one urban (St-Jean/Charmilles, within the city of Geneva) and one suburban (Champs-Fréchets in the town of Meyrin, close to Geneva).

The Scenario Workshop Method

Scenario workshops bring together local politicians, experts and civil servants, as well as civil society representatives and local residents. The first stages of the scenario workshop method—the critical phase and the visionary phase—aim at exploring future solutions without being constrained by temporal limits such as the next election. This allows participants to project themselves forward by one or two decades and imagine what one's neighbourhood might look like in the future (Photos 10.4 and 10.5). As more and more possible directions for progress are envisioned, different scenarios appear. This process is usually underpinned by a sustainable development or urban health mind-set, which can then be used to guide local policymakers.

Photo 10.4 Citizen Workshop

Photo 10.5 Public discussion and debate after presentation of the results

Participation as a Practical Tool to Assist Decision-Making

Once the visions of the future have been expressed, they are exposed to the harsh realities of limited funding, and their political and practical feasibility are assessed. In the so-called implementation phase, the sometimes lofty ideals expressed during the first two phases are concentrated into a handful of feasible, practical measures. Stakeholder participation is therefore used as a decision-making tool, as it gives decision makers not only an idea of what people desire, but also of the practical measures most acceptable and most likely to succeed.

In practice, scenario workshops often yield practical suggestions in the fields of sustainable transport (measures in favour of walking and cycling), urban planning (green spaces), and the reinforcement of social networks (intergenerational meeting places, improvement of social service provision).

Learning from Each Other and Political Legitimacy

The workshops enable all participants to learn from each other because the initially often very creative ideas are hardened down to the level of practical implementation by the same group of people, in a process where originality and feasibility are equally valued and important. The greater the consensus between participants, the greater the chances that suggestions emerging from the group will be taken up and implemented by the political and administrative system.

Participation as a Key Tool for Local Governance

It is important to sustain participation between stakeholders throughout the critical and visionary phases and into the implementation phase. In addition to reinforcing links between local administration and civil society, this also helps reinforce links between public-sector departments participating in the same project. In the case of the St-Jean neighbourhood in central Geneva, city-wide administrative departments collaborated with more local structures involved in sectors such as public health or social services.

Limitations

The true challenge of the scenario workshop method is to establish meaningful dialogue between elected officials, civil servants, experts, local residents and civil society organizations. Initially the idea was to place all participants at the same level; however, in practice it has often been difficult for participants to set aside their role in order to participate in brainstorming. For example, a senior civil servant felt unable to express her ideas fully during one of the meetings because her boss, an elected politician, was also present; and some local residents said they felt inhibited by being placed at the same level as politicians whom they otherwise read about in the newspapers.

Each person or group of people in a scenario workshop tends to have a level of information on a given topic which is different from that of another person or group. This knowledge mismatch is often a challenge for workshop organisers, who must help participants realise that technical knowledge and local, practical knowledge are all valuable and necessary for a fruitful discussion to take place. The method suggests having equal numbers of participants in each category, but it is common for some residents to feel themselves at a disadvantage in the presence of civil servants accustomed to give their opinion easily, and they sometimes remain silent.

Participation vs. Democracy

For politicians who have been democratically elected to their positions, it is not always easy to have their authority temporarily set aside during the participatory process. In practice it is often difficult for them to refrain from defending the status quo or from supporting projects that they are already supporting at the executive or legislative level.

Technical Expertise vs. Social Reality

Experts, whether or not they are part of the local administration, often have trouble accepting the usefulness of public discussion. Because they already have a high level of information and often know the local area very well, they may be tempted to simply 'inform' on what they believe to be the best solution rather than enter into bona fide discussion with local residents.

Limitations of the Future Scenarios Approach

It is difficult for local residents to suspend criticism in order to participate in creating a scenario, and then see it dashed or considerably reduced in the implementation phase. In our experience, the future scenarios approach is best suited for structuring discussions among people with similar professional backgrounds, e.g. architects and urban planners.

The Impossible Consensus

Differences in scale often prove impossible to overcome, particularly in the transportation sector where local residents may suggest reducing traffic in a certain area or street without taking into account knock-on effects elsewhere in the city or conurbation. Likewise, local residents often argue for 'more parking spaces' in their neighbourhood while complaining about the negative consequences of car traffic. This ties in with the NIMBY phenomenon (Not In My Back Yard)—many residents would rather divert traffic away from their neighbourhood than address their own transport behaviour. It follows that ideas put forward by local residents do not necessarily coincide with the principles of health promotion and sustainable development.

Limiting Scope

In Geneva the scoping was very wide, covering 'health promotion and quality of life'. This led to suggestions involving several administrative departments and all political levels (neighbourhood, municipal, cantonal, federal), some of which were not involved in the project. Because of the number and complexity of suggested actions, it was not possible to investigate the feasibility of all of them in a satisfactory manner.

Representativeness

The organization of the Make your voice heard! series of workshops has confirmed how hard it is to reach marginalised populations such as young people (of any age), the elderly, people with immigrant backgrounds, and those with limited socio-economic and cultural capital. In the case of Geneva, several visits were made to places where such persons were liable to be found, but this was found to be a poor investment of resources because such people are often not interested in participating in such processes.

Box 10.3: Recommendations Regarding the Participatory Method
Inform clearly on:

The political and administrative objectives and the participation format (choice of topics, area covered, etc.) while making the thematic and geographical limits very clear.

The framework of the participation, i.e. the amount of power devolved to the residents and local NGOs during the selection phase and the implementation phases. In rising levels of devolution these are information, consultation, delegation, partnership, and ultimately self-rule.

The role of participants (residents, experts and elected officials) and what is expected of them regarding knowledge and experience during the critical and visionary phases.

The defining of institutional resources, i.e. the amount of funding and manpower that the public sector is willing to engage for the implementation phase.

The general idea is to

- Limit the negative effects of passive or intrusive participation by specifying very clearly the roles, missions and mandates of each of the participants (e.g. through a charter specifying the rights and obligations of participants).
- Guarantee a fair allocation of speaking time among the various stakeholders, while preserving opportunities for discussion, interaction and confrontation between ideas and opinions.
- Favour constructive dialogue and inter-subjectivity by valuing the complementarity of local knowledge (residents), expert opinion (government employees and NGOs) and more strategic approaches (elected officials).

Results

Saint-Jean-Charmilles Neighbourhood, City of Geneva

In this central area of the city of Geneva the municipality had delegated responsibility for participatory events to a local group called the Saint-Jean-Charmilles Forum, which has gone on to create a quality of life and health group. This group brings together local residents and health professionals active in the local area.

Initially the Social Department of the city helped set up the Make your voice heard! project, but a change in political outlook following local elections made its involvement less of a priority. The quality of life and health group therefore analysed the results of the workshops itself. Four main directions with practical suggestions emerged: information and communication, health promotion (including education for health), administrative support, and 'towards a greener and more friendly environment'.

Champs-Fréchets Neighbourhood, Town of Meyrin

In the suburban town of Meyrin the municipal departments responsible for social services and the environment were the most involved in setting up the scenario workshops, and then used the results of the and helped shape them into implementable measures. However, changes in political and administrative personnel, and in particular the reorganization of the municipal social service, hampered implementation and weakened the link between the municipality and the local residents who participated in the process. The scenario workshops found themselves in competition with another local project entitled 'Living together in Meyrin'. Probably for this reason, no follow-up committee was created. The workshops were considered a limited study or project, rather than a process to be followed up upon. In due course the municipality distributed the final report to each of its administrative departments, with no particular guidance on how the proposals were to be integrated among their other activities.

One consequence was that the proposals, initially intended for the Champs-Fréchets neighbourhood, were extended to cover the whole of Meyrin. Some of the ideas were implemented, sometimes in a modified form, while others were abandoned. The strongest ideas to emerge from the workshops which took place between 2005 and 2009 concerned public lighting, sustainable mobility, traffic calming, waste management, assigned walking areas for dogs, social networks and respect for others, setting up a network for the exchange of experiences, and reinforcement of links between generations. As several town-wide projects have moved forward in each of these fields, it is difficult to estimate the specific impact of the Make your voice heard! However, locally, the project is widely considered to have acted as a positive stimulant.

Discussion

In suburban Meyrin the municipality took a stronger lead in the process than in the city of Geneva, where the Forum Saint-Jean-Charmilles played a central role. Although this forum was set up by the municipality, its members were local residents and not city personnel. Meyrin emphasised projects linked to infrastructural development and the improvement of the social and natural environment; in the City of Geneva there was a greater need for information and prevention.

The nature of citizen participation in these two projects leads us to accept the constructivist hypothesis, that people involved in participatory processes can become actors likely to implement actions in the future (either directly or through subsidiarity or delegation).

Box 10.4: Recommendations for follow-up on results

Empower the public sector stakeholders who took part in the process, so they can address the economic, social and environmental concerns of the population.

Ensure the medium- and long-term participation of local residents and NGOs in the everyday management of public health and quality of life.

Strive for equal amounts of resources to be available for the workshops and for follow-up, by setting up a multi-scale governance system and an institutional and financial monitoring scheme. An example is an ad hoc working group, which would be maintained over time by an inter-party agreement or a neighbourhood contract between the municipality and the local population.

Use the opportunity of the participatory and community-oriented process to intensify and reinforce links between municipal public services and civil society organisations, to help address local issues and implement projects.

Use the opportunity of the participatory and community-oriented process to intensify interdepartmental collaboration, to improve the focus and efficacy of public policy measures (e.g. setting up a community centre as a joint venture between the social services department and the urban planning department).

Take into account the different needs of the various sub-populations in the area, through actions addressing specific objectives in specific population groups.

Advocate a sustainable development and health promotion culture, integrating a community approach and new governance involving all stakeholders.

Conclusion

Geneva Healthy City developed for more than 15 years, surviving several political and administrative changes but keeping the same coordinator (the author of this contribution). During this period it produced several key initiatives for health promotion, some of which continue to be successful after the official termination (in 2007) of the Geneva Healthy City framework. Subsequently, the approach has continued to impregnate the work of the Health Department and of its partners.

The origin, lifetime, termination and legacy of the Geneva Healthy City experience are linked to changes in contexts, leaders and policies. Its successes, its limitations and its sustainability profile contain important learning that will be discussed in the last chapter on the critical evaluation of the Healthy Cities movement.

References

Adams, L. (1989). Healthy cities, healthy participation. *Health Education Journal, 48*(4), 179–182.

Antonovsky, A. (1996). The salutogenic model as a theory to guide health promotion. *Health Promotion International, 11*(1), 11–18.

Ashton, J. (1992). The origins of healthy cities. In J. Ashton (Ed.), *Healthy cities* (pp. 1–12). Buckingham: Open University Press.

Ashton, J., Grey, P., & Barnard, K. (1986). Healthy cities—WHO's new public health initiative. *Health Promotion International, 1*(3), 319–324.

Ashton, J., & Seymour, H. (1988). *The new public health: The Liverpool experience*. Philadelphia: Open University Press.

Aurillon-Van Den Broucke, N. (2013). Quels sont les freins à l'accès aux soins : le cas du quartier de Bellevue à Nantes, Colloque national du Réseau Français des Villes-Santé de l' OMS, Grenoble.

Baum, F. E. (1993). Healthy cities and change: Social movement or bureaucratic tool? *Health Promotion International, 8*(1), 31–40.

Berkeley, D., & Springett, J. (2006). From rhetoric to reality: Barriers faced by health for all initiatives. *Social Science and Medicine, 63*(1), 179–188.

Berridge, V., Christie, D., & Tansey, E. (2006). *Public Health in the 1980s and 1990s: Decline and rise?* London, UK: Wellcome Trust Centre for the History of Medicine at University College London. Retrieved November 29, 2013, from http://discovery.ucl.ac.uk/2054/.

Blackman, T., Harrington, B., Elliott, E., Greene, A., Hunter, D. J., Marks, L., … Williams, G. (2012). Framing health inequalities for local intervention: Comparative case studies. *Sociology of Health and Illness, 34*(1), 49–63.

Blackman, T., Hunter, D. J., Marks, L., Harrington, B., Elliott, E., Williams, G., … McKee, L. (2010). Wicked comparisons: Reflections on cross-national research about health inequalities in the UK. *Evaluation, 16*(1), 43–57.

Boonekamp, G. M. M., Colomer, C., Tomas, A., & Nunez, A. (1999). Healthy cities evaluation: The co-ordinators perspective. *Health Promotion International, 14*(2), 103–110.

Buck, M., & Taylor, J. (2009). *Liverpool healthy city: A 21st century approach*. Liverpool, UK: Liverpool Partnership Group.

City of Nantes. (2012). La santé des Nantais. Nantoscope thématique des besoins sociaux, 58p.

City of Nantes. (2013a). *European green capital*. Retrieved April 8, 2016, from http://www.nantes.fr/home/a-nantes-et-pas-ailleurs/decouvrir-nantes/la-metropole-des-possibles/nantes-capitale-verte.html.

City of Nantes. (2013b). Rapport d'activité de la mission santé publique de la ville de Nantes.

Clavier, C., & de Leeuw, E. (2013). Framing public policy in health promotion: ubiquitous, yet elusive. In C. Clavier & E. de Leeuw (Eds.), *Health promotion and the policy process* (pp. 1–22). Oxford: Oxford University Press.

CSDH. (2008). *Closing the gap in a generation: Health equity through action on the social determinants of health. WHO Commission on social determinants of health 2005–2008*. Geneva: World Health Organization.

Davies, J. K., & Kelly, M. P. (1993). Healthy cities: Research and practice. In J. K. Davies & M. P. Kelly (Eds.), *Healthy cities: Research and practice* (pp. 1–13). London: Routledge.

DHSS (Black Report). (1980). *Inequalities in health: Report of a research working group*. London Department of Health and Social Security.

De Leeuw, E. (2001). Global and local (Glocal) health: The WHO healthy cities programme. *Global Change and Human Health, 2*(1), 34–45.

De Leeuw, E. (2003). Nurturing healthy cities: Research responsibility and accountability. In T. Takano (Ed.), *Healthy cities and urban policy research* (pp. 138–161). London: Spon Press.

De Leeuw, E. (2011). Do healthy cities work? A logic of method for assessing impact and outcomes of healthy cities. *Journal of Urban Health, 89*(2), 217–231.

De Leeuw, E. (2013). Evaluating WHO healthy cities in Europe: Issues and perspectives. *Journal of Urban Health, 90*(Suppl. 1), S14–S22.

De Leeuw, E., & Skovgaard, T. (2005). Utility-driven evidence for health cities: Problems with evidence generation and application. *Social Science and Medicine, 61*(6), 1331–1341.

Dooris, M. (2005). Healthy settings: Challenges to generating evidence of effectiveness. *Health Promotion International, 21*(1), 55–65.

Dooris, M., & Heritage, Z. (2013). Healthy cities: Facilitating the active participation and empowerment of local people. *Journal of Urban Health, 90*(Suppl 1), S74–S91.

Dooris, M., Poland, B., Kolbe, L., de Leeuw, E., McCall, D., & Wharf-Higgins, J. (2007). Healthy settings: Building evidence for the effectiveness of whole system health promotion—Challenges and future directions. In D. V. McQueen & C. M. Jones (Eds.), *Global perspectives on health promotion effectiveness* (pp. 327–352). New York: Springer.

Duhl, L. (1986). The healthy city: Its function and its future. *Health Promotion International, 1*(1), 55–60.

Eriksson, C. (2000). Review article: Learning and knowledge-production for public health—A review of approaches to evidence-based public health. *Scandinavian Journal of Public Health, 28*(4), 298–308.

Gassiot, M. (2015, juin). Agir sur l'aménagement urbain: Création d'un marché et de stations gourmandes à Nantes; in 10èmes journées de l'INPES.

Giroult, E. (1989). Equity and the urban environment. *Health Promotion International, 4*(2), 83–85.

Goumans, M., & Springett, J. (1997). From projects to policy: 'Healthy Cities' as a mechanism for policy change for health? *Health Promotion International, 12*(4), 311–322.

Green, G. (2012). Intersectoral planning for city health development. *Journal of Urban Health, 89*(2), 247–257.

Green, G., Acres, J., Price, C., & Tsouros, A. D. (2009a). City health development planning. *Health Promotion International, 24*(Suppl 1), i72–i80.

Green, G., Price, C., Lipp, A., & Priestley, R. (2009b). Partnership structures in the WHO European Healthy Cities Project. *Health Promotion International, 24*(S1), i37–i44.

Guyard, F. (2016). *Un plan d'action pour la santé environnementale. La gazette Santé Social.fr.* Retrieved April 8, 2016, from http://www.gazette-sante-social.fr/25371/un-plan-daction-pour-la-sante-environnementale.

Hancock, T. (1993). The evolution, impact and significance of the healthy cities/healthy communities movement. *Journal of Public Health Policy, 14*(1), 5–18.

Hancock, T., & Duhl, L. (1986). *Promoting health in the urban context.* Copenhagen, Denmark: World Health Organization Regional Office for Europe. Retrieved November 24, 2013, from http://www.euro.who.int/__data/assets/pdf_file/0013/101650/E87743.pdf.

Heritage, Z., & Dooris, M. (2009). Community participation and empowerment in Healthy Cities. *Health Promotion International, 24*(Suppl 1), i45–i55.

Hunter, D. J. (2007). The policy context and growing importance of health. In D. J. Hunter (Ed.), *Managing for health* (pp. 8–27). Abingdon, OX: Routledge.

Katikireddi, S. V., Higgins, M., Smith, K. E., & Williams, G. (2013). Health inequalities: The need to move beyond bad behaviours. *Journal Epidemiology and Community Health,* Online First.

Kelly, M. P., Davies, J. K., & Charlton, B. G. (1993). Healthy cities: A modern problem or a post-modern solution. In J. K. Davies & M. P. Kelly (Eds.), *Healthy cities: Research and practice* (pp. 159–167). London: Routledge.

Kickbusch, I. (1989). Healthy cities: A working project and a growing movement. *Health Promotion International, 4*(2), 77–82.

Kickbusch, I. (1996). Tribute to Aaron Antonovsky—'What creates health'. *Health Promotion International, 11*(1), 5–6.

Kickbusch, I. (1999). Global + local = glocal public health. *Journal of Epidemiology and Community Health, 53*, 451–452.

Kickbusch, I. (2003). The contribution of the World Health Organization to a New Public Health and Health Promotion. *American Journal of Public Health, 93*(3), 383–388.
Kickbusch, I. (2007). The move towards a new public health. *Global Health Promotion, 14*(2 Suppl 9), 9.
Kickbusch, I. (2012). 21st century determinants of health and wellbeing: A new challenge for health promotion. *Global Health Promotion, 19*(3), 5–7.
Lindstrom, B., & Eriksson, M. (2009). The salutogenic approach to the making of HiAP/healthy public policy: Illustrated by a case study. *Global Health Promotion, 16*(1), 17–28.
Lipp, A., Winters, T., & de Leeuw, E. (2013). Evaluation of partnership working in cities in Phase IV of the WHO healthy cities Network. *Journal of Urban Health, 90*(Suppl 1), S37–S51.
Lorenc, T., Petticrew, M., Welch, V., & Tugwell, P. (2013). What types of interventions generate inequalities? Evidence from systematic reviews. *Journal Epidemiology and Community Health, 67*(2), 190–193.
Marmot, M., & Bell, R. (2010). Challenging health inequalities—Implications for the workplace. *Occupational Medicine, 60*(3), 162–164.
Milton, B., Attree, P., French, B., Povall, S., Whitehead, M., & Popay, J. (2012). The impact of community engagement on health and social outcomes: A systematic review. *Community Development Journal, 47*(3), 316–334.
Mittelmark, M. B., & Bull, T. (2013). The salutogenic model of health in health promotion research. *Global Health Promotion, 20*(2), 30–38.
Nutbeam, D. (1998). Evaluating health promotion—Progress, problems and solutions. *Health Promotion International, 13*(1), 27–44.
O'Dwyer, L. A., Baum, F., Kavanagh, A., & Macdougall, C. (2007). Do area-based interventions to reduce health inequalities work? A systematic review of evidence. *Critical Public Health, 17*(4), 317–335.
ORSLA. (2010). Observation de la santé des Nantais, Observatoire Régional de la Santé de Loire-Atlantique (ORSLA), 52p.
OSCAR. (2011). Plan alcool à Nantes: Charte de la vie nocturne. Agence Régionale de Santé Pays-de-la-Loire.
OSCAR. (2014). Campagne d'information, de dépistage et de remédiation du radon. Agence Régionale de Santé Pays-de-la-Loire.
Paton, K., Sengupta, S., & Hassan, L. (2005). Settings, systems and organization development: The healthy living and working model. *Health Promotion International, 20*(1), 81–89.
PNNS. (2011). Plan National Nutrition Santé. Ministère du travail, de l'emploi et de la santé, Paris.
Popay, J., Whitehead, M., & Hunter, D. J. (2010). Injustice is killing people on a large scale—But what is to be done about it? *Journal of Public Health, 32*(2), 148–149.
Rayner, G., & Lang, T. (2012). *Ecological public health: Reshaping the conditions for good health*. Abingdon, OX: Routledge.
Samson, X. (2015). Bains douches à Nantes; in Revue des 25 ans du réseau français des villes santé de l'OMS, Sept 2015.
Saraux, P. (2012). Citoyenneté et santé mentale, 1ère réunion des conseils locaux de santé mentales; in Points de secteur et Médiation, Elus santé et territoire, Paris janvier.
Seedhouse, D. (2004). *Health promotion: Philosophy, prejudice and practice* (2nd ed.). New York: Wiley.
Shareck, M., Frohlich, K. L., & Poland, B. (2013). Reducing social inequities in health through settings-related interventions—A conceptual framework. *Global Health Promotion, 20*(2), 39–52.
Smith, K. E. (2008). *Health inequalities in Scotland and England: The translation of ideas between research and policy*. Ph.D. Thesis. Edinburgh: University of Edinburgh. Retrieved December 8, 2013, from http://hdl.handle.net/1842/7203.
Smith, K. E., Bambra, C., Joyce, K. E., Perkins, N., Hunter, D. J., & Blenkinsopp, E. A. (2009). Partners in health? A systematic review of the impact of organizational partnerships on public health outcomes in England between 1997 and 2008. *Journal of Public Health, 31*(2), 210–221.
Stevenson, H. M., & Burke, M. (1991). Bureaucratic logic in new social movement clothing: the limits of health promotion research. *Health Promotion International, 6*(4), 281–289.

Tollec, L., Gall, A. R. L., Jourdren, A., Auffray, F., Jabot, F., Vidy, A., … Simos, J. (2013). L'Évaluation d'Impacts sur la Santé (EIS): Une démarche d'intégration des champs santé-environnement dans la voie du développement durable. Application à un projet d'aménagement urbain: La halte ferroviaire de Pontchaillou à Rennes. Juillet 4(2). Développement durable et territoires [En ligne]. doi:10.4000/developpementdurable.9815.
Tsouros, A. D. (1989). Equity and the healthy cities project. *Health Promotion International, 4*(2), 73–75.
Tsouros, A. D. (1995). The WHO healthy cities project: State of the art and future plans. *Health Promotion International, 10*(2), 133–141.
Tsouros, A. D. (2009). City leadership for health and sustainable development: The World Health Organization European Healthy Cities Network. *Health Promotion International, 24*(S1), i4–i10.
Tsouros, A. D. (2013). City leadership for health: Back to the future. *Journal of Urban Health, 90*(Suppl. 1), S4–S13.
Tsouros, A. D. (1990). *World Health Organization Healthy Cities Project: A project becomes a movement. Review of Progress 1987 to 1990.* Copenhagen, Denmark: World Health Organization Regional Office for Europe.
Webster, P., & Sanderson, D. (2013). Healthy cities indicators—A suitable instrument to measure health? *Journal of Urban Health, 90*(Suppl 1), S52–S61.
Werna, E., Harpham, T., Blue, I., & Goldstein, G. (1999). From healthy city projects to healthy cities. *Environment and Urbanization, 11*(1), 27–40.
World Health Organization Regional Office for Europe. (2012). *Health 2020: The European policy for health and well-being*. Retrieved April 29, 2015, from http://www.euro.who.int/en/health-topics/health-policy/health-2020-the-european-policy-for-health-and-well-being/about-health-2020.
World Health Organization Regional Office for Europe. (2013). *Phase VI (2014–2016) of the WHO European Healthy Cities Network: goals and requirements*. Retrieved April 28, 2015, from http://www.euro.who.int/__data/assets/pdf_file/0017/244403/Phase-VI-20142018-of-the-WHO-European-Healthy-Cities_Network-goals-and-requirements-Eng.pdf.
Ziglio, E., Hagard, S., & Griffiths, J. (2000). Health promotion development in Europe: Achievements and challenges. *Health Promotion International, 15*(2), 143–154.

Conference Proceedings

Produits chimiques, responsabilités et réponses des collectivités locales Forum national Santé Environnement 29–30 March 2007 Published by the City of Rennes.
Santé et environnement dans le ville. Conference proceedings of Rennes 6–7 June 1991-Published by Ecole Nationale de la Santé Publique 1992.
Urbanisme et santé : Projet urbain et politique santé l'expérience de Rennes- Projet urbain Rennes 2015 Hubert Chardonnnet, Colloque Institut des Villes "Villes, santé et développement durable" 2006- La documentation française Paris 2007.

City of Rennes Documents

Agir ensemble Plan Rennes Ville-Santé 2005–2008.
Contrat Local de Santé 2013–2015 Ville de Rennes ARS—City of Rennes.
Plan Local de Santé de la Ville de Rennes January 2013. Retrieved from http://metropole.rennes.fr/politiques-publiques/culture-education-vie-sociale/la-sante/.

Rapport d'activité (2012, December 4). Mission Promotion Santé Territoires, City of Rennes Direction Santé Publique Handicap, Mission Promotion Santé Territoires.
Rennes Healthy City Working together for Health and sustainable development, 3rd City Health Development Plan.
Rennes Ville Santé presentation brochure.
Urbanisme et Santé, Projet urbain et politique de santé 2006.

Rennes Academic Research

Erwan Le Goff. (2012). Les Villes-Santé en Bretagne: quels choix de gestion et d'aménagement des espaces: quels choix de gestion et d'aménagement des espaces ? Doctoral thesis defended on 4 December 2012 at Université de Rennes 2. Retrieved from http://www.villes-sante.com/?s=erwan+Le+Goff.
Moulin E and S2D. (2008). L'évaluation d'impact santé. Mise en place d'une méthodologie et application à un projet pilote: la crèche Colette de la ZAC de Beauregard.
Pascal Thébault. (1992). La révision du Plan d'occupation des sols par une analyse multicritère du projet urbain et du projet Ville Santé à Rennes Science and Technology Master's thesis 'aménagement et mise en valeur des régions' Université de Rennes 1.
Tollec L. (2011). Construction et mise en œuvre d'une démarche d'EIS pour un projet d'aménagement urbain : application à la halte ferroviaire de Pontchaillou et ses abords.

Chapter 11
The Diversity of Healthy Cities in Asia and the Pacific

Keiko Nakamura and Ai Chaobang John Ashton

Overview

Already 50 % of the population of Northeast Asia and 42 % of Southeast Asia live in urban areas, and it is estimated that the total urban population of these regions will increase from 1.1 billion in 2010 to 1.8 billion in 2050 (Asian Development Bank 2011). The quality of urban environments such as infrastructure, housing, economy, education, and cultural diversity; the efficiency of urban life and work; the attainment of optimum and equitable population health; and social policies for ageing populations are major concerns on the public health agenda in this rapidly urbanising region.

Tokyo was the first Asian city to respond to the Healthy Cities movement of the 1980s, establishing the Tokyo Citizens' Council for Health Promotion in 1991 (Nakamura and Takano 1992). A series of consultative meetings by the WHO Regional Office for the Western Pacific (WPRO) on urban health issues in developing countries merged into discussions on health in rapidly growing urban areas at the World Health Assembly in 1991 (Ogawa 2003). Following endorsement of the WHO's Western Pacific Regional Committee, ten countries (Cambodia, China, Fiji, Lao People's Democratic Republic, Malaysia, Mongolia, Papua New Guinea, Republic of Korea, the Philippines, and Vietnam) implemented the programme (WHO (WPRO) 2000). In 1995 the Ministers of Health of Pacific Island Countries also articulated a vision of Healthy Islands (Galea et al. 2000). Called the Yanuca Island Declaration, it presented an idyllic image: Healthy Islands would be places where children were nurtured in body and mind; the environment would invite learning and leisure; people would work and age with dignity; and ecological

K. Nakamura (✉) • A.C.J. Ashton
Department of Global Health Entrepreneurship, Graduate School of Tokyo Medical and Dental University; WHO Collaborating Centre for Healthy Cities and Urban Policy Research, Tokyo, Japan
e-mail: nakamura.ith@tmd.ac.jp

E. de Leeuw, J. Simos (eds.), *Healthy Cities*,
DOI 10.1007/978-1-4939-6694-3_11

balance would be a source of pride. Since then the Pacific Island Countries have applied this vision in areas such as environmental health, diabetes, obesity, mental health, human resource management, and elimination of lymphatic filariasis, tuberculosis, HIV/AIDS, and parasitic diseases.

By 1998 pilot projects, guidance on the ground and capacity-building programmes for cities in Asia, had led to more than 150 cities in the Western Pacific region coming to address health as a central issue in urban policy (WHO (WPRO) 2000). Following a series of meetings and consultations with researchers, practitioners, and city leaders working on Healthy Cities, the Regional Guidelines for Developing a Healthy City Project were published in 2000 (WHO (WPRO) 2000). These guidelines provided a generic approach framework and were a useful reference for innovative local and national Healthy Cities initiatives.

The WHO-led promotion of Healthy Cities found fertile ground in these regions, as there had been long-standing efforts to focus on health development at local, country, and subregional levels. One of these was the National Patriotic Health Campaign Committee established in 1952 in China. While it has adapted in response to changes in the country's social and economic development (Qifa 2003) its aims remain: to improve health and sanitation in both rural and urban settings, through national, provincial, and local networks. This long-term development became an important basis for the later Healthy Cities movement. In Japan a national government-led programme, Health Culture Cities, started in 1993, followed by the Healthy Japan 21 programme which since 2000 has emphasised planning for health at the local level, together with numerical target-setting and evaluation. A health promotion programme establishing a community leaders' committee had been practised since 1961 at most local administrative units (Yajima et al. 2001). Healthy Cities in Japan have been developed based on community-led initiatives like this and other public health development efforts, at local and community levels. In the Republic of Korea, following the National Health Promotion Act in 1995, local government and health centres became active in developing health promotion programmes for communities (Nam 2011). These locally led initiatives are regarded as a basis for Healthy Cities. In Malaysia, in both cities such as Kuching (Chi 2003; Edris 2003) and Johor Bahru (Rahim 2003) and in village settings (Kiyu et al. 2006), a Healthy Cities approach has been applied and widely accepted as a way of improving the health of the population.

In 2003 a regional consultation meeting of WHO WPRO recommended formulating the Alliance for Healthy Cities (AFHC), a network of cities and organisations (WHO (WPRO) 2003). Since its inauguration in October 2004, members of AFHC have developed their Healthy Cities following WHO guidelines, national and regional recommendations, and their respective steering groups. A vision to improve citizens' quality of life has been widely shared, and diverse programmes have been implemented. The AFHC provided momentum for the institutionalisation of Healthy Cities (WHO (WPRO) 2010a) and promoted a generic approach adaptable to local contexts. The AFHC network respects diversity and accommodates governments, communities, and private organisations from both non-profit and profit sectors, depending on social and political circumstances. The WHO Collaborating

Centre for Healthy Cities and Urban Policy Research, designated by the WHO in 1997 and stationed at the Department of Health Promotion and International Health at Tokyo Medical and Dental University, is providing technical support to this network. In addition, following the WHO Western Pacific Regional Committee's resolution in 2010 on healthy settings to advocate Healthy Cities and Healthy Islands as a multisectoral development agenda, country-specific guidelines, recognition, and evaluation have been developed.

Creative moves at regional, national, and local levels towards health promotion and development, combined with the conceptual framework of Healthy Cities, have led to the growth of Healthy Cities in this region, aligned with the social and economic development of its nations and its cities.

Structure

There are country-based, subregion-based, and international networks of Healthy Cities. The interactions of these networks and their support for each other have helped expand the Healthy Cities movement because they meet specific regional and national needs, and nurture a spirit of international information-sharing. Among those networks, the AFHC provides international forums for mutual learning from locally based initiatives, following WHO guidelines and advice for developing Healthy Cities (Nakamura 2011). The members of the AFHC include local governments, national agencies, NGOs, academic institutions, organisations from the private sector, and international agencies. Those members pay membership fees and participate in its programmes.

In October 2004, when the Inaugural Conference of the AFHC was held in Kuching, Malaysia, 26 cities and organisations joined AFHC as founding members. Since then the membership has steadily grown: there were 213 member cities and organisations in October 2014.

The goal and objectives of the AFHC are clearly stated in its charter (see box below). The general assembly of the AFHC serves as its main governing body. The steering committee of the AFHC, which consists of 12 elected members, serves as the policy-making arm of the AFHC. The convenor city of the General Assembly is elected for a term of 2 years and serves also as the chair city, hosting a global conference and the meeting of the General Assembly every 2 years. AFHC chapters are organised at country, sub-country, and inter-country level on approval of the general assembly; as of October 2014, five national chapters had been formed in Japan, China, Korea, Hong Kong, and Australia and others were set to follow. These chapters play key roles by disseminating and sharing information in local languages and contexts, expanding membership, and offering mutual mentoring. Collaboration with WHO allows members to learn from the global public health agenda and from urban development challenges.

Goal and Objectives of the Alliance for Healthy Cities (from the Charter of the Alliance for Healthy Cities, October 12, 2014)

Goal:

To promote and protect health and improve the quality of life of urban populations in a sustainable manner through the Healthy Cities approach.

Objectives:

A. To strengthen Healthy Cities initiatives and encourage the development of innovative plans and programmes to improve the quality of life and address the health challenge of specific settings and communities

B. To share experiences in improving the quality of life and addressing common health problems among members

C. To recognise and promote outstanding practices and innovations within Healthy Cities

D. To mobilise and optimise the use of all available resources to promote and support the adoption of the Healthy Cities approach among cities and other communities in the Western Pacific Region and other regions

E. To develop new knowledge and technology in collaboration with the academe, universities, and centres of learning and to package technical resources for the improvement of planning, implementation, and evaluation of Healthy Cities

Membership is granted upon recognition by the steering committee, on submission of documentation providing (1) a written policy statement in support of Healthy Cities; (2) future vision and goals; (3) a profile of the city (baseline data); and (4) analysis of priority health problems. These requirements are mandatory for the initial phase of membership and developments are expected to occur based on these fundamental elements. All member mayors and representatives sign the AFHC charter, which outlines the vision of its members built on the experiences of Healthy Cities worldwide.

The work by national government agencies and related bodies helps develop country-based guidelines and capacity-building programmes, and mobilises seed funds for Healthy Cities planning and programmes when applicable. Academic institutions play roles in consultation, evidence-gathering, providing training for workers, and conducting needs assessments for the planning and efficient implementation of a Healthy City project.

Common Characteristics

Comprehensive City Planning

The health of city dwellers largely depends on their living conditions and lifestyles. The determinants of health are defined by the WHO as 'the range of personal, social, economic and environmental factors which determine the health status of individuals or populations' (1998). After developing experience in interpreting the 'social determinants of health' in the Healthy Cities context (WHO 1998), a series of works by the Commission on Social Determinants of Health concluded that action was needed so that health equity could be achieved (WHO Commission on Social Determinants of Health 2008).

Studies on the determinants of health of city dwellers revealed the large contribution of socioeconomic and residential environmental factors on a population's health status, in addition to healthcare-related factors (Takano and Nakamura 2001). It also noted that individual determinants of health interact with each other, producing a need for an integration-oriented perspective in urban health policy interventions. As an empirical application of a 'health promotion' concept into specific contexts, the Healthy Cities Guidelines encourage integration of policies and programmes to avoid duplication of efforts, and the sharing of policies with multiple partners (WHO (WPRO) 2000). This is reflected in the experiences of Healthy Cities in the region, in which considerable variety exists between cities' preferred interpretations of 'health' as well as their priorities, capabilities, and circumstances, both human and environmental. Most take a broad conceptual approach to health and are attempting to address it on multiple fronts. As this involves action in many different urban settings and the participation of a range of stakeholders, many cities have explicitly incorporated the Healthy Cities mission or ethos into their planning. The most ambitious have developed master plans entirely dedicated to Healthy Cities objectives and encompassing many different public health concerns, as well as engaging with stakeholders responsible for improving them.

In the Directory of the Alliance for Healthy Cities, examples of comprehensive city planning in the Asia Pacific region are as follows (AFHC 2014).

Japan

Japanese examples are the most consistent. Kameyama is notable for declaring its Kameyama City Comprehensive Plan as a city plan of the highest level; the outline of its basic policy stipulates its guiding goal as 'Creation of a healthy environment with blessings of nature', and its Everybody Healthy Project in 2012 made a policy-wide commitment to supporting citizens' physical and mental health. Owariasahi's Healthy City programme is based on the 5th Owariasahi City General Plan; Yamato started its 10-year strategy of the eighth Total City Plan in 2009; and the New Healthy Obu 21 Plan, set to run for 10 years, gives its objective as 'Extend healthy life—healthy city to celebrate longevity'. It is noticeable that these plans tend to state their core goals extremely broadly; these often are realised in multiple pillars of health policy, which may themselves contain a multitude of specific goals and indicators—in Yamato's case, for example, there are seven basic goals, 23 sub-basic goals, and 1000 projects under the three headings of Human Health, City Health, and Societal Health.

People's Republic of China

In Wujiang, in Jiangsu Province of the People's Republic of China, the city releases regular 5-year plans on healthy city construction, the current being 'Wujiang's Patriotic Health Campaign and Healthy City Construction Plans' for 2011–2015. Its goals are to establish five systems for health improvement based mainly on water environment management and measures for medical provision, and to raise health standards through innovative ideas and methods. Another city, Wuxi, issued a 3-year plan in 2008 covering five major tasks and encompassing 21 working plans and 68 targets; since 2011 it has progressed into a second-round 5-year plan.

Republic of Korea

Comprehensive planning is also notable in Korean cities. One of the most distinct is Changwon, the first planned city in the nation: its infrastructure had been well established, especially in transport, green areas and ecosystems, walking courses, and public sanitation. The city of Siheung also stands out, not only for its Siheung Healthy City Model 21 but also for its development of the Baegot New Town project—the first such settlement in the nation to be developed under the Healthy Cities concept from its beginning, to consist of parks and eco-friendly schools, the sea on one side, a rich green environment, and direction by a steering team of experts and citizens from the design stage. The existence of plans at that scale is generally very evident in Korean cities, most of them highly comprehensive and clearly spelling out their visions and goals.

Taiwan

The same is true for Taiwan. Chiayi drafted a 3-year programme called the Chiayi City Healthy City Promotion Project, which underwent a systematic process of preparation and establishment (2006–2007), and then execution and implementation (2008–2010). Most of the city's public health efforts are conducted under this planning framework, including the installation of many facilities for social welfare, safety, and cultural expression, such as parks, museums, and support or service centres, as well as environmental renovation. Another city of note, Pingtung, has a similarly varied and wide-ranging framework which includes, among many other things, large-scale strategic rehabilitations of countryside villages and reconstruction of flood-prone areas.

Institutional Developments

Along with changes at the level of comprehensive planning have come new institutions, created or adjusted to fulfil these visions of improving health. In the first instance are the cities' Healthy City representative bodies themselves, which show very varied institutional character and structure. According to the directory of the Alliance for Healthy Cities (2014), the majority of the Healthy Cities of their members have been founded by city governments and local authorities, often in cooperation with other stakeholders, and key responsibilities in overseeing these bodies' functions remain in the hands of mayors' offices and public departments. Exceptions include Illawarra and Onkaparinga in Australia, whose representative bodies are non-governmental organisations.

Regardless of their composition, most Healthy Cities stress collaboration with a wide range of stakeholders and engagement with citizens. Many cities incorporate the stakeholders into their leadership structures to form multisectoral committees or other organs for a more consultative, inclusive direction. The core teams and steering committees typically consist of public officials, usually from mayors' offices or multiple government departments, but they coordinate and rely upon any number of collaborative units to plan and implement their work.

Legislative changes have accompanied some schemes. Logan City in Australia, for example, credits 'local government reform and changes in legislation' as laying the foundations for its initiative. In the Philippines, the city governments of Marikina and Parañaque passed ordinances to found their Healthy City programmes. A number of laws have been passed at local level to regulate or promote activities in specific areas. Examples of these include implementing anti-smoking measures, particularly in the restriction of smoking in public places and indoor venues, increasing cigarette taxes, devising public policies advantageous to health, promoting physical activity, promoting dental and oral hygiene, reducing risks and vulnerabilities of older persons by providing discounted services, introducing an insurance system for injured cyclists, preventing atopic diseases, implementing suicide prevention measures, and promoting community participation.

To fund Healthy Cities programmes, Korean cities receive benefits from their national government's Health Promotion Act, which stipulates that Korea's Health Promotion Fund, which receives its money from taxes on tobacco, provides money to health promotion programmes (Nam 2011). The money is used for both health promotion activities and Korea's National Health Insurance, but the allocation to health promotion increased from 17.8 % in 2002 to 43.8 % in 2009 (Nam 2011). The City of Wonju made a further decision to use local tobacco tax revenue for priority projects identified by the Plan of Healthy City Wonju (Nam et al. 2011).

Public–Private Partnerships

A most consistent theme of Healthy Cities in the region is the emphasis on citizen engagement and participation, community empowerment, and cooperation between as wide as possible a range of groups with stakes in public health. The leadership structures of most cities include teams and committees encompassing many different sectors, including both public and private institutions: local government, medical services, academia, health experts, public agencies, NGOs, industry and businesses, and residents' representatives. Examples of public–private partnerships in the directory of the Alliance for Healthy Cities (2014) are as follows.

Hong Kong

The districts of Hong Kong, which place public–private and cross-sectoral collaboration at the core of their approaches, emphasise partnerships. The involvement of businesses and private enterprises is consistently observed. The Kwai Tsing Safe Community initiative combines the joint efforts of over ten local public and private organisations; and Tsuen Wan's Safe and Healthy City Steering Committee was supported by 20 local public and private groups including businesses, NGOs, large enterprises, and government departments (AFHC 2014). Its Safe and Healthy Occupational Safety Charter Working Group is a good example of these partnerships. Similar collaborations are found in other districts in Hong Kong, and in most Taiwanese Healthy Cities.

Japan

Japanese cities also develop partnerships, but what perhaps stands out most is the involvement of academic, scientific, and medical institutions. Some cities are developing physical fitness programmes and community care for senior citizens in collaboration with a venture firm which began from a university project or a university-led consortium in the private sector. Most cities' Healthy City or health promotion committees include academics, medical and dental professionals, health promotion personnel, community representatives, and local business representatives. Hamamatsu has chosen to reflect its cross-sectoral collaboration in its logo design, which displays a 'kite' representing health as something raised up by the cooperation between many people and many varied groups.

Republic of Korea

Korean cities consistently develop partnerships, with a certain variety in their focus. Government–academic collaboration has involved academic personnel in baseline data collection, analysis of health profiles, and planning and evaluation in relation to a number of Healthy Cities. Siheung Healthy City runs business-oriented activities including the construction of Siheung Business Center, cooperation with retailers to differentiate retail strategies in four marketplaces, and creation of public sector jobs. A strong community focus is another characteristic of public–private partnerships: Guro-gu in Seoul, for example, emphasises community-led projects whereby residents of a community are in charge of planning, carrying out, and evaluating programmes for themselves without intervention from city officials, thus engaging and bringing citizens together and developing a close social network.

Malaysia

Kuching South in Sarawak has an organisational structure consisting of mayors, councillors, and divisional heads, but also community leaders, NGOs, and some prominent private companies. The importance of collaboration between these groups is well recognised by its steering group, which regards 'the spirit of Corporate Social Responsibility' as behind the success of the Kuching Healthy City programme. This programme has a long history, working together with Kuching North, the State Planning Unit, and the Health Department of Sarawak since 1994.

Evidence Gathering

The importance of research for Healthy Cities has been well recognised since the earliest stage of Healthy Cities in this region (Takano et al. 1992), and a series of research projects on indicators and determinants of health have been conducted (Takano 2003; Nakamura 2003). Participatory-style research (Takano and Nakamura 2004), jointly undertaken by community members and researchers, involves both groups in planning, implementation, analysis, discussion, dissemination, and evaluation, and gives communities access to evidence-based information that facilitates better planning and improved activities; this is encouraged in Healthy Cities.

According to the directory of the Alliance for Healthy Cities (2014), efforts among current members of the AFHC on evidence-gathering, planning cycles, or needs assessment and evaluation vary. Baseline city health profile information is collected in most Healthy Cities, and results from the monitoring of these profiles are reported in some. Unfortunately, examples of systematic evidence-gathering or reporting are limited. Further work is required to assess the sustainability of Healthy Cities plans and their suitability to the unique contexts of individual cities.

Gathering evidence on specific topics according to the priorities of individual programmes is evident in Hong Kong and Macau. Wan Chai conducted an ethnically and demographically disaggregated survey with interviews to investigate the mental health status of the community; in Sai Kung the focus was on accessibility, with two rounds of inspections of barrier-free facilities carried out in 2009 and 2012, to help define possibilities to improve them and alert the owners of buildings to their obligations to optimise the barrier-free facilities on their properties. Kwai Tsing, with its strong emphasis on physical and occupational safety, conducted surveys to assess the community's perceptions of safety issues. Kwun Tong carried out a community diagnosis in 2005 to investigate the health status and living habits of residents; the health promotion activities in the years that followed were designed in response to the results. Macau completed a city health diagnosis in its 2003–2004 preparatory and organisational stage, and the identified priorities formed the basis for the six health task forces set up in 2005.

In Taiwan, Taoyuan developed a project cycle based on a framework of 20 steps towards a Healthy City, which it has split into three stages: preparation, organisation, and action. In Hsinchu, evaluation responsibilities are organised across several bureaus. Chiayi set up a wide range of local Healthy City indicators for monitoring and improving its progress, and developed a specific evaluation mechanism; and the Daan district of Taipei carried out a health census in the community, creating a set of health indexes to identify residents' needs.

Some Korean cities carry out comprehensive analyses of their plans and activities. Yangcheon-gu has evaluation mechanisms by which projects' inputs, outputs, and results, as well as budget commitment and user satisfaction, are evaluated according to an index. Pohang composed an urban health index profile with five areas and 34 indexes in 2012, and analysed how healthy the public consider the city in different aspects. A number of these cities emphasise citizen involvement: Guro-gu's community-led health promotion project actively involves residents in planning, carrying out, and evaluating programmes with guidance from the community health centre; and Siheung and Dangjin offer further examples of citizen participation in evaluation. Jincheon holds an open competition for healthy city programmes, where qualified organisations submit their proposals and the best are selected. More generally, Dobong-gu designs businesses and projects on the basis of surveys of residents' health conditions, requests, and satisfaction, as well as on the results of previous years' businesses and projects. Wonju's efforts are based on a 'plan–do–see' cycle, and Yongsan-gu carries out evaluations to compare plans at the start of each year with their results at year's end, with achievements and residents' satisfaction determining which planning procedures go forward.

Most cities in Japan monitor health promotion indicators, set objectives for each project, and then compare achievements after a year by considering national guidelines and the cities' specific priorities. For example, Hamamatsu performed an evaluation in 2011 to re-evaluate initiatives up to that point and used the results to launch ‘Healthy Hamamatsu 21’ in 2013. Fukuroi’s health promotion plan undergoes evaluation by taking into account the changes in performance of projects over the years and comparing them with others in the country and prefecture. Owariasahi holds citywide surveys every year whose results showed an increase of citizens aware of the city’s Healthy City efforts from 26.2 % in 2006 to 56.8 % in 2012.

Case Studies of Developments in Specific Cities

The following case studies consider in more detail the Healthy Cities efforts in three cities in the Asia Pacific region, each of which exemplifies a certain key theme. These are Phnom Penh, Cambodia, which demonstrates a targeted Healthy Settings approach; Changwon, Republic of Korea, which has integrated its concern for sustainable public health into its approaches to social and economic development; and Owariasahi, Japan, a model of comprehensive citywide collaboration.

A Healthy Settings Approach: Phnom Penh, Capital City of Cambodia

Phnom Penh is the national capital of Cambodia, located in the south of the country at the confluence of the Mekong, Tonlé Sap, and Bassac rivers. Established in the fourteenth century, Phnom Penh became the permanent Cambodian capital and seat of government in 1865. Major urban infrastructure investments were made during the French colonial period in the early twentieth century and after independence in 1954, but the city was devastated following the rise of the Khmer Rouge in 1975–1978. Since the early 1980s, with the help of foreign investment and the international community, the city has been steadily rebuilding. It is now a significant political, commercial, cultural, and tourist centre in Southeast Asia, occupying an area of 678.46 km^2 and with a population of approximately 1.5 million people as of 2012.

Phnom Penh established a Healthy City Steering Committee in 1997. It has been especially notable for organising its Healthy City activities according to a settings approach: that is, an integration of activities to focus on distinct locations and their social contexts in which people interact daily, such as schools, workplaces, hospitals, or marketplaces (Secretariat of Phnom Penh Healthy City 2011). Such a focus can help enable complex interventions specifically suited to particular settings, whose effects then synergise in promoting health in the city as a whole. Within this framework, hygiene, sanitation, and waste management have been foremost priorities in a city which has faced rising urban poverty and challenges in access to latrines and potable water. Medical services have received significant attention, as have road safety and risk factors in the spread of non-communicable diseases.

At the start three settings were targeted: Bak Touk School as a Healthy School, Orussey Market as a Healthy Market, and Phnom Penh Municipal Referral Hospital as a Healthy Hospital. Since then a growing number of initiatives have been added to these efforts, but the settings approach has remained the foundation. In the 2010–2015 period, prototype health settings are being set up under several major categories: Healthy Workplaces (all Capital Hall departments, 7 restaurants, 10 factories, 5 guesthouses); Healthy Markets (2 markets); Healthy Commune Offices (12 communes); Healthy Hospitals (2 hospitals); and Healthy Houses. Healthy Schools remain on the agenda. An overview of these categories, each of which has a specified minimum package of commitments, may suggest the advantages of this approach in addressing the health concerns specific to each setting.

Healthy Workplaces, including factories and small industries, focus on setting up workplace policies such as health and safety training and smoke-free environments, as well as improving the physical environment such as air quality and ventilation, using safety equipment, reducing exposure to work-related hazards, and implementing fire prevention plans. This setting also calls for the provision of health services and education to staff, supporting healthy lifestyles with programmes on personal health skills such as nutrition and smoking cessation, and addressing concerns outside the immediate workplace such as preventing pollution from the external environment and playing a positive role in the community.

Healthy Markets receive clear political commitments and policy leadership from local authorities and market management, while national or provincial authorities promote the concept among governmental and non-governmental agencies. Improving the physical environment is a key concern, and includes building and path maintenance, and providing maps and signs, light and ventilation, and fire prevention plans. Availability of hygiene facilities, such as toilets, clean water, and soap for hand-washing, and water supply for drinking, washing, and disinfecting are high priorities, as are additional sanitation measures such as solid waste management and improvements to the sewerage system. Markets are further identified as a place to promote health education, with training in first aid and food hygiene provided to vendors and employees, and healthy practices in food handling, hygiene, non-smoking, and environmental protection communicated to the public.

Healthy Communes are also concerned with improving public health education, healthy lifestyles, and solid waste management, supported by commune policies. (Communes are a third-level administrative subdivision in Cambodia, into which districts are divided. Phnom Penh has a total of 812 districts comprising 96 communes.) Additional attention in improving the physical environment is given to paving roads, constructing walking lanes along main streets, planting trees along roads, providing gardens and green spaces, and improving sewerage systems and air quality. Security is also a priority, in terms of preventing crime and drug abuse, implementing village safety policies, and ensuring the absence of gangsters.

Healthy Schools implement improvements to infrastructure and health facilities as well as sanitation. The former includes sports playgrounds and safe play surroundings, fire extinguishers, furniture, and first aid kits; the latter involves providing an adequate number of toilets, and sinks for hand-washing with clean water and soap, as well as solid waste management, sewage improvements, and vector control. Naturally schools are also a site for the promotion of health education, including hand-washing, hygiene, nutrition, and healthy lifestyles.

Healthy Hospitals' commitments include implementing infection control, and physical improvements such as ensuring adequate light and ventilation for safe operations, the availability of basic hygiene facilities, and the establishment of treatment plants. Policy commitments from management boards establish conditions for developing hospitals as healthy workplaces, while education is provided on breastfeeding, maternal and child healthcare, and other areas.

Healthy Houses are committed to a general range of improvements such as access to safe drinking water and sanitary facilities, good ventilation, solid waste management and sewerage systems, healthy lifestyles, and a non-crowded and smoke-free environment.

In addition to these settings, Phnom Penh runs an Environmentally Sustainable and Healthy Urban Transportation (ESHUT) initiative (WHO (WPRO) 2010b). This seeks to address healthy and safe road transport and urban beauty; it promotes such varied activities as cycling to reduce traffic congestion and accidents and the construction of gardens and parks. As a next step in relation to health and transport, Capital Hall has introduced a bus system, and is expanding the number of bus lines from three in 2014 to ten in 2015 (Alliance for Healthy Cities 2014).

Taken together, Phnom Penh's approach has enabled it to target interventions effectively, in accordance with the requirements and conditions of specific settings, drawing on the knowledge and leadership of their primary stakeholders and enabling

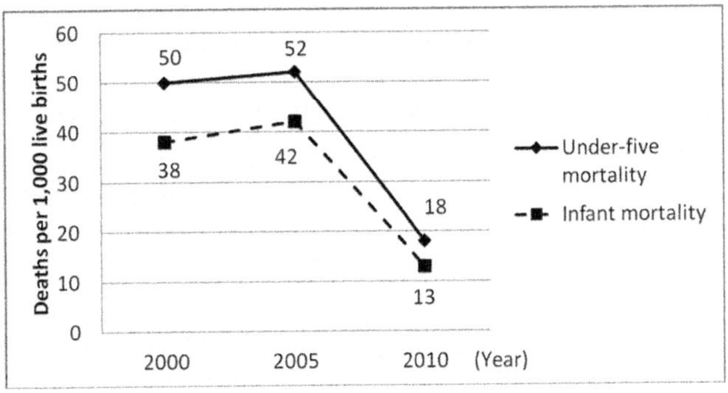

Fig. 11.1 Trends in childhood mortality in Phnom Penh (data source: 2000, 2005, and 2010 Cambodia Demographic and Health Survey.)

overarching priorities, such as sanitation and safety, to be implemented in ways most suitable to each environment (Secretariat of Phnom Penh Healthy City 2011). Although much work remains to be done, this approach has enabled Phnom Penh to make significant improvements under its Healthy City initiative: these have included reductions in infant and maternal mortality (Fig. 11.1), expansion of primary health service provision, improvements in road quality, and a decline in temporary settlements: by 2008, 86 % of households had access to safe water, electricity, and toilets compared with 66 % in 1998. The city thus stands as a leading example of a Healthy City initiative which takes a healthy settings approach to overarching public health challenges, and provides a valuable point of reference for other cities attempting to implement a settings-based approach, as well as a source of lessons for further development of the healthy settings concept itself.

Integration of Sustainable Health Development and Economic Development: Changwon, Republic of Korea

Changwon is a city in the southeastern Republic of Korea (ROK). It is the capital city of Gyeongsangnam-do (Gyeongsangnam Province), and is located close to ROK's second-largest city, Busan. The city, over 600 years old, was designated an Industrial Base Development Area in 1974. This involved constructing the Changwon Industrial Complex in the southern part of the city, and pursuing a development plan by which Changwon became the fourth planned city in the world and the first in Korea. In this capacity it has become an important centre of manufacturing, agricultural production, and knowledge-based high-tech industry, with a well-established public infrastructure in transport, sanitation, and parks. Rapid urbanisation and worker migration gave it a population of about 500,000 by 2008. In 2010 three neighbouring cities were merged with it, bringing its area to 736.34 km^2 and its total number of inhabitants to over one million.

Changwon's Healthy City initiative, Healthy Cities Changwon, has operated in the midst of this rapid economic development and transformation, and through it the city has sought to harness opportunities to improve the health and welfare of its citizens, and manage the drawbacks. Health promotion efforts have been carried out in the city since 1989, but since joining the Alliance for Healthy Cities as a founding member in 2004, it has consolidated these efforts into a set of systematic programmes under a comprehensive vision and organisational structure, with a settings approach, a broad and inclusive concept of health, and full engagement with communities and stakeholders to make programmes responsive to citizens' needs.

Changwon's principal health challenges are inseparably linked to its development journey over the last few decades, and the initiatives to address them have been grounded in that context. This may be illustrated by a brief exploration of some of its focus areas: urban planning, health equity, and communicable and non-communicable diseases.

First, Changwon has experienced very rapid structural, industrial, and population growth, particularly centred on its industrial complex. Although most of its factories are non-polluting, an increase in greenhouse gas emissions, mostly from rising motorised vehicle use, has contributed to climate change and air pollution. In response, Changwon has declared itself a 'cycling city', the first in Korea, placing the bicycle at the centre of an urban planning framework to promote healthy living and working conditions, lifestyles, accessibility, and social networks. It has identified cycling as a source of health benefits across the board: it is clean, relatively affordable, and efficient; it provides an opportunity for physical exercise and helps cultivate a sense of independence; it contributes to the goals of reducing pollution and congestion from car usage in a highly condensed population; and ultimately it creates an environmentally outstanding sustainable city. Under its Environmentally Sustainable and Healthy Urban Transportation (ESHUT) programme (WHO 2010b), a variety of resources and many new jobs have been created. A rigorous urban planning framework, with considerable budgetary support, has emerged to support pursuit of this vision with infrastructure plans such as improvements to cycle lanes and facilities, and social initiatives such as provision of bike insurance and bike hire systems, as well as numerous public events. To create a clean, beautiful urban landscape, the construction of pedestrian overpass bridges in the city has been avoided except for a single bridge between an existing elementary school and a renovated apartment complex, required for student safety.

Second, health equity in Changwon also reflects the city's transformation. A significant divide has emerged between its east side, consisting mostly of newly constructed offices with modern-style architecture and facilities, and its west side, the old town experiencing a sense of relative loss or deprivation that has deepened with the rapid population growth. The city has identified these socioeconomic inequalities—in income distribution, long-term unemployment, education, access to nutrition, clean water, and essential services—as a major source of adverse health outcomes and inequality, including the unequal presentation of non-communicable diseases, lifestyle-related issues like obesity and smoking, and overall life expectancy. A gendered dimension is also apparent in inequities arising from women's lower socioeconomic status relative to men. Changwon has taken these problems seriously. It has committed to the principle that health is a human rights issue and a development responsibility, and has acknowledged the need to increase the overall financial and resource pool, to allocate a bigger share of it to the most vulnerable groups, and to do so not only in primary healthcare provision but also in all policies and programmes relating to health, including those outside the immediate remit of the health sector. It has identified good governance, thorough data collection and analysis, and collaborative inter-sectoral partnerships as vital action points, and derived three principles of action on health equity: improving conditions of daily life; tackling inequitable distribution of power, money, and resources; and developing public awareness of the social determinants of heath. A new publicly funded health-promoting centre, in addition to public health centres, was established in the west-side old town in 2002. Critically, for a city whose modern history is predominantly told as a story of economic development, the city has asserted that economic growth, rather than a goal for its own sake, must be a means to health and health equity through appropriate social policies and action on the social determinants of health across all sectors of government and society.

Third, Changwon has experienced an increasing health burden from non-communicable diseases (NCDs) such as cardiovascular diseases, diabetes, cancer, arthritis, and obesity, which have disproportionately impacted on disadvantaged sectors. The city has identified NCDs as serious public health concerns and considers them closely linked to social and economic forces. Its strategies to address them have been synergetic: healthcare provision for NCD research, detection, and treatment is being strengthened and the health sector reformed, but the efforts to reduce risk factors have extended far beyond the health sector and affect public policy in areas like agriculture, pharmaceutical production, finance, trade, transport, education, urban planning, taxation, and sport. Cross-sectoral collaborative efforts have been established in a full range of settings. For example, to increase the physical activities of citizens, the city has encouraged the development of walking trails in apartment complex areas, and 85 % of its citizens now have access to these trails. To improve citizens' responses to health emergencies, the city provides instructions on cardipulmonary resuscitation. Automated external defibrillators have been installed in public places such as apartment complexes or shopping malls.

Fourth, Changwon, like much of Korea, experiences infectious diseases like autumn diseases, HIV/AIDS, and especially tuberculosis (TB) as leading causes of death. It suffered significantly from the swine flu pandemic of 2009. School-age and elderly persons have exhibited particularly high rates of TB infection, the former because of transmission between students sharing classrooms, and the latter because of endogenous reactivation of remote infections acquired when TB was more prevalent. Ethnic minorities also appear more vulnerable. In response the city has placed control of communicable diseases at the forefront of its urban development responsibility, and is implementing comprehensive awareness-raising initiatives, educational interventions including hand-washing programmes, and control measures on transmission in hospital settings, among others.

Changwon's Healthy City programme has infused the city's fundamental approach to development, re-framing the concept so that rather than being an end in itself, economic development is harnessed to improving the health and welfare of citizens, as a defining goal and primary public responsibility. This has given the city advantages in the ability to authorise and fund public policies to address health comprehensively: establishing wider social and economic challenges within the remit of public health, including those of fairness and equity, and implementing measures to address them in those terms. This gives Changwon tremendous potential as an international model, above all for cities where development narratives and frameworks are currently influential.

Healthy Cities Changwon has since evaluated its progress and identified six components contributing to its long-term sustainable development: (1) making the development of initiatives and their incorporation in existing community institutions a routine activity; (2) advocating policy change; (3) mobilising community residents; (4) changing community norms; (5) engaging stakeholders; and (6) providing leadership and political will.

Fig. 11.2 Owariasahi's guidelines (source: Owariasahi Healthy City Programme, December 2005)

Citywide Collaboration Towards a Healthy City: Owariasahi, Japan

Owariasahi is a city of approximately 83,000 inhabitants in Aichi Prefecture, central Japan. It is well connected by road and rail links, and its neighbourhood includes the major city of Nagoya and also Seto city, whose pottery and ceramic production is a key part of Owariasahi's industrial and cultural heritage. The city covers an area of 21.03 km^2 and has 49 m^2 of park space per capita, including the Aichi Prefectural Forest Park (Shinrin Koen). The city became a founding member of the Alliance for Healthy Cities in October 2004, has developed its own Healthy City programme, and has contributed to expanding the Healthy Cities in the region by showing steady progress through the teamwork of its city council, mayor's office, Healthy City Promotion Office, all divisions of the city, private organisations, and community groups (Fig. 11.2).

The Owariasahi Healthy City Programme began in 2005, and adopted three overarching guidelines: 'A city that prevents people from becoming bedridden'; 'A city people want to go out into'; and 'A city where people would always want to live'. Each guideline has been targeted through a set of policies: the first through strategies for physical, mental, and children's health; the second through strategies to infuse personal warmth into the city, make it easy to get around in, and offer many enjoyable activities; and the third through strategies to offer an attractive living environment where people consider and take good care of each other and their natural surroundings. These, through effective collaboration, have guided specific initiatives such as the Genkimaru health evaluation system, a muscular training course and walking programme, the Asapy City Bus system, and an annual Health Fiesta.

The city has continuously monitored progress, conducted evaluations, and updated leading plans. An initial evaluation was carried out in 2010, 5 years into the programme, and found general improvements across the board, particularly in the average length of periods of independence in people over 65 years old. It also identified issues requiring further work, such as mental and physical health, public infrastructure including pavements and train stations, and consideration for natural surroundings. A progress report in 2012 found further progress on elderly people's periods of independence, especially with their increasing use of the Asapy bus system. Indeed, when the city revised its plans and added new strategies in 2014, it cited elderly self-reliance, low costs of elderly nursing care, and a low percentage of elderly citizens requiring such care as among its proudest achievements. As of 2014 the city pursues seven leading plans: (1) cycling for health; (2) enhancing the health of the elderly; (3) making the city enjoyable for walking; (4) being refreshed; (5) providing food education for health; (6) developing an eco-garden city; and (7) renovating the whole city.

Owariasahi is exemplary of cities taking a comprehensive citywide approach to health promotion, with systematic collaboration between departments and sectors across many aspects of city life. This was envisaged from the outset, and since 2014 has been consolidated under the fifth Owariasahi City General Plan, which replaced the previous one that ran for 10 years. The new plan introduced and integrated the term and concept of the Healthy City into the core municipal policy framework. With regard to its central leadership, the programme is overseen by a subsection of the mayor's secretary section in the city's planning department, whereas an administrative evaluation system is provided by its Healthy City Promotion board of directors, which includes the mayor, the vice-mayor, the superintendent of schools, and the city's nine division managers. This provides strong political support at the core of the programme's coordinating structure.

Several additional structures are in place for cooperation between these officials and a wide range of citizens and stakeholders. Most prominently, a round-table conference, established to create a forum to incorporate the views of citizens, is held on a regular basis. Its 12 members include a university professor; professionals from the industrial, commercial, agricultural, sports, healthcare,, and education sectors; representatives of several community bodies and clubs; and two unaffiliated members of the public. In addition, the city has sought expert advice from numerous academic and professional specialists in formulating its policies, including the Tokyo Medical and Dental University. The city also supports the development of citizen leaders and their own projects and programmes: for example, the Asahi Health *Meister* recognition, started in 2008, commends and helps motivate citizens to further continuously engage in health promotion activities in communities, with efforts made on a citywide basis. Questionnaires every 2 years assess citizens' views on the achievement of policy targets, for use in continuing programme design. Recently, in preparing the fifth Owariasahi City General Plan, studies and discussions were undertaken with citizens, including groups such as junior high school students, to identify their needs and opinions.

This citywide engagement and collaboration extend into programme practice. Such processes are illustrated in three of its most prominent projects: the Asapy City Bus, the Asahi Health Fiesta, and Hotto Challenge Walking. The Asapy City Bus, one of the city's leading achievements, is a case in point. Rather than directly administering it, the city authorises an entrusted manager to manage the bus system with skill and flexibility, reducing costs and providing benefits for citizens. Citizen involvement and initiatives have been drawn on at all stages: a public petition in 1999 first put the idea of a city loop bus on the agenda; citizens were invited to choose the system's name in 2004, and were directly involved in discussions to develop the system and plan its routes to meet their needs and cover all required areas. A second example is the Asahi Health Fiesta, held on 29 April each year, which serves as a major event representing the city's efforts and promoting comprehensive health knowledge and services. The Health Fiesta has now grown to encompass a huge range of participating groups such as public divisions in all sectors, even those not immediately obvious in their relation to health; medical providers such as hospitals and health associations; corporations and consumers' groups; university and students' groups; and sports, art, and music organisations. A final example is the Hotto Challenge Walking activity, which since 2010 has held walking events and promotes the joy of walking: to this end, the organisers cooperated with a private business to develop a smartphone app, seeking to extend the activity's appeal to young people.

There are several additional examples of collaboration with the private sector. A company that runs a cafeteria helps supply a Healthy Lunch with low calories and salt content every Wednesday, supplemented by the city's special vegetables and fruit (*petit vert*—a new hybrid of Brussels sprouts and kale well suited to small-scale suburban intensive farming—and figs, respectively) contributed by local agricultural cooperatives. Businesses have also cooperated in the Company Cafeteria Lunch Tasting Tour to Promote Health since 2014, in which participants visit companies and sample health-conscious lunches in the employees' cafeterias. All five 24-h convenience store companies cooperated to set up defibrillators at their branches in the city, so that they are available within 5 min in 35 % of the city area, at any hour. An Agreement Regarding Cooperation in Health Promotion and Emergency Response was made with a leading pharmaceutical company in July 2014 to implement heat stroke and metabolic syndrome prevention measures, and set up food-vending machines with products available free of charge during disasters. The Green Curtain programme involved an experiment in which business operators gave away bitter gourd seedlings in exchange for tickets.

In other projects, administrators have worked with health promoters in offering muscular training, walking, and laughter activities; with neighbourhood associations in park conservation and development and in crime and disaster prevention; with local sports clubs in promoting enjoyment of sport; and with the local public transportation council, consisting of citizens, relevant experts, and transport business operators, in improving transportation services. From all of this it is clear that a great range of sectoral, professional, or neighbourhood bodies—committees, associations, clubs, volunteer groups, and so on—have become essential to the design, implementation, and community involvement in the city's projects.

Owariasahi also cooperates on an external basis. Nationally it has held discussions and lectures to explain its efforts and share information with other cities, encouraging them to join the AFHC network. Internationally, the city has actively provided information on how it developed its Healthy City Programme and the lessons it learned, through the exchange of experiences during visits by foreign professionals. The city has been an active AFHC member, engaging in global conferences and programmes, and has shared its experiences at other international events. Presentations at meetings and exchanges of experiences with people from other countries are brought back to Owariasahi's city office and shared, not only as learning opportunities for its representatives but also for the capacity building of all people committed to the city's work.

Owariasahi has developed a highly comprehensive approach to its Healthy City programme, encompassing a full spectrum of health concerns and constantly engaging with the public. It has been able to do this because of a citywide approach that emphasises cross-sectoral collaboration and participation at all levels in the city's day-to-day activities. Citizens' views and involvement are constantly sought, and communities given essential leadership roles in designing, implementing, and providing evaluation on projects that best suit their needs and can be delivered with maximum effectiveness.

Acknowledgements The authors appreciate Dr. Mean-Heng NGY, Deputy Director, Phnom Penh Municipal Health Department; Mr. Yoshinori Mizuno, Mayor of Owariasahi City; officers at Healthy City Promotion Office, Owariasahi; and Dr. Rhee Boo Ouk, Former Director of Changwon Health Center for their provision of materials and valuable comments.

References

Alliance for Healthy Cities. (2014). *Directory of the Alliance for Healthy Cities 2014*. Tokyo: The Alliance for Healthy Cities.

Asian Development Bank. (2011). *Asia 2050: Realizing the Asian century*. Manila: Asian Development Bank.

Chi, Y. S. (2003). Healthy City Kuching, intersectoral collaboration in healthy cities: Malaysia. In T. Takano (Ed.), *Healthy cities and urban policy research* (pp. 271–277). London: Spon Press.

Edris, A. (2003). Promoting quality of life through participation in healthy cities programme: Sharing the experiences of Kuching City. *Pertanika Journal of Social Science and Humanities, 11*, 165–172.

Galea, G., Powis, B., & Tamplin, S. A. (2000). Healthy islands in the Western Pacific—International settings development. *Health Promotion International, 15*, 169–178.

Kiyu, A., Steinkuehler, A. A., Hashim, J., Hall, J., Lee, P. F. S., & Taylor, R. (2006). Evaluation of the healthy village program in Kapit district, Sarawak, Malaysia. *Health Promotion International, 21*, 13–18.

Nakamura, K. (2003). Indicators for healthy cities: Tools for evidence-based urban policy formation. In T. Takano (Ed.), *Healthy cities and urban policy research* (pp. 76–103). London: Spon Press.

Nakamura, K. (2011). A network of healthy cities in Asia and the Pacific: The alliance for healthy cities. In T. Muto, T. Nakahara, & E. W. Nam (Eds.), *Asian perspectives and evidence on health promotion and education* (pp. 155–161). Tokyo: Springer.

Nakamura, K., & Takano, T. (1992). Image diagnosis of health in cities: Tokyo healthy city. In T. Takano, K. Ishidate, & M. Nagasaki (Eds.), *Formulation and development of a research base for healthy cities* (pp. 50–67). Tokyo: Kyoiku Syoseki.

Nam, E. W. (2011). Health promotion and healthy city projects in Korea. In T. Muto, T. Nakahara, & E. W. Nam (Eds.), *Asian perspectives and evidence on health promotion and education* (pp. 141–154). Tokyo: Springer.

Nam, E. W., de Leeuw, E., Moon, J. Y., Ikeda, N., Dorjsuren, B., & Park, M. B. (2011). Sustainable funding of health initiatives in Wonju, Republic of Korea via a tobacco consumption tax. *Health Promotion International, 26*, 457–464.

Ogawa, H. (2003). Healthy cities project in the Western Pacific. In T. Takano (Ed.), *Healthy cities and urban policy research* (pp. 21–40). London: Spon Press.

Qifa, G. (2003). Experience of national hygienic cities in China. In T. Takano (Ed.), *Healthy cities and urban policy research* (pp. 223–226). London: Spon Press.

Rahim, D. A. (2003). Towards a healthy city Johor Bahru, incorporating community action: Malaysia. In T. Takano (Ed.), *Healthy cities and urban policy research* (pp. 266–270). London: Spon Press.

Secretariat of Phnom Penh Healthy City. (2011). Phnom Penh Healthy City. Phnom Penh City Government, Phnom Penh, p. 62.

Takano, T. (2003). Examples of research activities for healthy cities. In T. Takano (Ed.), *Healthy cities and urban policy research* (pp. 162–189). London: Spon Press.

Takano, T., Ishidate, K., & Nagasaki, M. (Eds.). (1992). *Formulation and development of a research base for healthy cities*. Tokyo: Kyoiku Syoseki.

Takano, T., & Nakamura, K. (2001). An analysis of health levels and various indicators of urban environments for healthy cities projects. *Journal of Epidemiology and Community Health, 55*, 263–270.

Takano, T., & Nakamura, K. (2004). Participatory research to enhance vision sharing for Healthy Town initiatives in Japan. *Health Promotion International, 19*, 299–307.

WHO Commission on Social Determinants of Health. (2008). *Closing the gap in a generation. Health equity through action on the social determinants of health*. Geneva: World Health Organization.

WHO Regional Office for the Western Pacific (WPRO). (2000). *Regional Guidelines for Developing a Healthy Cities Project*. Healthy Settings Documents Series No. 2. Manila: WHO WPRO.

WHO Regional Office for the Western Pacific (WPRO). (2003). Report of a regional consultation meeting on scaling up and expansion of Healthy Cities. Manila: WHO WPRO.

WHO Regional Office for the Western Pacific (WPRO). (2010a). *Healthy Urbanization Regional Framework for Scaling Up and Expanding Healthy Cities in the Western Pacific 2011—2015*. Manila: WHO WPRO.

WHO Regional Office for the Western Pacific (WPRO). (2010b). *Environmentally Sustainable and Healthy Urban Transport: A strategic focus for urbanization and health*. Manila: WHO WPRO.

World Health Organization (WHO). (1998). *Health promotion glossary*. Geneva: WHO.

Yajima, S., Takano, T., Nakamura, K., & Watanabe, M. (2001). Effectiveness of a community leaders' programme to promote healthy lifestyles in Tokyo, Japan. *Health Promotion International, 16*, 235–243.

Chapter 12
Healthy Cities, Urbanisation, and Healthy Islands: Oceania

Evelyne de Leeuw, Anna Stevenson, Gwyn Jolley, Sandra McCarthy, and Erik Martin

For the purpose of this book, we see the region of Oceania as consisting of Australasia (New Zealand, Australia and its overseas territories such as Christmas Island, the Cocos Islands and Norfolk Island), Melanesia (with the exception of Indonesian territories in Papua as they are part of the Southeast Asian region of WHO), Micronesia and Polynesia (with the exception of Hawaii as it is part of the Americas region of WHO). The region is highly diverse in many respects, in terms of ecosystems, development levels and demographics. It comprises some of the most sparsely populated countries in the world (such as Australia with a population density of 2.7 people per km^2) and oddly—because of their relative isolation in the expanse of the Pacific Ocean—some of the most densely populated ones (such as Tuvalu—428.7/km^2—and the Marshall Islands—406.8/km^2; the full global range of population densities is between Macau at 21,190/km^2 and Greenland at 0.03/km^2). Most countries have been colonised over the last three centuries, and indigenous populations suffer disproportionately in terms of health equity.

Five of the countries in this region (Tonga, Fiji, Palau, New Zealand and Australia) rank above the world average of the Human Development Index (UNDP 2014) although the United Nations Economic and Social Commission for Asia and the Pacific (ESCAP) puts another seven Pacific sovereign territories in this category.

The region enthusiastically embraced the Ottawa Charter for Health Promotion following its adoption in 1986, and healthy settings projects and programmes quickly emerged. Some of the first Healthy Cities in the southern

E. de Leeuw (✉)
Centre for Health Equity Training, Research & Evaluation (CHETRE), University of New South Wales, South Western Sydney Local Health District, Ingham Institute, Locked Bag 7103, Liverpool, BC NSW 1871, Australia
e-mail: e.deleeuw@unsw.edu.au

A. Stevenson • G. Jolley • S. McCarthy • E. Martin
University of New South Wales, Sydney, NSW, Australia

E. de Leeuw, J. Simos (eds.), *Healthy Cities*,
DOI 10.1007/978-1-4939-6694-3_12

hemisphere were Noarlunga and the Illawarra region of Australia (Baum and Brown 1989). The Australian Federal Government embraced Healthy Cities enthusiastically and supported the establishment of pilot projects towards the end of the 1980s and early 1990s. Canberra, Australia's capital, for a while was designated a Healthy City. The fact that Healthy Cities appeared so visibly on the government agenda was certainly a sign of the times: many states and territories adopted local legislation mandating that local governments develop broad determinants of health-based health policy (see Chap. 16). New Zealand/Aotearoa followed suit, and the notion of Healthy Cities and Healthy Communities was embraced from the late 1980s, for instance in Aotearoa's North Shore (Abbema 1997). The settings idea in the region extended to schools, villages and marketplaces, and other initiatives in the smaller Pacific nations (Ritchie et al. 1998; Tagivakatini and Waqanivalu 2012).

The diversity of Oceania has had an impact on Healthy City development. Some of the smallest nations in this part of the world have the highest rates of urbanisation, although it may sound odd to speak of a 'highly urbanised' nation when we talk, for instance, of Kiribati, one of the most sparsely populated countries on earth (with vast expanses of ocean) but with one of the most densely populated capitals, South Tarawa.

After the initial success of Healthy Cities in this part of the world, the regional office for the Western Pacific Region of WHO decided to take a unique approach to Healthy Urbanisation (WHO 2011), embracing a mix of Healthy City-specific values with particular challenges found in this part of the world (Fig. 12.1). 'Healthy Urbanisation' as a concept casts the net beyond a model that views a Healthy City as the ultimate outcome of a local approach to health development, and embraces a dynamic perspective that has the potential to extend beyond the boundaries of the traditional city to the thriving network of Healthy Islands in the region.

For the future of Healthy Cities in the region, a series of consultations led to recommendations that WHO should not adopt a 'top-down approach' (possibly code for 'the WHO European Region process of designating and accrediting Healthy Cities and Healthy Cities National networks') but should support national and local governments to

- Strengthen institutional arrangements to promote healthy urbanisation
- Build capacity for action towards healthy urbanisation
- Strengthen the evidence base for governance, policies and actions, and value of investment for healthy urbanisation
- Advance national policies and action on healthy urbanisation
- Support city-to-city learning and networking within and across countries

The case studies in this section attest to the vibrancy of the settings approach in the region. Nearly two dozen Australian local governments and universities are members of the Alliance for Healthy Cities based in Japan which originally focused on the Western Pacific Region of WHO, and several states and territories in the country have statutory requirements to develop local health plans inspired by 'The New Public Health', as De Leeuw and Lin demonstrate. That same inspiration has extended to the smaller island states in the Pacific that have come together in a Healthy Islands programme.

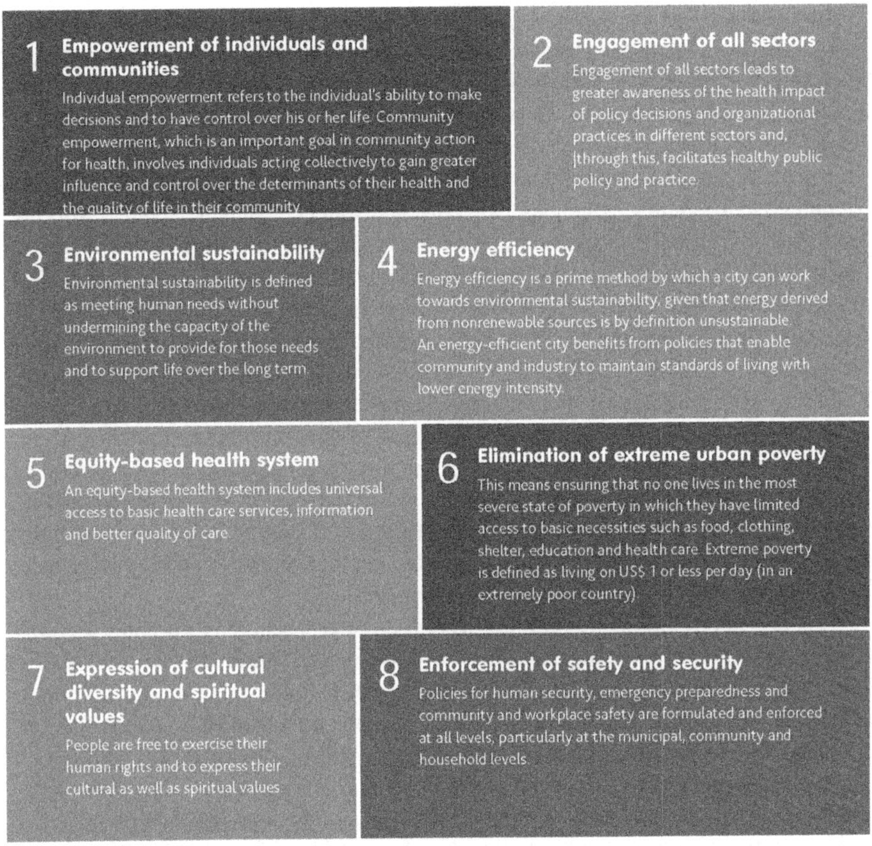

Fig. 12.1 Eight key domains for Healthy Cities in the Western Pacific Region of WHO (2011)

Christchurch: A Healthy Cities Case Study[1]

The City of Christchurch was built by British settlers in the 1840s and officially named in 1856; it is the oldest colonial city in New Zealand. It lies about one-third of the way down the eastern border of the South Island and is the main port and airport for the South Island. The area was known as Otatauhi by local Maori, who had been present in the area for several centuries prior to the arrival of European settlers. Christchurch has a population of approximately 360,000; currently it is New Zealand's third largest city. Compared with the rest of New Zealand it has an ageing population, with a median age of 37 years. The proportion of Maori is low, at just under 8 %.

An advisory group led by the Public Health Unit[2] and the city council was formed in 2001 to inaugurate Healthy Christchurch. This was a second attempt to create a

[1] Anna Stevenson.

[2] New Zealand has a comprehensive, publicly funded public health system. In 2001 regional public health units delivered services funded by the Ministry of Health. More recently, public health units have been located within 20 district health boards, which deliver and fund the full range of health services.

Healthy Cities interagency project locally; the failure of the first had been evaluated as resulting from the dual failure to engage at either a grass-roots or a chief executive level. In 2002 the Healthy Christchurch Charter[3] was developed, a formal document in which principles and protocols were agreed. The charter was the result of a wide-ranging engagement process to ensure that both high-level champions and grass-roots organisations were involved. By signing the charter, parties confirmed that they were committed to the values, principles and goals of Healthy Christchurch, and to working together to 'promote, protect and improve the health and wellbeing of the people of Christchurch' (in the process, becoming committed 'Signatories'). The incorporation of the Ottawa Charter for Health Promotion and Te Tiriti o Waitangi[4] (The Treaty of Waitangi, the founding document of New Zealand/Aotearoa) in the charter marked the intention of diverse signatories, from national government agencies to small NGOS, to work collaboratively to address the wider determinants of health.

By 2003, 88 signatories were actively working on nine projects together. These ranged from breast-feeding promotion to employment opportunities for Maori. In 2005 Healthy Christchurch made a submission to the local council, recommending that health impact assessments (HIAs) be utilised in the council policy cycle. Later that year the council and health board collaborated on a high-level policy HIA in the Greater Canterbury Urban Development Strategy.[5]

Capacity and relationship building were the focus when, in April 2008, Healthy Christchurch hosted an inaugural Healthy Cities short course. It was held over 4 days, with 30 people attending. The course focused on the importance and practicalities of embedding health and well-being in town planning, and included speakers from the fields of architecture, geography, policy, public health and health impact assessment. The guide for planners, *Health Promotion and Sustainability Through Environmental Design* (HPSTED),[6] was a joint production of the local council and the district health board, and was released just prior to the course.

In 2009 Healthy Christchurch focused on developing the Christchurch City Health and Wellbeing Profile,[7] using a combination of available health data, the signatories' views and widespread community consultation. The development process was based on an indigenous health promotion model known as Te Pae Mahutonga (Southern Cross Star Constellation)[8] (Fig. 12.2). Two questions were asked under each of the six headings: What is working well for you in this area? What do you need more of to make living a healthy life easier for you? Graphics (Fig. 12.2) were developed for each of the areas and these graphics have since been used to brand Healthy Christchurch activities, including its website. Consultation was completed in the week of the devastating September 2010 earthquakes, and the release of the final document was delayed until the following year (Fig. 12.3).

---

[3] http://www.healthychristchurch.org.nz/about/charter.aspx, accessed March 2016.

[4] http://www.treaty2u.govt.nz/the-treaty-up-close/treaty-of-waitangi/, accessed March 2016.

[5] http://www.greaterchristchurch.org.nz/background/related, accessed April 2016.

[6] http://resources.ccc.govt.nz/files/HPSTED.pdf, accessed March 2016.

[7] http://www.healthychristchurch.org.nz/city-health-profile.aspx, accessed March 2016.

[8] http://www.healthychristchurch.org.nz/media/9291/tepaemahutongatxtvers.pdf, accessed March 2016.

Fig. 12.2 Te Pae Mahutonga (the Southern Cross Star Constellation)—an indigenous model of health promotion developed by Prof. Sir Mason Durie Rangitane, Ngāti Kauwhata (Healthy Christchurch—http://www.healthychristchurch.org.nz/media/22388/te_pae_mahutonga.pdf)

A number of large and destructive earthquakes devastated Christchurch and the wider Canterbury region in 2010 and 2011, and more than 14,000 aftershocks led to ongoing community anxiety and difficult living conditions. Many Healthy Christchurch projects have focused on addressing the impacts of the earthquakes. At a community level these have included the River of Flowers,[9] a yearly event to commemorate those who died in the 2011 earthquakes; the Wellbeing Game,[10] a community project designed to support mental well-being in workplaces; and the world-leading social marketing campaign 'All Right?' which was designed to support population-level mental well-being.[11]

The planning guide referred to earlier (HPSTED) was reformatted and renamed the *Integrated Recovery Planning Guide*,[12] and training in how to use the guide and implement a broad range of actions on health determinants as an approach to planning and earthquake recovery was provided to many different groups.

The earthquakes exacerbated pre-existing issues in the region such as poor housing, low incomes, social and economic inequality, and accessibility to schools and employment opportunities. As with the rest of New Zealand, Canterbury faces rising rates of obesity and insufficient exercise levels and, linked to these risk factors, alarming levels of type two diabetes, cardiovascular disease and mental illness. The ongoing reduction in smoking rates in New Zealand (14.4% in Canterbury) is a

[9] http://www.healthychristchurch.org.nz/priority-areas/wellbeing-and-community-resilience/river-of-flowers.aspx, accessed March 2016.

[10] http://www.healthychristchurch.org.nz/priority-areas/wellbeing-and-community-resilience/the-wellbeing-game.aspx, accessed March 2016.

[11] http://www.healthychristchurch.org.nz/priority-areas/wellbeing-and-community-resilience/all-right-wellbeing-campaign.aspx, accessed March 2016.

[12] http://www.cph.co.nz/Files/IntegratedRecoveryGuideV2-Jun11.pdf, accessed March 2016.

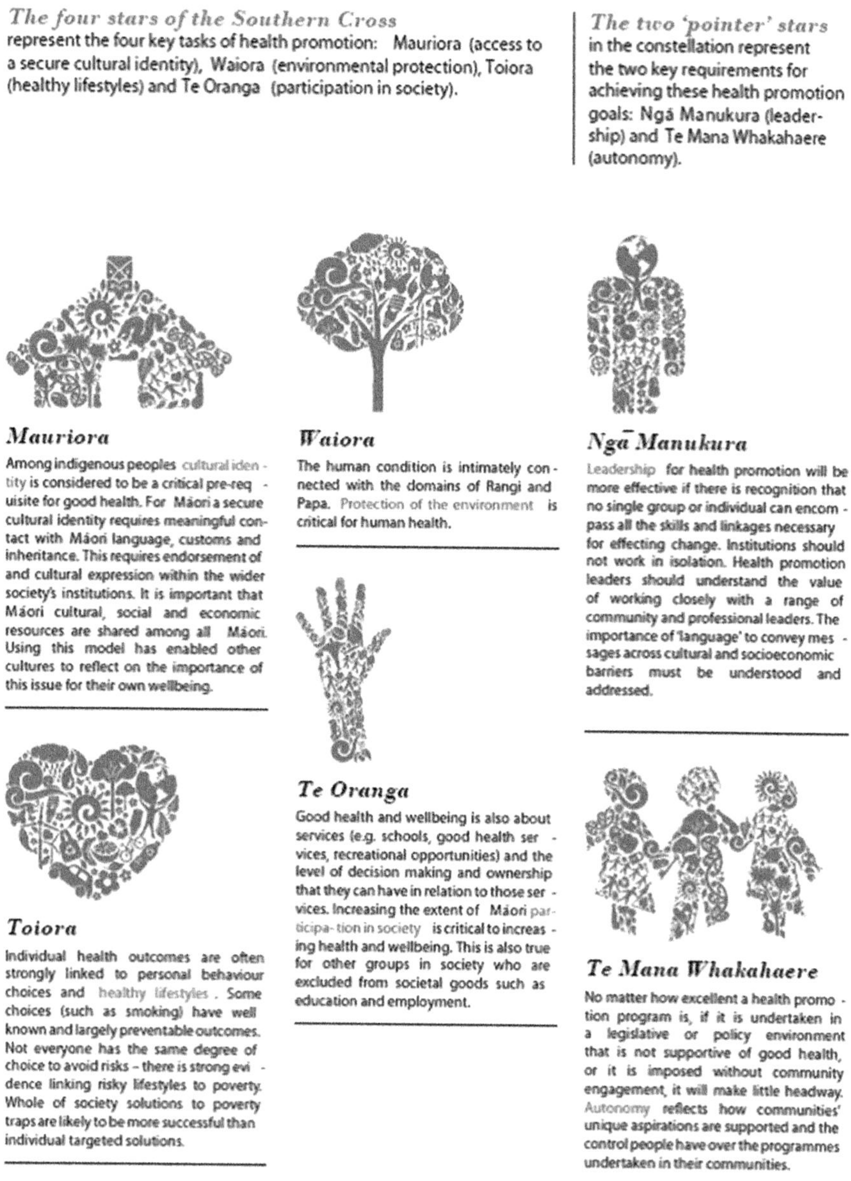

Fig. 12.3 A summary of Te Pae Mahutonga. From p.23 Christchurch City Health Profile http://www.healthychristchurch.org.nz/media/44655/chch_city_health_profile_2012.pdf

huge public health success and can be attributed in large part to a strong lead from central government. Post-earthquake Canterbury is experiencing challenges in dealing with the harmful use of alcohol, and Healthy Christchurch has supported the creation of a new position coordinating alcohol harm reduction activities.[13]

[13] http://www.healthychristchurch.org.nz/priority-areas/alcohol.aspx, accessed March 2016.

The public faces of Healthy Christchurch are 'champions' who give high visibility to its activities and work within their own organisations to support the Healthy Christchurch vision. They agree on specific priorities that they can champion within their own organisations while lending a voice to wider efforts. Champions are not individuals but representatives of organisations: the mayor of Christchurch is a champion, as is the CEO of the local health board and the kaiwhakahaere (chairperson) of the local tribe, Ngai Tahu. The champions have focused on alcohol harm minimisation for several years. They have contributed to central government alcohol policy, written editorial comments for local newspapers and advocated the safe use of alcohol within their organisations. They are supported by an advisory group that meets regularly to review the progress of Healthy Christchurch against its strategic plan.

Healthy Christchurch is currently serviced by the equivalent of 2.6 full-time staff, based at the local public health division of the district health board. This small team collates and distributes newsletters, maintains a Facebook page, arranges regular signatory meetings on topics of interest, maintains the Healthy Christchurch website and supports community projects as required.

The Canterbury Health in all Policies Partnership[14] is a subcommittee of Healthy Christchurch and has led a number of health impact assessments, joint organisational work plans and many other capacity-building projects that support a social determinants approach in sectors beyond the purview of the health board.

In 2014 there were well over 200 signatories to the Healthy Christchurch charter. A strategic plan is in place and, despite the challenges of the post-earthquake environment, the group is in good heart and making significant progress.

Onkaparinga: A Healthy Cities Case Study[15]

History and Context

Healthy Cities Noarlunga (HCN) was originally set up as one of the three Australian pilot Healthy Cities projects in 1987. Funded by the federal government for 3 years, the pilot project was modelled on WHO European Healthy Cities but adapted to the Australian context (Baum et al. 1990). Two evaluations of the first 3 years were conducted (Worsley 1990; Whelan et al. 1992), and found that the project had been successful in taking concrete action based on abstract principles, including the concepts of community participation and intersectoral collaboration.

Noarlunga is an outer suburb of Adelaide, the capital city of the state of South Australia. The population at the time was about 77,000. In 1997 the City of Onkaparinga was formed through the amalgamation of Noarlunga and two neighbouring areas. The population today is 165,000, with a higher proportion of children and youth than the average in Adelaide.

[14] http://www.healthychristchurch.org.nz/priority-areas/chiapp.aspx, accessed March 2016.

[15] Gwyn Jolley.

The early years of HCN built on a previous needs assessment and community visioning workshops. HCN was based in the then new primary health care service and community hospital, Noarlunga Health Services. In-kind support was received from the health service, the local council and other agencies. Following the loss of federal funding in 1991, local community activists reviewed the management of the project and decided to incorporate HCN as a non-government organisation. HCN changed its name to Healthy Cities Onkaparinga (HCO) in 2008 to reflect the new local government area.

Values

HCN was community driven from the start, with local community members and agency representatives joining to form the initial governance structures and to advocate incorporation. This community focus has continued, with community members having a majority of seats on the management committee. Community members are valued by being supported to attend training courses, forums and conferences relevant to Healthy Cities.

The importance attached to community participation demonstrates an equity approach that encourages all voices to be heard. Over the years, HCO has hosted many forums and planning sessions where members of the wider community have had an opportunity to discuss and vote on issues to be tackled. For example, increasing public transport and improving access to it has been a major area of action, culminating in an extension of rail links to newer suburbs.

A major achievement of HCO has been in fostering intersector activity. Various sectors, including health, local government, housing, education, policing, welfare agencies and small businesses, have contributed to HCO goals by providing small grants, in-kind support and representation on the target committee. This intersectoral approach has become a way of working in the region, as exemplified by a number of 'round tables' and collaborations that have evolved from HCO and that work across silos to improve services and support community well-being.

Political commitment to HCO has varied. Despite a generally positive evaluation, the pilot project was not extended beyond some 3-year 'networking' funding. Over the years South Australian state governments have occasionally shown interest in expanding the Healthy Cities approach, but this has not amounted to actual investment. HCO has striven to remain bipartisan in its relationships with changing governments. The local government (City of Onkaparinga) has been supportive throughout and has given in-kind support such as hosting events, providing a representative to the management committee, and giving access to competitive grants.

HCO has been sustained for over 25 years. This is a remarkable achievement given changes globally, nationally and locally in that time. The factors contributing to this sustainability have ensured the long-term nature of HCO and its evolution from a project to an approach: here for the long haul.

Triggers

The major contributor to the ongoing achievement of HCO can be linked to the passion and commitment of early champions. Fran Baum of Flinders University, Richard Hicks of Noarlunga Health Service and founding community members all fought hard to win the original pilot funding and establish Healthy Cities, and these early champions continue to support HCO today. Another long-standing supporter was Mr. Ray Gilbert, mayor in the 1980s and early 1990s.

The WHO Healthy Cities project has provided resources, guidance and networking opportunities. International awards from WHO have greatly increased the recognition of HCO in South Australia and nationally, and have given legitimacy to its approach. Baum and Hicks have both been formally recognised for their achievements in health promotion.

The ongoing connection with Flinders University has a two-way benefit, strengthening the academic focus within HCO and providing an opportunity for students and scholars to see a Healthy Cities initiative in action. Flinders University has run a Healthy Cities short course that attracted enrolments from home and overseas; it included a field trip to meet HCO members and see some of the practical outcomes the project had achieved.

Problems and Interventions

The main activity focus of HCO has been in advocacy for healthy public policy and bringing together community and other sectors to influence the social determinants of health. Examples of specific projects include the Onkaparinga Estuary Water Quality Group, community safety (e.g. eye safety in small business, body art), community action on drugs and public transport advocacy.

The Onkaparinga estuary is an important native bird and vegetation habitat and is a significant site for Aboriginal people dating from before white colonisation. From the mid-1800s to mid-1900s, vegetation was cleared and weirs and reservoirs were constructed to meet the growing demand for water from metropolitan Adelaide. In 1970 sludge drying ponds were constructed as part of the local waste water treatment plant. In 1989, following growing community and local government concern about high levels of pollution and health impacts from recreational use of the river, HCO was able to bring together the many agencies with responsibility for water quality, and organised public meetings to discuss pollution issues. From these the Onkaparinga Estuary Water Quality Group (OEWQG) was established (Baum 2002). Intense advocacy resulted in resources being made available by the state and federal governments, and regeneration began. In 1992 the first wetlands were constructed, and in 2013 the sludge ponds were rehabilitated and the Noarlunga Downs wetlands opened. The area now supports a diverse range of native fauna, and regenerating flora improve the visual amenity and extend the treatment time of storm water before it enters the estuary. The wetlands project was implemented with strong community engagement throughout. HCO and the OEWQG were influential in getting the water quality issue onto the government agenda, and active in arguing for resources that would provide a solution.

The OEWQG was supported by HCO from the beginning, demonstrating the way that issues could be addressed by drawing on community participation and intersectoral action. HCO was able to bring together partners, work with government, and provide access to the government and governance system and key people (Cooke 1995). The approach was to support a resident-driven project and enhance community empowerment. Many members of the OEWQG are still actively participating in environmental or other projects, so their advocacy effort has extended beyond the original project.

Conclusions and Lessons for the Future

In some ways HCO has done its work. Working collaboratively with community, business and government agencies has become the default: local government hosts intersectoral round tables and embraces the role of urban planning for health and well-being (Baum et al. 2006). But the impact of the global financial crisis and the dominance of a market-driven user-pays ideology mean that inequities continue, presenting ongoing challenges for the Healthy Cities approach.

HCO is at a turning point. In South Australia in the 2010s, government resources and support for health promotion are low (Government of South Australia 2013). There is an increased focus on clinical services and hospital demand management. Evidence-based medicine spills over to health promotion and demands 'hard' evidence that health promotion contributes to reduced hospital admissions. However, there are also opportunities in that the state government appears to be shifting some health promotion resources to local governments to support implementation of a new public health act which, for the first time, requires local governments to 'promote health' (South Australian Public Health Act 2011).

HCO is unique in having an incorporated structure rather than being based in local government. This brings both benefits, in terms of autonomy and independence, and challenges, in terms of being outside the system and facing uncertain funding; but as long as community members and agencies are committed to support HCO, it will continue to evolve to face the new challenges ahead.

Kiama Municipal Health Planning: Providing a Safe, Strong and Sustainable Community[16]

Introduction

The Kiama Local Government area is on the South Coast of New South Wales (NSW), Australia, approximately 120 km south of Sydney. It has an area of 256 km^2 and is essentially a rural community surrounded by beaches, rainforest and

[16] Sandra McCarthy.

farmland. Kiama Municipal Council has had a long-standing multidisciplinary commitment to the health of its community, through its operation of the Illawarra Home Nursing Service, which won a National Award of Excellence in Local Government and Blue Haven Aged Care Facility (nursing home, hostel, retirement facilities including a specialised Dementia Unit). It also provides a range of public health and health promotion services—the only Council in NSW to offer this quality of aged care.

The role, and subsequently the boundaries of responsibility, of local government is increasingly expanding and includes a focus on providing services for the health and well-being not only of the individual within a municipality but also of the community and environment. To address this role Kiama Municipal Council has embraced WHO's definition of health in recognition of the interaction between the health of the community and the wider environment. This socioecological definition of health is embodied in the development of local government initiatives which offer a holistic, collaborative approach to solving local issues.

Kiama was one of the first councils in NSW to develop and implement a municipal health plan. The plan came about by a chance opportunity: the Kiama District Health Watch was required to undertake community consultation using vision processes to identify priority health issues, and at the same time Healthy Cities Illawarra and Kiama Council were examining a proposal to develop a health plan. The development and implementation of the plan relied on the establishment of strong partnerships, particularly between Healthy Cities Illawarra, Kiama Council and the local health district, and with the support of the community and local organisations.

The Kiama Municipal Health Plan documents its holistic planning approach and outlines strategies to addresses the interaction between human and environmental health. It includes health, environment, social, strategic and land-use planning, and economic development. It is a whole-of-government approach which incorporates the principles of ecological sustainable development (ESD) and Agenda 21, and links international, commonwealth, state, regional and local issues, goals and programmes. Ultimately it attempts to encourage every citizen to become responsible for improving their own health and to preserve, protect and improve the local environment, and to provide opportunities to accomplish this.

The aims of the plan are:

- To improve the health, well-being and quality of life for all residents and visitors to the Kiama Municipality
- To improve, protect and enhance the environment for the benefit of present and future generations
- To promote ecological sustainable development that minimises hazards and maintains the local environment to a degree that human health and safety are not impaired and yet encourage development to proceed

The Process

The first Kiama municipal health plan of 2001–2004 had a strong focus on vision workshops to identify health issues within the areas of environment, community and economy. Participants were asked what their vision was for an ideal healthy municipality in 20 or 25 years' time. Groups participating included farmers, businesspeople, youth and tourists. Their visions included clean air and no pollution, land viable for food production, open spaces, a village atmosphere and a feeling of belonging and a sense of community. These workshops were supplemented with a community and tourist survey, a health needs assessment, research into and review of existing policies, goals and strategies being conducted by other agencies, and the construction of demographic and epidemiological profiles. The information from all of these was analysed to identify priorities in 12 theme areas: health, safety and injury, tourism, employment, transport, social and cultural, food and nutrition, leisure and recreation, pollution control, education and training, built and physical environment. Strategy workshops were then conducted with residents, community groups and government and non-government organisations.

Achievements of the Municipal Health Plan

The implementation of the Municipal Health Plan has led to significant programmes:

- A Safe Communities project.
- Kiamasphere, a sustainable living program: The objective of Kiamasphere projects is 'to manage, improve and protect the total environment to attain a sustainable and high quality lifestyle for present and future generations'. The logo incorporates the council's colours and represents the holistic approach required to achieve sustainable living.
- A Sustainable Living Expo, conducted annually to showcase the activities of the council and community groups and to introduce commercial products that promote a sustainable living environment.
- Sustainable Living grants which provide funding to community groups and schools to implement health and environmental strategies to create a more sustainable community. Grants have included the funding of tree planting, establishment of indigenous gardens, sun protection activities, building of a school chicken coop and development of community information resources.
- The development of a comprehensive physical activity programme including the provision of marked walking paths (the Heart Care Walks) to encourage regular physical activity and the development of a 12-km coastal walking track between Kiama Heights and Werri Beach.
- The establishment of the Kiama produce market, which came about through two significant issues: the lack of access to fruits and vegetables except through one large-chain grocery store which was very expensive, and the need for local

farmers to find uses for their land other than dairy farming after the deregulation of the dairy industry.
- The establishment of a community garden at Blue Haven Retirement Village, which is maintained by a wide range of community groups.

There are numerous other successful outcomes of the first plan, demonstrating how local government can lead and partner the community and local organisations to build a strong, healthy and safe community.

The Second Municipal Health Plan

Development of the second Kiama Municipal Health Plan 2005–2008 has involved widespread community participation through focus groups, interviews and surveys. Research and review of literature including local, state and national goals and programmes were conducted to identify key priorities for council; demographic, environment and epidemiological profiles were also prepared to assist in priority identification. The priorities identified through these processes were the following:

- Strengthening community by building partnerships and networks and community connections: The council undertook many public infrastructure projects, involving the community in planning and construction, to ensure the ongoing use of facilities and improved health outcomes. The projects included the Pavilion (town hall, community meeting place and home of agricultural show), redevelopment of the library to include the community college and a bridge club constructed on the AFL field so that senior citizens and young footballers share and socialise in a common public space.
- Providing supportive social environments for health, particularly through healthy lifestyle strategies, safety and injury prevention and high-quality health services.
- Protecting and enhancing the health of the environment, specifically through addressing issues related to the natural and built environments, waste and transport.

Kiama Healthy and Sustainable Cities Plan (2011–2017)

In the evaluation and review of the plan Kiama Council engaged with community focus groups to obtain vital feedback and recommendations. Overall there was broad consensus about the priorities identified by the council, and in turn the community delivered rich discussion and opinions which were incorporated in council decisions. A theme emerged: 'building partnerships, networks and community connections', acknowledging that positive relationships, social contact and a strong sense of belonging to and connection to the community in which one lives can greatly enhance health.

The priorities identified were the following:

- Building stronger partnerships with government and non-government organisations and community groups
- Addressing social isolation, particularly for new residents, young families and the aged
- Researching how residents in Kiama interact and how individuals become connected creating a harmonious and generous society
- Providing sustainable population and health equity programmes
- Bringing the community together with the celebration of food as a cultural experience
- Developing a new Resident's Kit to welcome new residents to the municipality, with the aim of empowering residents to form social connections and actively participate in the community

Triggers and Values

Kiama Council was a founding member of Healthy Cities Illawarra and still plays an active role in the organisation. In 2004 Sandra McCarthy, mayor of Kiama Council and president of Healthy Cities Illawarra, attended the inaugural meeting of the International Healthy Cities Alliance in Kuching. The meeting formed the beginning of strong, collaborative partnerships and friendships with global scholars, mayors and health and local government professionals. The key message of the conference was that if the Healthy Cities model were to be sustained, it had to be embedded in local governments, which were seen as the foundation stone or key component for community strategic planning, able to bring about behavioural change in policy and council organisation. On return from the conference, the mayor and council were active in working for the Kiama municipality to be recognised as Healthy City; this was achieved in 2008. That same year Kiama hosted the inaugural national forum of the Australian chapters of the Healthy Cities Alliance. International and local collaborations have continued to foster the exchange of information and experiences, from Taiwan, Singapore, Hangzhou in China and Muju City in South Korea, as well as all Australian members.

Long-term, sustainable community ownership of the plan was achieved due to the political commitment of the council. Significant contributions were made by the executive council officer and members of the health and sustainability committee.

Conclusion

Kiama's Health Plan has received international recognition from WHO. It is an evolving document which is regularly reviewed to ensure that it remains responsive to the local needs of a growing community. Encouraging the public to participate in the development, implementation and evaluation of the Health Plan is seen as essential to its continued success. The future in Kiama is positive: through the community's engagement in the Health Plan, it can look forward to better health and well-being, and to a safer, sustainable environment which ensures greater quality of life, empowerment, harmony and equity.

Healthy Islandsy[17]

One rather particular environment in which local government initiatives meet national interests is that of the small island state. Island nations face particular challenges. This was recognised as early as 1994 when the UN-sponsored Global Conference on the Sustainable Development of Small Island Developing States determined that sustainable development was the only option for the development of such nations. The conference subsequently adopted the Barbados Programme of Action for the Sustainable Development of Small Island Developing States (SIDS). Where the Barbados Programme of Action continues to work towards sustainable economic and ecological development (see sidsnet.org), the Yanuca Island Declaration (1995) squarely connects development with health. It adopts a settings-based health promotion agenda, recognising that (Healthy Pacific) islands are places where children are nurtured in body and mind, environments invite learning and leisure, people work with age and dignity, ecological balance is a source of pride and the ocean which sustains the people is protected. As a key action programme, the Yanuca Island Declaration advocates a strengthening of health care and the health care workforce, collaborative efforts between SIDS and the further development of expertise and technology to address the unique challenges faced by this group of countries.

The formal WHO Healthy Islands programme (Nutbeam 1996) governed by the Yanuca Declaration extends to nations and territories in the Pacific Ocean (American Samoa, Cook Islands, Fiji, French Polynesia, Guam, Kiribati, the Marshall Islands, the Federated States of Micronesia, Nauru, New Caledonia, Niue, the Commonwealth of the Northern Mariana Islands, Palau, the Pitcairn Islands, Samoa, Solomon Islands, Tokelau, Tonga, Tuvalu, Vanuatu, and Wallis and Futuna). There is considerable diversity among these islands, in terms of demography (most have Melanesian/Micronesian/Polynesian core populations, but there have been considerable European and Chinese influences), geography (some nations like Vanuatu are highly volcanic, others like Kiribati consist of numerous coral reefs and atolls and a few like Pitcairn or Nauru are singular rocky outcrops) and 'development' (Palau ranks 52nd on the Human Development Index—above Russia—and the Solomon Islands is 143rd) (Malik 2013).

All of these islands find themselves in the vastness of the Pacific Ocean, and those that occupy territorial areas of several million square kilometres (like Kiribati or the Cook Islands) suffer from an 'internal tyranny of distance' as much as the

[17] Evelyne de Leeuw and Erik Martin.

smaller ones do with their most direct neighbours. To travel from the outer islands to the administrative centre may well take weeks, and acute health issues are often hard to address promptly. This challenge, of course, is exacerbated by low population sizes and the smattering of expertly trained professionals, including public health and health promotion specialists.

Numbers alone may deceive, though. Pacific islanders have a rich cultural history, and they established systems of navigating the high seas through 'wayfaring' long before Chinese and European technologies enabled the colonisation of the area (Pyrek 2011).

The most serious threat to health and the very existence of these islands, however, is climate change. The International Panel on Climate Change forecasts that some of these 'may face serious threat of permanent inundation from sea-level rise. Among the most vulnerable of these island states are the Marshall Islands, Kiribati, Tuvalu, Tonga, the Federated States of Micronesia, and the Cook Islands' (Smith et al. 2001, p. 935). This effectively means the annihilation of human existence on those islands (Nunn 2013), and although some island governments have pre-empted this future, for instance by Kiribati purchasing land in Fiji for possible relocation of its residents (Office of the President 2014), there will still be profound impacts on the health of these nations.

Some of the more traditional health challenges in Pacific SIDS are summarised in Table 12.1. This material shows that some Pacific islands are surprisingly urbanised, and would merit a Healthy City rather than a Healthy Island programme on their own—for instance, as Suva on Fiji has declared, although there is very little documentation for 'Healthy Suva' (Fig. 12.4). Regardless, islands, and in particular SIDS, seem proverbial 'settings for health', although as Dooris (2004) has written, the range of settings (healthy marketplaces, health-promoting schools, healthy villages, health-promoting health services) could benefit from consistency in value systems across the board, given the extent to which they are places where people 'live, love, work and play' (WHO 1986) and create health. The Yanuca Island Declaration (WHO 1995) builds on a heritage of global conferences concerning health, sustainability and development in the later half of the twentieth century, and sees Pacific islands as places where

- Children are nurtured in body and mind
- Environments invite learning and leisure
- People work and age with dignity
- Ecological balance is a source of pride
- The ocean which sustains the people is protected

In 1997 health ministers of Pacific Islands confirmed their commitment to Healthy Islands (Galea et al. 2000), and stated in their Rarotonga Agreement, 'The Healthy Islands concept involves continuously identifying and resolving priority issues related to health, development and well-being by advocating, facilitating and enabling these issues to be addressed in partnerships among communities, organizations and agencies at local, national and regional levels'. The following priority areas and concerns are listed in the agreement:

Table 12.1 Pacific island nation/territory demographic information and health expenditure (WHO/WPRO, 2012—WHO Multi-Country. Cooperation Strategy for the Pacific 2013–2017)

Country or territory	Population			Per capita GDP		Health Expenditure			General government expenditure on health (% of total government expenditure)
	Year	Population (in '000s)	Urban population (%)	Year	US$	Year	Per capita US$	As % of GDP	
Melanesia									
Fiji	2010[e]	854.0	51.9	2009	2978.95	2009[p]	130.40	3.60	9.30
New Caledonia	2009[p]	245.6	57.4	2008	36,758.00	2008	3420.76	9.5	–
Solomon Islands	2009	515.9	18.6	2008	1014.00	2009[p]	71.84	5.30	16.80
Vanuatu	2009	234.0	25.6	2009	2685.10	2009	104.00	3.90	13.60
Papua New Guinea	2011[i]	677.4	12.5	2011	1767.24	2011[i]	75.98	4.28	12.75
Micronesia									
Federated States of Micronesia	2010[p]	102.6	22.7	2008	2223.00	2009[p]	333.33	13.80	20.60
Guam	2010[e]	180.7	93.2	2005	22,661.00	–	–	–	8.71
Kiribati	2010	103.5	48.3	2010	1307.40	2009[p]	204.80	12.20	8.70
Marshall Islands	2012	54.4	71.8	2007	2851.00	2009[p]	419.35	16.50	20.10
Nauru	2010[e]	10.0	100.0	2006–2007	2071.00	2009[p]	625.00	10.85	18.50
Northern Mariana Islands	2010[e]	63.1	91.3	2005	12,638.00	2007	–	–	25.40
Palau	2010[e]	20.5	83.4	2007	8423.00	2009[p]	1000.00	11.20	16.70

(continued)

Table 12.1 (continued)

Country or territory	Population			Per capita GDP		Health Expenditure			General government expenditure on health (% of total government expenditure)
	Year	Population (in '000s)	Urban population (%)	Year	US$	Year	Per capita US$	As % of GDP	
Polynesia									
American Samoa	2010[e]	65.9	93.0	2005	9041.00	2003	500.00	–	14.00
Cook Islands	2010[e]	23.3	75.3	2009	10,298.00	2009[p]	503.60	4.50	10.60
French Polynesia	2010[e]	268.8	51.4	2006	16,803.36	2008	3361.57	13.09	29.00
Niue	2010[e]	1.5	37.5	2006	8208.20	2009[p]	1866.55	16.94	15.81
Pitcairn Islands	2009	.05	n	–	–	–	–	–	–
Samoa	2010[e]	184.0	20.2	2009–10	2908.02	2009[p]	161.04	5.30	14.50
Tokelau	2006	1.5	0.0	2003	612.50	2001–09	3705.64	–	10.46
Tonga	2010[e]	103.4	23.4	2008–09	2988.00	2009[p]	161.04	5.30	14.50
Tuvalu	2010[e]	11.2	50.4	2002	1139.32	2009[p]	312.50	10.50	11.00
Wallis and Futuna	2010[e]	13.3	0.0	2004	3800.00	2008	–	–	24.00

e estimated, *p* provisional, *i* http://hiip.wpro.who.int/portal/countryprofiles/PapuaNewGuinea.aspx, *GDP* gross domestic product. Source: WHO Country Health Information Profiles (CHIPS) 2011

Fig. 12.4 Suva—a Healthy Fijian City

- Adequate water supply and sanitation facilities
- Nutrition, food safety and food security
- Waste management
- Housing
- Human resources development
- Communicable and non-communicable disease prevention and control
- Lifestyle and quality-of-life issues
- Reproductive and family health
- Promotion of primary health care
- Social and emotional well-being
- Population issues
- Ecological sustainability
- Information management
- Tobacco use
- Alcohol and substance abuse
- Environmental and occupational health.

These key remits were confirmed at a forum on Healthy Islands at WHO Headquarters in 2010, and a programme of recognition of successful and worthy projects in Healthy Islands was initiated.

The WHO Pacific division of the Healthy Island Recognition Programme explicitly connects with the visionary ideas emanating from the Ottawa Charter on Health Promotion and the work that was initiated by the Yanuca Declaration (Fig. 12.5).

Fig. 12.5 Ottawa Charter, Yanuca Declaration and Healthy Islands: complementary values

The WHO Healthy Island recognition programme was called for at the 2009 meeting of the Pacific Ministers of Health as part of the call for revitalisation of Healthy Islands. Since the conception of Healthy Islands, WHO has worked with its member countries and organisations to provide support for project and programme planning, implementation and evaluation, for the improvement of health and quality of life of Pacific Islanders. These initiatives fell on fertile ground, and a diversity of Healthy Island initiatives have been developed and maintained.

To encourage further development and building on this momentum, WHO member states decided that the sterling pioneering work needed formal recognition. With the main purpose of encouraging communities and countries to continue to innovate and demonstrate effective and efficient ways of promoting and protecting the health of their populations, a formal programme to recognise and award outstanding work in defined areas was developed. It focused on good practices or proposals for Healthy Islands, based on community actions or efforts, engaging across different sectors and with strong governmental support. Entries were open to all Pacific Island countries and organisations, and applications were to be submitted in the form of a report, either on work in progress or a completed project, for the Best Practice award. The Best Proposal category required the submission of a detailed proposal for intended work not yet initiated; the award monies could be used to start the proposed action.

Of the 19 applications received for assessment, 8 were for Best Practice and 11 for Best Proposal. Submissions came from the Federated States of Micronesia, Tonga, Solomon Islands, Samoa, Niue, Cook Islands, Tuvalu, Vanuatu, Papua New Guinea and Kiribati. The applications were of varying quality; some were quite exceptional in content and presentation, others needed extra work for presentation and a few did not align with the criteria. The broad range of quality and substance is not surprising, as some Pacific Islands have a proud record in health promotion: for instance, the Tonga Health Promotion Foundation has been recognised since its establishment in 2007 as a regional powerhouse of health innovation (Fotu et al. 2011) and others are still facing huge health and resource challenges.

A shortlist was made of the top five in each category. The committee engaged in detailed assessments against preset criteria based on Fig. 12.5, and scores on a number of categories were assigned. In discussion with representatives of Pacific Islands and WHO experts recommendations for awards were made to the Regional Director:

- Best Practice Community-based Effort (CBE): NCD Elimination in Aniwa and Aneityum (Vanuatu)
- Best Practice Government Sector: Healthy Workplace (Palau)
- Best proposal: Five Circles for Healthy Pacific Islands (NGO)
- Best Proposal: Adopt a Healthy Community (Chuk)

At the ninth Pacific Ministers of Health Meeting in 2011, the regional director of Western Pacific presented a recognition plaque and further monetary support for Best Practice projects, and monetary support only for Best Proposal, with the view to assess progress in ensuing years. Applications that did not get an award were referred to appropriate departments in WHO for further technical support. Everyone benefited.

A formal review of the process revealed that the countries applauded the recognition programme as a great measure of recognition of Healthy Island practices, which resulted in more and better quality proposals and reports being received for the second round which were awarded at the tenth Pacific Ministers of Health meeting in Apia, Samoa. The programme continues, with work towards the recognition at the 20th anniversary of Healthy Islands in 2015, to be held in Fiji where the first Pacific Ministers of Health Meeting was held in 1995.

Wrap-Up: Oceanic Values for Health Development

In this chapter we have highlighted the enormous diversity in approaches and governance models for Healthy Cities in Oceania. The Australian case studies show that—although local government is an important partner in these endeavours—Healthy Cities are essentially community-driven operations where NGOs (or, in the case of Onkaparinga—formerly Noarlunga—just *one* NGO) keep the momentum and sustain enthusiasm and vision. This is also the case in many Healthy Islands, as demonstrated by the initiatives that received WHO recognition across the Pacific.

It is not surprising that this emphasis on community presents itself so strongly in this region. Crawford et al. (2010) recognise and embrace the importance of diversity for healthy urban planning. Baum et al. (1990, 2006) consistently show evidence that community-based diversity in urban health (equity) planning approaches is a critical ingredient for success and sustainability.

The case study from Aotearoa (New Zealand) shows an inspirational perspective on how an indigenous vision of health promotion and Healthy Cities creates a powerful and appealing, as well as culturally safe and responsive, planning and development paradigm for urban health development. This is reaffirmed, more broadly, by Ryks et al. (2014) in their analysis of an inclusive urban health planning agenda in Aotearoa that builds on traditional patterns of settlement. It appears that embracing a value system for community living that is hundreds of years (and in the case of Indigenous Australia, tens of thousands of years) old in adaptation to 'modern' urban settlement could be an important way forward in dealing with equity matters in a culturally diverse Healthy City environment.

References

Abbema, E. (1997). *Twenty steps on North Shore: Healthy city development in New Zealand—A process evaluation*. Maastricht: Research for Healthy Cities Clearing House.

Baum, F. (2002). *The new public health* (2nd ed.). South Melbourne: Oxford University Press.

Baum, F., Cooke, R., Crowe, E., Traynor, M., & Clarke, B. (1990). *Healthy cities Noarlunga project evaluation*. Adelaide: Southern Community Health Research Unit.

Baum, F., Jolley, G., Hicks, R., Saint, K., & Parker, S. (2006). What makes for sustainable Healthy Cities initiatives?—A review of the evidence from Noarlunga, Australia after 18 years. *Health Promotion International, 21*(4), 259–265.

Baum, F., & Brown, V. A. (1989). Healthy Cities (Australia) project: Issues of evaluation for the new public health. *Community Health Studies, 13*(2), 140–149.

Cooke, R. (1995). Two healthy cities in Australia. In F. Baum (Ed.), *Health for all: The South Australian experience*. Adelaide: Wakefield Press.

Crawford, J., Barton, H., Chapman, T., Higgins, M., Capon, A. G., & Thompson, S. M. (2010). Health at the heart of spatial planning strengthening the roots of planning health and the urban planner health inequalities and place planning for the health of people and planet: An Australian perspective. *Planning Theory & Practice, 11*(1), 91–113.

Dooris, M. (2004). Joining up settings for health: A valuable investment for strategic partnerships? *Critical Public Health, 14*(1), 49–61.

Fotu, K. F., Moodie, M. M., Mavoa, H. M., Pomana, S., Schultz, J. T., & Swinburn, B. A. (2011). Process evaluation of a community-based adolescent obesity prevention project in Tonga. *BMC Public Health, 11*(1), 284.

Galea, G., Powis, B., & Tamplin, S. A. (2000). Healthy Islands in the Western Pacific—International settings development. *Health Promotion International, 15*(2), 169–178.

Government of South Australia. (2013). *SA health's response to the review of non-hospital based services*. Australia: Government of South Australia.

Malik, K. (2013). *Human development report 2013. The rise of the South: Human progress in a diverse world*. New York: UNDP Human Development Report Office.

Nunn, P. D. (2013). The end of the Pacific? Effects of sea level rise on Pacific Island livelihoods. *Singapore Journal of Tropical Geography, 34*(2), 143–171.

Nutbeam, D. (1996). Healthy Islands—A truly ecological model of health promotion. *Health Promotion International, 11*(4), 263–264.

Office of the President. (2014). *Kiribati buys a piece of Fiji. Kiribati Climate Change*. Retrieved from http://www.climate.gov.ki/2014/05/30/kiribati-buys-a-piece-of-fiji/.

Pyrek, C. C. (2011) *The Vaeakau-Taumako wind compass: A cognitive construct for navigation in the Pacific*. Doctoral dissertation, Kent State University.

Ritchie, J., Rotem, A., & Hine, B. (1998). Healthy islands: From concept to practice. *Pacific Health Dialog, 5*, 180–186.

Ryks, J., Howden-Chapman, P., Robson, B., Stuart, K., & Waa, A. (2014). Maori participation in urban development: Challenges and opportunities for indigenous people in Aotearoa New Zealand. *Lincoln Planning Review, 6*(1–2), 4–17.

Smith, J. B., Schellnhuber, H. J., Mirza, M. M. Q., Fankhauser, S., Leemans, R., Lin, E., … Yohe, G. W. (2001) Vulnerability to climate change and reasons for concern: A synthesis. In J. J. McCarthy, O. F. Canziani, N. A. Leary, D. J. Dokken, K. S. White (Eds.), *Climate change 2001: Impacts, adaptation, and vulnerability* (pp. 913–967). Cambridge: Cambridge University Press.

South Australian Public Health Act 2011. Retrieved March 26, 2014, from http://www.legislation.sa.gov.au/lz/c/a/south%20australian%20public%20health%20act%202011/current/2011.21.un.pdf.

Tagivakatini, S., & Waqanivalu, T. K. (2012). Whole school approaches to health promotion: The pacific journey. In N. Taylor, F. Quinn, M. Littledyke, & R. Coll (Eds.), *Health education in context: An international perspective on health education in schools and local communities* (pp. 5–15). The Netherlands: Sense.

UNDP, United Nations Development Programme. (2014). *Human development report 2014—'Sustaining human progress: Reducing vulnerabilities and building resilience'*. New York: Human Development Report Office.

Whelan, A., Mohr, R., & Short, S. (1992). *Waving or drowning? Evaluation of the National Secretariat, Healthy Cities Australia*. Bondi Junction, NSW: Australian Community Health Association.

WHO (1986) Ottawa Charter for Health Promotion. WHO regional Office for Europe. Copenhagen

WHO (1995) Yanuca Healthy island Declaration. WHO Regional Office for the Western Pacific. WPR/RC46/INF. DOC./1. Manila.

World Health Organization. (2011). *Healthy urbanization: Regional framework for scaling up and expanding healthy cities in the Western Pacific 2011–2015*. Manila: Regional Office for the Western Pacific Region of WHO.

Worsley, T. (1990). *National evaluation of healthy cities Australia pilot project*. Bondi Junction, NSW: Australian Community Health Association.

Part III
Grand Healthy City Challenges and Perspectives

Chapter 13
Wrap-Up: Values and Governance for Urban Health

Jean Simos and Evelyne de Leeuw

In part I of this book we looked at the history and development of what came to be known as Healthy Cities, and in Part II we presented a series of curated case studies from different parts of the world. In Chap. 6 we claimed that what sets Healthy Cities apart from other global urban networks is the strong and sustained set of commitments to a fairly revolutionary manifesto for change—Hancock and Duhl's (1986) 'Eleven Qualities'—a commitment to values. The material in Part II shows that each city, in each region and country, in each world region, commits to these values in different ways.

Jacques Ellul, the French philosopher, sociologist and theologian, was among the very first ones who issued the maxim 'Think globally, act locally' (Ellul 2006), and following this basic principle, Healthy Cities is a global concept intended to be implemented at the local level.

'Glocality'

A key term pertinent to this book is *glocal*. Glocal health (de Leeuw 2001; de Leeuw et al. 2006; Kickbusch 1999) is a term used to recognize and appreciate the intricate and inseparable interface between global developments (such as climate change or trade) and local responses (such as councils adopting building codes to address the

J. Simos
Institute of Global Health, University of Geneva, Campus Biotech, Geneva, Switzerland

E. de Leeuw (✉)
Centre for Health Equity Training, Research & Evaluation (CHETRE),
University of New South Wales, South Western Sydney Local Health District,
Ingham Institute, Locked Bag 7103, Liverpool, BC NSW 1871, Australia
e-mail: e.deleeuw@unsw.edu.au

E. de Leeuw, J. Simos (eds.), *Healthy Cities*,
DOI 10.1007/978-1-4939-6694-3_13

increased risks of flooding and heat islands or offering favourable opportunities for local entrepreneurs to engage in international forums). The glocalization dynamic is reciprocal: less desirable global developments may be mitigated—or exacerbated—by local action. For instance, the increasing number of local governments around the world adopting zero carbon emission policies (e.g. Koehn 2008) not only contribute to possible reductions in climate change risks, but also send the message to their colleagues at both local and higher levels that such actions are feasible and effective. On the other hand, the inability to address the consequences of local conflict locally (e.g. drug wars in Mexico, or religious fanaticism in the Middle East) impacts on the wider world. The diffusion of local policies (and their deliberate absence—as even no policy is a political choice) influences global change, and have led to the need for new words: for instance, analyses of local government effectiveness in the late 1980s and early 1990s led to the coining of the terms *glocal* and *glocalization* (Swyngedouw 1992). Virtually every development and phenomenon described in this book has glocal dimensions.

It is interesting to underline some glocal trends identified in Part II:

- Lack of government funding for Healthy Cities initiatives.
- Continuation of the global financial crisis, its negative impact and the dominance of a market-driven user-pays ideology, mean the perseverance of inequities and presenting ongoing challenges for the Healthy Cities approach.
- Intersection between Healthy City actions and a number of other city-focused initiatives, such as sustainable cities, safe communities, child-friendly or age-friendly cities, energy cities and so on. All of these 'movements' help to improve health and well-being or, more broadly, human development, and thus are complementary with Healthy Cities, but could also be showed as antagonistic, specifically in access to available financial resources.
- Consolidation of the link between the health of a community and urban planning, growing work in synergy between urban planners and public health professionals.
- Increase in the use of health impact assessments in Healthy Cities.

The Impossible Typology

In our original plans for this book we resolved to establish, from cases presented in Part II, a typology of the local implementation of Healthy Cities' overall concept. More specifically, we sought to identify 'clusters' and underlying patterns to classify these examples, using three parameters:

- The degree of involvement of the WHO regional office.
- The type of local institutional organization in charge of the programme.
- The importance accorded to community participation, often in opposition to local, regional or national health authorities' commitment.

Wrapping up the last chapter of Part II, we wrote, 'we have highlighted the enormous diversity in approaches and governance models for Healthy Cities in Oceania'.

This observation on diversity is valid across the board. It makes any methodological endeavour to achieve a typology irrelevant, given the numerous combinations encountered, the many shades of general schemes and the intermingling of theoretical archetypes. Thus, regarding the degree of involvement of the WHO regional office, we thought that if we took into consideration the involvement of each office, the result would be spread along a gradient ranging from 'low' (only general advice and general technical assistance provided to countries and cities in the region, such as one might have expected a priori for EMRO) to 'high' (management of a pilot network of Healthy Cities by the regional office like WHO/EURO), through 'mediumlow' (e.g. WHO/AFRO) or 'mediumhigh' (e.g. PAHO). But the reality has proven much too rich for such a grid: PAHO, for example, has a far more active role in Healthy City and Healthy Community networks in Latin America than in Canada or the United States, which act more autonomously. Similarly, WHO/AFRO is often limited to publishing educational materials and conveying incentives relayed by its National Focal Points (e.g. PointeNoire in Congo-Brazzaville), to carry out missions of expertise and support[1] and organize training seminars,[2] but it also has a strong commitment to promoting the specific issue of Healthy Markets and is sometimes more deeply involved with local Healthy Cities by providing additional funding (e.g. Sokodé in Togo). Moreover, WHO/WPRO, whose activity somewhat resembles the European experience, has a much stronger regional structure in charge of the programme's promotion and coordination, the Alliance for Healthy Cities, which was initially developed and strengthened independently of the regional office. WHO has often initiated an implementation process for Healthy Cities, especially with the World Health Days devoted to Healthy Cities or urban health in 1996, 2003 and 2010 (e.g. El Khroub). Finally, WHO/EMRO has evolved over the years, and in 2012 there was a strong push for the creation of a Healthy Cities and Villages network for the region.

Regarding the second parameter, the type of local institutional organization carrying out the programme, some cities have conducted an implementation process 'by the book', applying to the letter the procedures that are recommended by WHO (see, e.g. Athens), but most have proceeded by taking only some elements of the WHO guidelines and, for the rest, acting in response to specific contexts and local opportunities. Administrative solutions are numerous: attachment to the office of the mayor or deputy mayor (as in Liverpool), to a specific department (in Rennes and Nantes), creation of an association external to the administration structure (like the Canadian Institute of Planners, Woolwich), management at a higher administrative level like a region, province or Russian Oblast (as occurs in Chad, Mali, Togo, and EMRO countries), establishment of a public–private partnership (in Hong Kong and Japan) or of a non-governmental organization (in Kiama, Illawarra) or in another form that brings together the successful triad of research, training and intersectoral action. Other solutions adopt a more gradual engagement in the implementation process, as indicated by the examples coming from Latin America's networks, which have the status

[1] as in Algeria in 2009 (experts: Hawa Senkoro and Jean Simos).

[2] as in Libreville (Gabon), in 2008 (trainers: Paule Simard and Jean Simos).

of associated, full member, responsible or healthy municipality, or from the United States, where Healthy City or Healthy Community programmes do not even use these words. In the same vein, some Healthy Cities (e.g. Phnom Penh, WHO/AFRO) focus on, and associate with, particular settings: Healthy Markets, Healthy Schools, Health Promoting Workplaces, Healthy Hospitals, etc.

In the same way, a temporal dimension (that is, changes over time) must be taken into account because all initial forms evolve, some experiencing failures even after success in the beginning, as happened in Toronto, and later being reintroduced to a new experience of Healthy Cities such as the Canadian Healthy Communities Network.

Finally, regarding the third parameter, the importance accorded to community participation, virtually everywhere community participation was emphasized, although it could take drastically different forms. For instance, a systemic cultural difference exists between Anglo-Saxon and Latin countries, the first favouring direct civic participation, the second participation through representative organizations. Very often the case studies in Part II reveal that the implementation of Healthy Cities projects, and the original definition of local priorities, begin in a workshop or forum with very broad participation and representation, even when it is organized on the initiative of the Ministry of Health. Usually the following events, including any action plans and outcomes of the workshop, are managed or monitored by national or regional authorities (e.g. N'Djamena; Sokode), and sometimes with local authorities that give reluctant support (e.g. Mali). Sometimes community initiatives are the core drivers of change without any involvement of local authorities at the beginning or at any later stage of a healthy community (e.g. Mathare in Nairobi; Gha Massie in Accra).

From Local Political Culture … (or 'It Is the Local Political Culture, Stupid!')

The shape and practice that local Healthy Cities implementation takes depends heavily on the local (and sometimes regional or national) political culture. We can define *political culture*—after Braud (2006)—as a set of values, beliefs, traditions and strategies that enables people to make sense of their routine links with the leaders and groups that they use as identity references. This relationship can explain or predict how power is exercised or balanced in a specific country (or region or city). For instance, Rosanvallon (2004) shows that French political culture is characterized by a deliberate disregard for differences from the national standard, a way of thinking that can be cleanly traced back to the Jacobin era, with its vision of establishing a nation-state where any local differences would be eliminated.

This definition of political culture involves two specific dimensions (Braud 2006):

- A link to the past. It conveys a history and collective memory more or less developed and internalized.
- A projection into the future. Political culture enhances ideal modes of self-realization, legitimate expectations and hopes.

In other words, a political culture helps build social links through the continuous revitalisation of a shared history and shared values, while offering tasks to be accomplished together.

There are three functions to distinguish (Wildavsky 1987):

- Meaning function: allowing the identity of categories of thought that give meaning to things and events.
- Responsibility function: allowing standards to be set that define the level of responsibility of individuals in their behaviour.
- Boundary function: identifying patterns of behaviour or lifestyles perceived as unacceptable.

The literature proposes three different approaches of political culture that are useful in research. The first and the most traditional is the socio-historical approach, which is interested in the political dimensions of cultural traditions, including religion. Its purpose is to identify elements of overall belief that influence institutions and political life. Max Weber is the best-known supporter of this approach: he established a link between protestant culture and behaviours adapted to the development of capitalism, such as asceticism, saving, and concern for material success (Moscovici 1988). This approach allows one to reconsider governance while avoiding an ethnocentric belief that political requirements specific to social systems have always to be understood with the same connotations as in the Western world.

The second approach is neo-institutionalist, focused on the institutions (understood in a broad sense), standards and symbolic schemes framing actors' rationality. According to Berger and Luckmann (1967), persons and groups interacting in a social system create mental representations of each other's actions, which eventually become habituated and expressed into reciprocal roles played by the actors. When this phenomenon is generalized, these reciprocal interactions are said to be institutionalized, and their meaning is embedded in the society. Reality therefore is socially constructed, as knowledge and people's conceptions and beliefs of what reality is become embedded in the institutional fabric of society.

Such symbolic universes emerge to provide legitimation to the institutional structure. A symbolic universe is a set of beliefs 'everybody knows', which gives the institutionalized structure plausibility and acceptability for individuals who might otherwise not understand or agree with the underlying logic of the institution. As an ideological system, the symbolic universe 'puts everything in its right place'. It provides explanations for why we do things the way we do. Its function is to make objectively available and subjectively plausible the 'first-order' objections that have been institutionalized. Proverbs, moral maxims and wise sayings are common on this level. The symbolic universe also orders history: it locates all collective events in a cohesive unity that includes past, present and future.

The third approach is the anthropological approach, which consists of empirical investigations aiming to identify the values and representations that constitute the content of a political culture. According to Almond and Verba (1963) whose survey was based on statistics and interviews conducted among the populations of Germany, Italy, Mexico, United Kingdom and United States, there are three basic types of political culture explaining why people do or do not participate in political processes:

- The parochial political culture, associated with traditional and decentralized structures, where citizens are mostly uninformed and unaware of their government and take little interest in the formal political process.
- The subject political culture, where citizens are somewhat informed and aware of their government and occasionally participate in the political process.
- The participant political culture where citizens are informed and actively participate in the political process.

At the same time, Almond and Verba are describing three individual orientations to the political and global system (nation, history, constitution) and its structures and decisions:

- The emotional dimension, i.e. the daily perception of the nature of social relations between individuals in terms of trade, compromise or conflict, and the emotions aroused by public events (indifference or interest, attraction or rejection, etc.).
- The cognitive dimension, which reflects individuals' knowledge of the political system and governance rules, mainly as the product of education.
- The evaluative dimension, the most advanced form, which refers to the ability to make value judgments, enlightened or not, on the system's performance, what takes place there (e.g. legal/illegal, effective/ineffective, legitimate/illegitimate). The authors demonstrate a strong correlation between this dimension and individuals' level of education and exposure to the media.

Hofstede (2001) studied cultural differences based on two unique data bases. The first and largest covered 40 countries, and consisted of matched business employees' answers to the same attitude survey. The second consisted of answers to some of the same questions by his executive students, who came from 15 countries and a variety of companies and industries. Systematic differences between nations in these two data bases were identified, in particular in responses dealing with values. Values, in this case, are 'broad preferences for one state of affairs over others', and are mostly unconscious. The values that distinguish countries, rather than individuals, from each other group themselves statistically into four clusters. They dealt with four anthropological problem areas that different national societies handle differently:

- Ways of coping with inequality or 'power distance'. It is the extent to which the less powerful members of organizations and institutions (like a family) accept and expect that power is distributed unequally, indicating that a society's level of inequality is endorsed by the followers as much as by the leaders. Power distance scores attributed by Hofstede and colleagues are high for Latin, Asian and African countries and lower for Anglo and Germanic countries.
- Ways of coping with uncertainty, or 'uncertainty avoidance'. This deals with a society's tolerance for uncertainty and ambiguity. It indicates to what extent a culture programmes its members to feel either uncomfortable or comfortable in unstructured situations: that is, situations that are novel, unknown, surprising, and different from usual. Uncertainty-avoiding cultures try to minimize the possibility of such situations with strict laws and rules, safety and security measures,

and on the philosophical and religious level by a belief in an absolute truth: 'there can only be one Truth and we have it'. The opposite type, uncertainty-accepting cultures, are more tolerant of opinions different from what they are used to; they try to have as few rules as possible, and on the philosophical and religious level they are relativist and allow many currents to flow side by side. Uncertainty avoidance scores are higher in Latin countries, in Japan and in German-speaking countries, lower in Anglo, Nordic and Chinese cultures.

- The relationship of the individual with her or his primary group, or 'individualism versus collectivism'. It is the degree to which individuals are integrated into groups. The word *collectivism* here has no political meaning: it refers to the group, not to the state. On the individualist side we find societies in which everyone is expected to look after her/himself and her/his immediate family. On the collectivist side we find societies in which people from birth onwards are integrated into strong, cohesive in-groups, often extended families (with uncles, aunts and grandparents), which continue protecting them in exchange for unquestioning loyalty. Individualism prevails in developed and Western countries, while collectivism prevails in less developed and Eastern countries; Japan takes a middle position.
- The emotional implications of having been born a girl or a boy, or 'masculinity versus femininity'. Very often women's values differ less between societies than men's values; and men's values from one country to another contain a dimension from very assertive and competitive and maximally different from women's values on the one side, to modest and caring and similar to women's values on the other. The assertive pole has been called masculine and the modest, caring pole feminine. The women in feminine countries have the same modest, caring values as the men; in the masculine countries they are more assertive and more competitive, but not as much as the men, so that these countries show a gap between men's and women's values. Masculinity is high in Japan, in some European countries like Germany, Austria and Switzerland, and moderately high in Anglo countries; it is low in Nordic countries and in the Netherlands, and moderately low in some Latin countries (those speaking a Romance language such as Spanish, Portuguese, French, Italian or Romanian), like France and Spain, and Asian countries like Thailand.

In the early 1990s, the Canadian Michael Bond, working in Hong Kong at the time, re-ran Hofstede's questionnaires (first established in the mid-70s and largely replicated in six other cross-national studies between 1990 and 2002) and found significant differences on certain questions, leading to the addition of a fifth area: the 'Long-Term Orientation' dimension (Bond et al. 2004). Long-term oriented societies foster pragmatic virtues oriented towards future rewards, in particular saving/thrift, persistence/perseverance, ordering relationships by status, having a sense of shame, and adapting to changing circumstances. Short-term oriented societies foster virtues related to the past and present such as national pride, respect for tradition, preservation of 'face', and fulfilling social obligations (personal steadiness and stability, reciprocation of greetings, favours, and gifts). The influence of the teachings of Confucius is obvious in all these values. Long-term orientation scores are highest in East Asia, moderate in Eastern and Western Europe, and low in the Anglo world, the Muslim world, Latin America and Africa.

More recently, a sixth dimension has been added by Hofstede et al. (2010), based on Minkov's World Values Survey data analysis for 93 countries:

- 'Indulgence versus restraint'. Indulgence stands for a society that allows relatively free gratification of basic and natural human drives related to enjoying life and having fun. Restraint stands for a society that suppresses the gratification of needs and regulates it by means of strict social norms. Indulgence scores are highest in Latin America, parts of Africa, the Anglo world and Nordic Europe; restraint is mostly found in East Asia, Eastern Europe and the Muslim world.

Hofstede et al. (2010) argue that the groupings of country scores in their analysis points to some of the roots of cultural differences, which should be sought in the common history of similarly scoring countries. All Latin countries, for example, score relatively high on both power distance and uncertainty avoidance. Latin countries have inherited at least part of their civilization from the Roman Empire, which was characterized by a central authority in Rome and a system of law applicable to citizens anywhere. This established in its citizens' minds a value complex that we still recognize today: centralization fostered large power distance, and a stress on laws fostered strong uncertainty avoidance. The Chinese Empire also knew centralization, but lacked a fixed system of laws: it was governed by men rather than by law. In present-day countries that were once under Chinese rule, the mindset fostered by the empire is reflected in high power distance but medium to weak uncertainty avoidance. The Germanic part of Europe, including Great Britain, never succeeded in establishing an enduring common central authority, and their descendent countries show smaller power distance. Hofstede et al. found also that these six dimensions are statistically correlated with a multitude of other data about the countries. For example, power distance is correlated with the use of violence in domestic politics and with income inequality. Uncertainty avoidance is associated with Roman Catholicism and with the legal obligation in developed countries for citizens to carry identity cards. Individualism is correlated with national wealth and with mobility between social classes from one generation to the next. Masculinity is correlated negatively with the percentage of women in democratically elected governments. Long-term orientation is correlated with school results in international comparisons. Indulgence is correlated with sexual freedom and a call for human rights such as the free expression of opinions.

Proponents of the anthropological approach attempt to determine the values that are consensually shared. Post-materialism theory explains the long-lasting effects of childhood economic and social conditions, putting face to face two categories of values (Inglehart 1977):

- Materialist values, adapted to the logic of the market society.
- Post-materialist values, reluctant vis-à-vis money and social competition.

At the end of the twentieth century, the dominant 'materialist compromise' model ('growth is intrinsically a good thing' for all kind of ideologists) was widely disputed, as were its characteristic components: work, authority, religion, and sexual and social norms. New generations are promoting personal growth as an end in itself. Two fundamental reasons explain the development of post-materialist societies:

- The raising of the educational level, leading to the development of a system of values legitimizing critical thinking, personal autonomy and tolerance vis-à-vis other opinions.
- The increase of mobility, as migration allow emancipation from traditional cultural constraints which were not considered appropriate in circumstances other than those in which they emerged.

According to Almond and Verba (1963), modernization is a process of emancipation from parochial culture towards broader horizons. According to Braud (2006), the late twentieth century experienced a rebirth of identity claims based on political and religious particularities, territorial or regionalisms creation of subcultures. They are reactions to the transnationalization of economic and cultural exchanges and lead to 'a new world ethnicization' manifested by increased mistrust of migrants, a resurgence of separatist claims, and violent or latent forms of ethnic cleansing.

Other theories proceeding from this approach have paid attention to several particular elements:

- How political culture takes root and is transferred from generation to generation through political socialization (e.g. Seymour Martin Lipset's formative events theory, which describes the long-lasting effects of key events that took place when a country was founded).
- How European colonization has had long-lasting effects on countries and societies (Louis Hartz's fragment theory).

Even if the concept of political culture is relevant, encouraging researchers to overcome their own ethnocentrism when analyzing systems belonging to civilizations with different representations schemes and values, it is important to set down its limits. Among them we can mention (Bayard 1996):

- There is no cultural content specific to only one nation.
- There is always borrowing and reciprocal influences between cultures.
- Practices and values change over time: the weight of cultural determinism on the behaviour of individuals should not be overestimated, yet many 'culturalists' have a tendency to exaggerate determinism and to present the concept of political culture with a certain rigidity.

… to Global Values (or 'But It Is Also About Global Values, Stupid!')

Social values are like sociological institutions: they are the invisible glue to society and its cultures. This is not the place to provide a comprehensive primer on the practical and conceptual conundrums of defining and applying a value base to local health development. Like all value-based activities, Healthy Cities implementation is shaped by the local political culture. Earlier in the book we ascertained that Healthy

Cities strive for a value-based form of health development—and in Part II we have seen that they adopt and adapt this to their context. In returning to our original ambition of an analysis of Healthy Cities around the world, we can ascertain that their qualities align well with values as outlined in the previous section (Table 13.1).

From MDGs to SDGs and the 2030 Agenda…

The eight Millennium Development Goals (MDGs) formed a blueprint agreed to by all United Nations Member States and the world's leading development institutions. They galvanized unprecedented efforts to meet the needs of the world's poorest, and

Table 13.1 Eleven qualities and values of a Healthy City

Quality	Value
A clean, safe, high-quality physical environment (including housing quality)	Resilience
	Reliability
	(Environmental) justice
An ecosystem which is stable now and in the long term	Sustainability
	Systems view
	Ecology
A strong, mutually supportive and non-exploitative community	Caring community
	Sharing
	Inclusive
A high degree of public participation in and control over the decisions affecting one's life, health and well-being	Participation
	Control
	Empowerment
	Strategic development
The meeting of basic needs (food, water, shelter, income, safety, work) for all the city's people	Universal human rights
	Basic needs
	Responsive
Access to a wide variety of experiences and resources with the possibility of multiple contacts, interaction and communication	Equal access
	Equity
	Community control
	Solidarity
A diverse, vital and innovative city economy	Prosperity
Encouragement of connectedness with the past, with the cultural and biological heritage and with other groups and individuals	Unity in diversity
	Cultural respect and safety
A city form that is compatible with and enhances the above parameters and behaviours	Healthy urban planning
	Systems views
	Complexity
An optimum level of appropriate public health and sick care services accessible to all	(Primary) health
High health status (both high positive health status and low disease status)	Positive health
	Health equity

were very important for Healthy Cities located in developing countries. As their target date was at the end of 2015, the UN together with governments, civil society and other partners built on the momentum generated by the MDGs and carried on with a '2030 Agenda for Sustainable Development' adopted by world leaders in September 2015, a 'plan of action for people, planet and prosperity' which includes 17 goals (SDGs), 169 targets and 231 proposed indicators. The classic determinants of health such as poverty (SDG 1), food (SDG 2) and water (SDG 6) feature prominently, as do core public health concerns such as inequality (goal 10) or recent thinking on planetary health, like climate and ecosystems (SDGs 13–15), while governance, increasingly recognized as key to achieving health, features in SDG 16 (Moore et al. 2016).

Healthy Cities is explicitly linked to SDG 11, 'Make cities and human settlements inclusive, safe, resilient and sustainable', but in reality health promotion at the local level tends to focus on other SDGs, as per Table 13.2.

… Through the Global Charter for the Public's Health

In the light of and in conjunction with the SDGs, the World Federation of Public Health Associations (WFPHA) developed in early 2016 the *Global Charter for the Public's Health* (GCPH) to adapt public health to its global context. The Charter aligns with a number of other statements that were made in 2016—a pivotal year for urban health. The Quito New Urban Agenda (October 2016, formulated by participants in the Habitat 3 process), the statement emerging from the Ninth Global Conference on Health Promotion (Shanghai, November 2016) and the Kuching Statement (International Institute for Global Health 2016) all espouse a similarly broad value base for human development.

The core issue in the collaborative consultation of WFPHA with WHO was to investigate how the global public health community should position itself to influence all key actors across the entire spectrum of public health issues, in government, civil society and industry (Moore et al. 2016). A teleconference was organized of selected high-level public health stakeholders worldwide to reflect on the need to renew the meaning of public health in relation to the changing global context. The group also sought to identify terms of reference for a concept note to be discussed with the director-general of WHO. The meeting resulted in a follow-up qualitative study, based in pretested questions and an inductive content analysis (Lomazzi et al. 2016).

Findings included that the positive and negative impacts of globalization need to be better understood by public health professionals and more widely acknowledged by policymakers. Globalization is marked by increased interconnectedness and interdependence of peoples and countries, based on the opening of borders to increasingly fast flows of goods, services, finance, people and ideas, and changes in institutional and policy regimes at international and national levels to facilitate or promote such flows. It is recognized that globalization has both positive and negative impacts on health development (WFPHA 2016).

Table 13.2 Healthy Cities links to key Sustainable Development Goals (SDGs) (http://www.who.int/healthpromotion/conferences/9gchp/policy-brief1-healthy-cities.pdf?ua=1)

Healthy cities and the SDGs	
	Healthy cities can ensure access to safe, nutritious and sufficient food by adopting innovative policy measures than improve the food environment through: (1) increased access to healthy options (e.g. green markets); (2) empowering people with clear information to make healthier choices (e.g. calorie labelling at point-of-purchase, graphic labelling); (3) restricting or disincentivising te availability of unhealthy foods and beverages (e.g. economic zoning policies) and (4) helping end malnutrition by targeted delivery of nutrient-dense meals to the poor
	Urbanization promises efficiencies, better infrastructure, and technology. A health cities approach ensures that these efforts give due attention to increasing access to safe drinking water and improved sanitation for large segments of the population, as well as proper waste disposal, pollution management and good hygiene. Informal as well as formal settlements warrant attention for adequate sanitation standards
	A healthy cities approach views better housing and sanitation, reduced overcrowding and upgraded slums as public health priorities. Substandard housing and sanitation increases the risk of TB and other airborne illnesses, allowing malaria, yellow fever and now Zika to flourish, especially where there is stagnant water. Healthy cities also encourage better urban planning to prioritize increased access to safe transport systems, green and public spaces, and emergency responses to natural disasters, which together reduce road traffic deaths, improve air quality, promote physical activity and save lives from disasters
	Unsustainable consumption and production patterns that harm the environment also harm health, whether through air pollutants, contaminated water supplies or food losses. Healthy cities are, therefore, sustainable cities. They push transnational corporations, and support individuals, to adopt sustainable practices for the health of both the planet and its people
	A healthy cities approach recognizes that extreme weather events bear significantly on health, through disrupting food supply chains, spreading water borne illness, causing uprooting and migration, and resulting in physical injuries. Healthy cities aim to reduce carbon emissions, thereby improving air quality and promoting physical activity (e.g. replacing cars with walking and cycling) simultaneously. With large segments of the population, including poorer populations, now concentrated in cities, a healthy cities approach is a major pathway toward climate change mitigation
	Health cities are peaceful and inclusive. They help eliminate violence by providing safe places for people to live, work and play. They also pioneer the type of effective, transparent and accountable intersectoral governance needed to advance health, achieve universal health coverage, and progress other SDGs. Healthy cities, as natural place for pro-health social activism to start or gain momentum, also help ensure access to justice for all, including the marginalized and displaced. These are just a few of the reasons why health responses, from WHO's Healthy Cities project to the Urban Health and Justice Initiative to UNAIDS' Fast-Track Cities, have prioritized cities. UNDP's ART initiative helps countries to establish multi-actor, multi-sector and multi-level governance structures and systems to strengthen local governance and development

In parallel, the WFPHA conducted a literature review on global public health (Jenkins et al. 2016). This review revealed some important policy implications:

- The 'political landscape' needs to be understood by the global public health community in a broad sense, recognizing the proliferation of actors working within the field.
- The global public health workforce needs to recognize the importance of political and commercial determinants of health and strive to collaborate with and, where appropriate, challenge political and commercial actors to improve health outcomes.
- Global public health professionals as well as other actors in global public health should be trained to understand the glocal political context and work effectively within it. Professionals need to see the importance of connecting different fields that impact upon population health.

GCPH was officially launched at the May 2016 World Health Assembly in Geneva. It was discussed, as a basis for a policy tool-kit, in the Commonwealth Health Ministers Meeting that year in relation to SDG 3 (Commonwealth Secretariat 2016). At the time that the world adopts the SDGs, the intent of the Charter is to take the next step in providing a succinct and practical guideline to the implementation of public health associations to work with other NGOs, universities, civil society and governments to plan and implement strategies for better health outcomes across the globe.

The proposed GCPH provides 'a clear and flexible framework that can be applied globally and within individual countries' (Moore et al. 2016). It recognizes the need to challenge new threats to public health in a globalized world: these include the power of industries dealing in unhealthy commodities, challenges to public health posed by international treaties, and any ideology, peddled by powerful forces, that attacks any sensible regulation as an infringement on individual freedom or the creation of a 'nanny state'. The two institutions (Commonwealth Secretariat 2016; WFPHA 2016) see that the future of the public's health in a glocalized world depend on the establishment, maintenance and growth of services (the protection of health; prevention of disease; promotion of well-being; and people-centred efforts) and overarching functions (including governance; advocacy; capacity; and information or knowledge) (Fig. 13.1).

There are differences between the two institutional gazes: WFPHA uses the word 'information' where the Commonwealth speaks of 'knowledge'. WFPHA elaborates three services: protection, prevention and promotion; the Commonwealth adds people. These differences highlight the intractable political nature of this field as we outlined earlier in this chapter: there is no one global political health culture, although there may be generic value systems that drive the discourse. In our view this generic value system is illustrated in the compound version of the WFPHA and Commonwealth views—with three important 'tweaks': we see people-centred health dependent on a strong grounding in participatory and empowered communities (de Leeuw 2016; Löfgren et al. 2011); we know that 'information' and 'knowledge' are not value-free constructs but are shaped by the

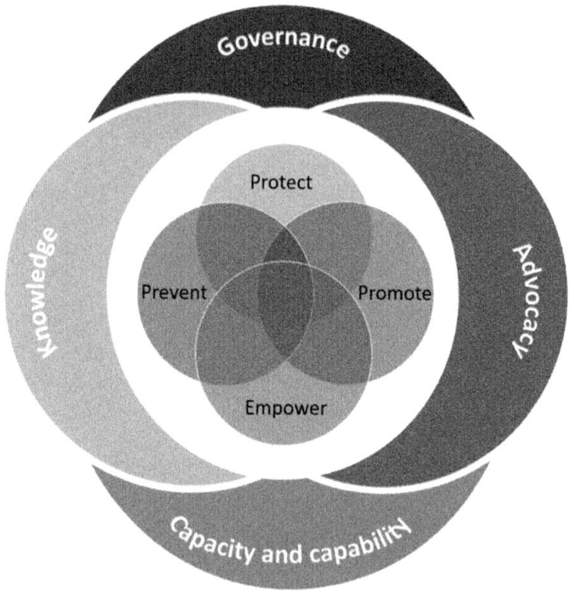

Fig. 13.1 A framework for strengthening the public's health, based on WFPHA (2016) and Commonwealth Secretariat (2016)

societies we live in (see Chap. 21); and similarly that 'capacity' must be framed in a political context that distinguishes between inherent human capacity and society-shaped and sanctioned human capability (e.g. Abel and Frohlich 2012; Nussbaum and Glover 1995).

There is overlap between all these services and functions, especially between health promotion, primary prevention and environmental health, which benefit from a cross-sector approach. Specific public health topics require components drawn from a range of services and functions (WFPHA 2016). These include:

- Governance: public health legislation; health and cross-sector policy; strategy; financing; organization; quality assurance (transparency, accountability and audit).
- Knowledge (or Information): surveillance, monitoring and evaluation; monitoring of health determinants; research and evidence; risk and innovation; dissemination and uptake.
- Protection: international health regulation and co-ordination; health impact assessment; communicable disease control; emergency preparedness; occupational health; environmental health; climate change and sustainability.
- Prevention: primary prevention: vaccination; secondary prevention: screening; tertiary prevention: evidence-based, community-based, integrated, person-centred quality health care and rehabilitation; health care management and planning.
- Promotion: inequalities; environmental determinants; social and economic determinants; resilience; behaviour and health literacy; life-course; healthy settings.

- People: primary health care; secondary health care; tertiary health care and rehabilitation.
- Advocacy: leadership and ethics; health equity; social-mobilization and solidarity; education of the public; people-centred approach (community engagement and empowerment); voluntary community sector engagement; communications; sustainable development.
- Capacity: workforce development for public health, health workers and the wider workforce; workforce planning: numbers, resources, infrastructure; standards, curriculum, accreditation; capabilities, teaching and training.

The development of global public health requires political engagement, use of social networks, identification of political leverage points, and steerage of public health agendas through the new societal and political environments. Importantly, global public health values should inform, be embedded within, and be used to assess political and policy decisions (WFPHA 2016).

Moving Forward

Value-based urban health development has strong evidence-based, human-centred credentials. Thousands of cities around the world, each in a unique context, adapt and adopt the evidence within their value base. In the remainder of this book we identify some real challenges that Healthy City operators, glocally, need to face and address—and we are aware, in line with the abstract high-level values that we have described above, that there is still an entire world to discover and develop in a spirit of glocal collaboration and respect. This book by definition will remain incomplete as we cannot replicate, meaningfully, the important and critical work that has been done in the areas of, for instance, Healthy Urban Planning (e.g. Barton et al. 2015). But there are a number of fields that we have identified as deserving further conceptual attention. In Part III we identify some of the grand challenges for Healthy Cities. In Chap. 15, Clavier and O'Neill take a stab at a political science perspective to frame and explain the role of communities in policy development and implementation in urban health. Henderson-Wilson and Townsend add another dimension to community health in the urban context in Chap. 16. They provide a systemic review that shows how access to nature is an ecological, health and well-being benefit to individuals, communities, and the sustainability of the urban landscape. In Chap. 17 we add yet another layer to these Healthy City values: the importance of governance systems for local government health planning and policy development. De Leeuw and Lin show how governance is not just an abstract concept that has been 'sold' to us (e.g. by Burris et al. 2007; Philippon and Braithwaite 2008; Wismar et al. 2012) but that it actually drives perspectives and priorities in local health policymaking. De Leeuw, in Chap. 18, illustrates how such policies can come about, and how policies are different from projects, programmes and actions. A number of glocal developments seem serendipitous and provide synergy—her chapter builds on the arguments in this chapter and shows that universal health coverage, complexity thinking and social movements around the world (to name a few) build momentum for lasting change. Jean Simos, in Chap. 19, takes us back to earth and shows that tools and technologies exist to guide these change processes. He describes the role of Health (Equity) Impact Assessment for urban health development. Bezold and Hancock, in Chap. 20, take this argument further: they show how thinking about the future is possible, feasible and necessary for better (urban) health world. Bringing all this together in a scholarly yet practical view, de Leeuw and Green in Chap. 21 demonstrate that in all this wicked, complex thinking, it is in fact possible to develop rigorous methodologies to monitor, assess and evaluate value-based health programmes such as Healthy Cities. They present a realist synthesis approach to research for, on, with and in Healthy Cities with a long case study of the inquiries that have been run in the context of the European network. Ultimately, however, developing Healthy Cities boils down to a strong and consistent, often courageous, political position. In the last chapter of this book Agis Tsouros illustrates how this is challenging, possible and rewarding.

References

Abel, T., & Frohlich, K. L. (2012). Capitals and capabilities: Linking structure and agency to reduce health inequalities. *Social Science & Medicine, 74*(2), 236–244.

Almond, G. A., & Verba, S. (1963). *The civic culture. Political attitudes and democracy in five nations* (xi-562 p.). Princeton, NJ: Princeton University Press.

Barton, H., Thompson, S., Burgess, S., & Grant, M. (Eds.). (2015). *The Routledge handbook of planning for health and well-being: Shaping a sustainable and healthy future*. London: Routledge.

Bayard, J.-F. (1996). *L'Illusion identitaire*. Paris: Editions Fayard. 310 p.

Berger, P. L., & Luckmann, T. (1967). *The social construction of reality: A treatise in the sociology of knowledge*. New York: Anchor Books Edition. 228 p.

Bond, M. H., Leung, K., Au, A., Tong, K. K., De Carrasquel, S. R., Murakami, F., … Boen, F. (2004). Culture-level dimensions of social axioms and their correlates across 41 cultures. *Journal of Cross-Cultural Psychology, 35*(5), 548–570.

Braud, Ph. (2006). *Sociologie politique*. Librairie Générale de Droit et de Jurisprudence—LGDJ (8th ed., 1st in 1992, 744 p.). Paris: Manuel.

Burris, S., Hancock, T., Lin, V., & Herzog, A. (2007). Emerging strategies for healthy urban governance. *Journal of Urban Health, 84*(1), 154–163.

Commonwealth Secretariat. (2016). *A systems framework for health policy. Advancing Global health security and sustainable well-being for all*. London: Commonwealth Secretariat. 31 p.

de Leeuw, E. (2001). Global and local (glocal) health: The WHO healthy cities programme. *Global Change and Human Health, 2*(1), 34–45.

de Leeuw, E. (2016). We need action on social determinants of health—But do we want it, too? Comment on 'Understanding the Role of Public Administration in Implementing Action on the Social Determinants of Health and Health Inequities'. *International Journal of Health Policy and Management, 5*(6), 1–4. doi:10.15171/ijhpm.2016.25.

de Leeuw, E., Tang, K. C., & Beaglehole, R. (2006). Ottawa to Bangkok: Health promotion's journey from principles to 'glocal' implementation. *Health Promotion International, 21*(Suppl. 1), 1–4.

Ellul, J. (2006). *Penser globalement, agir localement*. Paris (quoting the review he did in 1968): Pyremonde.

Hancock, T., & Duhl, L. (1986). *Promoting health in the urban context*. Copenhagen: WHO Regional Bureau for Europe. Retrieved June 14 , 2016, from http://www.euro.who.int/__data/assets/pdf_file/0013/101650/E87743.pdf.

Hofstede, G. (2001). *Culture's consequences: Comparing values, behaviors, institutions and organizations across nations* (2nd ed.). Thousand Oaks, CA: Sage. 596 p.

Hofstede, G., Hofstede, G. J., & Minkov, M. (2010). *Cultures and organizations: Software of the mind* (Revised and expanded 3rd ed.). New York: McGraw-Hill. 550 p.

Inglehart, R. (1977). *The silent revolution: Changing values and political styles among western publics*. Princeton, NJ: Princeton Legacy Library, Princeton University Press. 496 p.

International Institute for Global Health, United Nations University, Kuala Lumpur, Malaysia. *People, planet and participation: the Kuching statement on healthy, just and sustainable urban development*. Health Promot. Int. first published online July 1, 2016 doi:10.1093/heapro/daw046.

Jenkins, C., Lomazzi, M., Yeatman, H., & Borisch, B. (2016). Global public health: A review and discussion of the concepts, principles and roles of global public health in today's society. *Global Policy, 7*, 332–339. doi:10.1111/1758-5899.12302.

Kickbusch, I. (1999). Global + local = glocal public health. *Journal of Epidemiology and Community Health, 53*(8), 451.

Koehn, P. H. (2008). Underneath Kyoto: Emerging subnational government initiatives and incipient issue-bundling opportunities in China and the United States. *Global Environmental Politics, 8*(1), 53–77.

Löfgren, H., de Leeuw, E., & Leahy, M. (Eds.). (2011). *Democratising health: Consumer groups in the policy process*. Cheltenham: Edward Elgar Publishing.

Lomazzi, M., Jenkins, C., & Borisch, B. (2016). Global public health today: Connecting the dots. *Global Health Action, 9*, 28772. doi:10.3402/gha.v9.28772.

Moore, M., McKee, M., Borisch, B., & Ricciardi, W. (2016). The global charter for the public's health. *European Journal of Public Health, 26*(2), 207. doi:10.1093/eurpub/ckw013.

Moscovici, S. (1988). *La machine à faire des dieux*. Paris: Editions Fayard. 492 p.

Nussbaum, M. C., & Glover, J. (Eds.). (1995). *Women, culture, and development: A study of human capabilities*. Oxford: Clarendon.

Philippon, D. J., & Braithwaite, J. (2008). Health system organization and governance in Canada and Australia: A comparison of historical developments, recent policy changes and future implications. *Healthcare Policy, 4*(1), e168.

Rosanvallon, P. (2004). *Le modèle politique français: la société civile contre le jacobinisme de 1789 à nos jours*. Paris: Le Seuil. 462 p.

Swyngedouw, E. (1992). The Mammon quest: 'Glocalization', interspatial competition and the monetary order: The construction of new scales. In M. Dunford & G. Kafkalis (Eds.), *Cities and regions in the new Europe: The global–local interplay and spatial development strategies* (pp. 39–67). London: Belhaven Press.

WFPHA (2016). A global charter for the public's health—The public health system: Role, functions, competencies and education requirements. *European Journal of Public Health*. doi:10.1093/eurpub/ckw011.

Wildavsky, A. (1987). Choosing preferences by constructing institutions: A cultural theory of preferences formation. *American Political Science Review, 81*(1), 3–21.

Wismar, M., McQueen, D., Lin, V., Jones, C. M., & Davies, M. (2012). Intersectoral governance for health in all policies. *Eurohealth, 18*(4), 3–7.

Chapter 14
The Role of Policy Coalitions in Understanding Community Participation in Healthy Cities Projects

Carole Clavier and Michel O'Neill

Since it was first created under the auspices of the World Health Organization (WHO) in the late 1980s, the Healthy Cities project has become a model—some say a social movement—for municipalities seeking to 'put health on local agendas' (de Leeuw et al. 2013, S14), first in Europe and North America and increasingly in Oceania and Central America (Ashton 1992; de Leeuw et al. 2013; Petersen and Lupton 1996). Healthy Cities asks of local decision makers that they commit to making health a central concern of all their decisions, through partnership working with professionals from different policy sectors and through community participation. In other words, Healthy Cities are about creating health locally. But they are also about urban policymaking.

In this chapter we contend that the policy studies and urban studies literature provides insightful arguments to better understand and support community participation in Healthy Cities projects. After a brief review of the literature on community participation in the Healthy Cities movement, the second part of the chapter introduces the concept of 'policy coalition' as a way to understand how several actors work together towards governing cities and making public policies. Based on these findings, the final part of the chapter dwells on three issues that Healthy Cities should pay attention to, to promote and secure community participation: the formal instruments of community participation, the processes of actor inclusion and exclusion, and the legitimacy of Healthy Cities.

A caveat of this chapter is that it focuses on literature developed in high-income countries of Europe and North America. Studies have pointed out that the Healthy Cities in low and middle-income countries face specific constraints related to, for instance, the importance of donor and international aid practices, administrative

C. Clavier (✉)
Department of Political Science, Université du Québec à Montréal, Montréal, Canada
e-mail: clavier.carole@uqam.ca

M. O'Neill
Université Laval, Ville de Québec, Canada

E. de Leeuw, J. Simos (eds.), *Healthy Cities*,
DOI 10.1007/978-1-4939-6694-3_14

capacities and routines, the salience of health issues related to the environment, and so on (Harpham 2009; Harpham et al. 2001). Fully understanding the challenges of community participation and urban policymaking in low and middle-income countries would require drawing on this and specific policymaking literature (Gilson and Raphaely 2008).

The Community in the Healthy Cities Movement

A cornerstone of the Healthy Cities movement (Dooris and Heritage 2013), community participation is understood as the 'process by which people are enabled to become actively and genuinely involved' (World Health Organization 2002) in the process of building and sustaining healthy urban development. But who, precisely, is the community and what are the conditions for its participation in the policymaking process?

'Community' is a catchphrase that can encompass a diversity of meanings. In the context of the policymaking process, four ideal types of community can be identified: the local community as the place where people live; the organized community as in non-governmental organizations (NGOs); the community as a collective identity; and the community as a political or cultural whole (Vibert 2007, p. 13). In the case of the Healthy Cities, the community is often understood as the mass of individual citizens, including those who are 'affected' by Healthy City policies, who feel disempowered by institutions (de Leeuw 2012); as the organized groups of citizens such as community organizations; and as other stakeholders from the local business, intellectual and trade-union elites. In other words, in the Healthy City context, the organized community grants 'tangible reality' to the local community and identity (Vibert 2007, p. 114).

In the Healthy Cities context, the community is expected to participate in 'setting priorities and making and implementing decisions' and to take part in 'an ongoing programme of training/capacity building activities for health and healthy public policy' (Corburn 2009, pp. 8–9). In return, its actions should be supported by its local Healthy City project, in line with the priorities of the Ottawa Charter for Health Promotion. This reflects the Healthy Cities commitment to an understanding of health as being 'everyone's business', including the business of those who usually have little say in urban politics and policy development (de Leeuw 2012, p. 1306). The strong emphasis on the community, therefore, reflects both the Healthy Cities' priorities and values.

The last evaluation of Healthy Cities in Europe highlighted the diversity of community participation practices across all 78 member-cities of the network, ranging from episodic consultations to institutionalized participatory instruments (Dooris and Heritage 2013). Distinguishing between four types of community participation—namely providing information to the population, consulting local people, enabling citizen participation in decision-making and empowering citizens—the authors were able to confirm the Healthy Cities' commitment to community participation while also making apparent that it was subject to different interpretations across cities. Establishing the full range of community participation practices, the authors said, is one necessary step. But further studies are needed to understand why the community participates as it does in each city. Why are communities fully involved in building and sustaining Healthy Cities in some cases, and not in others? What are the conditions for their participation? What role do they play, and what role could they play?

De Leeuw and Clavier suggested elsewhere that health promotion has failed at effectively including the community in the making of health in all policies (de Leeuw and Clavier 2011). This, it was argued, was due to a lack of understanding of policymaking processes. In this chapter, and following our claim that theory is practical for health policymaking (Clavier and de Leeuw 2013), we propose to outline a way of addressing the puzzle of community participation in Healthy Cities from the perspective of public policy studies and urban studies. Our premise is that the Healthy Cities projects are little different from other urban development strategies: despite their focus on a particular set of values, namely health, well-being and social justice that sets them somewhat apart (Carpenter 2012), the Healthy Cities are one among other policy strategies that are subject to the urban governance rules of the game (Clavier 2011).

Both public policy studies and urban studies have made significant contributions to the understanding of how cities are governed and what they do, namely about how politics and policies occur in cities. In particular, they have shown the importance of policy coalitions to understand urban governance and urban policies (Dormois 2008; Mossberger and Stoker 2001; Pinson 2006; Stoker 2000). The literature provides numerous definitions of coalitions, some insisting on the fact that they bring together unlikely partners (Gamson 1961; Lemieux 1998; O'Neill et al. 1997), others on what brings them together (Sabatier and Weible 2007), and others still on their actual influence on decision-making (Stone 1989). The term 'coalition' is most often used in relation to electoral coalitions, that is, a formal and short-term agreement between parties whose results in the elections do not allow them to govern independently.[1] This is not the type of coalition that we are concerned with in this paper; rather, we are interested in the longer-term coalitions that are involved in public policymaking processes, including, but going way beyond, electoral gain. For now we shall loosely define a *policy* coalition as a formal or informal group of actors that participate in influencing and making public policy. This conception of policy coalitions insists on key findings from the literature on urban governance, namely the multiplicity of actors (Le Galès 2011) and the fact that sustained interactions between them go a long way towards explaining who exercises influence on urban decision-making and on urban policies (Mossberger and Stoker 2001; Pinson 2006).

[1] Lipp et al. (2013, p. S38) ascribed this electoral-oriented meaning to the term coalition in their paper on the evaluation of partnerships in Phase IV of the European Healthy Cities Network. A consequence of our adopting a broader definition of the *policy* coalition is that we include the partnerships of Lipp et al.'s study as but an instance of formal policy coalitions.

The Community, Policy Coalitions, and the Politics and Policies of Healthy Cities

The literature on urban governance focuses on *how* cities are governed, that is, on urban politics. By contrast, the literature on urban policymaking focuses on *what* cities do, that is, on making urban policies. Both sets of literature consider that governing cities and making policies is not the sole prerogative of a tiny elite but, quite to the contrary, involves a more or less broad range of actors gathered in policy coalitions. This section draws on both the urban governance and urban policymaking literature to better understand how the community may contribute to the politics and policies of Healthy Cities.

How Cities Are Governed: The Influence of Policy Coalitions on Urban Politics

There are several approaches to the study of how cities are governed. A classic approach focuses on the institutions of municipal government, its regulations and resources. As Stoker puts it, this literature is 'often descriptive and generally concerned with examining semi-constitutional questions and administrative doctrines about the appropriateness of local autonomy or how best politicians could hold bureaucrats to account without undermining the distinctive contribution of each' (Stoker 2000, p. 91). In the course of the 1960s, such normative questioning of local institutions was largely abandoned in favour of the study of actual power relationships in urban government. The 'big' debate was framed in terms of elitism vs. pluralism: are cities governed by a small, tight-knit elite of decision makers or by a large number of competing groups whose alliances and influence fluctuate over time? Marxism and the political economy perspective have fuelled this debate, leading to a renewed interest in the contribution of private actors and the economy to urban government (Stoker 2000).

Building on these findings, but leaning more strongly towards the pluralist thesis, urban studies then went back to their early concern with institutions, though with a decidedly more sociological twist. They included insights from the new institutionalism or, rather, the three new institutionalisms (Hall and Taylor 1996), a perspective that argued that both formal and informal rules influence collective action.[2] The concept of urban governance addresses precisely this shift, from a focus on

[2] Sociological and historical institutionalisms (Hall and Taylor 1996), in particular, are in line with urban studies' reframing of their concern for urban government in terms of formal and informal interactions between the full range of actors involved, both public and private. Sociological institutionalism brings ideas back into the understanding of institutions in that it considers that institutions are not the result of rational developments only, but also reflect cultural preferences. In that sense, institutions influence not only the rational behaviour of actors (their interests), but also their moral and cognitive templates. Historical institutionalism explains individual behaviour as the result of both rational interest and cultural preferences. It also places emphasis on the permanence of institutions, explaining how past choices constrain future choices along a specific path.

urban government as an administrative activity to a focus on urban government as a collective action dilemma (Le Galès 1995; Stoker 2000). It also reflects real-life changes towards a less central role for central governments and heightened responsibilities for cities: changes in the economic and social context of cities (the growing role of the market and globalization processes), greater decentralization of powers in line with the neo-liberal policy paradigm, and the increased complexity of urban problems such as poverty and inequalities (Le Galès 1995; Stoker 2000). In sum, governance is concerned 'with governing, achieving collective action in the realm of public affairs, in conditions where it is not possible to rest on recourse to the authority of the state' (Stoker 2000, p. 93).

In a context characterized by both fragmentation and interdependence between high numbers of institutional, private and community actors, the urban governance perspective calls for a study of the mechanisms that are able to clarify their relationships, hence the interest in policy coalitions (Dormois 2008). In the United States, the growth coalition thesis explained urban development policy through alliances between land and building owners on the one hand, and developers and municipal councillors on the other (Logan and Molotch 1987). Although this thesis has been influential, city government has also been analysed as the product of coalitions pursuing other objectives, such as environmental protection or simply maintaining the status quo (Stoker 2000).

Urban regime theory is another influential approach. Originally developed in the United States, it has been transferred with some success into European studies of urban governance (Dormois 2008; Mossberger and Stoker 2001). It focuses on the formation of governing coalitions and on their ability to orient decisions so as to solve governance issues. Stoker defines urban regimes as 'informal yet relatively stable coalitions' (Stoker 2000, p. 95) that have 'access to institutional resources that have a sustained role in making governing decisions' (Stone 1989, p. 4; cited in Stoker 2000, p. 95). Uniting public and private actors, these governing coalitions go well beyond electoral arrangements. They reflect the integration of the capacities for action of each public and private actor taking part in the coalition: it is the combination—or blending—of their capacities that makes governing cities possible (Dormois 2008; Le Galès 1995; Stoker 2000; Stone 1989). Regime theory, therefore, focuses on the following questions: who are the actors involved in the coalition? Who is excluded or absent from the coalition? Why and how do they cooperate? What is their ability to secure resources for action?

Although regime theory implicitly postulates that coalitions build themselves naturally, Stoker (2000) points out that governments often use steering instruments to encourage coalition or partnership-building between public, private and community organizations. Drawing on broader classifications of policy instruments (Howlett 2000; Lascoumes and Le Galès 2007), he mentions promoting partnership through cultural persuasion, setting up arenas for public discussion, providing funding to interest potential partners, monitoring the products of partnerships, and creating semi-autonomous bodies to achieve specific goals (Stoker 2000).

What Cities Do: How Policy Coalitions Contribute to Making Public Policies

Public policy studies have also used the concept of policy coalitions to make sense of the policymaking process, 'the manner in which governmental policies get formulated and implemented, as well as the effects of those actions on the world' (Sabatier 1991, p. 147). As Sabatier puts it, 'understanding the policy process requires knowledge of the goals and perceptions of hundreds of actors throughout the country involving possibly very technical and legal issues over periods of a decade or more while most of those actors are actively seeking to propagate their specific 'spin' on events' (Sabatier 2007a, p. 4).

There are, of course, several competing theories of the policy process, some placing emphasis on institutions, others on interests, ideas or actors, to explain why policies get elaborated and how (Sabatier 2007b; Schlager and Weible 2013). Among these theories, some start with the premise that policy actors and the ideas that they carry are drivers of policy action and change. This is the case of Sabatier's Advocacy Coalition Framework (Sabatier and Jenkins-Smith 1993; Sabatier and Weible 2007; Weible et al. 2011).

One of the most influential approaches to the policy process of the past 20 years, the advocacy coalition framework, was developed to explain public policy changes occurring even in the face of strong opposition from powerful interest groups—as was the case with environmental policies in the 1980s (Sabatier and Jenkins-Smith 1993). According to the framework, in every area of public action, policy actors coalesce based on their shared ideas and beliefs about the issue at hand (Sabatier and Weible 2007), as well as on their shared interests (Bergeron et al. 1998). Policy actors gathered into two to five coalitions form a policy subsystem, that is, a subset of the broader policy universe (Weible et al. 2010). These actors comprise not only elected officials and civil servants, but also private actors, NGOs, the media and citizens.

The advocacy coalition framework considers that public policies reflect the belief systems of the coalition that dominates the policy subsystem (Sabatier and Weible 2007). As a consequence, public policy changes because of transformations occurring within the belief systems of the advocacy coalitions. These can be minor changes to the ideas and beliefs of the dominant coalition, or perhaps a major change when the balance between coalitions shifts and a new coalition becomes dominant. For the latter type of change to occur, policy coalitions usually take advantage of external events such as major shifts in socio-economic circumstances, in public opinion, in political power, or in the policies of other policy subsystems.

The advocacy coalition framework has been widely used to understand the complexity of policy emergence and policy change, including in some cases public policies that may influence health (Breton et al. 2008; Gagnon et al. 2007; Kübler 2001; Montpetit 2011). Although the framework has been mostly used in reference to national policymaking processes, some have claimed it may be used with equal validity in reference to local or urban policymaking processes (Weible et al. 2011).

Policy Coalitions, or How the Community Contributes to Healthy City Politics and Policies

Thus, policy coalitions are ubiquitous in the study of urban governance and urban policymaking. Both fields recognize that making decisions and policies at the urban level is not the monopoly of formal municipal institutions but rests on the ability and willingness of public, private and community actors to agree, or not, to join forces—or blend capacities—to achieve particular goals. Public actors do have control over a variety of instruments allowing them to steer the structuration and activity of coalitions, but it would be a mistake not to consider the dynamics initiated by other, non-public actors in structuring and guiding the activities of coalitions. Indeed, while urban politics and policies are characterized by high fragmentation between public, private and community actors, there is also high interdependence between these actors. Thinking in terms of policy coalitions also encourages us to broaden the scope of research: it should not focus solely on those actors formally involved in making a particular policy, but also on other actors that might form a coalition contesting or supporting the efforts of the formal actors.

This brings us back to our interest in the role of the community as part of Healthy Cities. In the terms of the theories presented above, the community—in its various possible incarnations in a city—is better understood if included among a range of actors in a fragmented policy subsystem that may or may not share the goals of Healthy Cities. Although public authorities usually officially lead Healthy Cities, theirs is in practice a steering role rather than a commanding, hierarchical, role. Steering instruments offer other—private and community—actors a way into policymaking processes. Healthy Cities may form an advocacy coalition in the terms of the advocacy coalition framework, which would require bringing together actors that share a set of values, beliefs and interests compatible with the objectives of Healthy Cities. Its existence does not preclude that of at least one other coalition, defending another set of beliefs and values altogether.

Based on these insights from the literature on urban governance and policymaking processes, three issues are worth considering in more detail in order to understand how the community might better contribute to Healthy City politics and policies: the often informal processes of actor inclusion in, and exclusion from, policy coalitions, the ideas bringing policy coalitions together, and the formal and informal instruments of coalition practice.

Three Issues that Healthy City Analysts Should Pay Attention to

Studies of Healthy City processes often take the official Healthy City partnership as their unit of analysis (Dooris and Heritage 2013; Lipp et al. 2013; Stern and Green 2005, 2008). This makes a lot of sense as most of what Healthy Cities achieve results from the efforts of the partnership members. In the terms of coalition studies, these partnerships can be said to 'materialize' the Healthy City policy coalition. The third issue that we identify follows in this tradition by focusing on the instruments regulating interactions between partnership members. However, the first two issues discussed below adopt a broader, policy subsystem-wide perspective. Analysts and practitioners could benefit from looking beyond existing Healthy City partnerships to consider who becomes part of, or is not included in, the Healthy City coalition and what the other policy coalitions might be. Departing from a formal definition of the Healthy City coalition as *the* partnership opens the way to considering tensions and oppositions among the actors involved and with non-partnership members.

The Processes of Actor Inclusion and Exclusion

Considering the participation of the various community incarnations of Healthy Cities in terms of policy coalitions calls attention to the processes of their inclusion in, or exclusion from, policy coalitions. In this respect the distinction between insiders and outsiders may be less clear-cut than some of the literature suggests. Observations of processes leading to the emergence of public policies have shown that those making the policies are both from inside and outside public authorities (Dupuy and Halpern 2009). Although interest groups are often viewed as contesting public policymaking—e.g. are outsiders—they may very well be part of a coalition supporting the policies—e.g. insiders. Conversely, those opposing a particular public policy may be interest groups as well as actors from within the public authorities.

This reminder is perfectly in line with the theories of policy coalitions and does not change from what we have said so far. The reason we bring it up at this stage is to argue against the temptation to consider that public authorities and the community are homogeneous actors. On the contrary, we should look for fracture lines in the positions of the actors representing public authorities, as well as of those representing private interests or the community.[3] The distinction between those who are included in a coalition and those who are excluded should derive from empirical observations of their positions, beliefs and interests rather than from a priori expectations (Dupuy and Halpern 2009).

The recent evaluation of partnerships in phase IV of the European Healthy Cities Network (Lipp et al. 2013) indicates that almost half of the cities that answered the questionnaire had developed collaborations with the community—in their terms, the voluntary sector. The other partners involved in the policy coalition for Healthy

[3] In line with this argument, a recent study has established that residents of US communities have different expectations of what constitutes effective community representation into health policy making arenas (Chung et al. 2012). From there, there is only a short step to considering that community representation reflects, at least partly, these diverse expectations.

Cities[4] represented the health service sector, social services, education, urban planning, and, to a lesser extent, environmental protection, transport and economic development. This was not part of the focus of the article by Lipp et al. (2013), but we can make the hypothesis that most of the representatives of these other sectors were public actors, with a minority of private actors. The value of mutual support from the different sectors emerged as a lesson that partnership members drew from their experience. This evaluation gives very important insight into the types of partnership—or policy coalition—that support Healthy Cities in European cities. Large-scale evaluations, such as the one Lipp et al. conducted, do not lend themselves to the in-depth questioning necessary to understand who becomes a member of the policy coalition, what unites the members, and if there are other opposing coalitions. Building on the results of that evaluation, though, one could study the processes of partnership/coalition formation in two or three of these cities so as to bring to the fore potential fracture lines between the members of the community or of the public authorities. This might contribute towards explaining the varying intensity of partnership work[5] according to the cases.

The Ideas Bringing Healthy City Coalitions Together

As an example of how ideas contribute to shaping policy coalitions in the Healthy Cities context, let us look at the City of Toronto. In his account of the emergence of the Health Advocacy Unit in this city—the unit that triggered Healthy Cities there—Hancock (1988) tells about its commitment to document the causes of social and health inequalities in the Toronto area. This unit published the 'The Unequal Society Report', a document that raised 'a set of social, political, economic and environmental issues' (Hancock 1988, p. 216). For instance, it identified chemicals, and in particular chemicals used in the city's works and parks departments, as a cause of health problems and health inequalities. As a result, 'the Unit made enemies as well as friends. In particular, some of the "old guard" among the Department's management were offended by both its style and its implicit (and sometimes quite explicit) criticisms of their competence. Additional enemies were to be found among other City Departments that had been attacked by the Unit, the Provincial Ministry of Health and Conservative politicians and their supporters at the municipal and provincial level' (Hancock 1988, p. 216).

In 1985, the city council's board of health and the City of Toronto organized the 1985 'Beyond Health Care' conference, which for the first time put healthy public policy on the concept map of health promotion. In the wake of this event, the board

[4] It is worth reminding that, as per our definition, we consider official Healthy City partnerships as formal instances of policy coalitions.

[5] The intensity of partnership work is classified in five types of collaboration: no contact, share information, agreement to collaborate, collaboration resulting in agreed plans/strategies, and implementation of collaborative plans, projects or programmes (Lipp et al. 2013, p. S41).

of health established a planning subcommittee that was mandated to 'clarify what a healthy Toronto would be like' (Hancock 1988, p. 220). The subcommittee led a large number of consultations and public hearings with city officials, with professionals from several city departments, and with the communities. The concept of the healthy city emerged as a result of this yearlong, broad consultation, as did the strategy Healthy Toronto 2000.

For the purpose of our reflection on the community and policy coalitions, Hancock's thorough account of the processes leading to the emergence of Toronto's healthy city strategy raises two crucial elements. The first is that he identifies policy coalitions that are for and against the healthy city—although not in these terms. In favour of the healthy city were the environmental and social justice movements, some health groups with a non-biomedical focus, and the multicultural community. Against were shifting, issue-based coalitions that formed when their interests were threatened, especially in the fields of the environment and health care (Hancock 1988, p. 224). In line with what was said above, these coalitions cut across public, private and community boundaries, with some of the community constituencies supporting the healthy city coalition and others, presumably, not voicing a strong opinion.

The second crucial element is that these coalitions were divided according to values and knowledge, that is, ideas formed the basis of policy coalitions. For instance, the actors' positions on social justice seem to have been a dividing factor: committed municipal councillors from the New Democratic Party—Canada's incarnation of social democracy—actively supported the efforts of the Health Advocacy Unit and of the Board of Health. By contrast, conservative politicians were listed among opponents to the original healthy city idea. Knowledge about the social determinants of health and consequences of that knowledge for public action were also strong dividing factors among the coalitions, as illustrated by the controversy that surrounded the 'Unequal Society' report.

In the advocacy coalition perspective, knowledge and values are central to the process of public policy change. They are, in effect, the main drivers of change. In this vision, producing knowledge is not enough to change public policy. This knowledge has to be appropriated by policy actors and transformed into propositions for public action. Policy change is not the result of a linear process of knowledge transfer (de Leeuw and Skovgaard 2005; Jansen et al. 2012).

What does this mean for communities and Healthy Cities? For one thing, the analyst should consider not only the contribution of the community representatives within formal Healthy City partnerships, but also the support and opposition emanating from the (diverse) community groups to specific Healthy City strategies and interventions. In addition, shared values, beliefs and knowledge are opportunities for political, public, private and community actors to extend cooperation and seek to broaden their circle of allies, that is, to reinforce their coalition with new actors. Thinking in terms of opposing policy coalitions united around shared ideas does not erase power imbalances between the various categories of actor that are inherent to Healthy City partnerships. It may, however, clarify strategies for advancing cooperation within, and beyond, official partnerships, both for public authorities seeking to involve the communities in Healthy Cities, and for community constituencies trying to influence the agenda and implementation of Healthy Cities. Political, public, private and community actors may thus derive practical insights to guide their action from the theory of policy coalition (de Leeuw et al. 2013).

The Formal and Informal Instruments of Community Participation

It has been well documented that a major obstacle to collaborative endeavours is that partners have their own agendas, interests, (professional) cultures and ways of knowing. This has been especially emphasized in the literature on participatory research associating academics with community actors (Cargo and Mercer 2008; Gélineau et al. 2013; Mantoura et al. 2007; Tsai Roussos and Fawcett 2000). In the cases where such difficulties were overcome as part of Healthy Cities, it was demonstrated that strong support of community involvement on the part of public actors was clearly a condition for their enduring success (Baum et al. 2006). In this context, one possible avenue to understand how community involvement may be better supported is to investigate the instruments of community participation to policy coalitions—at least to the most formal of these coalitions, such as Healthy City partnerships.

We mentioned above that public authorities have several policy instruments at their disposal to steer urban governance. Could some of these instruments be favourable to community participation? Existing research suggests that it might well be the case. For the sake of clarity, we shall distinguish here between formal and informal instruments of governance. Starting with formal instruments, the presence of 'boundary workers' (Stern and Green 2005; Williams 2002), 'policy entrepreneurs' (de Leeuw 1999) and 'translators' (Bernier et al. 2006) has been associated with facilitated interactions within partnerships. As partners have such diverse agendas, cultures and ways of knowing, the presence of 'boundary workers' helps translate the expectations and interests of one actor into the languages of the others. Ultimately, boundary workers are considered valuable actors in a partnership or policy coalition because of their ability to align the interests of those present by finding a common focal point, and because of their ability to creatively suggest ways forward in their cooperation (Clavier et al. 2012). It is notable that the role of the boundary spanner needs not correspond with an official role within the partnership: although it may be the prerogative of partnership coordinators, the role may just as well be taken on by any other actor who is considered for his or her equitable attitude and appropriate set of skills.

The metaphor 'a foot in the door' (Stern and Green 2005, 2008) has sometimes been used to explain why community actors remained involved in Healthy City processes and partnerships together with local public authorities. Coalitions are fraught with tensions with regard to participants' objectives. Community actors sometimes choose to lower their expectations to make sure that at least some of what they want will get adopted. 'The foot in the door' metaphor thus refers to the fact the community strives to remain within the partnership with expectations of increased benefits in the future. The metaphor can also refer to formal instruments that partnerships/coalitions give themselves to secure their existence beyond the next crisis. For instance, some cooperation agreements stipulate that one partner shall give a year's notice before he actually withdraws from the cooperation (Clavier 2007). The dissociation in time between the crisis and the end of the coalition/partnership favours inertia, in that case maintaining the coalition/partnership is less effort than disbanding it. It is, indeed, difficult to draw political credit from slamming the door a year earlier, and such inertia may benefit the communities in the sense that coalitions/partnerships tend to be maintained.

As for informal instruments of governance, Stern and Green (2008) have shown that inequities in power and institutional support between public actors and the community are reflected in how the partners interact. They observed interactions between public actors and community members at partnership meetings and highlighted 'the struggle over the content and form of meetings' on the one hand, and the public actors' control over the 'rules of engagement' (Clavier 2007), that is, who chairs meetings and how, how to address each other so as to reduce status inequalities, etc. The authors' analysis shows both how the participants frame their participation in the decision-making process by being seated 'around the table' and how their interactions are heavily influenced by their respective institutional positions. This means that although community representatives have a chance to take part in decision-making, they may have little chance to actually influence the agenda of the partnership/coalition.

Studying the instruments of community participation calls for in-depth analysis of how community and public and private actors interact within Healthy Cities. The studies mentioned here focused on what happens inside the more formal partnerships set up towards Healthy City implementation. Building upon our knowledge of the processes of actor inclusion and exclusion, it would be interesting to question the instruments regulating access to these partnerships, as well as those regulating interactions with actors outside them.

Conclusion

Community participation is a core value, as well as a prominent challenge, for Healthy Cities. Following the saying of Kurt Lewin that 'there is nothing more practical than a good theory', this chapter has drawn on the fields of urban studies and policy studies to suggest a new reading of community participation to Healthy Cities. Thinking in terms of urban governance and policymaking as the result of conflict between opposing policy coalitions has led us to replace Healthy Cities within a broader policy subsystem, beyond the confines of Healthy City partnerships. Understanding and strengthening the role of community actors can benefit from a better understanding of the processes of actor inclusion in, and exclusion from, the Healthy City policy coalition, from a broader overview of the policy coalition actors' values and knowledge, and from a raised awareness of the instruments regulating interaction between the actors of a policy coalition. These issues call for a broadened perspective on the community and the Healthy Cities: not just a matter of community representation in one official instance, it is a matter of politics and policymaking.

Acknowledgments CC thanks Caroline Patsias for her insightful comments on an earlier version of the manuscript.

References

Ashton, J. (1992). The origins of healthy cities. In J. Ashton (Ed.), *Healthy cities* (pp. 1–12). Buckingham: Open University Press.

Baum, F., Jolley, G., Hicks, R., Saint, K., & Parker, S. (2006). What makes for sustainable Healthy Cities initiatives?—A review of the evidence from Noarlunga, Australia after 18 years. *Health Promotion International, 21*(4), 259–265.

Bergeron, H., Surel, Y., & Valluy, J. (1998). L'Advocacy Coalition Framework. Une contribution au renouvellement des études de politiques publiques? *Politix, 41*, 195–223.

Bernier, J., Rock, M., Roy, M., Bujold, R., & Potvin, L. (2006). Structuring an inter-sector research partnership: A negotiated zone. Reply to commentaries. *Sozial- und Praeventiv Medizin, 51*(6), 352–354.

Breton, E., Richard, L., Gagnon, F., Jacques, M., & Bergeron, P. (2008). Health promotion research and practice require sound policy analysis models: The case of Quebec's Tobacco Act. *Social Science & Medicine, 67*(11), 1679–1689.

Cargo, M., & Mercer, S. L. (2008). The value and challenges of participatory research: Strengthening its practice. *Annual Review of Public Health, 29*, 325–350.

Carpenter, D. (2012). Is health politics different? *Annual Review of Political Science, 15*, 287–311.

Chung, P., Grogan, C. M., & Mosley, J. E. (2012). Residents' perceptions of effective community representation in local health decision-making. *Social Science & Medicine, 74*(10), 1652–1659.

Clavier, C. (2007). *Le politique et la santé publique. Une comparaison transnationale de la territorialisation des politiques de santé publique (France, Danemark).* Doctoral dissertation. Université de Rennes 1. Rennes.

Clavier, C. (2011). La santé publique, un enjeu politique local? La politisation des politiques publiques en France et au Danemark. *Revue Internationale de Politique Comparée, 18*(4), 13–27.

Clavier, C., & de Leeuw, E. (2013). Framing public policy in health promotion: Ubiquitous yet elusive. In C. Clavier & E. de Leeuw (Eds.), *Health promotion and the policy process* (pp. 1–22). Oxford: Oxford University Press.

Clavier, C., Sénéchal, Y., Vibert, S., & Potvin, L. (2012). A theory-based model of translation in public health participatory research. *Sociology of Health and Illness, 34*(5), 791–805.

Corburn, J. (2009). *Toward the healthy city. People, places, and the politics of urban planning.* Cambridge, MA: The MIT Press.

de Leeuw, E. (1999). Healthy Cities: Urban social entrepreneurship for health. *Health Promotion International, 14*(3), 261–269.

de Leeuw, E. (2012). Healthy Cities deserve better. *The Lancet, 380*(9850), 1306–1307.

de Leeuw, E., & Clavier, C. (2011). Healthy public in all policy. *Health Promotion International, 26*(Suppl. 2), ii237–ii244.

de Leeuw, E., Keizer, M., & Hoeijmakers, M. (2013). Health policy networks: Connecting the disconnected. In C. Clavier & E. de Leeuw (Eds.), *Health promotion and the policy process* (pp. 154–173). Oxford: Oxford University Press.

de Leeuw, E., & Skovgaard, T. (2005). Utility-driven evidence for healthy cities: Problems with evidence generation and application. *Social Science & Medicine, 61*(6), 1331–1341.

Dooris, M., & Heritage, Z. (2013). Healthy Cities: Facilitating the active participation and empowerment of local people. *Journal of Urban Health, 90*(1), 74–91.

Dormois, R. (2008). Les coalitions dans l'analyse des politiques urbaines post-keynésiennes—Discussion à partir de la comparaison des politiques de régénération urbaine menées dans trois villes européennes en reconversion. *Métropoles, 4*, 46–82.

Dupuy, C., & Halpern, C. (2009). Les politiques publiques face à leurs protestataires. *Revue Française de Science Politique, 59*(4), 701–722.

Gagnon, F., Turgeon, J., & Dallaire, C. (2007). Healthy public policy. A conceptual cognitive framework. *Health Policy, 81*(1), 42–55.

Gamson, W. A. (1961). A theory of coalition formation. *American Sociological Review, 26*(3), 373–382.

Gélineau, L., Dupéré, S., Fradet, L., Landry, É., Beaulieu, M., & O'Neill, M. (2013). Une rencontre panquébécoise sur la recherche-action participative francophone en santé et services sociaux : origines, déroulement et principaux apprentissages. *Nouvelles Pratiques Sociales, 25*(2), 50–72.

Gilson, L., & Raphaely, N. (2008). The terrain of health policy analysis in low and middle income countries: A review of published literature 1994–2007. *Health Policy and Planning, 23*, 294–307.

Hall, P. A., & Taylor, R. C. R. (1996). Political science and the three new institutionalisms. *Political Studies, 44*(4), 936–957.

Hancock, T. (1988). From 'public health in the 1980's' to 'healthy Toronto 2000': The evolution of healthy public policy in Toronto. *Innovation: The European Journal of Social Science Research, 1*(2-3), 213–228.

Harpham, T. (2009). Urban health in developing countries: What do we know and where do we go? *Health & Place, 15*(1), 107–116.

Harpham, T., Burton, S., & Blue, I. (2001). Healthy city projects in developing countries: The first evaluation. *Health Promotion International, 16*(2), 111–125.

Howlett, M. (2000). Managing the "hollow state": Procedural policy instruments and modern governance. *Canadian Public Administration/Administration Publique du Canada, 43*(4), 412–431.

Jansen, M. W., de Leeuw, E., Hoeijmakers, M., & de Vries, N. S. (2012). Working at the nexus between public health policy, practice and research: Dynamics of knowledge sharing in the Netherlands. *Health Research Policy and Systems, 10*(33), 1–19.

Kübler, D. (2001). Understanding policy change with the advocacy coalition framework: An application to Swiss drug policy. *Journal of European Public Policy, 8*(4), 623–641.

Lascoumes, P., & Le Galès, P. (2007). Introduction: Understanding public policy through its instruments—From the nature of instruments to the sociology of public policy instrumentation. *Governance, 20*(1), 1–21.

Le Galès, P. (1995). Du gouvernement des villes à la gouvernance urbaine. *Revue Française de Science Politique, 45*(1), 57–95.

Le Galès, P. (2011). *Le retour des villes européennes. Sociétés urbaines, mondialisation, gouvernement et gouvernance*. Paris: Presses de Sciences Po.

Lemieux, V. (1998). *Les coalitions liens, transactions et contrôles*. Paris: Paris Presses universitaires de France.

Lipp, A., Winters, T., & de Leeuw, E. (2013). Evaluation of partnership working in cities in phase IV of the WHO Healthy Cities network. *Journal of Urban Health, 90*(Supp. 1), S37–S51.

Logan, J., & Molotch, H. (1987). *Urban fortunes: The political economy of place*. Berkeley, CA: University of California Press.

Mantoura, P., Gendron, S., & Potvin, L. (2007). Participatory research in public health: Creating innovative alliance for health. *Health and Place, 13*(2), 440–451.

Montpetit, É. (2011). Scientific credibility, disagreement, and error costs in 17 biotechnology policy subsystems. *Policy Studies Journal, 39*(3), 513–533.

Mossberger, K., & Stoker, G. (2001). The evolution of urban regime theory. The challenge of conceptualization. *Urban Affairs Review, 36*(6), 810–835.

O'Neill, M., Lemieux, V., Groleau, G., Fortin, J.-P., & Lamarche, P. A. (1997). Coalition theory as a framework for understanding and implementing intersectoral health-related interventions. *Health Promotion International, 12*(1), 79–87.

Petersen, A., & Lupton, D. (1996). *The new public health. Health and self in the age of risk.* London: Sage.

Pinson, G. (2006). Projets de ville et gouvernance urbaine. Pluralisation des espaces politiques et recomposition d'une capacité d'action collective dans les villes européennes. *Revue Française de Science Politique, 56*(4), 619–651.

Sabatier, P. A. (1991). Towards better theories of the policy process. *PS: Political Science and Politics, 24*(2), 147–156.

Sabatier, P. A. (2007a). The need for better theories. In P. A. Sabatier (Ed.), *Theories of the policy process* (pp. 3–17). Cambridge, MA: Westview Press.

Sabatier, P. A. (Ed.). (2007b). *Theories of the policy process*. Cambridge, MA: Westview Press.

Sabatier, P. A., & Jenkins-Smith, H. C. (1993). *Policy change and learning: An advocacy coalition approach*. Boulder, CO: Westview Press.

Sabatier, P. A., & Weible, C. M. (2007). The advocacy coalition framework: Innovations and clarifications. In P. A. Sabatier (Ed.), *Theories of the policy process* (2nd ed., pp. 189–220). Boulder, CO: Westview Press.

Schlager, E., & Weible, C. M. (2013). New theories of the policy process. *Policy Studies Journal, 41*(3), 389–396.

Stern, R., & Green, J. (2005). Boundary workers and the management of frustration: A case study of two Healthy City partnership. *Health Promotion International, 20*(3), 269–276.

Stern, R., & Green, J. (2008). A seat at the table? A study of community participation in two Healthy Cities Projects. *Critical Public Health, 18*(3), 391–403.

Stoker, G. (2000). Urban political science and the challenge of urban governance. In J. Pierre (Ed.), *Debating governance: Authority, steering, and democracy* (pp. 91–109). Oxford: Oxford University Press.

Stone, C. (1989). *Regime politics: Governing Atlanta 1946–1988*. Lawrence: University of Kansas Press.

Tsai Roussos, S., & Fawcett, S. B. (2000). A review of collaborative partnerships as a strategy for improving community health. *Annual Review of Public Health, 21*, 369–402.

Vibert, S. (2007). *La communauté au miroir de l'Etat. La notion de communauté dans les énoncés québécois de politiques publiques en santé*. Laval: Presses de l'Université Laval.

Weible, C. M., Pattison, A., & Sabatier, P. A. (2010). Harnessing expert-based information for learning and the sustainable management of complex socio-ecological systems. *Environmental Science & Policy, 13*(6), 522–534.

Weible, C. M., Sabatier, P. A., Jenkins-Smith, H. C., Nohrstedt, D., Henry, A. D., & deLeon, P. (2011). A quarter century of the advocacy coalition framework: An introduction to the special issue. *Policy Studies Journal, 39*(3), 349–360.

Williams, P. (2002). The competent boundary spanner. *Public Administration, 80*(1), 103–124.

World Health Organization. (2002). *Community participation in local health and sustainable development: Approaches and techniques European sustainable development and health series*. Copenhagen, Denmark: WHO Regional Office for Europe.

Chapter 15
Greening the City: The Health Evidence of Urban Nature

Mardie Townsend and Claire Henderson-Wilson

Introduction

> Everybody needs beauty as well as bread, places to play in and pray in, where nature may heal and give strength to body and soul. (John Muir, *The Yosemite*, 1912)

A profound connection exists between people and the natural environment and has done so for thousands of years, but until fairly recently this connection has been largely overlooked in the peer-reviewed literature. While John Muir (founder of the Sierra Club) and Frederick Law Olmsted (a nineteenth-century American landscape architect) identified this link, it is largely only recently that biologists, psychologists, psychiatrists and sociologists have begun to publish appreciations of the innate connection people have for the natural world and the health benefits derived from such a connection.

According to Harvard biologist Wilson (1984, 1993), there exists an innate, or genetic, emotional connection between people and living organisms, a view of the connection between people and the natural environment that led him to develop what is known as the biophilia hypothesis. According to this hypothesis, there exists 'a biologically based, inherent human need to affiliate with life and lifelike processes' (Kellert 1993, p. 42) that results in people being linked to the natural environment emotionally, cognitively, aesthetically and even spiritually (Wilson 1993). Yet modern city living, with its emphasis on increasing housing densities, may result in residents feeling a loss of connection to the natural environment. Suzuki (1997) believes that city living has created a divide between humans and nature, a

M. Townsend (✉) • C. Henderson-Wilson
School of Health and Social Development, Deakin University, 221 Burwood Highway, Burwood, VIC 3125, Australia
e-mail: mardie.townsend@deakin.edu.au; claire.henderson-wilson@deakin.edu.au

E. de Leeuw, J. Simos (eds.), *Healthy Cities*,
DOI 10.1007/978-1-4939-6694-3_15

view supported by researchers such as White and colleagues (2013, p. 920) who claim that the increasing incidence of mental health problems is 'associated with increased urbanization (Sundquist et al. 2004) and detachment from the kinds of natural environments people evolved in and are thus best adapted to (Kaplan and Kaplan 1989; Wilson 1984)'.

This chapter outlines the notion of 'greening the city', highlights the benefits flowing from such greening, describes mechanisms by which it can be achieved, and identifies barriers to be overcome.

Greening the City

The concept of urban greening, or greening the city, relates to the development of 'green resources in and around urban areas, such as woodlands, public and private urban parks and gardens, urban nature areas, street tree and square plantations, botanical gardens and cemeteries' (Konijnendijk 2014). The human need for urban nature has been evident since the Industrial Revolution, when separation from nature first became an issue with the 'massive migration of people from rural areas to urban' (Lenski et al. 1995, p. 277). Not until the nineteenth century was this need began addressed, through the development of urban public parks (Rohde and Kendle 1997). The famous American urban planner Frederick Law Olmsted stated in 1865 that natural scenery 'employs the mind without fatigue and yet exercises it; tranquilizes it and yet enlivens it; and thus, through the influence of the mind over the body, gives the effect of refreshing rest and reinvigoration to the whole system' (Kaplan 1995, p. 174).

In addition to its link to human well-being, urban greening is also important for its contribution to ecosystem services. In a report to the Victorian Environmental Assessment Council on the contribution of public land to Melbourne's liveability, West and colleagues (2009, p. 20) cited evidence showing that 'Public land resources such as the water catchments and treed parks support ecosystem functioning that enhance air and water quality' (Environment Protection Authority 2008; United Nations Department of Public Information 2002).

The Human Health Benefits of Nature in Cities

Urban green spaces have been associated not only with reduction of mental fatigue and stress, but also with an increase in physical activity and its associated benefits (Björk et al. 2008; Krenichyn 2006), a reduction in morbidity (Maas et al. 2009) and mortality (Mitchell and Popham 2008) and increased social integration and well-being (Peters et al. 2010; Seeland et al. 2009; Sullivan et al. 2004).

Benefits for Psychological Well-Being

The need for the restorative mechanisms alluded to by Olmsted (cited in Kaplan 1995) is even greater in twenty-first century urban environments than it was in the context in which Olmsted was speaking. While industrialization created detrimental living environments, modern city living is characterized by crowding, traffic, excessive stimulation, and overload of information, and can be considered very stressful (Han 2003). This was highlighted by Han, who noted that 'Feelings of stress, anxiety, and worry have been shown to have negative impacts on both physical and psychological well-being of humans' (Han 2003, p. 209).

Attention restoration theory (Kaplan 1995; Kaplan and Kaplan 1989) argues that an individual's concentrated or directed attention, defined as the type of attention used when a task requires deliberate and sustained attention (Herzog et al. 2003; James 1962), can be fatigued by overuse. Directed attention requires intellectual effort because the material attended to may not be attractive or interesting, and one maintains focus by inhibiting distractions (Herzog et al. 2003). Green spaces can alleviate mental fatigue and restore people's capacity for voluntary or directed attention in a number of ways:

- Soft fascination: features of the natural environment such as clouds, sunsets, snow patterns and the motion of leaves in the wind hold human attention effortlessly, while providing sufficient opportunity for the mind to focus on other things.
- Sense of being away: natural environments, including easily accessible natural environments within urban areas as well as environments such as the sea, mountains, lakes, streams, forests and other idyllic places, provide opportunities to gain distance from routine activities and thoughts (an escape from the pressures of daily life).
- Extent (sense of being part of a larger whole): the 'extent' of the natural environment can provide a scope or depth of experience in which one can become immersed so that the mind becomes engaged with it and gains rest from other concerns.
- Compatibility (satisfying innate needs and inclinations): as Wilson (1984, 1993) identified in his biophilia hypothesis, because people have an inherent affinity with nature, natural environments provide a setting that is compatible or well matched to human desires, which allows their attention to rest (Hartig et al. 1991).

Frances Kuo and colleagues examined the effectiveness of attention restoration theory in the inner city context. Their work focused on high-rise residents and the effects of nearby nature on a range of factors including the ability to cope with major life issues, attention deficit disorder and children's self-discipline. One of these studies (Taylor et al. 2002) examined the relationship between nearby views of 'green' nature and children's ability to concentrate, inhibit impulses and delay gratification. They found that the more 'green' a female child's view from her high-rise window was, the better able to concentrate and the more self-disciplined she was. Kuo (2001) also examined whether nearby nature affects high-rise residents' ability to cope with poverty and life issues. She found that residents with 'green' surroundings were able to pay attention more effectively and coped more effectively with major life issues than their counterparts from 'barren' environments. Other studies (Taylor et al. 2001; Kuo and Taylor 2004; Taylor and Kuo 2009) tested whether attention restoration theory could help children to deal with attention deficit disorder (ADD). In the 2001 study, children's attention levels were tested in a range of play settings; green settings were found to be most effective in enhancing attention. The authors concluded that the more 'green' a child's play setting, the less severe their ADD symptoms appeared.

Other research by Berman et al. (2008) has indicated that simple and brief interactions with nature can increase adults' cognitive control. Nielsen (2007), Maas et al. (2009), Grahn and Stigsdotter (2010) and Roe et al. (2013) have produced further evidence of the restorative qualities of natural environments. At a time when the gap between rich and poor is widening in many developed societies, when urban densification is increasing (in many cases threatening existing green areas in cities) and mental illness is on the rise, the evidence from Roe and colleagues' study of the relationship between green space and stress in deprived neighbourhoods of Scotland is timely. Their findings that 'higher levels of green space in residential neighbourhoods … are linked with lower perceived stress and a steeper (healthier) diurnal cortisol decline' (Roe et al. 2013, p. 4087) highlight the importance of greening the city.

Benefits for Physical Well-Being

In addition to their benefits for psychological well-being, urban green spaces have been shown to have measurable benefits for physical health and well-being. There is evidence that obesity is on the increase. The data from Australia (Table 15.1) is indicative of that trend.

One of the contributing factors to the rise in obesity is increasingly sedentary lifestyles, which see almost two-thirds of Australians fail to meet national physical activity guidelines. Similar patterns are evident elsewhere in the world.

In the context of rising obesity rates (Fig. 15.1) and declining levels of physical activity (Fig. 15.2), Australian research by Giles-Corti and colleagues (2005) finds that access to large, high quality public open spaces are associated with higher levels of walking, while evidence from a Swedish study by Björk et al. (2008) that shows clear links between access to high quality natural environments within 300 m of the home and time spent by residents on physical activity confirms the importance of greening the city. Other research has demonstrated links between green urban environments and health, in terms of morbidity and mortality. Research undertaken by Maas et al. (2009) drew on the medical records of 195 general practitioners serving almost 350,000 people in the Netherlands to explore links between proximity to green space and the annual prevalence of seven disease clusters (cardiovascular, musculoskeletal, mental, respiratory, neurological, digestive and mis-

Table 15.1 Overweight and obesity in Australia (AIHW 2013)

Overweight and obesity by the numbers	
3 in 5 Australian adults are overweight or obese (based on *BMI*). That's over 12 million people!	
5 % More *adults* are overweight or obese than in 1995	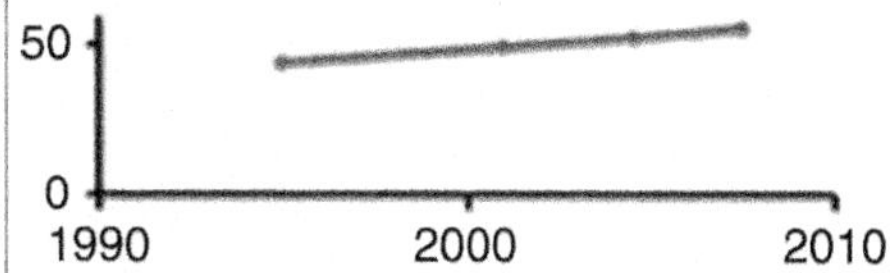
1 in 4 Australian *children* are overweight or obese	
Over 30 % More people living in *outer regional and remote areas* are obese than people living in major cities	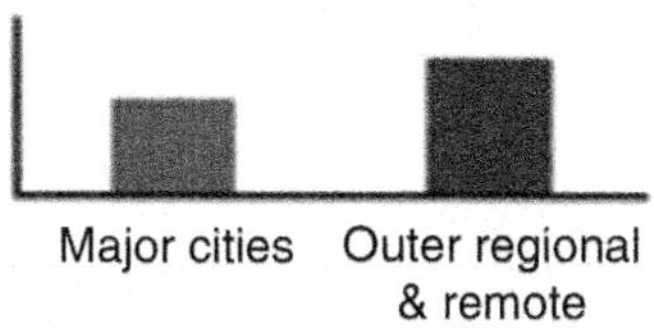
Third place Overweight and obesity is only beaten by smoking and high blood pressure as a contributor to *burden of disease*	

http://www.aihw.gov.au/overweight-and-obesity/

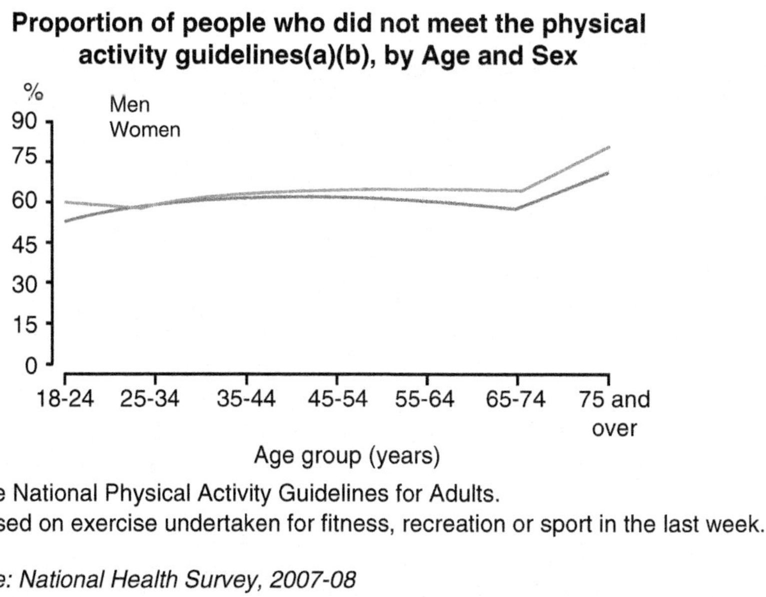

Fig. 15.1 Failure to meet physical activity guidelines in Australia (ABS, 2011)

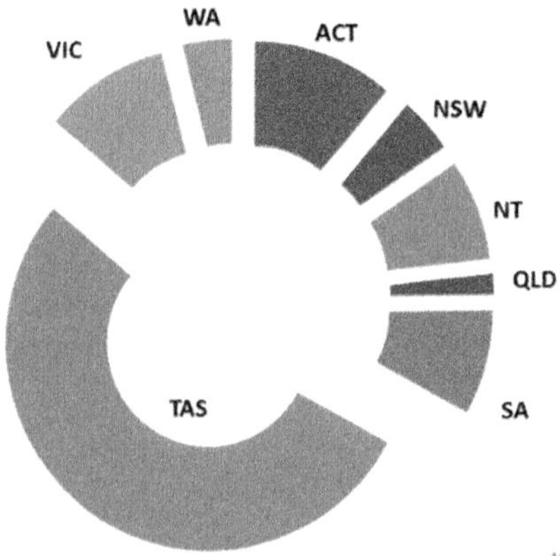

State	Gardens	2013 Population
NSW	121	7,381,100
VIC	216	5,713,000
QLD	28	4,638,100
WA	37	2,497,500
SA	51	1,667,500
TAS	103	512,900
ACT	16	381,700
NT	7	237,800

Fig. 15.2 Number of gardens per head of population in Australia: garden density by state

cellaneous conditions). The study (which controlled for demographic and socioeconomic characteristics) found that the prevalence of 15 of the 24 specific conditions within the seven categories was lower in residential areas within a kilometre of green space (Maas et al. 2009). UK research undertaken by Mitchell and Popham (2008) matched the records of more than 350,000 deaths with information about the level of greenness in the relevant residential areas, and found that exposure to green environments could reduce health inequalities based on income deprivation.

Benefits for Social Well-Being

In an era when social capital and social connectedness is perceived by many as declining, research seems to support the claim that greening the city can contribute to increased social connectedness. Sullivan et al. (2004) studied 59 outdoor spaces within a public housing development in Chicago to explore whether the level of vegetation within the spaces related to the level of social activity occurring there. They found 'systematic evidence that trees and grass help create vital neighbourhood spaces—spaces that not only bring neighbours together but that also support social interaction among them' (p. 698). Researchers from the Netherlands reported that parks in urban areas provide 'sites where different ethnic groups mingle and where informal and cursory interactions can stimulate social cohesion' (Peters et al. 2010, p. 93). These findings echo the results of Swiss research (Seeland et al. 2009) that urban green spaces play an important role in facilitating increased interaction between different social groups and improving social inclusion.

Other Benefits of Nature in Cities

A review of existing literature shows evidence of a range of environmental benefits from urban greening, especially related to, but not limited to, urban forestry, 'the art, science and technology of managing trees and forest resources in and around urban community ecosystems for the physiological, sociological, economic, and aesthetic benefits trees provide society' (Konijnendijk et al. 2006, p. 93).

Environmental benefits of urban greening include mitigation of the Urban Heat Island effect (UHI), protection from wind, reduction of rainwater run-off and wash-aways, increased biodiversity, noise reduction, improved water and air quality, and the storage of carbon dioxide, all of which provide benefits to the city and the humans living in it, economically as well as physically.

Mitigation of the Urban Heat Island Effect

The Urban Heat Island (UHI) effect is the name applied to the increase in temperatures in built-up areas, particularly overnight, in comparison with surrounding less built-up areas (Loughnan et al. 2010). This may exacerbate already high temperatures in urban areas, increasing morbidity and mortality rates. Evidence from the United States and Australia indicates that excessive heat exposure has contributed to more deaths in these countries over the past 100 years than hurricanes, lightning, tornadoes, floods and earthquakes combined (EPA 2011b; McKeon 2011). Increased daytime surface temperatures, reduced night-time cooling, and higher air pollution levels associated with UHI can affect human health by contributing to

- General discomfort
- Respiratory difficulties
- Heat cramps and exhaustion
- Non-fatal heat stroke
- Heat-related mortality (EPA 2011a)

Other negative effects of UHI include increased peak energy demand in summer and an associated increase in greenhouse gas emissions and air pollution, as well as increased air conditioning costs. There is evidence to suggest that, because of climate changes, the number of excessive heat events is increasing worldwide. Research in Indian by Hardin and Jensen (2007) showed that increasing leaf cover through urban forestry could have a substantial effect on surface temperatures, while in Arizona Silva and colleagues (2010) found that the use of vegetation is an effective strategy in mitigating UHI. Melbourne-based research by Loughnan et al. (2010) supports these findings, with evidence of inner city areas and the less tree-covered western suburbs experiencing higher temperatures than the more leafy eastern or southern suburbs.

Improving Air Quality

According to an Australian study by Denison et al. (2000, p. 4), in general 'air quality in Australian cities is much better than that observed in many American and European cities' but 'particle pollution is a concern during the autumn and winter' (p. 5). Another Australian study (Peach 1997, p. 7) notes that overseas research has linked the levels of particulates, sulphur dioxide and ozone with increased morbidity and mortality. Research in Santiago by Escobedo and colleagues (2008) has found that urban forestry may be effective in improving air quality, particularly in relation to removing atmospheric particulates. Another Australian study by Coutts et al. (2007) found that vegetation could limit CO_2 flux in the atmosphere, which results from urban traffic

Noise Reduction

'Noise, defined as any unwanted or undesirable sound, is a known environmental pollutant and health hazard' (Shankar et al. 2001, p. 244). The subject of adverse health effects from noise is complex. Sound is of major importance for human well-being, but sounds perceived as noise can have adverse effects on health, including noise-induced hearing impairment, interference with speech communication, disturbance of rest and sleep, psycho-physiological, mental-health and performance effects and effects on residential behaviour and annoyance, as well as interference with intended activities (Berglund et al. 1999). Studies show the high utility trees have in blocking noise, with Çobani and Aleksi (2012) reporting that vegetation barriers are not only effective in controlling noise pollution but also have social and cultural benefits through the reduction of impacts on sleep, annoyance and speech interference. Other research has demonstrated that roadside green belts in urban areas have benefits not only in reducing noise levels but in mitigating air pollution (Islam et al. 2012).

Biodiversity

The loss of biodiversity worldwide is of great concern as biodiversity plays a vital role in long-term ecosystem functioning (Groombridge and Jenkins 2002). Its effect on human health needs to be further investigated (Tzoulas and James 2003), but at the very least biodiversity loss has human health and well-being implications related to

- The incidence of infectious diseases, the potential for development of new treatments and medicines, and the quality of essential life support systems (Chivian and Bernstein 2004).
- Rates of exposure to neurotoxic effects of pesticides, etc. (Reid 1995).
- Sense of connection to place, personal identity and psychological well-being (Horwitz et al. 2001).

- Economic and social well-being (Speldewinde 2006).

It is also likely that the biodiversity benefits of urban forests and green spaces are influenced not only by the quantity but also by the quality of the vegetation. This was particularly evident in relation to human well-being as shown by Fuller and colleagues, who explored the psychological benefits people obtained from spending time in green spaces. Participants reported greater well-being benefits in parks that contained a greater diversity of plant species and habitat types, 'an effect that was not simply attributable to the area of the green space involved' (Fuller et al. 2007, p. 1). Although cities may not have the ability to preserve large areas of natural habitat, urban forests and green spaces have been shown to harbour an important percentage of species, even endangered ones, and species of high conservation value. The importance of urban forests and green spaces as reserves of biodiversity will continue to grow as urbanization increases and non-urban repositories of biodiversity decrease (Alvey 2006).

Mechanisms for Greening the City

The greening of cities and the provision of nature access occur in various ways, including through the provision of urban parks, community gardens and rooftop gardens, and the proximity to bodies of water. As Nankervis states, 'open space is one of the characteristics which define Australian cities and make them particularly liveable. It would not be too heroic a claim to say that urban open space is an integral part of Australian culture' (1998, p. 162).

A study by West and colleagues (2009) of public open spaces green spaces, and the contribution of public land to Melbourne's liveability, used the community well-being framework developed by Wiseman and colleagues (2006). The framework comprises five domains which, taken together, describe community well-being:

- Healthy, safe and inclusive communities.
- Dynamic resilient local economies.
- Sustainable built and natural environments.
- Culturally rich and vibrant communities.
- Democratic and engaged communities.

Table 15.2, taken from the report by West et al. (2009, p. 17), highlights the importance of urban public green spaces in achieving those well-being outcomes.

Urban Parks

One of the main forms of green space in cities is the urban park. According to Hamilton-Smith and Mercer (1991), urban parks appear to have existed ever since people gathered in cities, suggesting that they are considered vital for inner city

Table 15.2 Liveability and Melbourne's public land network

	Natural environment				Built environment	
	Parks and nature reserves, foreshores	Creeks and rivers utility easements Railway reserves	Sports fields and complexes Playgrounds	Roads and footpaths roadsides and road-related areas	Galleries and libraries school halls Health centres	Municipal offices Utility infrastructure police stations hospitals
Healthy, safe and inclusive communities	Open-air activities	Active and passive recreation	Active and passive sport	Communication Social interaction	Cultural and educational activities	Civic governance
	Physical and mental reinvigoration and preventative health measures	Physical and mental reinvigoration	Adult- and child-focused recreation	Active recreation, e.g. jogging, skating, cycling	Health and well-being	Water, sewerage, etc.
			Preventative health measures			Law and order
						Medical services
Dynamic, resilient local economies	Conservation-related employment	Communication and transport infrastructure	Recreation-related employment	Infrastructure Shopping strips	Community use venues and notice-boards	Local and regional services
				Al-fresco dining		
				Footpath trading		
Sustainable built and natural environments	Conservation of indigenous flora	Remnant vegetation and habitat	Water conservation	Urban vegetation	Heritage buildings	Recycling Transport
	Habitat for fauna	Biodiversity corridors		Community gardens		
	Landscape	Landscape				

Culturally rich and vibrant communities	Egalitarian and cross-cultural interactions	Revegetation	Sporting clubs	Street festivals	Events, exhibitions, social functions	Civic engagement
	Spatial autonomy (e.g. for teenagers)			Street art and sculpture		
	Educational areas					
	Sense of place					
Democratic and engaged communities	'Friends' groups	'Friends' groups	Committees of Management	Processions Rallies and protests	Cultural societies	Local government activities
	Conservation volunteers	Conservation volunteers and advocacy groups			Parents' groups Ethnic-based social clubs	

residents. In the nineteenth century, urban parks were considered to provide cities with 'green lungs', and more recently they have been considered 'optional amenities' and a contrast to the inner city landscape of concrete and glass (Kaplan and Kaplan 1989; McIntyre et al. 1991; Maller et al. 2008).

Today, life in the city is hectic and residents may seek solace by visiting urban parks. These may be the best access residents have to nature if they do not have their own garden (Maller et al. 2008). A number of researchers believe that if urban parks are within a 3- or 5-min walk of a resident's home, they are more likely to be visited (e.g., Giles-Corti et al. 2005; Kaplan and Kaplan 1989; Ward Thompson 1998, 2002, 2002; Ward Thompson et al. 2005).

Urban parks provide individuals with a range of positive health benefits with flow-on effects for the inner city community. For example, research conducted by SUPER Group (2001) on the economic benefits provided by use of Sydney's urban parks indicated that

- There may be decreased health care costs resulting from physical activity performed in urban parks.
- There may be saved capital investment as a result of using established facilities available in urban parks (e.g. barbeque facilities, seating, sporting equipment).
- Using urban parks as a source for rehabilitating minor offenders (through involving them in maintenance programmes) may reduce or avert crime.

Community Gardens

> Plants are so effective in eliciting human responses because their environment contrasts sharply with the social world in which we move. The garden is a safe place, a benevolent setting where everyone is welcome. Plants are non-judgemental, nonthreatening, and non-discriminating. (Lewis 1996, p. 64)

Lewis' statement highlights the important role gardens play in people's lives. Community gardens may bring together people from all ethnic backgrounds (Bartolomei et al. 2003; Saldivar-Tanaka and Krasny 2004; Shinew et al. 2004) and socio-economic groups to work on maintenance, alleviating social exclusion.

There is a range of health benefits that may be derived from participating in gardening, particularly inner city community gardening, ranging from the benefits of good dietary consumption and physical activity through to restorative and mental health benefits. In particular, a strengthening of community and social capital may be derived from participation in community gardening. Results of studies conducted by Armstrong (2000), Firth et al. (2011), Francis (1987), Glover et al. (2005) and Kingsley and Townsend (2006) indicate that participation in community gardens facilitates social networks, friendships and community development. Given these benefits, there is an increasing interest in the role of community gardens as a mechanism to strengthen social infrastructure (Bartolomei et al. 2003). This is shown by the number of community gardens that have been established throughout Australia (Australian City Farms and Community Gardens Network 2014).

Rooftop Gardens

As city councils strive to conserve energy, manage storm water runoff and reduce ozone levels and air pollution to comply with sustainability regulations, they are studying rooftop gardens as a way to satisfy these objectives (Knepper 2000). However, urban designers are recognizing the value of plants not only as important for balancing the polluting effects of cities but also for the aesthetic and emotional benefits they may provide high-rise residents (Ong 2003; Wong et al. 2005; Yuen and Wong 2005).

Rooftop greening is well established in Switzerland, Germany and Singapore, and is increasingly being adopted in Australia and the United States. It has been reported that in 2008, green roof projects in the States grew by an estimated 45 %, with 3.18 million square feet of new rooftops added (Litichevskaya 2010). One example is Brooklyn Grange, on the rooftop of the Brooklyn Navy Yard. At 65,000 square feet it is reported to be the largest rooftop farm in the world, totalling 2.5 acres and producing over 50,000 lbs (roughly 22,700 kg) of organically grown vegetables each year (Brooklyn Grange 2014).

There are economic, environmental, aesthetic and mental health benefits resulting from planting rooftop gardens on high-rise buildings (The Cardinal Group 2004; Knepper 2000; Leng and Sia 2001; Ong 2003; Wong et al. 2005). For example, a study by Leng and Sia (2001) of 333 high-rise residents in Singapore indicated that the residents perceived rooftop gardens to provide an opportunity for relaxation and social interaction. More than 60 % of the respondents perceived the benefit of rooftop gardens in providing a setting for them and their families to participate in physical activities (Leng and Sia 2001).

Bodies of Water

Water is a dominant feature in many environments and a highly prized element in the inner city landscape (Kaplan 1977; Pitt 1989). Throughout the world, building houses near bodies of water is growing in popularity. For instance, in the Netherlands developers are well aware of the value of water features, given the number of plans that include buildings located near water bodies (Luttik 2000). Likewise, Chicago, Vancouver, Paris and Seattle are examples of cities that have been built creatively around waterways (Carr et al. 1992). Marina Bay in Singapore is one example of an urban development built around a large body of water. This 360 ha development is planned to extend Singapore's downtown district and further support continuing growth as a major business and financial hub in Asia (Marina Bay Singapore 2014). Part of the Marina Bay precinct includes green elements: tree-lined pedestrian malls, city rooms, sky terraces and roof gardens. 'Gardens by the Bay', free outdoor gardens that are themed around plants important to Singapore's history and culture, are a key feature of the precinct. They were based on principles of environmental sustainability and a particular effort was made to plan and design for sustainable cycles of energy and water in the Bay South Garden, one element of the project (Gardens by the Bay 2014).

Living near water can enhance people's health and well-being. Research in England by Wheeler et al. (2012) suggests that living close to water is positively associated with good health. After adjusting for age, sex, greenspace density and socio-economic confounders, they found a gradient of increasing self-reported good health related to proximity to the coast. They also found that, like other analyses of green space accessibility, the health benefits of living near the coast were greater among more socio-economically deprived communities.

Barriers to Greening the City

Clearly, many strategies exist for increasing the greening of urban areas, amid a growing pool of evidence to support the claim that urban greening has positive benefits for human health and well-being and environmental improvement, as we have discussed. Other evidence indicates that greening has economic benefits to individual landholders, increasing the value of urban properties. For example, a study by ECOTEC Research and Consultancy reports that in urban areas of the United Kingdom, 'houses near parks were, on average, 8 % more expensive than comparable houses further away, with the highest increase seen at 36 %', and noted community-level economic benefits from urban greening:

> Green Infrastructure investment can act to encourage and attract high value industry, entrepreneurs and workers to a locality and region through the maintenance and creation of high quality, landscape sensitive, environmentally friendly living and working environments. (2008, p. 22)

Yet there is still significant resistance to the notion of urban greening. In part this arises from the norms and values of the community at large. Such values-based resistance is implied by Svedin (2011, p. 33) in the following statement:

> The basis for a possible stronger green urban direction is the strengthening of the connected values at large that would support such a development. Without such common and public support, it will be difficult to mobilise the resources needed and also to face some of the risks that the larger environmental challenges will most probably provide (e.g. a stress on a low carbon agenda due to climate change, water availability, etc.).

There appear to be other factors involved in resistance as well, including lack of recognition of the benefits of urban greening, perceptions that economic disadvantage (either individual or collective) might flow from such greening, limitations (especially in compact cities such as Hong Kong and Singapore) on the space available for new parks and urban green spaces, and population growth which results in pressures that threaten existing parks and green spaces.

Tian et al. (2012), in their article 'Challenges and Strategies for Greening the Compact City of Hong Kong', draw attention to physical, social and institutional constraints undermining urban greening. In terms of physical constraints, they note 'extremely limited availability of urban sites for greening', 'numerous cables, pipes, large chunks of concrete, and rocks in the topsoil as a result of underground facilities', and locality-specific problems associated with soil composition, terrain and climatic factors (p. 104). Williams and colleagues (2010) also note the importance of local conditions affecting the capacity of cities to adopt urban greening strategies used elsewhere, and suggest that specific research relevant to local conditions needs to be undertaken. According to Tian and colleagues (p. 104), social constraints such as residents' lack of awareness and understanding of the benefits of urban greening together with the 'overriding aspiration of the working-class population for economic survival and advancement … weakens urban residents' awareness of and desire for urban greening' and are key factors undermining urban greening. Conversely, research in Sweden (Ernstson et al. 2008) and in Australia (Moore et al. 2006) demonstrates that knowledge of the environment and its health benefits results in a commitment to its preservation, and may thus play a key role in fostering urban greening. Research by Mell and colleagues (2013) involving a survey of over 500 residents in Manchester, which found that almost three-quarters of those surveyed were willing to pay extra local council tax to provide for green infrastructure, suggests an awareness within the community of the benefits this might provide, and a commitment to its maintenance.

Bureaucratic factors may serve to undermine the execution of this commitment. Tian and colleagues, in their Hong Kong-based research, draw attention to institutional constraints such as the absence of a 'city-wide long-term landscape plan to describe the conceptual foundation for greening'. This lack is further compounded by 'the vague division of management responsibility for UGS between different departments', undermining the effectiveness of greening projects (p. 104).

Conclusion

The overwhelming evidence of the benefits of the greening of cities highlights that such places not only provide the 'beauty' or quality of life referred to by Muir, quoted above, but also the 'bread'—the physical and economic conditions that sustain our life both quantitatively and qualitatively.

In 1901 (p.1), Muir wrote:

> Thousands of tired, nerve-shaken, over-civilized people are beginning to find out that going to the mountains is going home; that wildness is a necessity; and that mountain parks and reservations are useful not only as fountains of timber and irrigating rivers, but as fountains of life.

Muir was not referring to twenty-first century, technology-dominated urban life but to the need of people at the beginning of the twentieth century to access nature in order to regain balance in their lives. How much more do we, who live in this fast-paced world, need to find ways not just to allow people to go out into nature, but to bring nature into our urban areas, so that all may access it, no matter who they are or what their resources. We all, as individual citizens, as communities, as policy and decision makers and as practitioners, need to work to remove the barriers to urban greening so that all may experience both beauty and bread.

References

Alvey, A. A. (2006). Promoting and preserving biodiversity in the urban forest. *Urban Forestry and Urban Greening, 5*, 195–201.

Armstrong, D. (2000). A survey of community gardens in upstate New York: Implications for health promotion and community development. *Health and Place, 6*, 319–327.

Australian Bureau of Statistics (ABS). (2011). *Physical activity in Australia: A snapshot, 2007–08*. Cat. No. 4835.0.55.001. Canberra: Commonwealth of Australia.

Australian City Farms and Community Gardens Network. (2014). Retrieved February 12, 2014, from http://directory.communitygarden.org.au/data.

Australian Institute of Health and Welfare (AIHW). (2013). *Overweight and obesity*. Retrieved February 13, 2014, from http://www.aihw.gov.au/overweight-and-obesity/.

Bartolomei, L., Corkery, L., Judd, B., & Thompson, S. (2003). *A Bountiful Harvest: Community Gardens and neighbourhood renewal in Waterloo*. Sydney: The University of New South Wales.

Berglund, B., Lindvall, T., & Schwela, D. H. (1999). *Guidelines for community noise*. Geneva: World Health Organization.

Berman, M. G., Jonides, J., & Kaplan, S. (2008). The cognitive benefits of interacting with nature. *Psychological Science, 19*(12), 1207–1212.

Björk, J., Albin, M., Grahn, P., Jacobsson, H., Ardö, J., Wadbro, J., … Skärbäck E. (2008). Recreational values of the natural environment in relation to neighbourhood satisfaction, physical activity, obesity and wellbeing. *Journal of Epidemiology and Community Health, 62*, e2.

Brooklyn Grange. (2014). *Farms*. Retrieved February 12, 2014, from http://brooklyngrangefarm.com/farms/.

Cardinal Group. (2004). *Public benefits of green roofs—Green roofs for healthy cities*. Retrieved February 11, 2004, from http://peck.ca/grhcc/public.htm.

Carr, S., Francis, M., Rivlin, L. G., & Stone, A. M. (1992). *Public space*. New York: Cambridge University Press.

Chivian, E., & Bernstein, A. F. (2004). Embedded in nature: Human health and biodiversity. *Environmental Health Perspectives, 112*(1), A12–A13.

Çobani, E., & Aleksi, L. (2012). Application of tree barriers as mitigation measures for noise pollution—Road '29 Nëntori', Tirana, Albania. *Journal of International Environmental Application and Science, 7*(2), 427–431.

Coutts, A. M., Beringer, J., & Tapper, N. J. (2007). Characteristics influencing the variability of urban CO_2 fluxes in Melbourne, Australia. *Atmospheric Environment, 41*, 51–62.

Denison, L., Simpson, R., Petroeschevsky, A., Thalib, L., Rutherford, S., Williams, G,… Streeton, J. (2000). *Melbourne mortality study: Effects of ambient air pollution on daily mortality in Melbourne 1991–1996*. Victoria: Environment Protection Authority.

Environment Protection Authority. (2008). *Melbourne mortality study: Effects of ambient air pollution on daily mortality in Melbourne 1991–1996*. Victoria: Environment Protection Authority.

EPA. (2011a). *Endangerment and cause or contribute findings for Greenhouse gases under section 202(a) of the Clean Air Act*. United States Environmental Protection Agency. Retrieved April 20, 2011, from http://www.epa.gov/climatechange/endangerment/comments/volume5.html.

EPA. (2011b). *Heat Island impacts*. United States Environmental Protection Agency. Retrieved April 20, 2011, from http://www.epa.gov/heatisland/impacts/index.htm.

Ernstson, H., Sörlin, S., & Elmqvist, T. (2008). Social movements and ecosystem services—The role of social network structure in protecting and managing urban green areas in stockholm. *Ecology and Society, 13*(2), 39ff.

Escobedo, F. J., Wagner, J. E., Nowak, D. J., De la Maza, C. L., Rodriguez, M., & Crane, D. E. (2008). Analyzing the cost effectiveness of Santiago. Chile's policy of using urban forests to improve air quality. *Journal of Environmental Management, 86*, 148–157.

Firth, C., Maye, D., & Pearson, D. (2011). Developing 'community' in community gardens. *Local Environment, 16*(6), 555–568.

Francis, M. (1987). Some different meanings attached to a city park and community gardens. *Landscape Journal, 6*(2), 101–112.

Fuller, R., Irvine, K., Devine-Wright, P., Warren, P., & Gaston, K. (2007). Psychological benefits of greenspace increase with biodiversity. *Biological Letters, 3*, 390–394.

Gardens by the Bay. (2014). *About the gardens*. Retrieved February 12, 2014, from https://www.gardensbythebay.com.sg.

Giles-Corti, B., Broomhall, M. H., Knuiman, M., Collins, C., Douglas, K., Ng, K., … & Donovan, R. J. (2005). Increasing walking: How important is distance to, attractiveness, and size of public open space? *American Journal of Preventive Medicine, 28*(2), 169–176.

Glover, T. D., Parry, D. C., & Shinew, K. J. (2005). Building relationships, accessing resources: Mobilizing social capital in community garden contexts. *Journal of Leisure Research, 37*(4), 450–474.

Grahn, P., & Stigsdotter, U. K. (2010). The relation between perceived sensory dimensions of urban green space and stress restoration. *Landscape and Urban Planning, 94*, 264–275.

Groombridge, B., & Jenkins, M. (2002). *World Atlas of biodiversity: Earth's living resources in the 21st century*. Berkley, CA: University of California Press.

Hamilton-Smith, E., & Mercer, D. (1991). *Urban parks and their visitors*. Melbourne: The Parks Division, Melbourne and Metropolitan Board of Works.

Han, K.-T. (2003). A reliable and valid self-rating measure of the restorative quality of natural environments. *Landscape and Urban Planning, 64*, 209–232.

Hardin, P. J., & Jensen, R. R. (2007). The effect of urban leaf area on summertime urban surface kinetic temperatures: A Terre Haute case study. *Urban Forestry and Urban Greening, 6*, 63–72.

Hartig, T., Mang, M., & Evans, G. W. (1991). Restorative effects of natural environment experiences. *Environment and Behavior, 23*, 3–26.

Herzog, T., Maguire, C., & Nebel, M. (2003). Assessing the restorative components of environments. *Journal of Environmental Psychology, 23*, 159–170.

Horwitz, P., Lindsay, M., & O'Connor, M. (2001). Biodiversity, endemism, sense of place, and public health: Interrelationships for Australian Inland aquatic systems. *Ecosystem Health, 7*(4), 253–265.

Islam, M. N., Rahman, K.-S., Bahar, M. M., Habib, M. A., Ando, K., & Hattori, N. (2012). Pollution attenuation by roadside greenbelt in and around urban areas. *Urban Forestry and Urban Greening, 11*, 460–464.

James, W. (1962). *Psychology: The briefer course*. New York, NY: Collier Books.

Kaplan, R. (1977). Down by the riverside: Informational factors in waterscape preference. In *Paper presented at the Symposium on River Recreation Management and Research, 24–27 January 1977*. St Paul, Minnesota: USDA Forest Service, North Central Forest Experiment Station.

Kaplan, S. (1995). The restorative benefits of nature: Toward an integrative framework. *Journal of Environmental Psychology, 15*, 169–182.

Kaplan, R., & Kaplan, S. (1989). *The experience of nature: A psychological perspective*. New York: Cambridge University Press.

Kellert, S. R. (1993). The biological basis for human values of nature. In S. R. Kellert & E. O. Wilson (Eds.), *The biophilia hypothesis* (pp. 42–69). Washington, DC: Island Press.

Kingsley, J., & Townsend, M. (2006). 'Dig In' to social capital: Community gardens as mechanisms for growing urban social connectedness. *Urban Policy and Research, 24*(4), 525–537.

Knepper, C. A. (2000). Gardens in the sky. *Journal of Property Management, 65*(2), 36–40.

Konijnendijk, C. C. (2014). *Aims and scope*. Urban Forestry and Urban Greening. Retrieved from http://www.journals.elsevier.com/urban-forestry-and-urban-greening.
Konijnendijk, C. C., Ricard, R. M., Kenney, A., & Randrup, T. B. (2006). Defining urban forestry—A comparative perspective of North America and Europe. *Urban Forestry and Urban Greening, 4*, 93–103.
Krenichyn, K. (2006). 'The only place to go and be in the city': Women talk about exercise, being outdoors, and the meanings of a large urban park. *Health and Place, 12*, 631–643.
Kuo, F. E. (2001). Coping with poverty: Impacts of environment and attention in the inner city. *Environment and Behavior, 33*(1), 5–34.
Kuo, F. E., & Taylor, A. F. (2004). A potential natural treatment for attention-deficit/hyperactivity disorder: Evidence from a national study. *American Journal of Public Health, 94*(9), 1580.
Leng, O. C., & Sia, A. (Eds.). (2001). *Handbook on skyrise greening in Singapore* (3rd ed.). Singapore: National Parks Board and National University of Singapore.
Lenski, G., Nolan, P., & Lenski, J. (1995). *Human societies: An introduction to macrosociology* (Internationalth ed.). New York: McGraw-Hill.
Lewis, C. A. (1996). *Green nature/human nature—The meaning of plants in our lives*. University of Illinois Press.
Litichevskaya, J. (2010). Reviving the world wonder: Why rooftop gardens should cover urban landscapes. *Rutgers Computer and Technology Law Journal, 37*(1), 58–92.
Loughnan, M. E., Nicholls, N., & Tapper, N. J. (2010). When the heat is on: Threshold temperatures for AMI admissions to hospital in Melbourne Australia. *Applied Geography, 30*, 63–69.
Luttik, J. (2000). The value of trees, water and open space as reflected by house prices in the Netherlands. *Landscape and Urban Planning, 48*, 161–167.
Maas, J., Verheij, R. A., de Vries, S., Spreeuwenberg, P., Schellevis, F. G., & Groenewegen, P. P. (2009). Morbidity is related to a green living environment. *Journal of Epidemiology and Community Health, 63*(12), 967–973.
Maller, C., Townsend, M., St Leger, L., Henderson-Wilson, C., Pryor, A., & Prosser, L. (2008). *Healthy parks, healthy people: The health benefits of contact with nature in a park context: A review of relevant literature*. Burwood: Deakin University.
Marina Bay Singapore. (2014). *Greenery, parks and open spaces*. Retrieved February 12, 2014, from http://www.marina-bay.sg/.
McIntyre, N., Cuskelly, G., & Auld, C. (1991). The benefits of urban parks: A market segmentation approach. *Australian Parks and Recreation, 27*, 11–18.
McKeon, G. (2011), *Living in a variable climate*. Australian Government, Department of Sustainability, Environment, Water, Population and Communities. Retrieved April 20, 2011, from http://www.environment.gov.au/soe/2006/publications/integrative/climate/box-2.html.
Mell, I. C., Henneberry, J., Hehl-Lange, S., & Keskin, B. (2013). Promoting urban greening: Valuing the development of green infrastructure investments in the urban core of Manchester, UK. *Urban Forestry and Urban Greening, 12*, 296–306.
Mitchell, R., & Popham, F. (2008). Effect of exposure to natural environment on health inequalities: An observational population study. *The Lancet, 372*, 1655–1660.
Moore, M., Townsend, M., & Oldroyd, J. (2006). Linking human and ecosystem health: The benefits of community involvement in conservation groups. *EcoHealth, 3*(4), 255–261.
Muir, J. (1901). Our National Parks. New York: Houghton, Mifflin and Company.
Muir, J. (1912). *The Yosemite*. New York: The Century Co.
Nankervis, M. (1998). Our urban parks: Suitable pieces of real estate? *Journal of Australian Studies, 57*, 162–163.
Nielsen, T. S. (2007). Do green areas affect health? Results from a Danish survey on the use of green areas and health indicators. *Health and Place, 13*, 839–850.
Ong, B. L. (2003). Green plot ratio: an ecological measure for architecture and urban planning. *Landscape and Urban Planning, 63*, 197–211.
Peach, H. G. (1997). *Air quality and human health*. Australia: State of the Environment Technical Paper Series (The Atmosphere), Canberra.

Peters, K., Elands, B., & Buijs, A. (2010). Social interactions in urban parks: Stimulating social cohesion? *Urban Forestry and Urban Greening, 9*, 93–100

Pitt, D. G. (1989). The attractiveness and use of aquatic environments as outdoor recreation places. In I. Altman & E. H. Zube (Eds.), *Public places and spaces* (pp. 217–252). New York: Plenum Press.

Reid, W. V. (1995). Biodiversity and health: Prescription for progress. *Environment: Science and Policy for Sustainable Development, 37*(6), 12–39.

Roe, J. J., Ward Thompson, C., Aspinall, P. A., Brewer, M. J., Duff, E. I., Miller, D., … & Clow, A. (2013). Green space and stress: Evidence from cortisol measures in deprived urban communities. *International Journal of Environmental Research and Public Health, 10*, 4086–4103

Rohde, C. L. E., & Kendle, A. D. (1997). Nature for people. In: A. D. Kendle, & S. Forbes (Eds.), *Urban nature conservation—Landscape management in the urban countryside*, (pp. 319–335), E. and F.N. Spon, London.

Saldivar-Tanaka, L., & Krasny, M. E. (2004). Culturing community development, neighbourhood open space, and civic agriculture: The case of Latino community gardens in New York City. *Agriculture and Human Values, 21*, 399–412.

Seeland, K., Dübendorfer, S., & Hansmann, R. (2009). Making friends in Zurich's urban forests and parks: The role of public green space for social inclusion of youths from different cultures. *Forest Policy and Economics, 11*, 10–17.

Shankar, N., Malhotra, K. L., Ahuja, S., & Op, T. (2001). Noise pollution: A study of noise levels in the operation theatres of a general hospital during various surgical procedures. *Journal of the Indian Medical Association, 99*, 244, 246–247.

Shinew, K. J., Glover, T. D., & Parry, D. C. (2004). Leisure spaces as potential sites for interracial interaction: Community gardens in urban areas. *Journal of Leisure Research, 36*(3), 336–355.

Silva, H. R., Phelan, P. E., & Golden, J. S. (2010). Modeling effects of urban heat island mitigation strategies on heat-related morbidity: A case study for Phoenix, Arizona, USA. *International Journal of Biometeorology, 54*, 13–22.

Speldewinde, P. (2006). Environmental degradation and human health in Western Australia. *Epidemiology, 17*(6 Supp.), 435.

Sullivan, W. C., Kuo, F. E., & DePooter, S. F. (2004). The fruit of urban nature—Vital neighborhood spaces. *Environment and Behavior, 36*, 678–700.

Sundquist, K., Frank, G., & Sundquist, J. (2004). Urbanisation and incidence of psychosis and depression. *British Journal of Psychiatry, 184*, 293–298.

SUPER Group (2001). *The social benefits of public open space* (pp. 1–41). Super Group, Sydney.

Suzuki, D. (1997). *The sacred balance: Rediscovering our place in nature*. St Leonards, NSW: Allen and Unwin.

Svedin, U. (2011). *Urban development and the environmental challenges—'Green' systems considerations*. European Commission, Directorate-General for Regional and Urban Policy.

Taylor, A. F., & Kuo, F. E. (2009). Children with attention deficits concentrate better after walk in the park. *Journal of Attention Disorders, 12*(5), 402–409.

Taylor, A. F., Kuo, F. E., & Sullivan, W. C. (2001). Coping with ADD: The surprising connection to green play settings. *Environment and Behavior, 33*(1), 54–77.

Taylor, A. F., Kuo, F. E., & Sullivan, W. C. (2002). Views of nature and self-discipline: Evidence from inner city children. *Journal of Environmental Psychology, 22*, 49–63.

Tian, Y., Jim, C. Y., & Tao, Y. (2012). Challenges and strategies for greening the compact city of Hong Kong. *Journal of Urban Planning and Development, 138*(2), 101–109.

Tzoulas, K., & James, P. (2003). Finding the links between biodiversity in urban green space and human well-being, *British Ecological Society Annual Meeting, 9–11 September 2003, Manchester Metropolitan University*.

United Nations Department of Public Information (2002). *Water: A matter of life and death, United Nations*.

Ward Thompson, C. (1998). Historic American parks and contemporary needs. *Landscape Journal, 17*(1), 1–25.

Ward Thompson, C. (2002). Urban open space in the 21st century. *Landscape and Urban Planning, 60*, 59–72.

Ward Thompson, C., Aspinall, P., Bell, S., & Findlay, C. (2005). 'It Gets You Away From Everyday Life': Local woodlands and community use—What makes a difference? *Landscape Research, 30*(1), 109–146.

West, S., Jones, C., Townsend, M., Henderson-Wilson, C., Maller, C., & Rosenberg, K. (2009). *The contribution of public land to Melbourne's liveability*. McCaughey Centre, University of Melbourne, NiCHE Nature and Health Research Group, Deakin University and The Centre for Design, RMIT University.

Wheeler, B. W., White, M., Stahl-Timmins, W., & Depledge, M. H. (2012). Does living by the coast improve health and wellbeing? *Health and Place, 18*, 1198–1201.

White, M. P., Alcock, I., Wheeler, B. W., & Depledge, M. H. (2013). Would you be happier living in a greener urban area? A fixed-effects analysis of panel data. *Psychological Science, 24*, 920–928.

Williams, N. S. G., Rayner, J. P., & Raynor, K. J. (2010). Green roofs for a wide brown land: Opportunities and barriers for rooftop greening in Australia. *Urban Forestry and Urban Greening, 9*, 245–251.

Wilson, E. O. (1984). *Biophilia*. Cambridge, MA: Harvard University Press.

Wilson, E. O. (1993). Biophilia and the conservation ethic. In S. R. Kellert & E. O. Wilson (Eds.), *The biophilia hypothesis* (pp. 31–41). Washington, DC: Island Press.

Wiseman, J., Heine, W., Langworthy, A., McLean, N., Pyke, J., Raysmith, H., & Salvaris, M. (2006). *Measuring wellbeing, engaging communities: Developing a community indicators framework for Victoria*. The final report of the Victorian Community Indicators Project (VCIP).

Wong, N. H., Wong, S. J., Lim, G. T., Ong, C. L., & Sia, A. (2005). Perception study of building professionals on the issues of green roof development in Singapore. *Architectural Science Review, 48*, 205–214.

Yuen, B., & Wong, N. H. (2005). Resident perceptions and expectations of rooftop gardens in Singapore. *Landscape and Urban Planning, 73*, 263–276.

Chapter 16
Local Health Planning and Governance

Evelyne de Leeuw and Vivian Lin

Introduction

Municipal health planning has increasingly adopted a whole-of-society approach in its embrace of Healthy Cities principles and strategies. The political science literature shows that policy and planning considered too complex or controversial to deal with at state or national levels seems more easily developed at a local level (De Leeuw and Polman 1995), as Corburn et al. (2014) have found in the United States development of Health in All Policies. Givel (2006) describes how the policy momentum in US tobacco control has shifted from the federal to the state level, and Bulkeley and Kern (2006) show how local governments in the UK and Germany, despite statutory and governance differences, take leading roles in the formulation, adoption and implementation of climate change policies. An important aspect for the greater commitment—and possibly success—of local governments in developing policy responsive to community needs is that spatial and cognitive conditions for engagement in planning processes at the local level allow more immediate and relevant feedback between stakeholders.

Planning is an important participatory process as well as a way for translating vision into action (Laverack and Labonte 2000). It is a technocratic exercise and a policy and management tool, but also a core component of governance and stewardship at all levels. Echoing the Greek poet Cavafis' work 'Ithaka', we argue that planning is more about the journey (the participatory process) than about arrival (the plan).

E. de Leeuw (✉)
Centre for Health Equity Training, Research & Evaluation (CHETRE),
University of New South Wales, South Western Sydney Local Health District,
Ingham Institute, Locked Bag 7103, Liverpool, BC NSW 1871, Australia
e-mail: e.deleeuw@unsw.edu.au

V. Lin
College of Science, Health and Engineering, School of Psychology and Public Health,
Department of Public Health, La Trobe University, Victoria, Australia

E. de Leeuw, J. Simos (eds.), *Healthy Cities*,
DOI 10.1007/978-1-4939-6694-3_16

Ithaka
As you set out for Ithaka
hope the voyage is a long one,
full of adventure, full of discovery.
Laistrygonians and Cyclops,
angry Poseidon—don't be afraid of them:
you'll never find things like that on your way
as long as you keep your thoughts raised high,
as long as a rare excitement
stirs your spirit and your body.
Laistrygonians and Cyclops,
wild Poseidon—you won't encounter them
unless you bring them along inside your soul,
unless your soul sets them up in front of you.

Hope the voyage is a long one.
May there be many a summer morning when,
with what pleasure, what joy,
you come into harbors seen for the first time;
may you stop at Phoenician trading stations
to buy fine things,
mother of pearl and coral, amber and ebony,
sensual perfume of every kind—
as many sensual perfumes as you can;
and may you visit many Egyptian cities
to gather stores of knowledge from their scholars.

Keep Ithaka always in your mind.
Arriving there is what you are destined for.
But do not hurry the journey at all.
Better if it lasts for years,
so you are old by the time you reach the island,
wealthy with all you have gained on the way,
not expecting Ithaka to make you rich.

Ithaka gave you the marvelous journey.
Without her you would not have set out.
She has nothing left to give you now.

And if you find her poor, Ithaka won't have fooled you.
Wise as you will have become, so full of experience,
you will have understood by then what these Ithakas mean.

C.P. Cavafy Collected Poems. Translated by Edmund Keeley and Philip Sherrard. Edited by George Savidis. Revised Edition. Princeton University Press, 1992

This chapter considers the challenges of urban health governance by a review of the municipal public health planning experience. In particular, the experience of an early innovator, Victoria, Australia in legislating for health planning at the local governance, is assessed.

Planning and Governance

Planning as a policy instrument was adopted widely in post-World War II reconstruction, although the history of urban planning extends to a much earlier time in western civilizations (see Chap. 2). In its classic incarnation, planning was a top-down blueprint for resource allocation, often driven by technical possibilities and limitations. With the success of those forms of planning (for instance, the dramatic innovation of assembly-line production of automobiles by the Ford Motor Company in 1913), the notion of planning started to pervade wider society. Friedmann (1998) recounts how around the middle of the twentieth century steps were made toward theorizing planning, and how pioneers straddled fields such as philosophy, economy, political science, engineering and urban studies. No wonder that a single definition of the idea has been hard to settle on. In the arena of health and well-being, planning may become a process for articulating preferred approaches to problem-solving for health service delivery agencies or public health services; but 'planning' is embraced, too, as a credentialed process in more complex social systems like urban development, and in the marketplace when it comes to product development and placement.

Friedmann (1998) sees six kinds of planning: (1) applied rationality, (2) societal guidance, (3) a behavioural (positivist) approach, (4) a communicative practice, (5) social learning or (6) radical planning or emancipatory practice. An underlying assumption about planning, regardless of these perspectives, is that it is core to the governance of institutions and governments and a means for managing the path from the present to the future. The core can be mainstream and aligned with political priorities at government level, but can also take the shape of countervailing or radical pronouncements from, or on behalf of, populations on the fringe of political or social life—and anything in between, and in widely different forms. For instance, participatory budgeting and policymaking and deliberative democracy (Abers 2003) may be fully integrated in the institutional structures of a government, but also be found in social movements engaging with or opposing official policies.

Planning for health emerged in the 1960s as part of community moves to shift the focus in health sector planning to a concern for health outcomes, community participation and the voices of the disadvantaged. The major proponents of this approach were Laframboise in Canada (1973), Blum in the USA (1974) and Van der Werff in the Netherlands (1976). They contributed to major shifts in planning paradigms through their input on national policy documents (the Lalonde report in Canada), development of a public health profession strongly grounded in local health planning (spreading from California across the United States), and 'futuring' for health in the context of broad social targets for WHO/EURO member states (Brouwer and Schreuder 1988; Van Herten and Gunning-Schepers 2000). These approaches share a conviction that by collecting sufficient knowledge patterns of action may be determined and controlled.

Even before these important moves to community health there was a strong connection between community health planning and urban planning. Many participatory planning models found in public health and health promotion demonstrate this joint development: Arnstein's (1969) 'ladder of participation' and Davidson's (1998) 'wheel of participation' both connect community development, health and well-being, and urban planning processes and outcomes, and current practice in local health planning is built on these models.

Governance

Geoff Green (1998) was the first to analyse comprehensively patterns of local governance for health in WHO/EURO. To our knowledge there have been no similar attempts elsewhere or in other policy domains. He mapped responsibilities for health care delivery, public health and health promotion, and management of the social determinants of health at all levels of government in member states of the European Region of WHO. He did so for good reason. There is a profound connection between governance and health (e.g. Marmot et al. 2008; Plochg et al. 2006; Vlahov et al. 2007). In a foundation report for WHO/EURO Health 2020, Kickbusch and Gleicher (2012) build on Green's evidence to argue that there is a difference between health governance and governance for health: health governance is the administration and strengthening of a health system, while governance for health is any action by health or non-health sectors, public or private sectors, and community groups or individual citizens, for a common health cause. Kickbusch and Gleicher define governance for health as 'the attempts of governments or other actors to steer communities, countries or groups of countries in the pursuit of health as integral to well-being through both whole-of-government and whole-of-society approaches'.

Cairney (2012) explains why the idea of governance has replaced the imperative of government, essentially seeing the starting point for this development in the Thatcherism and Reaganism of the 1980s. With the assumption that market mechanisms would deliver 'value for money' more than government bureaucracies, public policy authority devolved from (supra)national to local government levels, and from government to quasi-government (QUANGO) and private sectors. This led to what some called a 'hollow state' (Milward and Provan 2000): 'a metaphor for the increasing use of third parties, often non-profits, to deliver social services and generally act in the name of the state'. With its multitude of state and non-state, individual and institutional actors, the health arena is a particular case of hollow state governance.

Kickbusch and Gleicher (2012) assert that

> many of the current health challenges could be better resolved through whole-of-society approaches, which include civil society and the private sector as well as the media.

> Health 2020 can support health ministries and public health agencies in reaching out to people within and outside government to find joint solutions. It can propose new programmes, networks and initiatives to engage many different stakeholders and, above all, citizens throughout Europe and explore new incentive mechanisms. Stakeholders could jointly identify and implement new means for assessing accountability and health impact, such as the contribution to a European health footprint. The WHO European Healthy Cities Network would be an excellent laboratory for such an innovation.

The responsibilities of local governments in governance for health are determined by statutory pronouncements and legal codes, and as Green (1998) shows for WHO/EURO, these are widely—and wildly—diverse. In Table 16.1 we present a selective (and perhaps biased) sample of statutory statements on local government responsibilities and commitments to health development. Some local governments are directly responsible for running health services (including their financing from local revenue), including public health and health promotion. Others are directed in

Table 16.1 Local government responsibilities for health in four jurisdictions

Nation	Local government responsibilities
United Kingdom	UK responsibilities follow the legal principle of *ultra vires*: local councils are able to do only what they are statutorily permitted to do. Their rights and competences are not general but specific (Wilson and Game 2011)
Germany	The right of 'self-government': local authorities have responsibility for all matters relevant to the local community (örtliche Gemeinschaft), but within existing legislation (Grundgesetz, article 28, section 2)
Victoria, Australia	(1) The primary objective of a Council is to endeavour to achieve the best outcomes for the local community having regard to the long term and cumulative effects of decisions
	(2) In seeking to achieve its primary objective, a Council must have regard to the following facilitating objectives:
	(a) to promote the social, economic and environmental viability and sustainability of the municipal district
	(b) to ensure that resources are used efficiently and effectively and services are provided in accordance with the Best Value Principles to best meet the needs of the local community
	(c) to improve the overall quality of life of people in the local community
	(d) to promote appropriate business and employment opportunities
	(e) to ensure that services and facilities provided by the Council are accessible and equitable
	(f) to ensure the equitable imposition of rates and charges
	(g) to ensure transparency and accountability in Council decision making (Local Government Act, 1989)
The Netherlands	Public Health Law: municipalities have the administrative responsibility to create, sustain and coordinate efforts in public health. Their task is to prevent, protect and promote the health of their populations. Apart from core responsibilities in public health, infectious disease control and the health of young people, municipalities need to enable intersectoral collaboration and the establishment of coherence between public health and cure. A Health care Inspectorate monitors compliance and has the authority to direct change (Steenbakkers 2012).

such responsibilities through national legislation that requires strict compliance and reporting, often without direct control over the flow of resources or of priorities. Some local governments do not have—or do not feel they have—any say over the health of their populations, including the provision of, and control over, the quality of housing, sanitation, public transport, roads and social support mechanisms.

Municipal Public Health Planning in Victoria: A Case Study

We now turn to a local case study, looking at municipal public health planning in Victoria, Australia. Australia has a federalist system, and the constitution designates the state as the level with primary responsibility for service delivery. Local governments are created by state legislation and their roles, therefore, vary widely across the country. More than any other state, Victoria has delegated responsibilities for health and other services to local government.

The responsibility for local government to develop a municipal public health plan (MPHP) was legislated in 1987 amendments to the Health Act of the State of Victoria, at a time when there were some 220 local governments. The implementation of this legal requirement was initially under the direction of one officer in the State's health department, working with the Municipal Association of Victoria. Councils were mandated to prepare a new plan every 3 years and review them annually. The plans had to identify and assess actual and potential public health dangers, and outline the programs and strategies the council would pursue to minimize these and enable people to achieve maximum well-being.

The initial response from local councils was one of caution (Wills 2001). There was the question of cost and capacity, the scope of the task, and a sense of a top-down imposition by the state government. There was confusion about this idea of a public health plan with its core in local government authority: was it a corporate (or 'business') plan for those environmental health officers working locally already, or was it to be a corporate plan for all local government services which had some influence on health? Or was it supposed to be a plan for the health of the local population? These initial doubts and questions were grounded in the fuzziness of the idea of 'public health': Is it 'the public's health', or 'public sector health', or 'rats and drains'?

As with anything newly mandated, there was some suspicion about the government's intent. Was the responsibility to formulate a municipal public health plan simply a devolution and decentralization of tasks, lip service to the emerging idea of 'government close to the people'? How would local governments without much experience or professional expertise be supported in framing and implementing their plans for social model of health?

Despite such initial concerns, political momentum for 'the new public health' built. Also in 1987, the Victorian Tobacco Act was passed, enabling the establishment of the Victorian Health Promotion Foundation (VicHealth), a world first: using revenues from tobacco taxation. VicHealth began to support development of 'healthy localities' and this helped local communities and governments understand what a health plan could entail. However, there were challenges in implementing planning priorities which touched the core of local political economies, as when a local community wanted to address pollution from the main local employer.

Plans and the planning experienced proved to be highly variable across the state, reflecting differing organizational capacity and culture (Bagley et al. 2007). There was agreement that the legislatively mandated process had improved local planning, but often the focus remained on the development of a plan rather than on implementing the identified priorities and strategies. Nonetheless, there now was a minimum standard for public health planning. For some councils, municipal public health planning was the testing ground for developing strategic planning and alliances, and the beginning of a whole-of-community approach to health (Wills 2001). It became evident in the early years that local government's role in public health could shift from hazard surveillance to active agents in the development of healthy communities.

In 2001, after more than a decade of experimentation, the government of Victoria adopted its Environments for Health (E4H) policy framework (Department of Human Services 2001). E4H provides evidence-based guidance for the development of local policies that address the social and environmental determinants of health in the overlapping domains of the social, built, economic and natural environments. E4H explicitly embraces a social model of health, and the policy package provides local governments with a comprehensive evidence base, assistance in building capacity in local health bureaucrats and communities, and exemplars of policy action. Five years after its adoption an evaluation of E4H assessed the extent to which it had

- been incorporated by local governments in their policies and practices.
- contributed to greater consistency and quality in the scope and approach of municipal public health planning across the state.
- led to the integration of municipal public health plans with other council plans.
- increased the level of understanding among appropriate local government staff of the impact of the social, economic, natural and built environments on health and well-being.
- created additional opportunities for health gain through strengthened intersectoral partnerships to address the social determinants of health.
- been supported effectively by the Department of Human Services and other stakeholders (De Leeuw et al. 2006).

The evaluation objectives were the outcome of negotiations between a range of stakeholders, including the Department of Human Services, local governments and research sector representatives. A number of political theories were hybridized in the evaluation framework (Pawson and Tilley 1997), notably policy diffusion theory (Lindblom 1959), implementation theory (Mazmanian and Sabatier 1983) and multiple streams theory (Kingdon 1984). It also drew on a range of data collection strategies:

- Analysis of local government authorities' municipal public health plans (62 plans).
- 73 individual and group interviews with key stakeholders in municipal public health planning.
- An online survey of relevant individuals, including councillors, council staff, non-council organizations and community members (108 respondents).

- Five community forums to present preliminary findings and obtain input from other stakeholders and groups.

The evaluation (Department of Health 2014a) found that E4H had substantially changed the way local governments thought about health, improved the way local governments planned for health, and had led to the start of sectoral integration. However, developing a plan was still frequently seen as a means in itself, and implementation often lagged. The Department of Health consequently launched programmes for implementation, knowledge co-creation, capacity-building and networking at the local level, case models for environments for health development, and political skills with stress on economic environments. These aligned closely with an update in 2008, the *Victoria Health and Wellbeing Act*, which mandated the fuller integration of an approach based on social determinants of health with local governments' daily practices and policies, setting the stage for the 'Healthy Together Victoria' prevention system, which acknowledges more fully the responsibilities and opportunities of a broad range of institutions and actors (Sylvan 2013; Department of Health 2014b). The 2008 Act represents a securing of intersectoral action as mainstream practice. The journey from the initial legislative requirement to undertake municipal public health plans to the present is a successful evolutionary pathway from creating an initial authorizing environment to creating a sustainable enabling environment for intersectoral action on local health (Lin 2013).

Lessons and Governance Challenges

From Victoria municipal public health planning diffused to other states. Queensland is now another exemplar in Australia. Interestingly, this diffusion contrasts with the expansion of Healthy Cities in Australia, which began in the late 1980s. While Victoria legislated municipal public health plans, Healthy Cities projects were initiated in New South Wales (Illawara/Kiama), ACT/Canberra and South Australia (Noarlunga) largely on the initiative of the novel health promotion sector, and on the back of the second WHO Global Conference on Health Promotion, held in Adelaide in 1988 with the theme of 'healthy public policy'. While a number of Australian cities have joined the Alliance of Healthy Cities, there is not the same widespread acceptance of the 'brand' as for municipal public health plans. This may be because planning has always been a core function for local government, and public health planning is not a completely foreign concept, and because a legislated mandate will be carried out, enabling public health planning to become an internalized function of local government. Moreover, a health promotion initiative may still be seen as a health sector project rather than as a core function of urban governance. In other words, municipal public health plans may be seen as an instrument of local governance and more readily accepted than Healthy Cities.

The challenge for local government is to remember that public health planning is not about producing a blueprint, in the way that land use or transportation or other statutory plans might be produced. Bagley et al. (2007) find that the Victorian councils who found the planning framework beneficial were those who took a strategic approach to public health plans, who made reference to the social model of health, who had high levels of community involvement and who had formal processes in place for implementation and evaluation. In a sense, these councils practised planning in a way consistent with Burris et al. (2005) of governance, 'the management of the course of events in a social system'.

Future of Local Health Governance and Environments for Health

Municipal public health planning is an illustration of what good governance and intersectoral action for health could be. Good governance is typically characterized as participatory, fair, accountable, transparent and sustainable (UN-Habitat 2002). At any level of governance in a complex social system the problems are how to mobilize dispersed knowledge, capacity and resources for the public good, and how to leverage local knowledge and capacity to influence policy outcomes. In a local urban setting where the scale is smaller and the social networks pervasive, poor governance may be more transparent while good governance may deliver desired outcomes; thus, municipal public health planning is a device for promoting good governance as well as improving health in the community.

Local councils around the world have widely varying responsibilities for and involvement with health. Depending on the political agendas of leaders and the capacities of local civil society, there is an opportunity for innovative approaches to address local health concerns, and particularly to tackle social determinants of health with a health-in-all-policies approach at the local level. Healthy Cities exemplifies both the potential and the uncertainties of governance strategies for health (Burris et al. 2007), given the diversity of contexts and leadership capabilities. However, municipal public health planning can also be a mechanism of intersectoral governance, with the prioritizing of problems to be addressed, the involvement of community interests and the tools chosen to implement solutions.

Cities across the world share many challenges. There are few, if any, places not experiencing global economic integration and climate change, and with them some degree of increased social division and social inequality. These forces ultimately produce a myriad of health challenges, including infectious and emerging diseases, risk factors for non-communicable diseases, violence, and other social dysfunction. Good urban governance for health can link governmental and civil society institutions, link local with regional and global communities of interest, create safe or trusting spaces for diverse interests to interact, and integrate solutions for health problems with solutions for other urban concerns.

For municipal public health plans to be successful, a variety of good urban governance practices appear to be important. These include building institutions to increase participation and network governance to strengthen connections across interests and sectors, ensuring technical competence, providing forums for creativity and ensuring responsive government (Burris et al. 2007). By recognizing local health issues as the manifestation of complex social interactions, and situating local health governance within an understanding of global forces, tools such as municipal public health plans can bring together diverse knowledge and practices through citizen engagement and intersectoral governance.

References

Abers, R. (2003). *Deepening democracy: Institutional innovations in empowered participatory governance* (Vol. 4). Verso.

Arnstein, S. R. (1969). A ladder of citizen participation. *Journal of the American Institute of Planners, 35*(4), 216–224.

Bagley, P., Lin, V., Sainsbury, P., Wise, M., Keating, T., & Roger, K. (2007). In what ways does the mandatory nature of Victorian's municipal public health planning framework impact on the planning process and outcomes? *Australia and New Zealand Health Policy, 4*, 4.

Blum, H. L. (1974). *Planning for health: Development and application of social change theory*.

Brouwer, J. J., & Schreuder, R. F. (Eds.). (1988). Scenarios and other methods to support long term health planning: Theory and practice. *Steering committee on future HealthScenarios*. Leidschendam: Ministry of Welfare, Health and Culture.

Bulkeley, H., & Kern, K. (2006). Local government and the governing of climate change in Germany and the UK. *Urban Studies, 43*(12), 2237–2259.

Burris, S., Drahos, P., & Shearing, C. (2005). Nodal governance. *Australian Journal of Legal Philosophy, 30*, 30–58.

Burris, S., Hancock, T., Lin, V., & Herzog, A. (2007). Emerging strategies for healthy urban governance. *Journal of Urban Health, 84*(1), 154–163.

Cairney, P. (2012). *Understanding public policy: Theories and issues*. Palgrave Macmillan.

Corburn, J., Curl, S., Arredondo, G., & Malagon, J. (2014). Health in all urban policy: City services through the prism of health. *Journal of Urban Health, 91*(4), 623–636. doi:10.1007/s11524-014-9886-3.

Davidson, S. (1998). Spinning the wheel of empowerment. *Planning, 1262*(3), 14–15.

de Leeuw, E., Butterworth, I., Garrard, J., Palermo, J., Godbold, T., & Tacticos, T. (2006). *Evaluation of the environments for health framework*. Melbourne, VIC: Deakin University and the University of Melbourne.

de Leeuw, E., & Polman, L. (1995). Health policy making: The Dutch experience. *Social Science and Medicine, 40*(3), 331–338.

Department of Health. (2014a). *Local government planning for health and wellbeing—Resources*. Melbourne: Department of Health. Retrieved February 4, 2014, from http://www.health.vic.gov.au/localgov/resources.htm.

Department of Health. (2014b). *Healthy together. Prevention and population health*. Melbourne: Department of Health. Retrieved August 15, 2014, from http://www.health.vic.gov.au/prevention/healthytogether.htm.

Department of Human Services (2001). *Environments for health: Promoting health and wellbeing through built, social, economic and natural environments*. Municipal Public Health Planning Framework. Melbourne, VIC: Department of Human Services.

Friedmann, J. (1998). Planning theory revisited. *European Planning Studies, 6*(3), 245–253.

Givel, M. (2006). Punctuated equilibrium in limbo: The tobacco lobby and US state policymaking from 1990 to 2003. *Policy Studies Journal, 34*(3), 405–418.

Green, G. (1998) Health and governance in European cities: A compendium of trends and responsibilities for public health in 46 member states of the WHO European Region. *European Hospital Management Journal.*

Kickbusch, I., & Gleicher, D. (2012). *Governance for health in the 21st century*. Copenhagen: World Health Organization Regional Office for Europe.

Kingdon, J. W. (1984). *Agendas, alternatives and public policies*. Boston: Little, Brown and Co.

Laframboise, H. L. (1973). Health policy: Breaking the problem down into more manageable segments. *Canadian Medical Association Journal, 108*(3), 388.

Laverack, G., & Labonte, R. (2000). A planning framework for community empowerment goals within health promotion. *Health Policy and Planning, 15*(3), 255–262.

Lin, V. (2013). Conclusions. In Lin, V. (Ed.), *Australian social determinants of health casebook.*

Lindblom, C. E. (1959). The science of 'Muddling Through'. *Public Administration Review, 19*, 79–88.

Marmot, M., Friel, S., Bell, R., Houweling, T. A., & Taylor, S. (2008). Closing the gap in a generation: Health equity through action on the social determinants of health. *The Lancet, 372*(9650), 1661–1669.

Mazmanian, D. A., & Sabatier, P. A. (1983). *Implementation and public policy*. Glenview, IL: Scott Foresman.

Milward, H. B., & Provan, K. G. (2000). Governing the hollow state. *Journal of Public Administration Research and Theory, 10*(2), 359–380.

Pawson, R., & Tilley, N. (1997). *Realist evaluation*. Thousand Oaks: Sage.

Plochg, T., Delnoij, D. M. J., Hogervorst, W. V. G., Van Dijk, P., Belleman, S., & Klazinga, N. S. (2006). Local health systems in 21st century: Who cares?—An exploratory study on health system governance in Amsterdam. *The European Journal of Public Health, 16*(5), 559–564.

Steenbakkers, M. (2012). *Lokaal integraal gezondheidsbeleid: Realistische uitdaging of utopie. Een onderzoek binnen gemeenten naar mogelijkheden tot intersectorale samenwerking*. PhD thesis. Maastricht University, Academic Collaborative Centre of Public Health Zuid Limburg.

Sylvan, L. (2013). Chapter 5: The challenges ahead. In *'State of Preventive Health 2013'*. Canberra: Australian National Preventive Health Agency.

UN-Habitat. (2002). *The global campaign on urban governance: Concept paper*. Nairobi: UN-Habitat.

Van Der Werff, A. (1976). *Organizing health care systems: A developmental approach*. Eindhoven, Greve Offset.

Van Herten, L. M., & Gunning-Schepers, L. J. (2000). Targets as a tool in health policy: Part I: Lessons learned. *Health Policy, 53*(1), 1–11.

Vlahov, D., Freudenberg, N., Proietti, F., Ompad, D., Quinn, A., Nandi, V., & Galea, S. (2007). Urban as a determinant of health. *Journal of Urban Health, 84*(1), 16–26.

Wills, J. (2001). Case study: Municipal public health planning. In P. Garrett, K. Eagar, & V. Lin (Eds.), *Health planning: Australian perspectives*. Allen and Unwin: Sydney.

Wilson, D., & Game, C. (2011). *Local government in the United Kingdom*. Palgrave Macmillan.

Chapter 17
From Urban Projects to Healthy City Policies

Evelyne de Leeuw

Introduction

The urban health world is rife with projects—to alleviate poverty, empower communities, provide better roads, increase access to education, secure primary care, etc. Many of these initiatives, rightly, have a fixed life. A new school needs to be built; upon completion it requires staff and maintenance: these are entirely different things that can be managed well on a project basis. But a proper policy would take a longer term perspective that includes not just the building and maintenance of the infrastructure. Such a policy would set parameters to how infrastructure relates to continued access, how both of these relate to the delivery of a curriculum, and how to undertake regular reviews of accomplishments.

Goumans and Springett (1997) have identified that many Healthy Cities suffer from projectism. Interestingly, although this term (alternated with 'projectitis') has become an integral part of the vernacular of the—critical—government bureaucrat, we have been unable to ascertain an authoritative definition for it. In development studies, Pareschi (2002) and Little (2005) refer to projectism when they describe the tendency of the international development community (the 'do-gooders' (Christensen 2004) in development aid) to impose a project format with defined beginning and end, contained resources (money, people) in time and space, on daily activities undertaken by indigenous communities that by their very nature are organic and ongoing, e.g. the defence of territory, production of food, and political organization. Such project 'containment', they assert, can provoke major changes in cultural values, leadership patterns, time conceptions, organizational structure, and political relations in affected indigenous communities. True as this may be, the

E. de Leeuw (✉)
Centre for Health Equity Training, Research & Evaluation (CHETRE),
University of New South Wales, South Western Sydney Local Health District,
Ingham Institute, Locked Bag 7103, Liverpool, BC NSW 1871, Australia
e-mail: e.deleeuw@unsw.edu.au

E. de Leeuw, J. Simos (eds.), *Healthy Cities*,
DOI 10.1007/978-1-4939-6694-3_17

popularity of projectism/itis suggests that the phenomenon may have detrimental effects in high-income, industrialized nations as well. A definition that reaches beyond indigenous culture into (Western) bureaucratic and political values might read 'The reflex, intention and action to contain organisational and community efforts in terms of resource allocation, time, space and conceptualisation purely for the sake of accountability and management purposes, not because the effort at hand necessarily lends itself to such limitation'.

Goumans and Springett (1997) have investigated the potential of Healthy Cities to move beyond projects and projectism, and found that ample opportunities exist to move into longer term programmes or policies that embed the value base and strategic outlook of the approach. In this chapter we take a look at the conditions that may facilitate such a longer term endeavour.

Policy Diffusion and Glocalization

Some research suggests that achieving policy innovation, which is required for introducing systemic and sustainable intersectoral perspectives across society, cannot be achieved at the national level, or not at that level alone. Policy diffusion researchers (e.g. Shipan and Volden 2008) argue that local governments drive policy innovation and diffusion of novel policies horizontally to other local governments, and vertically to regional and national governments. For example, policy diffusion facilitated the Netherlands' efforts to develop a broad Healthy Public Policy in the 1980s (de Leeuw and Polman 1995). Global commitments, such as the Kyoto Protocol for climate change adaptation and mitigation or the Framework Convention for Tobacco Control, can be seen as crucial benchmarks for the need to develop new policy types. Policy innovation does not happen exclusively bottom-up or top-down, but must be characterized as happening through a process called 'mixed scanning' (Etzioni 1967, 1986) in which systems of incremental and reciprocal checks and balances between governance levels create opportunities for change.

A key term that has been encountered throughout this book is 'glocal'. 'Glocal health' (de Leeuw 2001; de Leeuw et al. 2006; Kickbusch 1999) is a term used to recognize and appreciate the intricate and inseparable interface between global developments (e.g. climate change or trade) and local responses (e.g. councils adopting building codes that account for the increased risks of flooding and heat islands, or offering favourable opportunities for local entrepreneurship to engage in international forums). This glocalization dynamic is reciprocal. Less likely or desirable global developments may be mitigated—or exacerbated—by local action. For instance, the increasing number of local governments around the world adopting 'zero-carbon emission' policies (e.g. Koehn 2008) not only contribute to possible reductions in climate change risks, but also send signals to their colleagues, at local as well as higher government levels, that such actions are feasible and effective. Through policy diffusion these local policies impact on global change. In fact, analyses of local government efficacies in the late 1980s and early 1990s led to the introduction of the terms 'glocal' and 'glocalization' into our vocabularies (Swyngedouw 1992). Virtually every development and phenomenon in Healthy Cities has glocal dimensions.

Health as a Social Resource and Created in All Policies

So why would we make policy at all? Healthy people are an important resource for society. Healthy communities are thriving communities, not just in economic terms (because they may more comprehensively contribute to building their common resources) but certainly also in terms of social development and the resilience to cope with shifts and challenges in their social and natural environments. Societies and communities with high levels of positive health are resilient. They can face adversity better.

A firm expression of the nature of such a health perspective is often found in its definition as engrained in the Constitution of the World Health Organization (1948):

> Health is a state of complete physical, mental, social [and spiritual, Larson 1996] well-being and not merely the absence of disease or infirmity.

In spite of this broad framing of health, in many countries the health service delivery (or 'sick care') sector is not fully embracing these views and their consequences. Most healthcare establishments focus on individual treatment and disease prevention, and are challenged to adopt a full social model of health. Around the world, the health delivery industry has become a dominant economic sector in its own right and efforts to involve it in actions to promote community health (rather than cure and prevention of disease) face strong beliefs that individual-focused interventions are better, quicker, or more effective.

The microbiologist-philosopher Dubos (1959) recognized the profound interface between individual and social health and defined health as

> the expression of the extent to which the individual and the social body maintain in readiness the resources required to meet the exigencies of the future.

The key to appreciating this definition is the notion of 'the social body': it refers to community as well as society and its institutions. The institutions can be seen as tangible 'hardware' (hospitals, transport services, bodies of government) but also in a more sociological sense. The formidable Ahrendt (1970) saw an institution as 'a body of people and thought that endeavours to make good on common expressions of human purpose'. This idea of an institution (as in 'the institution of marriage' rather than 'the hospital institution') has intimate relations to concepts of government and governance.

The ways in which local governments are shaped are functions of both the philosophical and structural views of institutions. In democratic traditions, the assumption is that local government can directly represent constituents and respond to individual, family, community, and neighbourhood needs. But that assumption is firmly based in other assumptions about representation and eligibility of people to partake in the communal and political processes leading to the values that pervade governance, and the resulting shape of government.

A key aspiration of modern glocal government is to deliver justice. As Spinoza (1670) said,

> the ultimate end of the State is not dominion, nor restraint by fear, nor the exaction of obedience; on the contrary, its end is to free every man from fear, so that he may live securely.

contrast this with Ronald Reagan's definition of government as

> like a big baby—an alimentary canal with a big appetite at one end and no sense of responsibility at the other (Adler 1981, p. 30)

and the clash of political ideologies will be clear.

Governments can secure and facilitate different forms of justice (e.g. Ruger 2005):

- Procedural justice—decision-making about policy, programme, service design, and delivery—making the composition of decision-making bodies more descriptively representative of the community (in cultural, socioeconomic, gender, etc. senses); and strengthening communities' power to define 'agenda' items independently of the 'dominant culture'
- Substantive justice—influence—putting items on the agenda, influencing discussion and debate on all agenda items, and influencing the outcomes of decision
- Distributive justice—ensuring that the population has equitable opportunities to access social resources including high-quality health care, but also preventive services and education, employment, transport, etc.

We assert that local government is an expression and instrument of priority setting for shaping the resources for health that Dubos describes, and creating the forms of justice that allow people to fully participate. This happens through policy development and the management of social and environmental assets. The growing body of evidence, over recent decades, on the social, political, and commercial determinants of health may well enable local government better than other levels of government and governance to take decisive action. Evidence from the other chapters in this book shows that local governments (and especially Healthy City ones) are in closer contact with their constituents and would purportedly be able to respond more effectively and quickly to needs expressed. Clearly this is an ideal-type description: not all local governments are transparent and accountable, and not all people may be, or may feel, represented. This is particularly true for slum dwellers. Sometimes urban inequities are literally hidden—in many Third World cities the slums and their informal populations are located in gullies and ravines. But in others, the favelas rise up high on the slopes surrounding affluence. It appears that technology can come to a degree of rescue, whether it is enabling social connectedness in Nairobi slums (Corburn and Karanja 2014) or physical connectedness through novel public transportation solutions in Medellin, Colombia (Díez et al. 2015).

Local government also has the potential to address the wider determinants of health and health equity. The determinants of health extend far beyond the workings of the health care system, and include the provision and levels of education, the availability of work and employment and standards, the quality of the built and natural environment, the existence of intangible things like a sense of community and solidarity expressed in 'social capital', and the apparent immutability presence of general social gradients between those at the highest and lowest ends of the socioeconomic spectrum.

Families and communities, and their elected representations in local governments, most directly suffer and enjoy the negative and positive consequences of their decisions on how their lives are shaped in all these domains. Complex and connected issues require complex and integral responses. Local government does not stand alone in this—it can respond (and has responded, e.g., through the Healthy Communities and Healthy Cities networks) more efficaciously to population needs; but at the same time it is bound by regional (provincial, state) and (inter)national contexts. Horizontal and vertical collaboration and synergy can and should be sought.

Moving from Singular to Complex

Analyses of the workings of modern society and its institutional structures (governance, democracy, leadership, etc.) have shown that traditional sectoral and vertical (top-down) responses may yield short-term success but may not address the systemic and complex causes of problems. The consequence of such analyses has been a call for better integration of (and within) problem formulation, policy development, and comprehensive action. Such integration would assume equitable access of highly heterogeneous stakeholders to all elements of enormously multifaceted systems (anyone should have access at any time wielding the same influence over the process, no matter who and where they are—a utopian ideal). It is no wonder that solving this issue has eluded politicians, scholars, practitioners, and communities.

At an *abstract* level, the solution has been found in concepts such as 'systems thinking', 'complexity science', and identification of problems as being 'wicked', 'messy', or 'fuzzy'. For *policy-making*, those terms have translated into perspectives on 'whole of government', 'joined-up government', 'integral government', and 'horizontal government' (Carey 2016; Carey and Crammond 2015; Pollitt 2003). There is a strong argument to be made that these perspectives play out best at the local level because it is there that cooperation between state, market, and civil society actors is considered most likely to produce coordinated planning and action (Christensen and Lægreid 2007). The search for whole, joined-up, integral, or horizontal local government approaches achieved momentum, some scholars and politicians claim, since the (perhaps overly zealous) adoption of 'new public management' (NPM) principles from the 1980s. In NPM citizens are viewed as customers, and public servers/administrators are considered managers of product and service delivery.

The assumption of NPM was that marketization of public goods would yield greater efficiencies. However, vulnerable, socially excluded, marginal, and under-represented populations in particular often cannot claim a voice of influence and power in this pseudo-economic discourse. Governments have tried to repair the resulting gaps in the system with the application of (often cunningly rhetorical) tools that go by monikers such as 'new social partnerships' and 'empowered clients'. In many cases a new balance between complete state control (the 'nanny state' caring for everyone 'from cradle to grave', cf. Rivett 1998) and full dissolution of services to commercial sectors is yet to be struck.

In the health field, the recognition of 'health' as an issue across social and government sectors has led to the launch of *policy* perspectives such as 'Healthy Public Policy' and 'Health in All Policies'. In *action* terms (that is, for specific intervention development) we have seen the emergence of terms like 'strategic', 'comprehensive', 'multisectoral', or 'intersectoral' action.

In the scientific literature we see important efforts to distinguish between all these terms. Analysts also suggest ways in which they interrelate. A Canadian publication (Gagnon and Kouri 2008) starts this discussion with a description stemming from Australia of 'integrated governance':

> the structure of formal and informal relations to manage affairs through collaborative (joined-up) approaches which may be between government agencies, or across levels of government (local, state and Commonwealth) and/or the nongovernment sector.

This describes the overarching principles driving both policy and intervention responses to complex systems issues in health development: managing health, health development, and health equity through collaborative approaches. The current perspective on Health in All Policies (HiAP) finds a basis in the call to develop Healthy Public Policies in the Ottawa Charter (1986).

Around the world, governments at all levels have experimented with integrated health policies. Some of these actually inspired the pronouncements of the Ottawa Charter, e.g. the Norwegian Farm–Food–Nutrition policy, the Chinese 'barefoot doctors' programme, and women's health initiatives in the Americas. Two initiatives from opposite ends of the world started the developmental process of what now is called HiAP. One came from Finland during its presidency of the European Union in 2006: Finland, building on its experience in the long-running North Karelia project (labelled a 'horizontal health policy'), urged other members of the Union to engage in

> a horizontal, complementary policy-related strategy contributing to improved population health. The core of HiAP is to examine determinants of health that can be altered to improve health but are mainly controlled by the policies of sectors other than health. (Ståhl et al. 2006)

The other came almost simultaneously from the government of the state of South Australia, which identified opportunities for a broad policy programme to invest in the health of its people:

> Health in All Policies aims to improve the health of the population through increasing the positive impacts of policy initiatives across all sectors of government and at the same time contributing to the achievement of other sectors' core goals (Ståhl et al. 2006, quoted in Baum et al. 2015; Rudolph et al. 2013).

These provided impetus for the organization of the Eight Global Conference on Health Promotion where a statement and framework were adopted that expressed HiAP as follows:

> Health in All Policies is an approach to public policies across sectors that systematically takes into account the health implications of decisions, seeks synergies, and avoids harmful health impacts in order to improve population health and health equity. It improves accountability of policymakers for health impacts at all levels of policy-making. It includes an emphasis on the consequences of public policies on health systems, determinants of health and well-being. (World Health Organization 2014a, b)

In different countries and jurisdictions the emphases of the different dimensions of HiAP vary. Consistently, values (in **bold**) associated with the concept centre around the importance of **collaboration** between sectors of **public policy-making** in good **partnership**. Other aspects where less coherence exists between the different jurisdictions include **health equity**, the attainment of **synergy**, HiAP leading to or driven by **accountability**, the character of **innovation**, ways of **integration**, and the very **nature of policy**, e.g.:

> Health in All Policies is a collaborative approach that integrates and articulates health considerations into policy making across sectors, and at all levels, to improve the health of all communities and people.—US Association of State and Territorial Health Officers. (ASTHO)
>
> Health in All Policies is a collaborative approach to improving the health of all people by incorporating health considerations into decision-making across sectors and policy areas. (California Health in All Policies Task Force)
>
> Health in All Policies is the policy practice of including, integrating or internalizing health in other policies that shape or influence the [Social Determinants of Health (SDoH)] … Health in All Policies is a policy practice adopted by leaders and policy makers to integrate consideration of health, well-being and equity during the development, implementation and evaluation of policies. (European Observatory on Health Systems and Policies)
>
> Health in All Policies is an innovative, systems change approach to the processes through which policies are created and implemented. (National Association of County and City Health Officials) (from Rudolph et al. 2013)

Shifts like these don't just happen. They are the result of, and embedded in, intricately connected webs (Kickbusch (2012) identified those for the health promotion endeavour as rhizomes) that grow, dissolve, reconnect, and interact between people, agents, and events. Callon (1984) and Latour (1987) based their actor-network theory (ANT) on such a view of social, policy, and technology development. This is not the place to engage in a fully fledged ANT analysis; instead we will list six keystone developments that made the current momentum for HiAP possible.

Primary Health Care

One foundation and lasting contribution to our thinking about Health in All Policies has been the drive towards primary health care (PHC). At the WHO/UNICEF conference on Primary Health Care in Alma Ata in 1978, the concept was defined (UNICEF and WHO 1978) as

> essential health care based on practical, scientific and socially acceptable methods and technology. It is made universally accessible to individuals and families in the community through their full participation and at an affordable cost to the community and country.

In its further development, PHC became more fine grained, and two perspectives emerged. One was a horizontal (comprehensive, systems-driven) approach aligned with a set of strong values around equity, participation, and community-driven bottom-up action for health and well-being, and the other a vertical (disease and health care-driven) approach aligned with the need to address specific (burdens of) disease in many countries, and grounded in existing institutions and patterns in the delivery of clinical interventions. Ideology-inspired debates have raged, contrasting the superiority of each approach. Reviews show that vertical programmes, particularly those targeting infectious disease morbidity, may yield short-term and specific health gains (e.g. from vaccination campaigns), but that long-term population health development (e.g. aiming at the reduction of the incidence of non-communicable disease [NCD]) does not unequivocally benefit from such selective approaches (e.g. Magnussen et al. 2004). Vertical programmes work for particular threats, and horizontal programmes contribute to general well-being. In particular, addressing health equity and NCDs does not align well with a selective, vertical approach. Evidence has emerged that, depending on the existing health profile and management of (social) determinants of health in different communities and countries, an appropriate balance between the two should be struck. Building on a mix between vertical and horizontal primary health care, the aspiration should be to engage in the development of comprehensive health strategies accessible to all (Rasanathan et al. 2011).

Community Development and Asset Thinking

In North America, planning emerged as a discipline early in the twentieth century. Initially the planning professional focused on urban development, but soon social planning and other areas such as health and environmental planning were added to the repertoire of the planner. Considering the 'best' ways of planning, experts before long found that the full participation of people in planning considerations was important. What 'full participation' entailed was (and perhaps continues to be) a matter of debate, and Arnstein's 'ladder of participation' as well as Davidson's 'wheel of participation' have contributed significantly to insights into the circumstances and degrees of public participation in the planning endeavour. These views have also made a significant contribution to public health and health promotion practice around the world (Wallerstein 2006), in the Americas (Wallerstein and Duran 2006), and in European Healthy Cities (Boulos et al. 2015; Green and Tsouros 2008).

A second tradition in this arena was driven by Paulo Freire's work in the area of community development through new forms of education, famously called 'the pedagogy of the oppressed' (originally published in 1968 in Portuguese, translated into English in 1970; his *Politics of Education* (1985) gives a good reflective overview). The views espoused by Freire and others in this tradition hinge on a philosophy that all in society should be able to engage with personal and social development equitably, through open forms of democracy and decision-making. In order to attain such a capacity, empowerment was, and remains, a key strategy in (local) (health) development.

Others have taken this important work as a starting point for, for instance, asset-based community development (recognizing that the people in particular social contexts are an important resource for change), deliberative democracy, and a particular form of the latter, participatory budgeting. The 'father of asset-based community development' is John McKnight. He sees community assets as all the potential resources in a community—not only financial, but also the talents and skills of individuals, organizational capacity, political connections, buildings and facilities, and so on (Kretzmann and McKnight 1993). Some authors (e.g. Page-Adams and Sherraden 1997) criticize such a broad conceptualization as assets might be taken to mean 'all good things', and in order to make assets more tangible they prefer to frame them in a more economic manner. Such a view denies, in our view, the fact that social and health equity both depend on much more than financial and resource capability, and also involve culture, history and heritage, and context (Wilkinson and Pickett 2010).

The asset model presented by Morgan et al. (2010) aims to redress the balance between evidence derived from the identification of problems to that which accentuates the positive capability to identify problems jointly and activate solutions, and so promotes the self-esteem of individuals and communities and leads to less dependency on professional services. This can lead to an increase in the number and distribution of protective/promoting factors that are assets for individual- and community-level health. The asset approach should be seen as the 'shiny' side of the coin. The deficit approach remains valuable in responding to acute crises (at individual, community or societal levels), but in evidence terms at least, the asset model may help to further explain the persistence of health and well-being inequities despite increased efforts to do something about it.

Harrison et al. (2004) have defined health assets as resources that individuals and communities have at their disposal and that protect against negative health outcomes, or promote health status. These assets can be social, financial, physical, environmental, or human resources (e.g. education, employment skills, supportive social networks, natural resources) (Harrison et al. 2004). As such, a health asset can be defined as any factor or resource which enhances the ability of individuals, groups, communities, populations, social systems, or institutions to maintain and sustain health and well-being and to help to reduce health inequities. These assets can operate at the level of the individual, group, community, or population as protective (or promoting) factors to buffer against life's stresses. Obviously a balance needs to be struck between the intangible assets (skills, knowledge, intents, and aspirations) and the hardware assets of a community (schools, work, infrastructure, etc.). Even when both are available there may still be a disconnect between the two: individuals, families, and communities may want to improve their health, but insidious factors such as (health) literacy, culture, sexism, and racism may stand in the way of full and equitable access and use. An asset-based health approach should carefully take into account all elements of a complex individual, social and ecological environment.

Effectively mobilizing and empowering communities for their health, health equity, wealth, and well-being is an inherently political enterprise and may upset the status quo. Not all governments, locally or nationally, may see the full benefits of participation and empowerment. The maturity of government and governance styles as well as patterns of accountability, transparency, and responsiveness to need may not always allow for the full mobilization of community assets. We will return to these challenges later in this chapter.

The Ottawa Charter: A Lasting Foundation of the New Local Public Health

Due to a growing recognition that health lifestyle change through traditional behavioural (health education) interventions had limited efficacy, and needed to be embedded in broader social change, the World Health Organization with Health Canada and the Canadian Public Health Association organized the first international conference on 'the move toward a new public health' in Ottawa, in 1986. The conference, followed by a series of global health promotion conferences, culminated in the adoption of the Ottawa Charter (World Health Organization et al. 1986). The Charter defined health promotion as 'the process to enable individuals, groups and communities to increase control over the determinants of health and thereby improve their health'.

The conference and its Charter saw a responsibility to enable, mediate, and advocate a broad view of health and health action in four areas:

- To reorient health services towards a broader, participatory, and health-promoting position in society at any level
- To create supportive social, economic, natural, and built environments to create and sustain health promotion and to address the determinants of health equitably
- To invest in personal skills and community action to drive and complement these actions
- To build healthy public policy, recognizing that health is created across many sectors in society that all have the potential to enhance institutional, community, and personal health

Reviews of the accomplishments of the Ottawa Charter have found that substantial progress has been made in our understanding of the drivers of success in each of these fields. Our understanding of the complex nature of natural, social, political, and commercial determinants of health has increased, as has our appreciation of the impact of policies on all of these. Great advance has been documented in linking ('enabling, mediating, and advocating') individual and community health potential with systemic action on environments for health. The only area where success has been lagging is the reorientation of health services (Ziglio et al. 2011).

The global community of health promoters continues to work on the basis of these principles and advances, and implements these especially in the context of 'healthy settings'—a concept that the Charter launched:

> Health is created and lived by people within the settings of their everyday life; where they learn, work, play and love. Health is created by caring for oneself and others, by being able to take decisions and have control over one's life circumstances, and by ensuring that the society one lives in creates conditions that allow the attainment of health by all its members.

Economic Development and the Role of the World Bank

Health and economic development go hand in hand, although the interface between the two can best be described as 'fuzzy', or in terms of complex systems policy development 'wicked'. The fact that investment in health is a sound economic strategy started to gain traction from the late 1980s and achieved prominence for the first time in the 1993 World Bank's world development report *Investing in Health*. A strong case was made of the importance to national economies and local communities of addressing health and disease factors that impeded full economic development (Jamison 1993). The report was criticized for espousing New Public Management and neoliberal principles of outsourcing and privatizing health as a public good (including, e.g., the supply of safe drinking water) and quantifying the impact of disability on economic development through a measure called the 'disability-adjusted life year' (DALY) (e.g. Navarro 2007). However, it succeeded in placing health promotion and public health management on global and local agendas as legitimate strategies for development. The argument for HiAP, also at the global level between international bodies, has evolved in the past 20 years with the family of UN agencies, including World Bank, UNDP, and WHO, now mobilized for NCD action.

The argument has been developed and refined over the years; for instance, in the Jeffrey Sachs-led WHO Commission on Macro-economics and Health. More recently the WHO Commission on Social Determinants of Health (the Marmot Commission, WHO Commission on Social Determinants of Health (2008)) forcefully indicted unequal economic conditions and pervasive poverty as one of the most critical drivers of health inequity around the world. Impressively, the global Marmot Report has had a number of regional (Europe), national (e.g. Brazil, England), and local (Malmø) reincarnations, highlighting the opportunities and benefits of political action on the social determinants of health. In recent years there has also been a move to take the discourse further, with some starting to address commercial and political determinants of health. Recently WHO and UNDP issued *Guidance Note on the Integration of Noncommunicable Diseases into the United Nations Development Assistance Framework* (2015), an expression of the joint-agency work that was an outcome of the high-level meeting at the UN in which NCDs were given utmost priority. In the *Guidance Note* the vicious cycle of poverty and health is described with great insight into the implications of this perspective for local government action.

Health Equity

The recognition that health is unequally distributed across populations is not new to the twenty-first century. The terminology used for this phenomenon is possibly as political as its causes and consequences. Various terms are pertinent to this discourse, including 'health disparities' and 'health differences' (scholars of the unfair distribution of resources and its consequences in society claim that these are deliberately 'value-free' functional descriptors to obscure the political nature of the issue) and 'the social gradient' (the statistical slope between those at the top of a socioeconomic spectrum and those at the bottom) upon which most health and disease expressions can be mapped. **(In)equality**, some say, is purely a description of that social health gradient, whereas **(in)equity** conveys a view of the moral and social injustice of such differences in society. Wilkinson and Pickett (2010) describe how equitable societies provide and create better opportunities for health for all, including enhanced economic development, sustainability, and educational attainment. Striving for equity is not necessarily a requirement or prerogative of national government alone—it depends and thrives on a vibrant civil society and its political representation, extending from local action to global policy and the other way around.

Equity is a driving concept in various global strategies, including those on climate change, sustainable development, and gender. Particularly in the health domain, the work by the Marmot Commission has been instrumental. Its report reviews the causes and consequences of health inequity, and demonstrates that it is possible to close the gap within a generation. Policy and action at every level are required to mitigate the possible negative influences of globalization on equity; some authors, however, also allude to the significant potential that global connectedness through new social media may have on an equity agenda.

Globalization, the Rise of the Local, and Governance for Health

The idea that we live in a globalized world has become a mainstream perspective in the twenty-first century. Goods, capital, and knowledge travel, sometimes with the speed of light, around the world. Globalization goes beyond the bounded role of the traditional nation-state. Indeed, although countries continue to collaborate and expand their vision in the globalized world, the phenomenon to no small extent is driven by commercial interests—but also by a new global civil society. The latter includes NGOs like Greenpeace, Médecins Sans Frontières, Amnesty International, Human Rights Watch, and the Peoples' Health Movement.

The actions of this variety of actors on the global scene have made the traditional borders of sovereign states more permeable. No country can thrive without interaction, not just with its neighbours but across the globe, and not just with other countries but with such ‘non-state actors’. In discussions about global health governance, experts agree that a new architecture for managing health and health systems in this context is very important. At the same time, new technologies and social media offer new opportunities for knowledge development and community mobilization (de Leeuw et al. 2013).

Local governments around the world see the dissolving integrity of the nation-state as an opportunity to take action. The challenges to the sovereign nature of the nation-state have become prominent during (and in the aftermath of) the SARS epidemic; authors such as Fidler (2007) argue for a new architecture of global health governance (De Leeuw 2013). NCD control, Ebola, HIV/AIDS, the Zika and Chikungunya viruses, and other health issues have become a global health concern, and new options for policy development at the interface between global and local need to be developed. This has happened through the creation of networks of cities around themes such as climate change and sustainability, age-friendly cities, and knowledge and creativity. Assessments of these networks show that such contacts benefit the quality of policy development and actions to improve the quality of life of their citizens. These developments interface with a current discourse about governance.

Governance

Geidne et al. comprehensively review the emergence of the concept of governance as relevant for local health development. They explain that a focus on governance, as complementary to studies of government, derives from a more refined understanding of the scope and nature of the welfare state. This understanding has led to a convergence of ideas that ‘government directed by sovereign politicians is not necessarily the most rational arrangement’ (Geidne et al. 2012, p. 307). Stoker (1998) argues that, despite there being no unequivocal definition of governance, a consensus exists that it refers to the development of governing styles that blur the boundaries between, and within, the public and private sectors. This makes governance a multidimensional and contextually relevant approach to local arrangements for health development, but also a phenomenon that can be construed as a messy research problem (e.g. Sinkovics and Alfoldi 2012), and ‘evidence’ for it must be generated in ways beyond the epidemiological paradigm of (quasi-)experimental studies.

There is a profound connection between governance and health (e.g. Marmot et al. 2008; Plochg et al. 2006; Vlahov et al. 2007). In a foundation report for the WHO European Region Health 2020 strategy, Kickbusch and Gleicher (2012) build on this evidence and argue that there is a difference between health governance and governance for health: (1) the governance of the health system and the strengthening of health systems is called health governance; and (2) the joint action of health and non-health sectors, of the public and private sectors, and of citizens for a common interest is called governance for health. The definition of the latter they propose is

the attempts of governments or other actors to steer communities, countries, or groups of countries in the pursuit of health as integral to well-being through both whole-of-government and whole-of-society approaches.

In many reports and pronouncements the concepts of health governance, health policy, and health action are used interchangeably, especially when they deal with complex intersectoral endeavours. It is useful to distinguish between these, particularly as there is value in seeing an overlapping.

Intersectorality and Governance

Intersectoral governance can be defined as

> the sum of the many ways individuals and institutions, public and private, manage the connections of their common affairs. It is a continuing process through which conflicting or diverse interests may be accommodated and cooperative action may be taken. It includes formal institutions and regimes empowered to enforce compliance, as well as informal arrangements that people and institutions either have agreed to or perceive to be in their interest. (Commission on Global Governance 1995)

In the European Region of WHO, from the early stages of the programme, a commitment to intersectoral governance has been a criterion for designation as a Healthy City. From phase II onwards, cities needed to submit evidence that they had established an intersectoral steering committee (ISC) that would oversee policy and intervention development (Heritage and Green 2013; Lipp et al. 2013). There are no specific requirements to the design or architecture of such ISCs, as they are often driven by unique local contexts and requirements. Whether cities lived up to the expectation beyond their formal application commitments was ascertained via annual reporting templates. Virtually all members of the network reported that they did establish an ISC, although the frequency with which this body met was variable. In some cities it met only once a year, and in others more regularly, up to monthly. In cities where the ISC met annually, the role of the body was more at a systems and regulatory level, such as driving and approving policy development and monitoring of intersectoral deliverables; ISCs that met more regularly tended to engage more directly in the operational aspects of partnership development, such as allocation of resources and direct supervision of working relationships.

Both the strategic and the operational aspects of intersectoral governance are important. In their multiple governance framework, Hill and Hupe (2006) show these different dimensions of governance as complementary requirements for effective and transparent policy development and implementation (Fig. 17.1). Intersectoral governance moves between, and encompasses, an architecture in which implicit and explicit rules at a systems level ('institutional design' in Fig. 17.1) explicitly connect to the way in which individuals in collaborative pro-

Scale of action situations	Constitutive Governance	Directive Governance	Operational Governance
		Action level	
SYSTEM	Institutional design	General rule setting	Managing trajectories
ORGANISATION	Designing contextual relations	Context maintenance	Managing relations
INDIVIDUAL	Developing professional norms	Situation-bound rule application	Managing contacts

Fig. 17.1 Multiple governance framework (Hill and Hupe 2006)

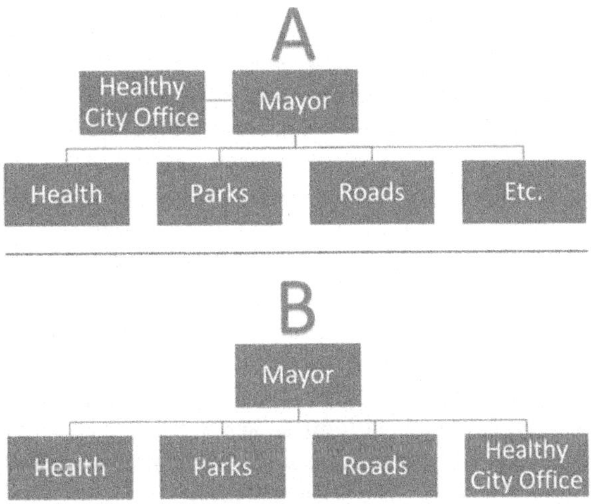

Fig. 17.2 Coordinating Healthy Cities: a staff function (**a**) or line responsibility (**b**)

cesses manage their contacts. Since Healthy Cities in Europe have been deliberately considered a natural laboratory of health policy innovation at the local level (Tsouros 1995), in hindsight it has been appropriate that the specific terms of reference of ISCs have never been spelt out in great detail. This flexibility has allowed the emergence of all types of governance, and an evolution of praxis in which these different levels and types of governance have been tried, tested, and connected.

Regarding the actual architecture of intersectoral governance arrangements in Healthy Cities, all designated cities are required to have a coordinating office. Similar to the flexibility in terms of reference for the ISCs, WHO has not set specific expectations regarding the organizational positioning of this office. There has been an ongoing debate whether this coordinating body should be directly associated with the local government executive office, that is, as a staff unit appended to the Mayor's office (Fig. 17.2, model A), or should be a line unit at a relatively high hierarchical level able to coordinate efforts within government (Fig. 17.2, model B). Both models can serve a distinctive purpose, depending on the nature and maturity of the Healthy City. The evaluation of phase V revealed another type of governance architecture whereby Healthy Cities increasingly integrate and devolve the responsibility for intersectoral action for health throughout both the government and civil society.

Intersectorality and Action

Intersectoral action is the engagement of relevant sectors, both within and outside the public policy arena, in the implementation of activities, programmes, and projects that have a multidimensional nature. Obesity, for instance, has lifestyle-choice dimensions but must also be addressed through structural interventions in the obesogenic environment (Kirk et al. 2010), e.g. in public transport, food security, and community development. Ideally this requires a policy and managerial context that embraces the values of HiAP described below, and it is important that different sectoral stakeholders collaborate effectively.

Lipp et al. (2013) show that from phase II through phase IV of the European Healthy Cities programme, intersectoral action has expanded and strengthened. For example, the 31 cities participating in both phase III and phase IV increased the extent of partnership working in all sector studies: health services, social services, education, urban planning, voluntary, environmental protection, transport, and economic development. For phase V, Farrington et al. (2015) show that Healthy Cities, in trying to address the prevention of non-communicable disease, also make explicit efforts to work intersectorally in distal determinants of health. European Healthy Cities, they find, recognize that to make healthier choices easier requires appropriate structuring of the upstream determinants of health. For example, interventions in the built environment to make active living an easier choice include investment in city sports and exercise facilities, investment in cycling infrastructure, and redesignating streets for pedestrians only.

Successive European Healthy Cities evaluations therefore show that local governments are not only embracing intersectoral work through the creation and maintenance of appropriate governance architectures, but are also increasingly deploying resources in terms of operational action to deal with complex problems in dynamic partnerships. Following the programme logic of realist synthesis, this suggests that the social determinants of health are being addressed more effectively and sustainably.

Intersectorality and Policy

Elsewhere I have argued, with Clavier and Breton, that ‘policy’ can mean different things to different actors at the same time. A bureaucrat may use the word ‘policy’ to indicate ‘standard operation procedures’, a community activist may mean by it ‘a dictate coming from above’, and a politician may use the term to denote ‘an intention to change'. Not surprisingly, the same diversity is found in the field of political science. For the purpose of this chapter, and to distinguish meaningfully between governance, action, and policy, we define the latter as

> the expressed intent of government to allocate resources and capacities to resolve an expressly identified issue within a certain timeframe (De Leeuw et al. 2014).

Such an approach clearly distinguishes between the policy issue, its resolution, and the tools or policy instruments that should be dedicated to attaining that resolution.

Thinking about intersectoral health policy has evolved over the years. Healthy Cities engage enthusiastically—and beyond mere rhetoric—in the development of health and health equity in all policies. Building on a strong foundation in the various political statements on Healthy Cities over the years and most recently in the Athens Declaration (Tsouros 2015), local governments work with diverse stakeholders from the public and civil society sectors to develop such policies. The nearly three decades of Healthy City development are clearly leaving a legacy, in that Healthy Cities manage the politics and logistics of interorganizational work effectively. This is clearly dependent on strong yet flexible governance arrangements and demonstrated commitments to the action component of intersectorality (McQueen et al. 2012a, b)

In the evidence on intersectoral policy development and implementation compiled for European Healthy Cities (de Leeuw et al. 2015), there was an interesting mix between more traditional health approaches, such as a programme on active living in Izhevsk, Russian Federation, and initiatives where the health sector has more peripheral ownership, such as a programme on sustainability in Amaroussion, Greece. This is precisely the message for effective HiAP development—that the health sector has the capacity to share, redistribute, and even disavow ownership of policy initiatives beyond its traditional remit. Healthy Cities show that such actions do not compromise but strengthen the integrity of health sector policy-making capacity.

The Consequence: Health in All Policies

The above developments have created a strong historical footing for the development of Health in All Policies. They are, however, often seen as abstract global concepts and aspirations rather than operational local inspirations. In this second decade of the third millennium there are, nevertheless, many reasons why local governments and their communities in particular should be inspired to make a real difference. We compile five themes that drive further action.

The Health Promotion Evidence Base

It is important for society and its communities to spend their resources where they matter. Although it can be easily contested what 'where it matters' actually means in different contexts (for instance, a national re-election campaign of a politician based in a megacity would probably not recognize the needs of rural and remote communities to their fullest magnitude), this idea has driven the development of evidence-based (health) policy. Substantial impact on this broader aspiration was made by the evidence-based medicine mantra that has its foundation in the work of Archibald Cochrane. He found that many medical practices were not firmly rooted in evidence of effectiveness (whether something produces the intended result) or efficiency (how well it produces that result). The consequence of this position was that decision-makers, both in policy and in practice, invested in approaches to demonstrate the effectiveness of medical procedures.

This effort has had its influence on policies that espouse a broad social model of health and health promotion, both globally and locally. The methods to generate evidence of effectiveness on this arena are, naturally, different from the often controlled circumstances under which clinical procedures can be tried and tested. Where in clinical environments an assumption is that an experimental group can be matched with a control group, is it much harder in reality to find the perfect experimental match for, for example, a barrio in Medellin, in order to test the effectiveness of social investment.

Yet very good progress is being made in demonstrating the effectiveness and efficiency of health policy and health promotion. Evaluation efforts around Healthy Cities show that it is easier to achieve public participation and good governance for health at the local level. Equity is a concept close to the heart of many local politicians. International research shows that health and health equity impact assessments are not just highly effective tools for measuring the consequences for population health of broad social, environmental, and economic change, but also have a significant impact on the quality and sustainability of policy development and implementation. Concepts like Healthy Urban Planning that embrace a wider view of transport and mobility show not just health, but far broader social improvement.

The Ottawa Charter for Health Promotion also launched the ideas of settings for health ('where people live, love, work and play') as a critical aspect of health development. Significant evidence has been accumulated on the efficacy and health impacts of initiatives beyond Healthy Cities, for instance in Health Promoting Schools (globally the most significant network of settings for health with tens of thousands of participating primary and secondary schools, currently expanding into kindergarten environments), Health Promoting Market Places, Healthy Islands (notably in the Pacific through the Yanuca Declaration, linked to the Barbados Programme of Action), Health Promoting Universities, Health Promoting Prisons, and Healthy Transport.

This evidence continues to be compiled by international organizations like WHO, UNDP, IUHPE (the International Union for Health Promotion and Education), and other global agencies, but also through networks of civil society like international city networks (e.g. C40 and Healthy Cities) and academia. There is, in fact, 'meta-evidence' that networking for evidence generation enhances the quality, relevance, and responsiveness for glocal action.

Universal Health Coverage

The enthusiasm and vigour that were originally part of the Alma Ata Declaration on Primary Health Care were rekindled a few years ago when the World Health Assembly formally re-endorsed the broad social nature of the Declaration. It was further sustained by a global campaign to work towards universal health coverage (UHC) at all levels of governance and health system operation. It is defined as

> ensuring that all people can use the promotive, preventive, curative, rehabilitative and palliative health services they need, of sufficient quality to be effective, while also ensuring that the use of these services does not expose the user to financial hardship (World Health Organization 2016).

In some instances UHC is conceived as an exclusively financial issue that requires fiscal programmes and discipline to redistribute key social resources. The evidence shows that, in fact, the monetary dimension is maybe the least problematic to address. Moving from divisive health (delivery) services towards inclusive ones requires much more than the reallocation of resources.

UHC has many benefits and creates ample win–win situations, apart from the obvious health gain. They secure a (human) rights-based perspective on population health, have the potential to organize and rally communities for social and economic development, and have the strong potential for higher quality health information collection and management, thus adding to more bespoke evidence-based local health policy.

Local governments may not always have control over fiscal opportunities and the management of health facilities and professionals. Often these are organized and financed at higher levels of governance, and partly for good reason: not every town needs highly specialized neurosurgeons and expensive fMRI scanners. But the essential population-based 'first point of contact' with the health system, i.e. primary care, is by its very nature integrated in local communities – even where there may be no doctor. Community health workers and local health posts play critical roles in maintaining and integrating universally accessible and appropriate health and social support; they are also the natural champions of (local) community development. Even when there are no formal governance arrangements for local government institutions (and in slum areas may even have an informal nature), these professionals and their operational bases are very much part of the social and political landscape of local government.

UHC at point of delivery is therefore a concern for local action, whether it has been formalized as a local government remit or not. Experiences from the Americas, e.g. of people-centred programmes in Mexico and Brazil, show that UHC is possible and yields significant dividends, not just for population health but more broadly for social development (Quick et al. 2014). Evidence suggests that success of UHC schemes depends on the presence of (1) the strength of organized progressive groups in local communities; (2) the potential of mobilizing adequate economic resources; (3) the absence of significant societal divisions; (4) a weakness of institutions that might oppose it (such as for-profit hospital enterprises); and (5) a skilful identification and opening up of windows of opportunity by (local) policy entrepreneurs (McKee et al. 2013).

Determinants of Health

The description of the social gradient in health (that is, the fact that health parameters like mortality, morbidity, and life expectancy follow the patterns of the distribution of wealth, prestige, status, and education in society) has moved from a mere epidemiological curiosity to a political issue. Increasing numbers of governments endeavour to place health equity and its causes high on their political agendas. This happens with varying degrees of success.

There have been arenas of governance with such a strong belief in their equitable nature that a debate around the sheer existence of health inequity in those societies and communities has been unimaginable. There are also cases where existing inequity is attributed to personal lifestyle choice rather than to broader determinants of health. This so-called lifestyle drift can be inspired either by uninformed behaviourist tendencies (assuming that all human behaviour is entirely within the control of the individual) or by political ideologies like conservative liberalism (assuming that the fate of societies can be entirely attributed to the resourcefulness of its individual members). The evidence, however, demonstrates that individual choice is determined by social, environmental, cultural, economic, natural, and built environments. Clearly these interact in extremely intricate ways. They are also the result of political preference, and of commercial interest.

Both England's report *Fair Society Healthy Lives* and the Swedish *Socially Sustainable Malmö* stress the interrelation between policies that aim to

- Give every child the best start in life
- Enable all children, young people, and adults to maximize their capabilities and have control over their lives
- Create fair employment and good work for all
- Ensure a healthy standard of living for all
- Create and develop healthy and sustainable places and communities
- Strengthen the role and impact of ill-health prevention

A Reflection on Silos: How Did We Get into Them, and How to Move Beyond Them

One of the most prominent challenges in establishing cross-cutting policies and actions is to move outside traditional disciplinary and sectoral boundaries—silos. How did we end up with, and in, those silos?

The classic ideal of a good citizen was that of the Renaissance man, of whom perhaps the best example is Leonardo da Vinci (1452–1519), the Italian polymath, painter, sculptor, architect, musician, mathematician, engineer, inventor, anatomist, geologist, cartographer, botanist, and writer. Typically Da Vinci would not regard himself as any of these 'professions'—he accomplished all of this without silos. And certainly he was not unique—many advances to modern society have been made by men and women who branched out across scientific disciplines and the arts.

This comprehensive integration of the sciences and the arts, not just in one person but in a world view, was challenged in the period of Enlightenment, in the sixteenth and seventeenth centuries when the world view evolved towards one of a separation of body and mind, and of distinctly different disciplines arguing that the observed world could be understood mainly through rigorous analysis, that is, deconstructing it into its unique parts. Scholars started to focus on particular bodies of knowledge and developed strong theories for each. When in the nineteenth century medicine, as one of these disciplines, became highly professionalized (influenced by the industrial revolution and a growing upwardly mobile middle class) these disciplines started to specialize even further. The process is sometimes called 'hyperspecialization' and today can be witnessed in the proliferation of academic journals focusing on quite particular areas of interest.

Hyperspecialization is one reason that modern societies operate in management and policy silos. Professionalization is another. This is the process of establishing acceptable qualifications, a professional body or association to oversee the conduct of members of the profession, and some degree of demarcation of the qualified from unqualified amateurs. The process creates a hierarchical divide between the knowledge authorities in the professions and a deferential citizenry and creates strong patterns of inclusion and exclusion: building a bridge requires an engineering professional, taking someone to court needs legal professionals, and treating disease must involve qualified medical professionals.

Specialist and professional segregation are continuously challenged. In the early twentieth century, for instance, debate raged in North America whether public health was within the remit of the medical profession. The matter was resolved with the publication of the Flexner Report (Flexner et al. 1910), urging a proper 'scientific' approach to clinical medicine teaching, thus excluding public health. In Europe—and in countries that followed a European model of health professionalization—medical education continued to include public health matters under the banner of 'social medicine'.

Specialization and professionalization created formidable commercial and political forces to maintain and protect their status quo. Even when the evidence base concerning social determinants of health rationally dictates collaboration and integration of efforts, these forces often prevent successful and effective action and policy development.

There is a growing body of rhetorical and evidence-based knowledge that addresses these problems. Effective partnering for health starts with the recognition that the capacities of a discipline or specialty in isolation are insufficient to make a difference. The process that enables such a recognition requires the presence of leadership, communication and analytical skills, and something that can be called social entrepreneurship (the capacity to advocate, mediate, and manage opportunities and differences in diverse communities of policy and practice). Firm pronouncements by executive offices (e.g. a mayor, CEO, or spiritual leader) in support of reaching out to other sectors are indispensable. Reliable and sustainable grounding of such positions in community action helps maintain momentum.

Such approaches to removing the walls of silos play out at a relatively high level of abstraction; a workforce that is receptive to interdisciplinary work and has been trained to reach out to others is, of course, vital, too. Increasingly we see programmes and curricula across primary, secondary, and tertiary education that do in fact embrace such values.

Health in All Policies: State of the Art and Local Opportunity

Policy and Action

The terms *intersectoral action* (sometimes *intersectorial action*) and *multisectoral action* have been part of the rhetorical repertoire of public health and health promotion since the mid-1970s. The terms achieved credence through the Alma Ata Declaration, the Ottawa Charter, and a series of other pronouncements by global bodies including WHO. The international discourse has also included arguments and evidence around variations of ideas about working together for health on the spectrum networking–coordinating–cooperating–collaborating (see Lipp et al. 2013 for a brief discussion). Although there may be conceptual shades of grey around the interpretation of these terms, this focus of public health and health promotion clearly hinges on the noun *action*.

Agencies, individuals, groups, and communities may come together to jointly *act* on health concerns or determinants of health—but this does not necessarily mean that these actions are either *driven by policy* or *result in policy*. A series of case studies, however, are starting to build an evidence base that demonstrates that successful intersectoral action may inspire the need for HiAP. HiAP, however, may not necessarily lead to intersectoral action: for instance, policies to limit lead (Pb) content in paints and gasoline are singularly industrial–economic in nature, and apart from commitments required by industry do not necessitate the deep involvement of other government sectors.

Considering the importance of successful intersectoral action for the development of HiAP, it may be worthwhile to reflect on the reasons why it appears such a challenge to break the walls of the silos and move beyond pithy interests. Irwin and Scali (2010), at the request of WHO, assessed the reasons for the failure of intersectoral action and policy to become an 'easy', 'mainstream' effort. They show that intersectoral action for health failed because (1) it was driven by the health sector alone; (2) the intersectoral rhetoric was effectively challenged by the absence of supporting empirical evidence and research programmes to establish such evidence; (3) public health was 'messed up' by New Public Management ideologies that moved health responsibilities out of government into private and civil society spheres and complicated matters; and (4) international donors and healthcare agencies achieved rapid success with single disease-focused vertical delivery programmes.

It appears that, with the resurgence of primary health care, the strengthening of UHC, and an increasing commitment to equity around the world, the seesaw with neoliberalism and free market principles on one side and deeper human values on the other have become balanced again, and that the political climate for successful intersectoral action initiatives is more positive. This is expressed in the commitment to HiAP formation and implementation, but the limitations and challenges in the comprehensive embrace of integral action will remain and need to be addressed.

This discussion on the critical connection between action and policy raises the question of what the process to attain and sustain Health in All Policies will entail, and which actors need to be engaged. McQueen et al. (2012a, b) describe various governance models for HiAP. These have been mapped onto the different elements of the policy process (Fig. 17.3) and hinge on seven best practice models for HiAP implementation. Different (groups of) government and non-government agencies can play different roles during the HiAP process. Figure 17.3 describes some of the governance parameters for positioning HiAP development within government structures. In addition to this, we have also identified eight institutionally different structural interaction patterns (Fig. 17.4) that describe the linkages between the health care system and its public policy agencies (e.g. a Ministry of Health at the national or provincial level; or a public servant within a local government agency with a public health remit), other public sector agencies and executives (the office

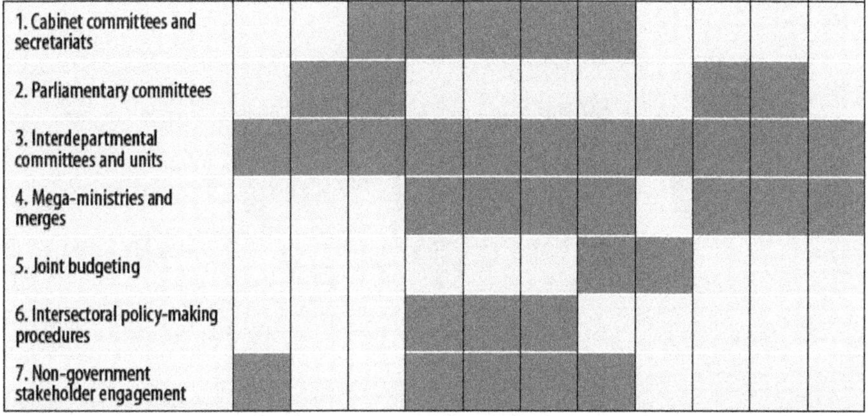

Fig. 17.3 Actors engaged in governance for HiAP. WHO (2015), based on McQueen et al. (2012a, b)

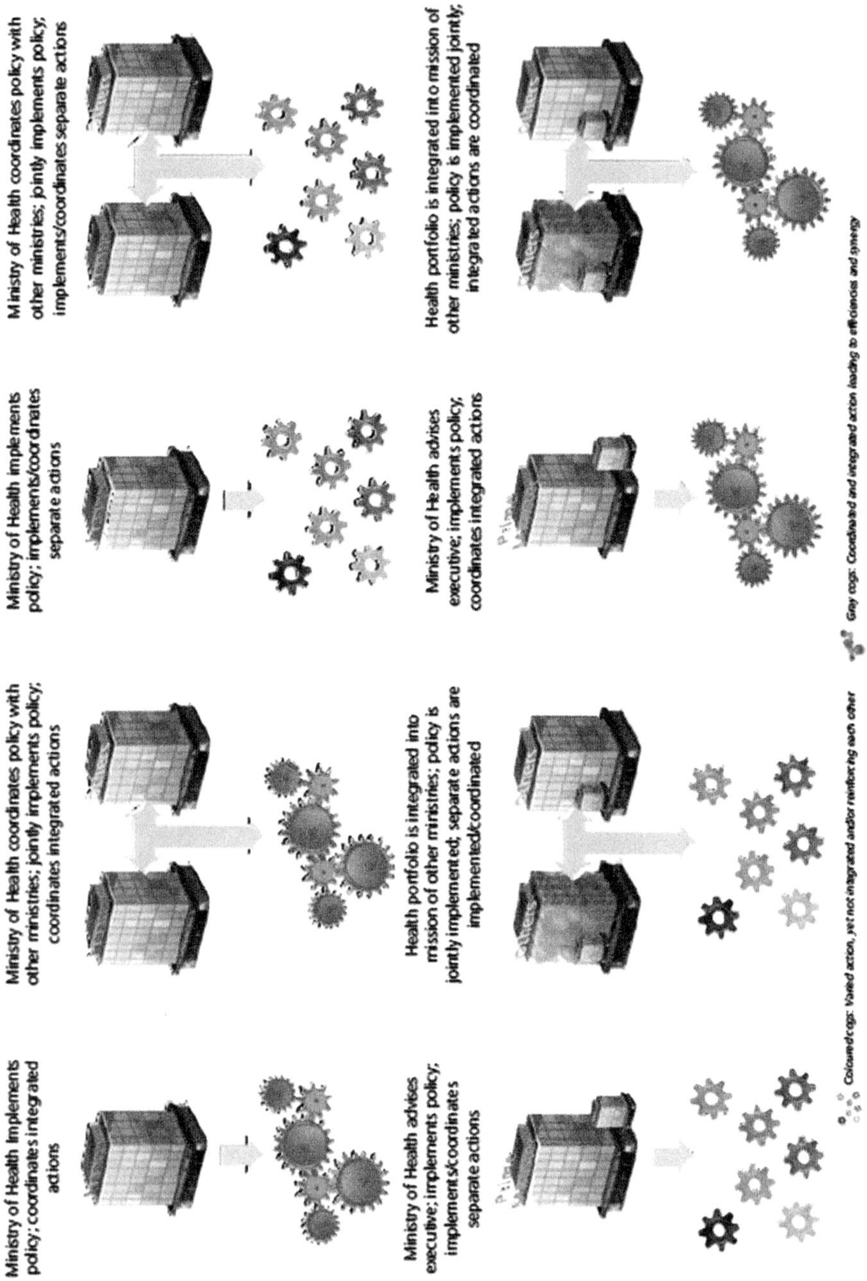

Fig. 17.4 Eight ways to coordinate between sectors, and implement HiAP in integrated or separate action

of the president or mayor, for instance), and the shape that intersectoral action for health and health equity may take.

It would be an illusion to think that inter-agency integration or collaboration will automatically lead to integrated action. There are many cases where collaboration still leads to a multitude of singular projects without a lot of systemic and synergy consequences. An example might be inter-agency collaboration on road safety: even where there may be agreement on the nature of a road safety issue (e.g. child fatalities) and action is taken, those actions may not be coordinated and sometimes may be counterproductive. Road design, improvements, trauma response, safety communications (signage), and behaviour communications (promoting seatbelts and helmets) should be jointly assessed and developed. If they are not, the whole of the roadscape may be messy and confusing and the total effectiveness of interventions significantly decreased.

On the other hand, we know examples where the health sector successfully drives systemic and sustainable intersectoral action in cases where the sector is given the opportunity to engage with local communities. An example would be the integration of health checks, childcare, and (health) literacy training in 'casas de cultura' (Latin America) or community hubs.

Key to the success of *any* approach is the assessment of win–win opportunities, playing to the strengths of each sector and community, 'going with the flow' rather than against it, demonstrating co-benefits to those involved (and that goes beyond government sectors), avoiding turf wars, and a more comprehensive appreciation of different forms of evidence that are generated and applied beyond the health system alone. This includes the exploitation of successful inter/multisectoral action driven by stakeholders outside the health and public sectors.

Partner for Purpose

Intersectoral action and HiAP must not happen for their own sake. Collaboration without joint ownership and outcomes, and integrated policy addressing one-dimensional issues, is senseless. Many lessons have been learned from the integrated partnership agenda in health promotion, particularly in Healthy Cities (e.g. Lipp et al. 2013). Planned action to connect, integrate, and scope the integral policy agenda needs to address the following evidence-based stages:

- Map and recognize organizational mission and resource capacities and acknowledge the boundaries of the traditional organizational footprint.
- Describe organizational challenges in addressing issues and populations that permeate and move beyond the organization's legitimate area of concern.
- Map and include organizations that cover the same, similar, or different issues and populations, or share the same, similar, or different approaches and interventions to deal with these.
- Recognize the legitimate potential of other stakeholders to be involved in intersectoral action or integral policy development and strive for transparency in sharing these views.

- Scope the dimensions of probable and possible collaboration and factors that may stand in the way of respectful joint action.
- Involve real authorities and decision-makers, including organization executives as well as street-level bureaucrats (frontline implementation personnel who deal with inter-sectoral action challenges on an everyday basis), in shaping the joint agenda.
- Formalize and celebrate each of these stages, as far as possible including individuals, communities, and neighbourhoods that are at the 'pointy end' of the implementation of action and policy outputs.
- Make all stakeholders in these processes, as far as is culturally and organizationally possible, accountable for their actions, but apply the 'Chatham House Rule' (full confidentiality of sensitive and strategic considerations) wherever necessary.

Identify Existing Supportive Structures and Processes and Agendas for Their Development

In this book we have seen that in many local government areas there are already effective structures and processes that would further facilitate the development of inter-sectoral action for health and a strongly associated integral policy development potential. Such structures and processes include

- An engaged and empowered community
- Successful experience in deliberative democratic and participatory processes
- Successful experience in partnerships and collaboration for health and well-being
- A broad recognition of the urgency of NCD strategies, supported at executive and council levels
- A broad recognition of the 'causes of the causes' of ill health, supported at executive and council levels
- An existing agenda to strengthen or move towards universal health coverage
- Existing role models and examples of inter-sectoral action and HiAP in other local governments in the countries, for instance, connected through Healthy Communities networks
- Vertical integration of governance models for inter-sectoral action and HiAP between different levels of government
- Existing evidence of social, economic, and sustainability win–win situations, and ongoing connections with local and national agencies and structures that may support the creation and maintenance of such evidence (e.g. local and national universities and NGOs)

Build Lasting Capacity

Addressing the complexity of modern health and health equity issues requires a lasting, continuous process. The establishment and implementation of one Health in All Policy in one place cannot be considered the end point. It is a stage in an evolutionary practice: the policy needs to be reviewed, adapted, and renewed to meet the exigencies that it has created. The context, and local stakeholders, in which this happens will constantly change. Political shifts may require a renewal of executive commitment; evolving community concerns will dictate ongoing participatory consultative action; and technological advances may inspire new solutions.

The local government apparatus will require a firm grounding in flexible understanding of the foundations of inter-sectoral action and HiAP, and the processes required to maintain and develop its potential and impact. The above steps, when documented and conscientiously applied, form a local basis for sustained capacity to address new complex health issues through HiAP and inter-sectoral action. A form of 'corporate memory' is required to keep such lessons on the radar, and a public repository (virtual or real) can be such a resource.

Various organizations, often at the interface of policy and practice, offer capacity-building programmes that engage with real-life environments and aim to integrate new understanding and improved potential for sustained change in local health development. One example is the Learning by Doing programme (Harris-Roxas and Harris 2007; Pennington et al. 2015) that makes local government agents across sectors engage in, and reflect on, health impact assessments. Another example is the efforts of the Victoria Health Promotion Foundation (Australia) to build capacity for local operators to include broad determinants of health thinking (Environments for Health) in actual processes towards the development of compulsory municipal public health plans through its 'Leading the Way' programme (VicHealth 2002).

Conclusion

We started this chapter by asserting that many local initiatives merit a project approach with dedicated temporal and resource dimensions. But to drive those projects in an integrated vision they should transcend themselves, transforming into a perspective that connects an overall vision for the future of a glocal issue or environment. We have described how such vision has been given momentum by developments both local and global, firmly grounded in a set of coherent values including equity, participation, sustainability, and accountability. It is not just preferable to work towards policy for health: it is the *only* thing to do to respond to the needs of communities, and through integrated connections between sectors, in Health in All Policies.

References

Adler, B. (Ed.). (1981). *The Reagan Wit.* Ottawa, IL: Caroline House Publishers.

Ahrendt, H. (1970). *On violence.* New York: Harcourt, Brace and World.

Baum, F., Lawless, A., MacDougall, C., Delany, T., McDermott, D., Harris, E., & Williams, C. (2015). New norms new policies: Did the Adelaide Thinkers in Residence scheme encourage new thinking about promoting well-being and Health in All Policies? *Social Science and Medicine, 147*, 1–9.

Boulos, M. N. K., Tsouros, A. D., & Holopainen, A. (2015). 'Social, innovative and smart cities are happy and resilient': Insights from the WHO EURO 2014 International Healthy Cities Conference. *International Journal of Health Geographics, 14*(1), 3.

Callon, M. (1984). Some elements of a sociology of translation: Domestication of the scallops and the fishermen of St Brieuc Bay. *The Sociological Review, 32*(S1), 196–233.

Carey, G. (2016). *Grassroots to government. Creating joined-up working in Australia*. Melbourne: Melbourne University Press.

Carey, G., & Crammond, B. (2015). What works in joined-up government? An evidence synthesis. *International Journal of Public Administration, 38*(13–14), 1020–1029.

Christensen, J. (2004). Asking do-gooders to prove they do good. *New York Times, 3*, 7.

Christensen, T., & Lægreid, P. (2007). The whole-of-government approach to public sector reform. *Public Administration Review, 67*(6), 1059–1066.

Commission on Global Governance. (1995). *Our global neighbourhood: The report*. Oxford University Press.

Corburn, J., & Karanja, I. (2014). Informal settlements and a relational view of health in Nairobi, Kenya: Sanitation, gender and dignity. *Health Promotion International, 31*(2), 258–269.

de Leeuw, E. (2001). Global and local (glocal) health: The WHO healthy cities programme. *Global Change and Human Health, 2*(1), 34–45.

de Leeuw, E. (2013). Global health governance. *Australian Journal of Political Science, 48*(2), 251–253.

de Leeuw, E., Green, G., Spanswick, L., & Palmer, N. (2015). Policymaking in European healthy cities. *Health Promotion International, 30*(Suppl. 1), i18–i31.

de Leeuw, E., & Polman, L. (1995). Health policy making: The Dutch experience. *Social Science and Medicine, 40*(3), 331–338.

de Leeuw, E., Tang, K. C., & Beaglehole, R. (2006). Ottawa to Bangkok—Health promotion's journey from principles to 'glocal' implementation. *Health Promotion International, 21*(Suppl. 1), 1–4.

de Leeuw, E., Townsend, B., Martin, E., Jones, C. M., & Clavier, C. (2013). Emerging theoretical frameworks for global health governance. In C. Clavier & E. de Leeuw (Eds.), *Health promotion and the policy process*. Oxford: Oxford University Press.

de Leeuw, E., Clavier, C., & Breton, E. (2014). Health policy—Why research it and how: *Health political science. Health Research Policy and Systems, 12*(1), 1.

Díez, A. E., Alzate, P., Restrepo, V., & Bohórquez, A. (2015). The reintroduction of grid-connected alternatives into the planning of mass transportation systems in Colombia. *Energy and Sustainability VI, 195*, 293.

Dubos, R. (1959). *Mirage of health: Utopia, progress, and biological change*. Doubleday.

Etzioni, A. (1967). Mixed-scanning: A 'third' approach to decision-making. *Public Administration Review, 27*(5), 385–392.

Etzioni, A. (1986). Mixed scanning revisited. *Public Administration Review, 46*(1), 8–14.

Farrington, J., Faskunger, J., & Mackiewicz, K. (2015). Evaluation of risk factor reduction in a European City Network. *Health Promotion International, 30*(Suppl. 1), i86–i98.

Fidler, D. P. (2007) Architecture amidst anarchy: Global health's quest for governance. *Global Health, 1*(1).

Flexner, A., Pritchet, H., & Henry, S. (1910). *Medical education in the United States and Canada bulletin number four* (The Flexner Report). New York, NY: The Carnegie Foundation for the Advancement of Teaching.

Freire, P. (1985). *The politics of education: Culture, power, and liberation*. Greenwood Publishing Group.

Gagnon, F., & Kouri, D. (2008). *Terms relating to integrated governance*. National Collaborating Centre for Healthy Public Policy. Institut national de santé publique du Québec (INSPQ), Québec. Retrieved from http://www.ncchpp.ca/docs/GouvIntGlossaireAN.pdf.

Geidne, J., Fröding, K., Montin, S., & Eriksson, C. (2012). Implementation structure and participation at neighbourhood level-a multiple case study of neighbourhood development in Sweden. *Systemic Practice and Action Research, 25*(4), 305–322.
Goumans, M., & Springett, J. (1997). From projects to policy: 'Healthy cities' as a mechanism for policy change for health? *Health Promotion International, 12*(4), 311–322.
Green, G., & Tsouros, A. (Eds.). (2008). *City leadership for health. Summary evaluation of Phase IV of the WHO European Healthy Cities Network.* Copenhagen: World Health Organisation Regional Office for Europe.
Harrison, D., Ziglio, E., Levin, L., & Morgan, A. (2004). *Assets for health and development: Developing a conceptual framework.* Venice: European Office for Investment for Health and Development, World Health Organization.
Harris-Roxas, B., & Harris, P. (2007). Learning by doing: The value of case studies of health impact assessment. *New South Wales Public Health Bulletin, 18*, 161–163.
Heritage, Z., & Green, G. (2013). European National Healthy City Networks: The impact of an elite epistemic community. *Journal of Urban Health, 90*(1), 154–166.
Hill, M., & Hupe, P. (2006). Analysing policy processes as multiple governance: Accountability in social policy. *Policy and Politics, 34*(3), 557–573.
Irwin, A., & Scali, E. (2010). *Action on the social determinants of health: Learning from previous experiences.* Social Determinants of Health Discussion Paper 1 (Debates). Geneva: World Health Organization.
Jamison, D. (1993). Investing in health. *Finance and Development, 30*(3), 2.
Kickbusch, I. (1999). Global+local=glocal public health. *Journal of Epidemiology and Community Health, 53*(8), 451.
Kickbusch, I. (2012). Understanding the rhizome effect: Health promotion in the twenty-first century. In *Health promotion in Canada: Critical perspectives on practice* (pp. 308–313). Toronto: Canadian Scholars' Press.
Kickbusch, I., & Gleicher, D. (2012). *Governance for health in the 21st century.* Geneva, Switzerland: World Health Organization.
Kirk, S. F., Penney, T. L., & McHugh, T. L. (2010). Characterizing the obesogenic environment: The state of the evidence with directions for future research. *Obesity Reviews, 11*(2), 109–117.
Koehn, P. H. (2008). Underneath Kyoto: Emerging subnational government initiatives and incipient issue-bundling opportunities in China and the United States. *Global Environmental Politics, 8*(1), 53–77.
Kretzmann, J. P., & McKnight, J. L. (1993). *Building communities from the inside out. A path toward finding and mobilizing a community's assets.* Evanston, IL: Northwestern University, Center for Urban Affairs and Policy Research.
Larson, J. S. (1996). The World Health Organization's definition of health: Social versus spiritual health. *Social Indicators Research, 38*(2), 181–192.
Latour, B. (1987). *Science in action: How to follow scientists and engineers through society.* Harvard University Press.
Lipp, A., Winters, T., & de Leeuw, E. (2013). Evaluation of partnership working in cities in phase IV of the WHO Healthy Cities Network. *Journal of Urban Health, 90*(1), 37–51.
Little, P. E. (2005). Indigenous peoples and sustainable development subprojects in Brazilian Amazonia: The challenges of interculturality*. *Law and Policy, 27*(3), 450–471.
Magnussen, L., Ehiri, J., & Jolly, P. (2004). Comprehensive versus selective primary health care: Lessons for global health policy. *Health Affairs, 23*(3), 167–176.
Marmot, M., Friel, S., Bell, R., Houweling, T. A., & Taylor, S. (2008). Closing the gap in a generation: Health equity through action on the social determinants of health. *The Lancet, 372*(9650), 1661–1669.
McKee, M., Balabanova, D., Basu, S., Ricciardi, W., & Stuckler, D. (2013). Universal health coverage: A quest for all countries but under threat in some. *Value in Health, 16*(1), S39–S45.
McQueen, D. V., Wismar, M., Lin, V., Jones, C. M., & Davies, M. (2012b). *Intersectoral governance for health in all policies—Structures, actions and experiences.* Observatory Studies Series No. 26. Copenhagen: WHO Regional Office for Europe.

McQueen, D., Wismar, M., Lin, V., & Jones, C. M. (2012a). Introduction: Health in All Policies, the social determinants of health and governance. In *Intersectoral governance for health in all policies. Structures, actions and experiences*. Copenhagen: World Health Organization Regional Office for Europe.
Morgan, A., Ziglio, E., & Davies, M. (Eds.). (2010). *Health assets in a global context: Theory, methods, action*. Springer Science and Business Media.
Navarro, V. (Ed.). (2007). *Neoliberalism, globalization, and inequalities: Consequences for health and quality of life*. Baywood Pub.
Page-Adams, D., & Sherraden, M. (1997). Asset building as a community revitalization strategy. *Social Work, 42*(5), 423–434.
Pareschi, A. C. (2002). *Desenvolvimento sustentável e pequenos projetos: Entre o projetismo, a ideologia e as dinâmicas sociais*. Ph.D. Thesis. Anthropology Department, Univ. of Brasilia.
Pennington, A., Dreaves, H., Scott-Samuel, A., Haigh, F., Harrison, A., Verma, A., & Pope, D. (2015). Development of an urban health impact assessment methodology: Indicating the health equity impacts of urban policies. *The European Journal of Public Health*, ckv114.
Plochg, T., Delnoij, D. M. J., Hogervorst, W. V. G., Van Dijk, P., Belleman, S., & Klazinga, N. S. (2006). Local health systems in 21st century: Who cares?—An exploratory study on health system governance in Amsterdam. *The European Journal of Public Health, 16*(5), 559–564.
Pollitt, C. (2003). Joined-up government: A survey. *Political Studies Review, 1*(1), 34–49.
Quick, J. D., Canavan, C. R., & Jay, J. (2014, October) People-centered health systems for UHC. *Strengthening Health Systems*. doi:10.7196/shs.9.
Rasanathan, K., Montesinos, E. V., Matheson, D., Etienne, C., & Evans, T. (2011). Primary health care and the social determinants of health: Essential and complementary approaches for reducing inequities in health. *Journal of Epidemiology and Community Health, 65*(8), 656–660.
Rivett, G. (1998). *From cradle to grave: Fifty years of the NHS*. London: King's Fund.
Rudolph, L., Caplan, J., Ben-Moshe, K., & Dillon, L. (2013). *Health in all policies: A guide for state and local governments*. Washington, DC/Oakland, CA: American Public Health Association and Public Health Institute.
Ruger, J. P. (2005). Health, capability, and justice: Toward a new paradigm of health ethics, policy and law. *Cornell Journal of Law and Public Policy, 15*, 403.
Shipan, C. R., & Volden, C. (2008). The mechanisms of policy diffusion. *American Journal of Political Science, 52*(4), 840–857.
Sinkovics, R. R., & Alfoldi, E. A. (2012). Progressive focusing and trustworthiness in qualitative research. *Management International Review, 52*(6), 817–845.
Spinoza, B. (1670). Tractatus Theologico-Philosophicus. In J. Van Vloten, & J. P. N. Land (Eds.). (1914). *Benedicti de Spinoza Opera—quotquot reperta sunt* (3rd ed., Vol. 2). The Hague: Martinus Nijhoff.
Ståhl, T., Wismar, M., Ollila, E., Lahtinen, E., & Leppo, K. (2006). *Health in all policies. Prospects and potentials*. Helsinki: Finnish Ministry of Social Affairs and Health.
Stoker, G. (1998). Governance as theory: Five propositions. *International Social Science Journal, 50*(155), 17–28.
Swyngedouw, E. (1992). The Mammon quest: 'Glocalization', interspatial competition and the mone-tary order: The construction of new scales. In M. Dunford, & G. Kafkalis (Eds.), *Cities and regions in the New Europe. The global–local interplay and spatial development strategies* (pp. 39–67). London: Bel-haven Press.
Tsouros, A. D. (1995). The WHO Healthy Cities Project: State of the art and future plans. *Health Promotion International, 10*(2), 133–141.
Tsouros, A. D. (2015). Twenty-seven years of the WHO European Healthy Cities movement: A sustainable movement for change and innovation at the local level. Health Promotion International. In E. de Leeuw, A. Tsouros, & G. Green, (Eds.), *Intersectoral governance for health and equity in European cities—Healthy cities in Europe*, Supplement 30(Suppl), i3–i7.
UNICEF and World Health Organization. (1978, September). *Primary health care: Report of the international conference on primary health care*. Alma-Ata, USSR.

VicHealth. (2002). *Leading the Way: Councils creating healthier communities. VicHealth resource designed to increase the level of understanding across local government about how social, economic and environmental factors can impact upon health and wellbeing*. Victorian Health Promotion Foundation, Carlton. Retrieved February 23, 2016, from https://www.vichealth.vic.gov.au/media-and-resources/publications/leading-the-way-councils-creating-healthier-communities.

Vlahov, D., Freudenberg, N., Proietti, F., Ompad, D., Quinn, A., Nandi, V., & Galea, S. (2007). Urban as a determinant of health. *Journal of Urban Health, 84*(1), 16–26.

Wallerstein, N. (2006). *What is the evidence on effectiveness of empowerment to improve health?* Copenhagen: Health Evidence Network, World Health Organization Regional Office for Europe.

Wallerstein, N. B., & Duran, B. (2006). Using community-based participatory research to address health disparities. *Health Promotion Practice, 7*(3), 312–323.

WHO Commission on Social Determinants of Health. (2008). Closing the gap in a generation: Health equity through action on the social determinants of health: Commission on social determinants of health final report. In World Health Organization (Ed.), World Health Organization.

Wilkinson, R., & Pickett, K. (2010). *The spirit level: Why equality is better for everyone*. London, UK: Penguin.

World Health Organization. (2014a). The Helsinki statement on health in all policies. *Health Promotion International, 29*(Suppl. 1), i17–i18.

World Health Organization. (2014b). Health in all policies (HiAP) framework for country action. *Health Promotion International, 29*(Suppl. 1), i19–i28.

World Health Organization. (2015). *Health in all policies: Training manual*. Geneva: WHO.

World Health Organization. (2016). *Health financing for universal coverage. What is universal coverage*? Retrieved February 18, 2016, from http://www.who.int/health_financing/universal_coverage_definition/en/.

World Health Organization & United Nations Development Programme. (2015). *Guidance note on the integration of noncommunicable diseases into the United Nations development assistance framework*. Geneva: WHO Press. Retrieved February 23, 2015, from http://www.undp.org/content/undp/en/home/librarypage/hiv-aids/guidance-note-NCDs-UNDAF.html.

World Health Organization, Health and Welfare Canada, & Canadian Public Health Association. (1986). *Ottawa Charter for health promotion: An international conference on health promotion—The move towards a new public health, Nov. 17–21, Ottawa*. Geneva, Switzerland: World Health Organization.

Ziglio, E., Simpson, S., & Tsouros, A. (2011). Health promotion and health systems: Some unfinished business. *Health Promotion International, 26*(Suppl. 2), ii216–ii225.

Chapter 18
A Supportive Tool for Urban Health: Health Impact Assessment at the Local Level

Jean Simos

What Is a Health Impact Assessment (HIA)?

HIA has emerged as a privileged tool for the implementation of health in all policies (HiAP). It is a vehicle to help health practitioners inform societal decision-making and can be very effective in the decision-making process as a decision-aid tool.

The Gothenburg Consensus defines HIA as 'a combination of procedures, methods and tools by which a policy, programme or project may be judged as to its potential effects on the health of a population, and the distribution of those effects within the population' (WHO 1999). This definition was completed in the mid-2000s by the following sentence: 'HIA identifies appropriate actions in order to manage those effects' (IAIA/WHO 2006).

HIA identifies how and through which pathways a decision can influence a wide range of health determinants. Pathways can be direct (e.g. increases in respiratory illnesses occur when air pollution increases) or indirect (e.g. dense vehicle traffic reduces walking and cycling) (Dahlgren 1995). HIA aims to improve the quality of decisions by maximizing positive impacts and minimizing negative impacts (WHO 1999). It follows that recommendations for action are inherent in HIA (Quigley et al. 2006), which is predictive (evaluating future consequences of decisions on health) and informative, and can be used for advocacy. HIA also seeks to define the distribution of effects within populations, thereby raising the issue of equity (Kemm 2013).

Health impact assessment (HIA) emerged in the context of development projects in the early 1990s, and guidelines were first published in 1992 (Birley and Peralta 1992; Birley 2011). HIA was popularized at first in the UK (Scott-Samuel 1996) and is now implemented in almost all highly developed countries (Kemm 2013).

J. Simos (✉)
Institute of Global Health, University of Geneva, Campus Biotech, Geneva, Switzerland
e-mail: jean.simos@unige.ch

E. de Leeuw, J. Simos (eds.), *Healthy Cities*,
DOI 10.1007/978-1-4939-6694-3_18

HIA has built on the methodology of the environmental impact assessment (EIA): it uses a similar screening stage to decide whether an evaluation should take place, followed by scoping which defines the perimeter, methodology, management and participants of the process (see Fig. 18.1). The assessment itself then produces results and recommendations. After the decision about the assessed by HIA proposal is taken and implemented, its effects are monitored to examine whether predicted impacts have occurred (Cantoreggi et al. 2007). This approach has much in common with the policy appraisal process (Scott-Samuel et al. 2001), which HIA strengthens by allowing a systematic review of health consequences (Kemm 2013; Milio 1986).

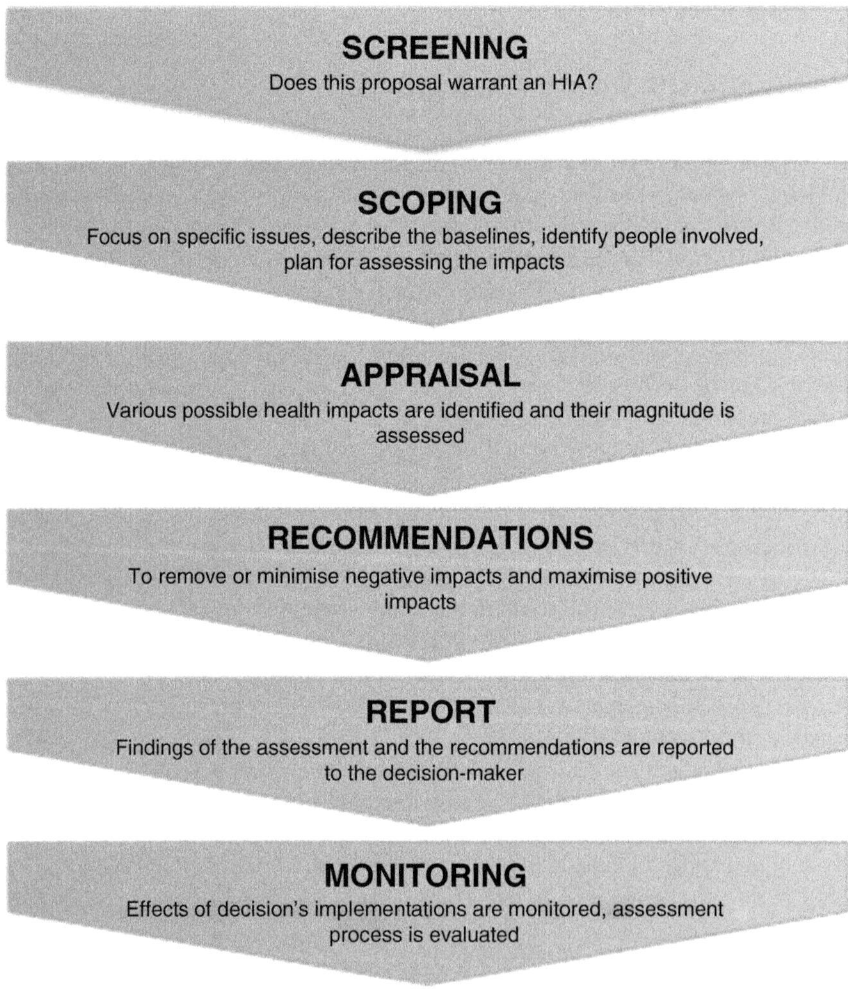

Fig. 18.1 Methodology steps in Health Impact Assessment

HIA as a Health Promotion Process at City Level

Underpinning the Ottawa Charter is the realization that health depends to a great extent on environmental factors which are beyond the influence of individuals (Dever 1976). To be effective, health promotion must influence the environment in which people live. This can only succeed if the needs of health promotion can be factored into political and administrative decisions beyond the boundaries of health care, in areas such as urban planning, transport and the building sector. According to the Ottawa Charter, the very first step in health promotion is to build up healthy public policy, and HIA is an adequate tool to make public policies healthier (Diallo 2015).

Nevertheless, it remains a relatively novel tool for intersectoral planning, so HIA practitioners can expect to face bureaucratic, political and methodological challenges in its application (see Table 18.1). At the administrative level, institutional roles and remits are almost always fragmented, there is a lack of shared responsibility for health in urban planning, and social and environmental determinants of health are poorly understood and often challenged by professionals outside the health promotion sector. At the methodological level, tools and practices for health impact assessment are underdeveloped. At the political level, new forms of impact assessment are commonly perceived as entailing new, costly regulations and requirements for development projects. This may generate suspicion and even preemptive opposition from those whose interests may be affected. HIA is not currently viewed as a pro-active, preventative and supportive force in advancing city planning. Given the historic use of environmental impact assessment and more recent calls for social, strategic and sustainability impact assessments, health and HIA must maximize added value, focus on unanswered questions, serve collaborations and coalitions, promote synergies (e.g. with EIA) and even compete for resources and constituencies (Corburn 2009; Corburn and Bhatia 2007).

Table 18.1 Potential barriers to the acceptance, use and integration of HIA in urban planning

Bureaucratic	Methodological	Political
Lack of understanding of the social determinants of health and lack of knowledge of the purpose and value of HIA	Lack of accepted health-relevant environmental indicators	Lack of understanding of the social determinants of health and lack of knowledge of the purpose and value of HIA
Lack of working relationships and trust between planners and health professionals	Lack of appropriate analytic tools for forecasting	Lack of HIA mandate and constituency
Coordinated timeframes for project and plan design		Contested policies
Capacity for additional responsibilities		Fears that HIA will prevent development or make it more costly

What are the effective strategies to overcome these challenges? And have these potential barriers actually been observed in Healthy Cities that have implemented HIA?

HIA in Phase IV of the WHO European Healthy Cities Network

HIA was introduced into the European Healthy Cities Network during Phase IV (2003–2008) as one of four core themes (Ison 2012). The objectives were to

- Raise awareness and create a common understanding of HIA.
- Provide leadership and strengthen capacity.
- Share results and evidence from HIA practice with other European cities.
- Provide evidence of HIA's contribution to areas such as healthy urban planning and healthy ageing.
- Work towards mainstreaming HIA as a framework for integrating health and well-being concerns into all urban policies and projects.

The implementation of HIA methodology at local level has been observed across Europe in over 30 countries with widely differing economies and administrative or sociopolitical backgrounds. The main barriers to and facilitators of the introduction and implementation of HIA identified during the evaluation of Phase IV of the Network are presented below (Ison 2012). These elements are consistent and complementary with those presented in Table 18.2.

The evaluation in Phase IV showed that the majority of city representatives did not feel that they had the resources, knowledge or experience to achieve these objectives. Nevertheless, gains from introducing and implementing HIA during Phase IV included improved understanding of the social determinants of health, placing health on the agenda of policymakers and decision makers, initiating or improving intersectoral working on health, and influencing decision-making in favour of health (Ison 2012).

Table 18.2 Identified barriers and main facilitating factors to the implementation of HIA in the WHO European Healthy Cities Network during Phase V (2004–2008)

Barriers	Facilitating factors
Lack of skills	Receiving training in HIA
Lack of knowledge and experience of HIA	Collaboration with an academic/public health institution or local health agency
Newness of the HIA concept	A pre-existing culture of intersectoral working
Lack of a legal basis for implementation	Political support
Lack of political support	Supportive national policy context
	Access to WHO materials about HIA or access to expertise in HIA
	Membership of the European Network, HIA Sub-network, or a national network

HIA in Phase V of the WHO European Healthy Cities Network

Although it was no longer a core theme during Phase V of the European Network (2009–2013), HIA remained linked with the overarching theme of health and health equity in all local policies and a requirement regarding capacity building (WHO 2009). This viewpoint is founded on the Adelaide Statement on Health in All Policies (HiAP), which promotes HIA as one of the most effective tools for HiAP operationalization (WHO and Government of South Australia 2010). One part of the Phase V evaluation focused on verifying the truthfulness of this assertion and on the sustainability of introducing HIA into Phase IV. The methodology for this evaluation was driven by a multi-method approach, allowing triangulation of data and information from various sources (general evaluation questionnaire, case study templates, NVivo software text analysis, telephone discussions with local Healthy Cities coordinators to clarify anything which seemed unclear in written documents) and was based on the realist synthesis approach (Simos et al. 2015).

One of the first interesting outcomes obtained by this specific evaluation was the confirmation of the flexibility of HIA as a tool and as a process. The ways in which European Healthy Cities have used HIA include the following:

- Full HIA process applied to a project.
- Integration of HIA into an EIA of a project (or in addition to EIA).
- Full HIAs conducted as part of a wider set of activities.
- A series of pilot HIAs to develop an HIA approach to planning policy decisions.
- Principles and elements of HIA used to develop indicators and checklists for regeneration proposals.
- HIA used as a screening tool for policies.
- Incorporation of HIA or elements of HIA into other impact assessment tools for use on policies.
- Use of elements of HIA as part of a Health Equity in All Policies framework.
- Furthermore, the evaluation revealed some common threads:
- Support (and funding) from Healthy Cities politicians is often decisive for the initiation of the HIA process.
- International influence to initiate HIA was present in all examined case studies.
- The health 'ideology' underlying the work—promotion of the social model of health and the social determinants of health—was present in all cities involved in the evaluation.
- HIA implementation is to be seen as a strategic action for Healthy Cities, especially regarding governance and partner and policy action.
- As the process concerns the use of a new tool, learning from HIA implementation was an important part of the process. The same can be said for the generalization of this (often first) experience and the impact that HIA may have in the future.

- As HIA treats a wide range of objects, there is considerable diversity around thematic action, in particular as regards supportive environments and healthy living.
- Healthy urban planning and healthy urban design remain favourite fields for HIA.
- As HIA is promoted as a tool leading towards HiAP, this issue was mentioned in a majority of the case studies.
- Focus remains on vulnerable groups in the population and reduction of the sources of health inequalities and inequities.
- The status of HIA is as a voluntary and non-statutory requirement: when other, statutory, impact assessments exist, then HIA can be perceived by decision makers as an added burden.
- Financial requirements were covered by resources allocated to Healthy Cities in the various municipalities except in two of the case studies, where some funds came from external sources.
- In terms of human resources, cities worked on a partnership basis and often received support from several municipal departments. A few cities mentioned the use of external consultants to support their work.
- Capacity-building activities included training courses for municipal technicians and project officers, sometimes also to Healthy Cities administrators or to stakeholders external to the city administration, and far more rarely to politicians.
- A variety of social mechanisms were used in the case studies, e.g. establishing a steering group to oversee the work described in the case study, or holding workshops, focus groups or other types of participative meetings.
- Some factors increased the *acceptability*[1] of HiAP interventions:
  - The benefits, added value, cost-effectiveness, or other resource-saving aspects of the HIA framework.
  - Responsiveness and flexibility in the development or use of the tool/framework (willingness to adapt the tool following stakeholder feedback/adapting the framework to the organizational needs of different stakeholders/responding to requests for tool development, etc.).
  - Existence of an evidence base for the work, whether in the published literature or in the form of a good practice.
  - Engagement of stakeholders such as citizens from the municipality.
  - The use of simple, clear language when engaging with stakeholders.
  - The achievement of Healthy Cities accreditation and the profile of being a member of the European Network.
  - The 'local' relevance of the tool.
  - The focus on improving health and reducing inequalities.
  - The fact that policy authors were able to retain ownership of the policy.
  - The capacity of the tool to help policy authors identify interdependent outcomes and new links across policies.

[1] Acceptability is defined as 'are sectors willing to collaborate on health and equity?'

- Some factors increased the *feasibility*[2] of HiAP interventions:
  - Support at a strategic level or from senior politicians.
  - Support from internal and external partners or stakeholders.
  - Support from WHO expert advisers.
  - Involvement or participation of partners and stakeholders at the start of the process, or the introduction and use of the framework or tool early on in the process.
  - Active participation of staff always a key to implementation.
  - Investing time in the process, or recognizing that it is a long-term agenda.
  - The use of pilots/piloting.
  - Facilitating ownership of the intervention.
  - Effective and regular communication (information dissemination and explanation, especially of complex concepts, with a focus on key issues and clear messages) with councillors, partners (internal and external) and stakeholders, including the community.
  - Systematic nature of the HIA framework.
  - Fitness-of-purpose of HIA process and tools to support the work.
  - Use of the HIA framework to support integrated work on health equity issues.
  - Other factors increased the *sustainability*[3] of HiAP interventions:
  - Designing an inclusive process from the start.
  - Incorporating health and equity issues (including access to health services) and health improvement into policy and strategy documents, and decision-making for different types of proposals from policies to infrastructure projects.
  - Gaining trust among partners during the process.
  - Incorporating HIA in urban planning policies.
  - Improved intersectoral action on health in general.
  - Having partners who express interest in addressing issues of equity.
  - Having of a long-term vision.
  - Having the opportunity to take part in corporate-level discussions.
  - Developing a comprehensive impact assessment tool, and using it to enable the achievement of Healthy Cities aims and objectives for Phase V.
  - Using HIA as a health 'filter' by the municipality: the additional factor influencing sustainability was that the aims of each professional discipline and sector in the municipality now included health.
  - Establishing an HIA review service at the local university.
  - Enabling planners to gain ownership of the healthy urban planning agenda (in some cases planners became the strongest local advocates for public health).

---

[2] Feasibility is defined as 'do sectors have the capability to collaborate on health and equity?'

[3] An indicator of sustainability would be the description of successful or completed HiAP interventions (Shankardass et al. 2014). Acceptability and feasibility of HiAP interventions may be necessary pre-conditions for sustainability (Morestin et al. 2010).

Conclusion

The role of Healthy Cities and their networks in promoting and supporting innovations such as HIA is a central idea of the WHO Healthy Cities concept. It is encouraging to see that HIA implementation in many Network cities in Phase V was at least as vigorous as in Phase IV, despite the withdrawal of the HIA Sub-network from WHO-Euro and the discontinuation of HIA as an explicit target.

Work with HIA requires training and skills, political support, expertise in networking, and motivation for intersectoral collaboration (Saint-Pierre et al. 2014). The evaluation of Phase IV showed how difficult it was to introduce and implement HIA in countries without a history of HIA or without a supportive national context. Exceptions were cities that had made a strong commitment to the Phase IV core theme by joining the HIA Sub-network, which was an effective vehicle for spreading knowledge and supporting new local HIA leaders. Evidence for this is that all non-British Healthy Cities included in the Phase V evaluation were in countries that were members of the HIA Sub-network: France, Hungary, Italy and Spain.

The main barriers and facilitating elements for HIA implementation at city level are now well identified. The findings regarding the factors that increase the feasibility of HIA interventions within the European Network are convergent and complementary with the results of other recent surveys found in the scientific literature (Bourcier et al. 2015, Rhodus et al. 2013). The achievement of Healthy Cities accreditation and the raised profile of becoming a member of the WHO European Healthy Cities Network increased the acceptability of developing a comprehensive impact assessment tool (Simos & Cantoreggi 2008).

HIA pursues its development as a tool for decision makers who want to promote health across policy sectors. Already mentioned in the Adelaide Statement in 2010, HIA has a position in the Health 2020 Strategy (WHO 2012), for which Healthy Cities are an important vehicle for local-level implementation.

References

Birley, M. (2011). *Health impact assessment. Principles and practice*. London: Earthscan.

Birley, M., & Peralta, G. (1992). *Guidelines for the health impact assessment of development projects*. Asian Development Bank Environment Paper, Manila.

Bourcier, E., Charbonneau, D., Cahill, C., & Dannenberg, A. L. (2015). An Evaluation of Health Impact Assessments in the United States, 2011–2014. *Preventing Chronic Disease, 12*, 140376.

Cantoreggi, N., Frei, K., Simos, J., Casabianca, A., Litzistorf Spina, N., & Diallo, T. (2007). HIA in Switzerland: Considerations concerning the experience of the Cantons of Geneva, Jura and Ticino. *Italian Journal of Public Health, 4*(3), 169–175.

Corburn, J. (2009). *Toward the Healthy City: People, places and the politics of urban planning*. Cambridge: MIT Press.

Corburn, J., & Bhatia, R. (2007). Health Impact Assessment in San Francisco: Incorporating the social determinants of health into environmental planning. *Journal of Environmental Planning and Management, 50*, 323–341.

Dahlgren, G. (1995). *European health policy conference: Opportunities for the future*. Vol. 11: Intersectoral Action for Health. Copenhagen: WHO Regional Office for Europe.

Dever, G. (1976). An epidemiological model for health policy analysis. *Social Indicators Research, 2*, 453–466.

Diallo, T. (2015). *L'évaluation d'impact sur la santé, l'outil le plus efficace pour évaluer l'impact sur la santé des politiques urbaines de réduction des gaz à effet de serre*. PhD thesis (Thesis no 4879), University of Geneva.

International Association for Impact Assessment (2006). *Health impact assessment international best practice principles*. IAIA Special Publications, Series No. 5.

Ison, E. (2012). Health impact assessment in a network of European Cities. *Journal of Urban Health: Bulletin of the New York Academy of Medicine*, 90(1).

Kemm, J. (Ed.). (2013). *Health impact assessment. past achievement, current understanding and future progress*. Oxford: Oxford University Press.

Milio, N. (1986). *Promoting health through public policy*. Ottawa: Canadian Public Health Association.

Morestin, F., Gauvin, F. P., Hogue, M. C., & Benoit, F. (2010). *Method for synthesizing knowledge about public policies*. Montreal, Canada: National Collaborating Centre for Healthy Public Policy.

Quigley, R., den Broeder, L., Furu, P., Bond, A., Cave, B., & Bos, R. (2006). *Health impact assessment international best practice principles*. Fargo: International Association for Impact Assessment.

Rhodus, J., Fulk, F., Autrey, B., O'Shea, S., & Roth, A. (2013). *A review of health impact assessments in the U.S.: Current state-of-science, best practices, and areas for improvement* (National Exposure Research Laboratory, Office of Research and Development, U.S). Cincinnati, OH: Environmental Protection Agency.

Saint-Pierre, L., Lamarre, M.-C., & Simos, J. (2014). L'évaluation d'impact sur la santé: une démarche intersectorielle pour l'action sur les déterminants sociaux, économiques et environnementaux de la santé. *Global Health Promotion, 21*(Suppl. 1), 7–14.

Scott-Samuel, A. (1996). Health impact assessment: An idea whose time has come. *British Medical Journal, 313*, 183–184.

Scott-Samuel, A., Birley, M., & Ardern, K. (2001, May). *The Merseyside guidelines for health impact assessment* (2nd ed., p. 20.). International Health Impact Assessment Consortium.

Shankardass, K., Renahy, E., Muntaner, C., & O'Campo, P. (2014). Strengthening the implementation of health in all policies: A methodology for realist explanatory case-studies. *Health Policy and Planning, 2014*, 1–12.

Simos, J., & Cantoreggi, N. (2008). Mise en œuvre des EIS à Genève (Suisse) et utilisation de leurs résultats dans le processus législatif. *Téléscope, Revue d'analyse comparée de l'Observatoire de l'administration publique, Ecole nationale d'administration publique, 14*(2), 51–61.

Simos, J., Spanswick, L., Palmer, N., & Christie, D. (2015). The role of health impact assessment in Phase V of the Healthy Cities European Network. *Health Promotion International, 30*(S1), i71–i85.

WHO (1999). *Health Impact Assessment: Main concepts and suggested approach*. The Gothenburg Consensus paper. Copenhagen: WHO Regional Office for Europe.

WHO (2012). *Health 2020: A European policy framework supporting action across government and society for health and well-being*. World Health Organization Regional Office for Europe, Regional Committee for Europe, Sixty-second session, Malta.

WHO Euro. (2009). *Phase V (2009–2013) of the WHO European Healthy Cities network: Goals and requirements*. Copenhagen: World Health Organization Regional Office for Europe.

WHO and Government of South Australia (2010). *Adelaide statement on health in all policies*. Retrieved February 15, 2016, from http://www.who.int/social_determinants/publications/en/.

Chapter 19
Futures Thinking and Healthy Cities

Trevor Hancock and Clem Bezold

> *Thinking about the future is only useful and interesting if it affects what we do and how we live today.*
>
> —*James Robertson*

The Healthy Cities and Communities initiative is an exercise in what some call 'futuring'. From the start, the concept of a healthy city brings with it a key question: what *is* a healthy city? Answering that is an exercise in futures thinking, and more explicitly in visioning, because there is no such thing as a fully healthy city—nor is there ever likely to be. It will always be a vision of what might and should be, an aspiration to which we strive rather than a 'most likely' extension of the present.

These few short sentences touch on many of the key aspects of futures thinking. This chapter will cover them in more detail, beginning with an overview of futures thinking and followed by a discussion of health futurism based on the work we have done together in this field over more than 30 years, before we move to focus more specifically on futures thinking in the context of Healthy Cities.

Parts of this chapter are based on *An Overview of the Health Futures Field* (Bezold and Hancock 1993).

T. Hancock (✉)
Professor and Senior Scholar, School of Public Health and Social Policy, University of Victoria, Victoria, BC, Canada
e-mail: thancock@uvic.ca

C. Bezold
Chairman and Senior Futurist, Institute for Alternative Futures, Alexandria, VA, USA
e-mail: cbezold@altfutures.org

E. de Leeuw, J. Simos (eds.), *Healthy Cities*,
DOI 10.1007/978-1-4939-6694-3_19

How far away is the future?
The belief that the future will be shaped by human decisions and actions is a characteristic that most futurists share. They recognize that while the immediate future (the next 5 years) will be shaped largely by decisions previously taken (while recognizing that 'discontinuities' such as the 1973 oil shock or the 1989 fall of the Berlin Wall can dramatically alter the future very swiftly), the medium-term (5–20 years) and long-term (20–50 years) future will be substantially shaped by the decisions we make today.

Beyond 50 years, the future is so far removed as to make thinking about it extremely difficult. For example, how much of what we accept as commonplace today could have been anticipated in 1965? And how little of what was commonplace in 1965 would have been anticipated in 1915?—most of the mechanical and electrical equipment of the 20th century? This does not make thinking 50 years ahead irrelevant, but it frees us to make use of science fiction literature, which can be very helpful in thinking about dramatic changes which may surprise us in the coming decades.

Thinking About the Future

The comment that opens this chapter by James Robertson, a leading British expert on alternative futures, contains the essence of futurism. Good futures work is to some extent concerned with attempting to forecast the future, but more importantly it is concerned with thinking about the future and in helping people who are not futurists to think and act more wisely about our tomorrows. As Edward Cornish, founder of the World Future Society, puts it, 'futurists take historical fact and scientific knowledge and add human values and imagination to create images of what may happen in the future' (Cornish 1977).

James Robertson is also making another important point: thinking about the future is relevant to the decisions we make today. Futurists don't actually study the future, since it does not yet exist. Rather, they study ideas about the future, most often in the belief that the future is 'plastic' and can be shaped. Futures thinking is a tool for wise action that stimulates the imagination, encourages creativity, identifies threats and opportunities, and allows us to relate possible future choices and consequences to our values. While some experts argue that futures research should simply be able to develop accurate predictions—a dubious proposition in the first place—and that it is not the role of futurists to be normative and to work in the area of vision, good futures work includes both futures research and vision. These in turn contribute to strategic management by aiding in anticipation, setting direction and securing commitment.

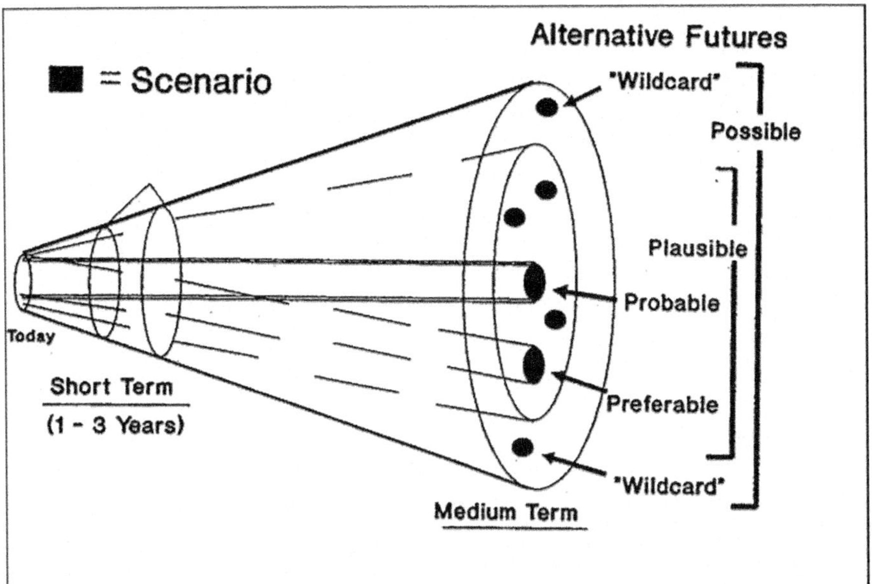

Source: from Clem Bezold and Trevor Hancock, An Overview of the Health Futures Field for the WHO Consultation, July 19 - 23, 1993.

Fig. 19.1 The futures cone. Source: C. Bezold and T. Hancock, An Overview of the Health Futures Field for the WHO consultation, July 19–23, 1993

Four Sorts of Future

We tend to talk about 'the future' as if there was just one, yet we know that there are many possible futures facing us, and that to some extent we create them—which is not to say that a large asteroid or a massive volcanic eruption or a war might not change everything. One way of categorizing the futures that we face has been proposed by Henchey (1978), a Canadian futurist. He suggests that we think about the future in four ways, presented in Fig. 19.1:

- Possible, i.e. what may happen.
- Plausible, i.e. what could happen.
- Probable, i.e. what will likely happen.
- Preferable, i.e. what we want to have happen.

There are two 'zones' in this model: the outer zone of possibilities, which is very large, and the narrower zone of plausible futures. Within this narrower inner zone there can be a number of scenarios; the probable future is one of these, but others can be described. The preferable future is often different from the probable future and is usually—but not necessarily—within the plausible zone. The futures cone makes clear that all futures start from where we are today, then diverge. The closer they are to today, the harder it is to tell them apart, but over time it becomes more clear that choices made now can have dramatic effects later.

Possible futures are all the things we can possibly imagine, no matter how unlikely. They may include 'science fiction' futures that transgress presently accepted laws of science (for example, we may learn how spiritual healing works and will be able to treat people by using the energy aura of the healer). Also included here are 'wildcard' futures; these are typically low-probability but high-impact events. Having a sense of what type of wildcards might arise is useful, although most planning efforts are barely able to deal effectively with the zone defined by the more plausible futures. Yet the fall of the Berlin Wall and the transformation of the Soviet Union remind us that dramatic, seemingly implausible change can occur very swiftly, while awareness of the possibility of a large meteor strike or a massive global mega-volcano eruption may give us pause for thought—if only to recognize how small and frail we humans and our civilizations might be. We need to be flexible enough to deal with surprises when they occur.

Plausible futures represent a narrower scope because they are those futures that seem to make sense, given what we know today. Plausible futures can be discrete forecasts of individual trends or a set of scenarios which combine differing trends and together describe a range of alternative futures. For example, common plausible futures include high technological growth; a 'green' or sustainable society; environmental, economic and social decline; a more controlled society; and a 'high spirit' or transformational scenario. For each of these we can describe a quite detailed societal scenario, including a description of, for example, a city, or health status, or health care system consistent with it. (The health care system, for example, will reflect the society of which it is a part, not the other way around, which is why we need to begin with the societal scenario.)

The probable future is the future we think is most likely to happen, based on the examination of our present situation and the appraisal of likely trends and future developments; it is a subset of the plausible future and is sometimes referred to as 'business as usual'. This consideration of likely or plausible futures is often called descriptive forecasting. Most people see the future as an extension of the present with little significant change; likewise much government and business planning assumes that the probable future will be a straightforward extrapolation of the present. Ironically, history has shown that 'the most likely future' turns out to be the most likely *not* to occur: as Jim Dator, one of the world's leading futurists, likes to say, 'the probable future is unlikely'. Descriptive forecasts which are based solely on recent trends can preclude futures that are different; they also often turn out to be the future we don't want! As Dator has also remarked, 'trends can take you with unerring accuracy to where you don't want to be'.

The preferable future is the one future we would like to have happen, and is sometimes called prescriptive futurism, or normative forecasting; this is where *vision* becomes important because, 'vision is values projected into the future' (Bezold). Visions, or preferable futures, generally begin by identifying and trying to create a future that does not yet exist. Vision moves reality beyond the present toward the best that can be; it should include visionary scenarios that move well beyond present realities. Creating a shared vision of a preferred future health care system or of a healthy community can be a powerful technique for mobilizing an organization or community around a common purpose (for more on vision, see Hancock 1993; Bezold 2009).

In our work, we usually use a combination of alternative future scenarios and visioning (or other ways of defining a preferable future). We may ask people to rate the probability and desirability of various alternatives and, in the case of a preferred future, its feasibility.

The value of the alternative futures approach is that it enables us to compare a range of quite plausible options and to choose among them. In making our choice we also make explicit the values that we hold, individually and collectively. There are other benefits of alternative scenarios. They can make us examine futures we don't want, which may lead us to seek ways to prevent them from happening, or at least to prepare for them. They can help us compare the future we think we will get with the future we would like to get, and if they are not the same, to think about how we can change our path to move closer to a desired future. And they can be a way of engaging people in a rich discussion about their future and what they intend to do about it.

The Future We Don't Want

None of us like to contemplate an unpleasant future, but just because it is undesirable does not mean it is implausible. So one or more of the scenarios—which, we need to recall, are based on existing driving forces—should be a plausible alternative future that involves ecological, economic or social decline or collapse—or all three. These are not scenarios that we like to explore, any more than we like to explore what it would mean to have cancer; but, as with cancer, ignoring it does not make it go away.

Indeed, a failure to consider negative scenarios may actually make them more likely, because if we don't take them seriously we will not plan to avoid them or figure out how to cope with them. As will be discussed later, there are some very plausible 'bad news' scenarios for some, perhaps many cities; we had better learn to think about these alternative futures and either figure out how to avoid them or plan how to manage them.

Probable vs. Preferable Futures

There is often a marked difference between the future we think is likely to happen and the one we would prefer to have happen. It is important to stress the value of examining both the probable and the preferable futures, a process which can be very liberating. Too often our image of the future is the image of what we think will most likely happen; if we don't like the way we think things are going, this may bring with it an awful sense that the light at the end of the tunnel is a train coming toward us! The probable future is something that seems to happen to us, something over which we have little or no control, and sometimes—perhaps often—something we don't like very much.

But the process of contemplating a future we find likely but undesirable, rather than a future we consider desirable but less likely, can be a great motivator for action; if we don't want the future we seem headed for, what can we do (together) to change the path we are collectively on?

The Preferable: An Empowering Future

If futures thinking focuses too much upon the probable, which it has a tendency to do (after all, planners—be they politicians, civil servants or private sector—like to know what to plan for, as do ordinary people) then it runs the risk, perhaps inadvertently, of disempowering people and denying them choice. If we are told 'this is the probable future', then the only choices left for us to deal with are how to prepare for it and if need be, how to brace for it. Such a passive way of thinking about the future inevitably results in apathy, in a feeling of impotence and lack of control.

The preferable future, on the other hand, is a liberating and empowering future, especially when it touches our more creative capacities. It not only enables but encourages us to say 'this is the future that we value and that we want to create' (the emphasis being on we; this should be a collective process). The energy and creativity released in a 'preferable futures' process can be quite astonishing (see the case study of Woolwich Healthy Community in Chap. 10; and earlier cases in Alternatives for Washington (Stilger 1978)).

Creating the Future We Prefer… Together

Clearly, if futures work is to help in creating a future that we find desirable, it must make policymakers and individuals better able to create the future they want. This brings us to two important and related points about futures thinking that were identified by Roy Amara, founder of the Institute for the Future, in a series of articles in *The Futurist* in 1981. The first point is that the futures field is concerned with creating new images of what is possible; the second, that good futures work increases people's participation in thinking about and creating their preferable future.

Futurists by and large take a proactive stance. If they are not actively involved in seeking to create change themselves, they are certainly aware that their presentations and 'predictions' will form the basis for others to seek to create change.

At its best, futurism is—to use a phrase coined by Alvin Toffler, author of *Future Shock* and *The Third Wave*—a form of 'anticipatory democracy', helping people to decide what sort of future they want and how they might achieve it. It is the very essence of democracy. Futures techniques that involve the use of complex and sophisticated technical models, professional expertise and language, and that limit their process to a small inner circle are anti-democratic. We must have techniques of anticipatory democracy (the basis for the establishment of the Institute for

Alternative Futures) and they must become the standard form of futures work. We need techniques that work at the community level and with 'ordinary' people. Vision workshops and other approaches can meet these needs, as well as goals programmes and other efforts, and these have been present in a variety of settings (see 'Anticipatory Democracy Updated' (Bezold 2006).

Futures Methodologies

Just as there is a range of alternative futures, there is a range of methodologies for exploring the future. Some are more suited for exploring one sort of alternative than another, although most can be adapted to explore different sorts of future. For example, although vision workshops are particularly well suited to exploring a preferable future, there is no reason why they cannot be used to explore a probable future; similarly, the Delphi method (an iterative polling technique, usually done with experts) can be used to explore a range of futures, given the right questions.

The point is that the range of futures methodologies each has strengths and weaknesses:

- **Forces and trends:** By identifying the major forces in the social, political, economic, environmental, technological, ethical and other sectors that will likely affect our city, organization or group, and by 'scanning' the scientific and public media, we can often identify emerging trends of importance. John Naisbitt popularized this approach in his 1984 book *Megatrends*.
- **Models:** Sophisticated computer models can be used to explore various options and to develop 'what if' scenarios. Most commonly used in the economics field, they are also used to predict the impact of global warming, as 'global models' to examine the combined effect of environmental, social and economic change, and in many other situations.
- **Delphi methods:** Various versions of this technique (which consults a number of people without convening them, thus protecting against the influence of powerful personalities) can be used to assess the feasibility or probability of certain events. For example, a panel of scientific experts might be asked to assess the likelihood of a particular set of medical advances being in place by 2030.
- **Cross impact matrices and 'futures wheels':** These and related methods are used to understand the interaction of key forces and trends, and to explore possible unanticipated and unintended consequences. For example, an unintended consequence of a society strongly focused on wellness might be to victimize and discriminate against the unwell. Sometimes sets of forces are mutually reinforcing, and at other times they may cancel each other out. Trying to understand this helps in the construction of scenarios and the anticipation of the impact of various forces.
- **Scenarios**: These are coherent 'stories' incorporating a large number of trends and forces, or plausible events, in a way that shows their interaction and their implications IAF recommends that they consider expectable or most likely

futures, a challenging scenario that explores 'what could go wrong', and visionary or surprisingly successful paths to the future (see Bezold 2009). Often presented as narratives, scenarios may take the form of a 'report' from the future and should include tables that compare forecasts for key elements in each of the scenarios.

- **Visioning**: This process creates a preferred future by projecting values and aspirations into the future and then seeking to describe that future succinctly, in a powerful phrase or sentence, and sometimes as a scenario as well. One technique of visioning is guided imagery (inviting participants to imagine their preferred future), which can evoke powerful images of the future. Visions are inspiring: they are for the heart (scenarios are for the head) and help us to be the best we can be.

The method—or better, if time and resources permit, the combination of methods—that we choose depends on what we are trying to do. In the context of a city, especially a large city, much of the technical work of scanning forces and trends, and perhaps even producing models, may be somewhat routine. Such cities may have several universities they can look to for expertise and advice, but be less able to engage citizens at the neighbourhood level in participatory processes and visioning. Small communities may lack the technical expertise and resources to undertake the technically intense work that large cities can handle, but may be effective in bringing local people together on a volunteer basis for a scanning exercise, a futures wheel and a vision workshop; they may also find participatory processes easier to establish and maintain.

Thinking About the Future of Health

Before discussing futuring for healthy cities, we recall a distinction we have already touched on: futurist thinking about a healthy city is part of what we call 'health futurism'. It is important to make a distinction between health futurism as we mean it here and what is often called health futurism but is in reality 'medical care futurism'. We need to distinguish between health and medical care, because otherwise we run the risk of falling into the trap of believing that medical care equates with health. The importance of health futurism is that it requires us to look beyond our mistaken preoccupation with medical care and to examine the real determinants of future health status.

Health futurism is concerned with the future state of health of a population—either a local population or the global population—and all the environmental, social, economic, political, cultural, biological and technological factors that will affect its health now and in the future. Inherent in those social, economic, political, cultural and technological factors is the way any given society or community chooses to organize the provision of medical care; but medical care is only one determinant of health, and not necessarily a very important one. It has been estimated that medical care explains only 10–20 % of health over the life-course (McGovern et al. 2014), meaning that 80–90 % is determined by other factors. Health futurism asks questions such as: Will we be more or less healthy in the future? What will be the major influences upon health in the future? What values will we attach to health and what role will health play in our decision-making in the future? What will be the future relationship between health and medical care?

A great deal of health futurism is thus concerned with societal futures in general, and particularly those aspects that will affect health and well-being. This includes such issues as wealth and poverty, future living and working conditions, the sustainability of the environment, the future state of social networks and social support, the extent of participation and empowerment and other issues that affect personal and collective health and well-being. Health futurism also considers the subject of healthy public policy, in other words, what future policies might be developed if achieving health and well-being were a prime determinant of public policy.

The value of health futures in the futures field more generally is that it serves to focus our attention on human and ecological ends, rather than on the economic and technological means that dominate so much of the work of corporate and governmental futurists and that is reflected in the field of medical care futurism. Thinking about the future in this way should lead us to examine more closely what we do and how we live today, and to ask what it takes to create a healthier future for all.

Healthy Cities and Futures Thinking[1]

So what does all this have to do with Healthy Cities? As noted at the outset, Healthy Cities is inherently an exercise in health futures thinking at the local level. It tends to be an exercise in what one might call optimistic futuring or aspirational futures. It envisions a preferred future for the city and then figures out how to achieve it. But what sometimes gets overlooked or ignored, as noted earlier, is the dark side of the future. While we all want a healthier future, there are many societal forces, not to mention environmental changes, that might push us in the opposite direction, so it is important to spend time examining the driving forces and the less attractive alternative futures that might result, rather than being focused only on the future we prefer.

Unhealthy Cities in the Future and in the Past

There are reasons to believe that cities in the future—or at least, many cities in many places—may be very unhealthy. For example, there are high levels of social inequity in many parts of the world, and in many cases it is growing. It results in significant inequalities in health, as documented in the report of the WHO

[1] For an application of these concepts to the Healthy Cities and Communities movement itself, see Bezold and Hancock 2014.

Commission on the Social Determinants of Health (2008) and the WHO/UN Habitat report *Hidden Cities* (WHO/UN Habitat 2010). In addition, the ecological changes resulting from human-induced climate change—which are accelerating—hold many threats. For cities on coasts, sea-level rise and more frequent storms will be a growing problem, while others, especially in tropical and equatorial areas, will face more prolonged and severe 'urban heat events'. All cities are likely to face more severe weather events; many will face shortages of water and of food caused by changes in both agro-ecosystems and ocean systems, and some will face growing numbers of eco-refugees (Intergovernmental Panel on Climate Change 2014, Chap. 8).

These and other problems arising from ecosystem change will add to and compound social inequity. In general, the poor live downhill, downwind and downstream, in low-lying land, in the most polluted parts of a city with the fewest public services and amenities. They are far more vulnerable to the impacts of climate change, and will have access to far fewer resources to help them cope. As resources such as food and water become more scarce, they will be unable to afford them. This combination of environmental and social inequity, together with climate change (and more broadly, global ecological change), poses a significant threat to the health of cities and their citizens in the twenty-first century (Friel et al. 2011; Hancock et al. 2015).

All of this is not inevitable, but it is plausible, and may have quite a high probability of occurring, especially in the often more vulnerable cities of the Global South that may have fewer resources and less capacity to respond. While decline and collapse is certainly not a desired future, it is a future that cities need to consider, to plan to avoid if they can, and to manage if they happen. In fact, thinking about how to manage decline—what might be thought of as a long, slow disaster—is something that we need to do in a systematic way, not only in our cities but as a society (Hancock 2015).

One place we can turn to when faced with such daunting challenges is history. It is a truism that good futurists need to be good historians: as George Santayana remarked, 'those who cannot remember the past are condemned to repeat it'. In thinking about the unhealthy future for cities described above, we can be helped by considering the state of cities in industrializing countries in the nineteenth century. To read descriptions of, say, Manchester in the 1850s (e.g. Girouard 1985, Chap. 12) is to realize that the situation cities faced then is very much like the situation that many cities in the Global South face today. They had appalling environmental conditions, with little or no sewage management, and less treatment; dirty and dangerous water supplies; atrocious housing and working conditions, dreadful air quality, and many other environmental and social ills. The average age at death among the labouring classes in some of the industrial cities in England in 1842 was as low as 15 (Liverpool), 16 (Bethnal Green in London) or 17 (Manchester), according to Chadwick, while for the 'gentry' in those communities it was 35, 45 and 35 years, respectively (Rule 1986, p. 89). Today, while far from perfect, these cities are vastly improved, and those improvements largely came in a period of 50–70 years—two generations or so. They happened for a variety of reasons: the threat of revolution, the pressure for democratic and social reform, the enlightened self-interest of the wealthy, the championing of good governance, charitable good works and other factors. While those may not be the same forces that operate today, and while the twenty-first century is not the nineteenth century, there are lessons to be learned and hope to be gained from knowing our history.

Towards a Healthier City

Beyond just dealing with crises, there is a need for cities and their citizens to dream about and plan for and work to create the healthier, more just, more sustainable future they desire. It is no coincidence that a key recommendation of the WHO Commission on the Social Determinants of Health (2008) was to 'place health and health equity at the heart of urban governance and planning', noting that 'within cities, new models of governance are required to plan cities that are designed in such a way that the physical, social, and natural environments prevent and ameliorate the new urban health risks, ensuring the equitable inclusion of all city dwellers in the processes by which urban policies are formed' (p. 6).

The need for what has come to be called 'healthy urban governance' is growing (Barten et al. 2011; Kickbusch and Gleicher 2012). The crucial point here is that this is—or needs to be—an exercise in anticipatory democracy. One of the key characteristics of a healthy city or community is that people are involved and engaged, that they participate. But what does this mean in the twenty-first century, with the widespread and swiftly growing use of social media and other forms of internet connectivity? What does a participatory and anticipatory democratic process mean, what does it look like and how can it work—and how does it work in very different settings, from megacities to small municipalities, from large cities in the Global North that are richly endowed with resources to small villages in the Global South?

Clearly, when faced with such diversity, there are no 'one size fits all' approaches that can work; but we hope we have shed some light on the potential use of a variety of futures methodologies to explore together the futures we might inhabit, to prepare us for and to help us avoid the 'bad news', and to have the conversations that will help us to find our way forward, together, to the healthier future for our cities and our citizens that we all desire.

The path from here to there is not clear, easy or straightforward. Futures thinking is not about predicting the future, mapping out that future and then developing and implementing a clear plan to get us there. Rather, on our journey we will make false starts, we will make mistakes, we will sometimes lose ground, we might even fail. We will battle opposition and be buffeted by the changing circumstances of our world, our society and our city. But that is the nature of change: it is like a sailing trip.

One lesson we have learned above all others from our work in the futures field over the past 30 or 40 years is that getting from here to there requires the engagement of as many as possible in the development of a clear understanding and expression of our shared values, the expression of those values in a shared vision and a detailed description of a preferred future scenario, and the engagement of people in

Sailing to Success

Nautical Tips for Navigating the Seas and Shoals of the Healthy Community Process

Dr Trevor Hancock, © 2001

I am a great believer in the 'vision-directed muddling through' approach to planning and action. Whether we are trying to create change in a community, a work place or any other setting, we need to have a very clear—and shared—vision of where we want to be and what it would look like if we achieve what we hope to achieve. But the process of getting from here to there is never a simple one, never amenable to simplistic and fixed plans. It is a process of muddling through, taking advantage of every opportunity that gets us a step closer to our goal.

As a student of sailing, I have found it useful to think of this process as analogous to sailing against the wind and through the reefs from one harbour to the next. Unlike the captains of power boats, who can just open up the throttle and go in a straight line from one harbour to the next (and who can end up in deep trouble if there is a mechanical failure or if they run out of fuel), change agents, like sailors, have to take sensible precautions and follow simple rules. For those of a nautical bent, such as myself, the following hints may be helpful.

- When sailing against the wind, it is necessary to tack back and forth: sometimes you can sail a long way and yet make little progress towards your goal.
- Sailors must be alert to shifts in the (political) winds: if the wind shifts (an election), you may have to change your course completely.
- Sometimes, the only way to get around a reef is to sail back the way you came—which is usually when they evaluate you!
- The 'crew' are involved in deciding what the destination will be; be prepared for the destination to change halfway there.
- If the wind gets too strong, you may have to run for a safe harbour and wait for the storm to die down.
- If the wind drops and you are becalmed, you just have to sit there and wait for the wind to pick up again. Be prepared for the journey to take a lot longer than you thought it would!
- If things get really nasty, the boat may start to sink. Make sure you have a life raft and some emergency rations!
- You need an experienced captain and crew, a team that works well together even in the roughest of storms.
- The skipper has to be respectful of the crew, or risk a mutiny.

Finally, but perhaps most importantly, navigation hinges on two things: knowing exactly where you are and knowing where you want to be. This is equally true of the community development process. We need to know exactly where we are in terms of our community's environmental, social, political, economic, health and cultural conditions, and we need to know both the good news and the bad news. We also need at least a pretty good picture of where we want to be, even if we don't have the precise coordinates and the destination may change. And remember, much of the point of sailing is to enjoy the trip, getting to your destination is an added bonus!

Good luck and happy sailing!

A video of TH telling this story is available at https://www.youtube.com/watch?v=GPfVygl8IN0

the process of 'muddling through' or 'sailing' towards that preferred future. This is the challenge facing the world's cities. How well we do that will determine the fate of our cities, our civilization and our citizens, as well as the fate of many other species with whom we share this planet.

References

Amara, R. (1981). 'The Futures Field' (3 part series). *The Futurist, 15*(1), 25–29.

Barten, F., Akerman, M., Becker, D., Friel, S., Hancock, T., Mwatsama, M., … & Stern, R. (2011). Rights, knowledge, and governance for improved health equity in urban settings. *Journal of Urban Health, 88*(5), 896–905.

Bezold, C. (2006, July). Anticipatory democracy revisited. In *Democracy and futures* (pp. 38–51). Helsinki: Parliament of Finland, Committee for the Future. Retrieved from http://www.eduskunta.fi/fakta/vk/tuv/democracy_and_futures.pdf.

Bezold, C. (2009). Aspirational futures. *Journal of Futures Studies, 13*(4), 81–90.

Bezold, C., & Hancock, T. (1993). *An overview of the health futures field.* (Background paper for the WHO Health Futures Consultation, Geneva, July 1993).

Bezold, C., & Hancock, T. (2014). The futures of the healthy cities and communities movement. *National Civic Review, 103*(1), 66–70.

Intergovernmental Panel on Climate Change. (2014). *Climate change 2014: Impacts, adaptation, and vulnerability* (Chapter 8: Urban Areas). Retrieved April 28, 2014, from http://www.ipcc.ch/report/ar5/wg2/.

Cornish, E. (1977). *The study of the future.* Washington, DC: The World Future Society.

Friel, S., Akerman, M., Hancock, T., Kumaresan, J., Marmot, M., Melin, T., … GRNUHE members (2011). Addressing the social and environmental determinants of urban health equity: Evidence for action and a research agenda. *Journal of Urban Health 88*(5), 860–874.

Girouard, M. (1985). *Cities and people.* Newhaven, CT: Yale University Press.

Hancock, T. (1993). Seeing the vision. Defining the role. *Healthcare Forum Journal.*

Hancock, T. (2015). Managing decline: Global change requires local action. In Butler, C., Dixon, J., & Capon, T. (Eds.), *Healthy work, health places, health planet* (A Festschrift publication in honour of Prof. Tony McMichael) Canberra: ANU Press.

Hancock, T., Spady, D. W., & Soskolne, C. L. (Eds.). (2015) *Global change and public health: Addressing the ecological determinants of health: The report in brief.* Retrieved from http://www.cpha.ca/uploads/policy/edh-brief.pdf.

Henchey, N. (1978). Making sense of future studies. *Alternatives, 7*, 24–29.
Kickbusch, I., & Gleicher, D. (2012). *Governance for health in the 21st century*. Copenhagen: WHO Europe.
McGovern, L., Miller, G., & Hughes-Cromwick, P. (2014). Health policy brief: The relative contribution of multiple determinants to health outcomes. *Health Affairs, 21*, 2014.
Naisbitt, J. (1984). *Megatrends*. New York: Warner Books.
Rule, J. (1986). *The labouring classes in early industrial England, 1750–1850*. Routledge.
Stilger, R. (1978). Alternatives for Washington. In C. Bezold (Ed.), *Anticipatory democracy* (pp. 88–99). New York: Vintage Books.
WHO Centre for Health Development and UN Human Settlements Program. (2010). *Hidden cities: Unmasking and overcoming health inequities in urban settings*. Geneva: WHO and Nairobi: UN Habitat. Retrieved from http://www.hiddencities.org/.
WHO Commission on the Social Determinants of Health. (2008). *Closing the gap in a generation: Health equity through action on the social determinants of health* (Final report). Geneva: WHO.

Chapter 20
The Logic of Method for Evaluating Healthy Cities

Evelyne de Leeuw and Geoff Green

Introduction

Developing and sustaining healthy cities is a complex business. Evaluating the journey undertaken by city administrations and their partners is equally challenging. Compared with traditional epidemiological paradigms, we must address socio-economic, political and spatial context, identify prerequisites for success, distinguish multiple and concurrent interventions in many domains of city life, account for multiple outcomes, and attribute wider impacts. This chapter reviews how our evaluation methodology has evolved over five phases spanning 25 years, based always on a social model of population health and keeping abreast of emerging scientific evidence on distal determinants. Our most sophisticated fourth generation evaluation (4GE) of phase V deploys realist synthesis methodology, utility driven by the requirements of city policy communities while retaining the WHO requirement of robust generation of scientific evidence. We conclude with a call for a fifth generation evaluation methodology.

E. de Leeuw (✉)
Centre for Health Equity Training, Research & Evaluation (CHETRE),
University of New South Wales, South Western Sydney Local Health District,
Ingham Institute, Locked Bag 7103, Liverpool, BC NSW 1871, Australia
e-mail: e.deleeuw@unsw.edu.au

G. Green
Emeritus professor, Sheffield Hallam University, Sheffield, United Kingdom

E. de Leeuw, J. Simos (eds.), *Healthy Cities*,
DOI 10.1007/978-1-4939-6694-3_20

Utility Driven Evidence

Nested within the primary modus operandi of the World Health Organization, evidence-driven policy and robust evaluation is an enduring feature of the European Healthy Cities Network since its inception in 1986. In its early phases (I and II 1986–1997) the social model of health was strongly adopted to supplement the dominant clinical and bioepidemiological and socioepidemiological theoretical positions identified by Eriksson (2000). Now, the generation of evidence within a policy-oriented framework of supportive environments—the final phase identified by Eriksson—is in harmony with Network perspectives. Kickbusch (2007), Kickbusch and Gleicher (2012) and Davies et al. (2014) highlight broad governance arrangements and the role of city and neighbourhood communities in the co-production of population health. Each phase of the Network and each new wave of scientific focus have not merely 'superseded' earlier ones, but have built on a perspective of health that has necessarily become more complex and certainly has started to include social and political dimensions of health.

This complexity has led to the explicit introduction of expanded multidisciplinary bodies of knowledge to explain health phenomena. It is not an easy task for evaluators. There are major challenges in calibrating and synthesizing all the evidence and ensuring good relations between public health, decision-making, research, knowledge generation and evidence. The European Advisory Committee on Health Research highlights Healthy Cities as a prime challenge in the amalgamation of evidence:

> a legitimate concern is that research in many areas of 'the new public health' aims at actions that are difficult to evaluate, such as those in health promotion. For example, what is a 'healthy city' and what are the general and specific outcomes sought? Because of these difficulties, decisions that are mainly determined by good evidence of effectiveness would favour interventions with a medical rather than a social focus, those that target individuals rather than communities and populations, and those that focus on the influence of proximal rather than distal determinants of health. This would clearly be unsatisfactory for population health activities. (Banta 2004, p. 566)

Birckmayer and Weiss (2000) have demonstrated that theory-based evaluation yields better research information on various elements of success and failure in health promotion programmes. It expects researchers and programme directors to spell out assumptions to a micro-theoretical level so that outcomes are not just made evident but can also be explained. In their theory of realist evaluation, Pawson and Tilly (1997) formulate the mechanism of change (M) as the central intervention component in the sequence $C \rightarrow (M) \rightarrow O$, taking account of context (C) in the resulting outcome (O). Recently we reaffirmed the importance of this perspective, stating that 'the selection of an appropriate theory would provide answers to questions that ask *why* things are (not) happening beyond a mere description *that* they are (not) happening' (de Leeuw et al. 2014a, b). This perspective offers opportunities to integrate intra-generational 'prevention projects' such as Healthy Cities, drawing heavily on the approaches that Eriksson calls socioepidemiological, and environment and policy-oriented, and thus unravel and analyse their various components.

Next is the issue of utility. These perspectives give indications of *how* evidence is to be produced, but not *to what purpose*. Good evidence should be useful evidence. The notion of utility-driven evidence (de Leeuw and Skovgaard 2005) is based on the observations that

- The generation of evidence serves a purpose beyond mere intellectual curiosity (McQueen and Anderson 2001).
- (Health) policymaking takes place in complex interaction between stakeholders (McDougall and de Leeuw 2006, Chap. 14).
- The application of evidence in decision-making argumentation may transform 'facts' into 'beliefs', and 'beliefs' into 'facts' (de Leeuw 1989). Cummins and Macintyre (2002) have described this phenomenon as leading to 'factoids'.

Eriksson (2000) has endeavoured to identify relevant evaluation strategies for each (increasingly complex) intervention type. In his perspective, the amalgamation of relevant and appropriate evaluation strategies and their outcomes would lead to compelling evidence for decision-making. Tones (1997) has indeed argued that evidence is multidimensional, and that measures of success are an assembly (ideally a synthesis) of different types of evidence, such as witness accounts, expert testimony or lab tests. In short, Eriksson has established an evidence heuristic, and Tones an evidence generation approach. Neither, however, is too explicit about the question, 'to which purpose is evidence generated?' We argue for an overarching utilitarian evidence paradigm. Whether taking a social or purely scientific perspective, the producers of evidence should take into account how their products may be used in shaping good practice and guiding policy choices (de Leeuw et al. 2008).

However, policy decision-making is a messy affair, described by some as 'muddling through' (Lindblom and Cohen 1979), 'wicked' (Hunter 2013) or 'negotiated' space (Stone 1997) within the realpolitik of city decision-making. We have described this process by using a juggling metaphor (de Leeuw et al. 2014a, b) in which various elements of the policy process can happen at the same time and a disciplined view of the whole is required to keep all the balls in the air. In terms of policy process analyses, however, a view of the elements as discrete events or opportunities is more appropriate. Kingdon (1995) has demonstrated that windows of opportunity for policy decision-making are created when policy entrepreneurs have applied a process of 'alternative specification' in which different representations of the same 'truth' are presented to stakeholders. This implies that the evidence used in alternative specifications may take different shapes for different stakeholders.

The perspective is shared by Weiss (1979, 1998) and Vedung (2000). In their work on research utilization (or, in our terms, the application of evidence for decision-making purposes) they maintain that research is put into action through different strategies. Six models are proposed:

1. The knowledge-driven model: New knowledge will lead to new applications, and thus new policies. An example could be fundamental research into nuclear resonance signals leading to the development of NMR and MRI scanners, the emergence of which led to medical technology assessments to assist governments in deciding where and how such costly new technology could be implemented.

2. The problem-solving model: Research findings are actively sought, and used for pending decisions. In their ideal form, health impact assessments (HIAs) are an instrument in this model: HIAs supposedly are commissioned to guide decision-making related to proposed profound environmental and social change operations.
3. The interactive model: Here incremental policy change is interactively driven back and forth by emerging research outcomes. The current Swedish national health policy is an exemplary application of this model, and has taken some 20 years to establish.
4. The political model: This leads to research being used to support partisan political positions. Debates around the acceptability of nuclear power demonstrate the different political connections to different research perspectives.
5. The tactical model: The fact that research is being undertaken may be an excuse for delaying decisions or may deflect criticism.
6. The enlightenment model: Concepts and theoretical perspectives that social science research has engendered permeate the policymaking process, rather than single studies or research programmes having a discernable impact on policy priorities.

The compelling conclusion is that social and academic evidence, even when produced 'properly' (i.e. following established scientific 'rigour'), may not have a significant impact on policy. We thus take the position that evidence, if it is to be used by policy entrepreneurs, should be utility-driven, drawing on Weiss' models two, three and six. Its generation should—again, within the parameters of scientific rigour—take into account how it can be used to influence the judgment of policy stakeholders on how success is to be defined. As this axiom is fundamental to Healthy Cities evaluation, the remainder of this chapter will turn to the nexus of research, policy and practice and the emerging meta-theory that may guide future evidence generation for complex health promotion interventions.

Framing the Methodology of Healthy Cities Evaluation

Before we explore the evidence base relevant for Healthy Cities, we need to say something about Healthy Cities methodology. The procedures for gathering evidence are ultimately framed by assumptions about society, the values and goals of cities and their decision makers, and a meta-theory of how these goals will be achieved. In scientific terms, 'methodology' is the logic of method. Methods (from Greek Μεθοδος 'guide') are procedures that guide the collection of data. Methodology—as the logic of choosing and applying methods—has a relation with a frame of reference (or 'theory'). Applied to the Healthy Cities movement, this means we will have to review some of its conceptual foundations, and more specifically, the assumptions that guide its strategy, programmes, processes and projects. To summarize the ascending elements of this hierarchy: (1) data collection requires the deployment of an appropriate (2) method, shaped by (3) a methodology which

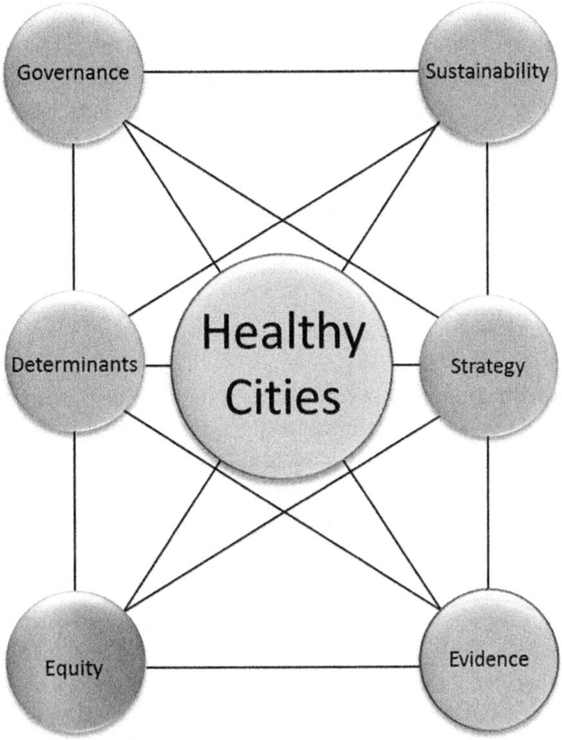

Fig. 20.1 A framework for Healthy Cities values and evidence generation

assumes (4) a theory linking intervention to outcome and (5) a framework or meta-theory about the dynamic of cities and the development of population health.

Sometimes this frame of reference is implicit, and so embedded in culture and value systems that it is difficult to formulate a rational justification why one method is preferred over another. Over six programme phases and 30 years, the WHO Regional Office for Europe has produced over 100 documents to support the Healthy Cities movement. Here we select and summarize six key imperatives which continue to shape cities' operational frameworks. As the network has evolved with the changing social and economic landscape of Europe, new knowledge and greater experience has consolidated our understanding of how cities should organize and act to secure better health for all their citizens. The six interactive elements of the framework are illustrated in Fig. 20.1.

City Governance for Health

From the very beginning WHO put local governments centre stage in promoting the health of city populations. This is innovative because in many European countries national and regional governments are formally responsible for health services

(Green 1998). This leads to a second innovation: partnerships between the many sectors responsible for the social determinants of health. City mayors and councillors are encouraged to provide leadership, orchestrating the contributions of many agencies. Healthy cities pioneered this new form of governance which evolved into the mainstream of European policy with the publication of *Governance for Health in the Twenty-First Century* (Kickbusch and Gleicher 2012). It is a whole-of-government and whole-of-society approach to health.

Equity

Health equity is the core value of the Healthy Cities Network. Though the concept is sometimes difficult to translate, it is defined in the English language as reducing 'avoidable and unfair inequalities in health' (WHO 2005). This fundamental value is intimately linked to justice, solidarity and widespread participation. Equity was the very first of the *Targets for Health for All* set by the WHO European Region in 1985, prior to creating the Network. Over 30 years later it remains the key strategic goal of *Health 2020.* Its attainment is the responsibility of all levels of government, and city mayors have consistently acknowledged their role. Their *Athens Declaration* (*WHO* 2014) which launched phase VI of the Network endorsed 'strengthening urban leadership for health, health equity and well-being for all'. All committed their local authorities and partners to a series of practical steps toward its 'effective implementation'. By providing guidance in *A framework for action,* WHO shows how best *Healthy Cities tackle the social determinants of inequities in health* (World Health Organization 2012).

Evidence

Public health has a strong tradition of evidence-based decision-making; local government less so. Operating on the frontier between these two cultures, the Network encourages healthy cities to adopt a scientific approach. They should (1) import evidence on the prerequisites of population health and (2) rigorously evaluate their own projects and programmes. Synthesis of global evidence is one of the key roles of WHO, together with producing policy frameworks and mandates. WHO provides tools, guidance and expertise, often tailored to the requirements of Network cities. Cities should use it freely to strengthen the case for action. An engaged and constructive Health Impact Assessment (Ison 2013) of non-health actions is very important for mayors who must persuade potential partners from many sectors of city life of their impact on health. Critical success factors are highlighted by the more general evaluation of all Network cities commissioned by WHO at the end of each phase (de Leeuw et al. 2014a, b).

Determinants of Health

This term refers to the range of personal, social, economic and environmental factors that determine the health status of individuals and city populations. From the beginning, healthy city partnerships were built on this model of health. Our scientific knowledge was enhanced with the 2003 publication of *Social Determinants of Health* in the Solid Facts series for the WHO European Healthy Cities Network (Marmot and Wilkinson 2003a, b). Co-author Michael Marmot then chaired the Global WHO Commission on Social Determinants of Health: *Closing the Gap in a Generation* was published in 2012 (WHO Commission on Social Determinants of Health 2008). According to Ilona Kickbush and colleagues, many health challenges require a unique mixture of structural and behavioural change and of agency and political action. Co-production of health involves local governments and partners and also a new role for citizens and civil society (Kickbusch and Gleicher 2012).

Sustainability

Three key dimensions of sustainability have guided the evolution of the Network: social, environmental and economic. First, the Primary Health Care conference meeting in Alma Ata in 1978 suggested that over-investment in expensive hospital services was unsustainable in most countries, and declared an intersectoral strategy of preventive primary care as the best way forward (WHO 1978). This thread is developed by 'active ageing' to expand disability-free life, and supported by the environments of *Age-friendly Cities and Communities* (WHO 2007). Second, environmental sustainability was the focus of the ground-breaking 1992 Rio Declaration of the United Nations (United Nations 1992). The local dimension (Agenda 21) of this global imperative was developed innovatively by Network cities utilizing the concept and practice of 'healthy urban planning' (WHO 1992). This strategy promotes 'human-centred' sustainability. Third, and often a priority for mayors, is economic sustainability. *Health 2020* (WHO 2014) utilizes more recent evidence to propose that 'good health underpins social and economic development'.

Strategy

The pioneering cities of the first phase prioritized demonstration projects—small initiatives which illustrated an innovative way forward. Indeed, a typical Healthy Cities project was often a pilot. If proved successful it might then be upscaled and 'mainstreamed' within the city administration. Later phases were characterized by a more strategic approach to policy within city governments (Price and Tsouros 1996).

Demonstration projects remained important, but in 'magic combination' with city health plans in phase II and city health development plans in phase III (WHO 2001). These involved all sectors with an influence on health and, according to Jostein Rovik, previous mayor of Sandnes (Norway), 'tell you where your city wants to be and how to get there' (Green et al. 2009). City health development plans or their equivalent remain a requirement for phase VI (WHO 2013); 'equivalent' because a lesson of phases IV and V is that potential partners have differing plans, priories and timescales. An overarching goal for phase VI is now 'health and equity in all local policies' (Sihto et al. 2006).

Member cities of the WHO European Healthy Cities Network have applied these elements differentially depending on when they joined, on socio-economic context (areas with a population with overall higher social-economic status tend to approach community action differently from areas with more vulnerable populations), on pre-existing political configurations (conservative governments tend to choose other models than social-democratic governments), demographic parameters and social movement traditions (Nordic welfare states traditionally involve communities in decision-making in radically different ways from most countries in Central and Eastern Europe). For each unique local and national context there is a variety of (theoretical) conceptualizations. It is important to be as explicit as possible in determining such variability in concepts and contexts between localities in Healthy Cities research. We will return to this later in our description of realist evaluation and realist synthesis methodologies.

Formalizing a Meta-Theory for Healthy Cities Evaluation

Underlying their complexity and unique contexts is a common meta-theory driving Healthy Cities forward. Specific theories of change find a place in such a comprehensive meta-theory, which is 'a theory about theory'. It reflects all the elements of the framework summarized in the previous section (and more), with a focus on the dynamic of health development. Over the years and through the different phases of the WHO European Network these elements and this meta-theory have been refined, from a broad 'demonstration project' approach in the first phase (1987–1992) to a much more strategic approach. Designation criteria (from the third through sixth phases, 1998–2018) have placed more emphasis on broad, inclusive, institutional and governance values that were reciprocally reflected in a pan-European strategic health approach (WHO 2014).

However, it is our position that the meta-theory for Healthy Cities has significance beyond specific project parameters established for the different phases of the European Network. It should also be relevant for Healthy Cities outside the 'official' scope of the project: it has global relevance. In Fig. 20.2 this meta-theoretical perspective is visualized. In the figure three main components are identified: proximal and distal determinants of health, proximal and distal interventions for health, and 'known impact'.

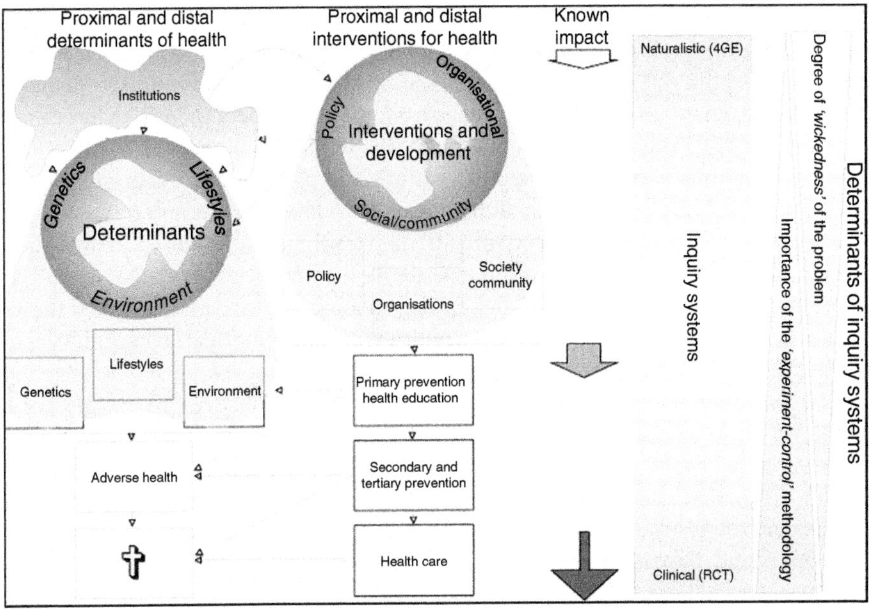

Fig. 20.2 The PADDIH model (proximal and distal determinants and interventions for health, de Leeuw 2009)

In the determinants field, illustrated in the first column of Fig. 20.2, the important elements are the determinants circle and the overarching notion of 'institutions'. The previous section summarized how the healthy cities approach gives greater weight to 'wider', 'upstream' or distal determinants of health relative to 'downstream' proximal determinants of lifestyle or genetics/human biology. However, for each unique dimension of population health in its unique city context, there is a unique mix of proximal and distal determinants of health. And to a significant extent, the observed and valued degree of impact of a determinant is the result of existing institutions and value systems in the given context.

The use of the word 'institution' merits some clarification. Basically, institutions are systems of order: social, political and administrative structures and processes that create, act on, preserve and legitimize complex forms of common knowledge. Given the task of stabilizing the identity of a society, institutions emerge from what Edmund Burke calls an act of constitution, that is, institutions enact norms necessary for social problem-solving. Hannah Arendt (1969) said that an institution is a body of people and thought that endeavours to make good on common expressions of human purpose.

Figure 20.2 illustrates how the 'deadweight' of pre-existing institutions (column 1), often firefighting adverse health, may be transformed (fundamentally or incrementally) to positively developing population health (column 2). Interventions by city governments are critical because of their formal competences for the distal socio-economic and environmental determinants (Green 1998). However, as highlighted in the previous section, Healthy Cities have pioneered forms of 'governance' involving many partners, including agencies with formal responsibility for shaping the 'causes of the causes' of health, and community organizations which may co-produce health by stimulating local action addressing proximal lifestyle issues. In our meta-theory the transformation from column 1 to column 2 requires these institutional structures and processes to adopt ideals and perspectives, such as equity in health, sustainable development and communicative qualities, in individual and organizational relations. The value of a meta-theory is demonstrated here: on an aggregate level Healthy Cities agree that these are important aspects of their work, but in the specific context for each city these institutions may take different shapes, employ different (political) paradigms, and guide different operational actions.

The third column in the graphic representation of our Healthy City meta-theory is 'known impact'. Here we have to draw upon the increasing body of knowledge in evidence-based health promotion. The Cochrane Collaboration in Public Health and Health Promotion (2002), the reviews by Britain's Health Development Agency (2002), and the instrument developed by Netherlands' National Institute for Health Promotion and Disease Prevention (2002) all provide lists of characteristics of effective interventions. The International Union for Health Promotion and Education (2000a, b) produced a report assessing 20 years of evidence of the health, social, economic and political impacts of health promotion.

Generally, the findings from these reviews demonstrate that distal interventions for health in their appropriate 'mix' provide a broad and sustainable effect on population health, whereas proximal interventions for health (health and patient education, health care) yield focused health gains (often disease-, gender-, and age group-specific) against relatively high cost (Vos et al. 2010). However, the further up the scale (the more distal the determinant and intervention mix), the more complex the associated evaluation methodology. This creates problems both for an effective argumentation of Healthy City actions ('What does policy development effectively do for health?') and for establishing a methodology for Healthy City research.

The logic of this impact model is mirrored by the evolution over five phases of evaluations commissioned, sponsored and supported by the WHO Regional Office for Europe over the five phases of the European Healthy Cities Network. In the first two phases, health education and more traditional health promotion endeavours, and their associated assessments, typically focused on the lower end of the model. Characteristically, lifestyle and environmental health issues were high on the agenda of local health services and government agencies. Throughout the phases of the European Healthy Cities Network there has been an endeavour to shift attention more to the upper end of the model. The charge of the development of city health plans (in phase II) was obviously to integrate 'lower end' perspectives and interventions into a more comprehensive package; city health development plans (phase III) were to address that integrated package through more institution-focused intervention frameworks. In subsequent phases there was much more explicit attention to important, yet fuzzy, institutional arrangements like equity, governance, sustainability and participation (de Leeuw et al. 2014a, b); clearly this has presented complex theoretical and methodological challenges to evaluation endeavours.

Healthy City Research Methodology

We can now summarize our pioneering methodologies for evaluating the European Network. In this section we give a generic perspective of research methodology for Healthy Cities, irrespective of their global location or association with local, national or international legal or moral imperatives. In some countries, national or regional authorities require local governments to publish health plans that set unique parameters for associated Healthy Cities research. These requirements differ from place to place, dependent on existing competences and traditions. Some WHO regions have stronger sets of guidance parameters for Healthy Cities than others. For instance, in the Western Pacific region there is a strong member state recognition of the challenges of urbanization, but the development and implementation of particular responses is left to local governments (WHO 2011). In the European region, designation requirements for joining the Network intersect with the competences and powers of local government and partners.

In order to illustrate a generic Healthy City research methodology, we emphasize the fourth column in Fig. 20.2. Inquiry systems range from our preferred fourth generation evaluation (4GE) to the randomized control trials traditionally associated with bioepidemiological subjects. Summarized below are the strengths and limitations of both systems and of intermediary quasi-experimental or naturalistic methodologies.

Orthodox research paradigms address relatively simple problems with relatively simple theories by relatively simple methodologies. The question of whether a new pharmaceutical product is effective (the straightforward theory that 'pill X cures disease Y') is addressed through a methodological approach which has become the gold standard in health sciences: the randomized controlled trial (RCT). RCTs establish two groups from matched populations (an 'experimental' and a 'control' group). The supposedly effective intervention is administered to one group, and the other group receives an intervention which is—supposedly—known to be ineffective. Neither researchers nor the groups are aware of which receives which type of intervention (this is called a 'double-blind' design). Any significant test results can be attributed to the effectiveness of the intervention, as all 'confounding factors' (outside factors that might influence measurements and effectiveness) have been cancelled out by the research design. Appropriate application of the RCT methodology is based on a number of assumptions: that the experimental and control groups are homogeneous (often 'healthy men between 18 and 60') and test conditions have been randomized completely (any factors that might influence the test procedures are distributed randomly in the populations) to allow for statistical analysis.

Across WHO regions there are many examples of quasi-experimental research designs which strive to adopt certain features of RCTs. Commissioners and investigators with roots in this tradition have taken advantage of interventions planned by national and regional governments, municipalities, and health authorities and their partners or contractors. Their evaluations are located higher than RCTs in the 'inquiry systems' column of Fig. 20.2, moving up toward a more naturalistic evaluation approach. Like RCTs, these studies generally focus on a specific intervention. For example, we earlier evaluated the impact on crime and health (Green et al. 2002) of housing regeneration in Liverpool (Critchley et al. 2004) and in Udine the local university evaluated the sensitization of politicians and administrators to a public health perspective using the Kirkpatrick protocol (Whitfield et al. 2013) and changes in the cognitive functioning ability of older people as a result of the 'Move your Mind Programme' (Green et al. 2015). These designs feature measurements at various points in time (a T_0 measure before the intervention, and T_{1-n} measures during and after the intervention) and in different natural settings with a high degree of similarity (e.g. neighbourhoods with comparable demographic profiles). Although investigators recognize that context cannot always be randomized and that 'real' populations are not as homogeneous as in the RCT approach, some studies (for example, the Liverpool study) have deployed control groups to account for 'deadweight'—the changes which would have occurred anyway because of external factors beyond the focused intervention.

Much more challenging for RCTs and their quasi-experimental variants are the complex dynamics of a whole Healthy City as the subject of evaluation (Green and Tsouros 2007). The conceptual frameworks and meta-theory guiding the research endeavour are far more complicated and intricate, as illustrated by Fig. 20.1. The sociopolitical dynamic of a city does not sit easily with these established methodologies. For example, would the control setting for a city like Horsens, if one could be found, be a neighbouring city in Denmark with a similar-sized population, with a similar socio-economic composition and trajectory? Suppose this 'match' is valid—there then may be elections, and political priorities may change during the inquiry period. The economy may experience an upswing, a new factory may open and unemployment fall. The housing authority may already have a long-term programme to redevelop a neighbourhood, generating an influx of people with entirely different characteristics from those assumed by the research design. This is context: the first column in the meta-theory illustrated by Fig. 20.2. Add to the challenge multiple and concurrent interventions illustrated by the second column. Even when at the starting point of an evaluation the environment of interest (i.e. the city—in our example Horsens) may be matched with a fully comparable environment without the 'experimental condition' (e.g. the city of Vejle, which is not part of the Healthy Cities network), as soon as we start following the two over time the context will start to diverge and comparisons will become increasingly challenging.

McQueen and Anderson (2001) in their 'Evaluation in Health Promotion' chapter—which should be required reading for anyone engaging in Healthy Cities research—eloquently describe the methodological problem:

> Unfortunately, many health promotion researchers put the cart before the horse when choosing research methods. They let research methodology drive the investigation, rather than allowing theory and models to provide the conceptual underpinnings for the advancement of knowledge. With such conceptual understanding, investigators can then seek appropriate methods. For instance, many researchers inappropriately use randomized controlled trials in health research. (p.73)

Our preferred methodology is fourth generation evaluation (4GE 'forge') nested within the Healthy Cities Framework elaborated in Part I of this book and the meta-theory of context and change illustrated by Fig. 20.1. The acronym 4GE ('forge') is appropriately chosen, as 4GE is a participatory, dialectic, post-modern scheme of reference ultimately leading to consensus on evaluation parameters, their use and expected outcomes. The 4GE methodology is not unique, extremely innovative, or past any current paradigm. Boutilier et al. (1997) describe what they call 'community reflective action research' that incorporates stakeholder perspectives in policy development. Fourth Generation Evaluation assumes the following steps in the evaluation project development process: (1) contracting, (2) organizing, (3) identifying stakeholders, (4) developing within-group joint constructions, (5) enlarging joint stakeholder constructions through new information/increased sophistication, (6) sorting out resolved claims, concerns, and issues, (7) prioritizing unresolved items, (8) collecting information/adding sophistication, (9) preparing agenda for negotiation, (10) carrying out the negotiation, (11) reporting and (12) recycling.

Within a negotiated 4GE evaluation remit, and appreciating the dynamism of context, our preferred logic of method has become grounded in 'realistic evaluation', and specifically in 'realist synthesis of evidence' (Pawson et al. 2005) to enlighten municipal policymakers and decision makers on interventions to maximize the health of their city populations. Evaluators have highlighted the methodological difficulties of gauging the impact of multiple interventions across many domains and sectors within the complex context of Healthy Cities, and delineating and scaling the causal pathways to health. We (de Leeuw 2009) earlier proposed applying a realist approach to such evaluations, as encapsulated in the Context (C) + Change Mechanism (M) = Outcome (O) model in the seminal work by Pawson and Tilley (1997). According to their elaboration, 'realist synthesis is an approach to reviewing research evidence on complex social interventions, which provides an explanatory analysis of how and why they work (or don't work) in particular contexts and settings' (p iv). Key steps in a realist review are to (1) clarify the scope, (2) search for evidence, (3) appraise primary studies and extract data, (3) synthesize evidence and draw conclusions and (4) disseminate, implement and evaluate.

Like Boutillier et al., Pawson and associates recommend a healthy two-way dialogue with the policy community, from the initial expert framing of the problem to their final judgment on what works. 'The tasks of identifying the review question and articulating key theories [of change] to be explored cannot meaningfully occur in the absence of input from practitioners and policymakers' (Pawson et al. 2005, p. 31). This is the imperative we take into our evaluation of the European Healthy Cities Network.

Towards Utility-Driven Evidence for Healthy Cities

Over the whole span of the European Network, member cities have asked, 'Do Healthy Cities make a difference?' or more precisely two related questions: 'Are we set on the right course toward the goal of a healthy city?' and 'Does our membership of the European Healthy Cities Network help us along this journey?' Obviously policymakers and decision-takers want answers, and they look to evaluators to provide them. The evidence that we accumulate may be utilized to shape policy and guide their investment decisions. To summarize, we aim for our evidence to be utility-driven.

However, a big constraint on robust evaluation is limited—human and financial—resources, particularly acute in a period of austerity and pressure on funding agencies. The task has also become more demanding of time and effort as the Network has developed from simple demonstration projects in phase I through strategic planning in phases II and III to governance for health in phases IV and V. Focused demonstration projects are more amenable to the relatively simple quasi-experimental methods referred to earlier. Much more demanding are expansive evaluations of the context and trajectory, a comprehensive picture of Healthy Cities, or even of their principal components and wicked problems. City health development plans illustrate the issue. They are emblematic of the switch in emphasis from demonstration projects to strategy. They implicitly adopt the $C \rightarrow (M) \rightarrow O$ model of realist evaluation, taking account of socioeconomic context, specifying a set of interventions from intersectoral partners and identifying outcomes as a precursor to predicted impact on health and health equity (Whitfield et al. 2013). A central question for one of the key deliverables is 'Are cities which explicitly adopt a city health development plan (CHDP) more or less likely to achieve greater health equity in and for their populations?'

The key issue is 'deadweight': would the outcome and impact have happened anyway, without adopting a CHDP? Answering this question could lead to a very elaborate set of methods. These include a review of historical factors leading to inequities in health in a number of selected (possibly matched) urban settings, social epidemiological data-gathering or compilation of relevant data from existing city sources (carefully scrutinizing an appropriate use of existing indicators or the application of standardized indicators), a review of urban policies or procedures explicitly or implicitly addressing the equity issue (the conceptual literature would indicate that explicit policies for the reduction of inequities in health might be as effective as general urban socio-economic policies which would thus be addressing the problem implicitly), a selection of cities or their neighbourhoods where an impact may be expected from these implicit or explicit policies versus settings where this might not be expected, process evaluations of the extent to which these policies are factually implemented (including the factors which impede or facilitate development and implementation of these policies), assessments to review the possible influence of participation of cities in Healthy City programmes (or similar grander schemes such as, for instance, in Britain the Health Action Zones or in the Netherlands, the

Ministry of the Interior's Urban Policy on disadvantaged neighbourhoods), evaluations of subjective and objective benefits of the programmes, and, ultimately, the attempt to attribute any changes in health inequities to any of the procedures developed.

Such research is grandiose and realistically untenable, even apart from the fact that the associated meta-theory and its logic of methods can hardly be considered central to the funding considerations of main national and international research grant organizations. So, even though we have made a compelling argument for appropriate research, it is at the same time likely to be unfeasible research.

Realist Synthesis

Though this 'grandiose' method for evaluating the efficacy of CHDPs is within the methodological paradigm of realist evaluation, its application in most cities is *un*realistic. It would require an enormous logistical effort and colossal resources. Even if an agency were found to fund such an effort, the organization of the evaluation project and the placement of mechanisms for the continuous monitoring of research quality (including measures of reliability and validity) would be almost beyond comprehension.

Consequently we have turned to realist synthesis as our preferred methodology, developing parameters for evaluation of phases III and IV and refining our methods for the evaluation of phase V. The WHO designation requirements for phase III (WHO 1997) provided an opportunity to update the meta-theory of change and develop a series of performance measures (de Leeuw 2009). To ensure utility (within a 4GE approach) evaluators consulted representatives of the Network at a series of business meetings, and advice from the WHO Healthy Cities Evaluation Advisory Committee, led us to operationalize four components of the designation criteria into three fields of inquiry and many questions to be asked of each city. The matrix is summarized in Table 20.1.

Three innovative features characterize our monitoring, assessment, reporting and impact (MARI) protocol. First it is a theory-based exercise in which the theory has been constituted by the normative, causal and final relationships assumed to govern Healthy City operations and development (cf. Milewa and de Leeuw 1995). The $C \rightarrow (M) \rightarrow O$ of realist evaluation is *operationalized* by the sequence (C) presence of policies, adherence to principles, and involvement of actors, (M) processes of change and (O) results, impact, outcomes and outputs. Second it has the objective of assembling then synthesizing the evidence generated by city responses to gain a picture of how they are functioning, indicating causal links between interventions and outcomes. Third an annual reporting template (ART) allows for at least 4 years' worth of information captured over the full phase, adding to the evidence and permitting a time-series analysis of Healthy City dynamics.

This ambition was not fully realized in the evaluation of phases III and IV. Momentum was lost because of limited funding from WHO and low response

Table 20.1 MARI annual reporting template (de Leeuw 2009)

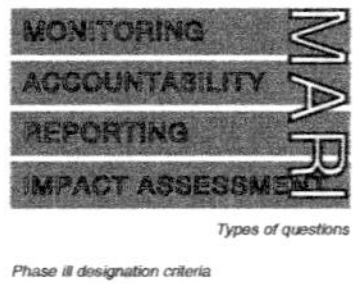	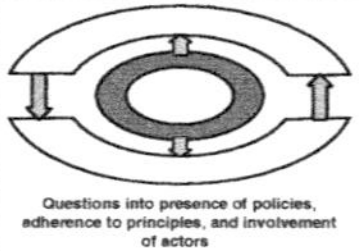	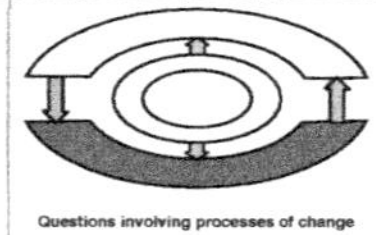	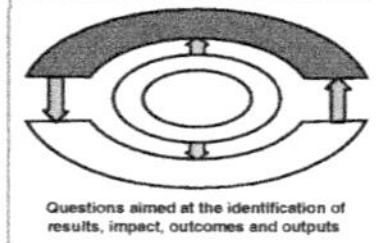
A Endorsement of principles and strategies	A1: Is commitment to the Healthy City project ensured?	A2. Have there been any political changes (elections, Mayor, composition of steering group), locally and nationally that have changed Healthy City practices?	A3. What are existing and new obstacles in the implementation of Healthy City principles? Are there are (new) facilitative factors for Healthy Cities?
B Establishment of project infrastructures	B1: What is the composition of the Project Team, of the Intersectoral Steering Group, and have there been any reviews (management, membership) of their functioning?	B2: What decisions have been taken by the Council, the Project Team and the Steering Group? Distinguish between policy decisions (general direction) and activity decisions (what challenge to take up by action)	B3: How do the activities of Project Team and Intersectoral Steering Group relate to practice in communities, with the private sector and in health and sustainable development in general?
C Commitment to specific goals, products, changes, outcomes	C1: Can you provide an activity list over the past year including at least the following for each activity: project title; name person in charge; professional affiliation of person in charge; objective(s) of the activity; time-frame of the activity; evaluation means associated with the activity	C2: Which activities have started this last year? Why were they started (i.e. who initiated the activity?)? How are activities supported financially and by communities/private businesses?	C3: How are separate activities linked with Phase III designation requirements? What impact do activities have on municipal policies and practice? Have you measured the results of the activities (qualitatively and/or quantitatively)?
D Investment in formal and informal networking and cooperation	D1: Can you provide a list of networks and collaborative efforts? Distinguish between: ad.hoc permanent; formal informal; public private; local national international	D2. How does networking and cooperation take shape (meetings, letters, e-mail, visits, common project)? Are you satisfied with these means? What could improve?	D3: What is the added value of networking and cooperation? To Project Team, Steerign Group, municipality, national network, communities, other local settings (schools, workplaces, hospitals, etc.)?

rates for ART caused by survey exhaustion and data overload. However, phase V (2009–2013) presented a new opportunity to build on MARI and refine our realist synthesis approach to evaluation. Discussions about the evaluation started early in the phase and in good time to present the results after its conclusion. Reflecting the development of earlier MARI templates, city coordinators and lead politicians wished to ensure the utility of the evaluation by linking it to the themes and designation criteria of individual Healthy Cities and the accreditation criteria of national Healthy Cities networks. Further methodological development and research into feasibility was undertaken in a series of conceptual notes, aided by the EU-funded DECiPHEr project (which involved four leading city partners from the Network, Whitfield et al. 2013). The final model for the evaluation is summarized in Fig. 20.3.

The programme logic for phase V implicit in the goals and requirements for membership of the Network (WHO 2008) is made explicit in the construction of the evaluation framework. In discussions with Healthy City stakeholders it was important to develop a conceptual model that distinguishes context and mechanisms in order to provide strong indications of likely outcomes. Our dynamic model shows a pathway linking prerequisites (leadership, vision, strategy, structures, processes and networks) through activities (policies, programmes, projects) which should 'make a difference' to city status in the core areas of caring and supportive environments, healthy urban environment and design, healthy living and health outcomes. Although the arrows in the model suggest linear and causal relationships between the three elements of the programme, an earlier evaluation (de Leeuw and Skovgaard 2005) shows these relationships are reciprocal and dynamic.

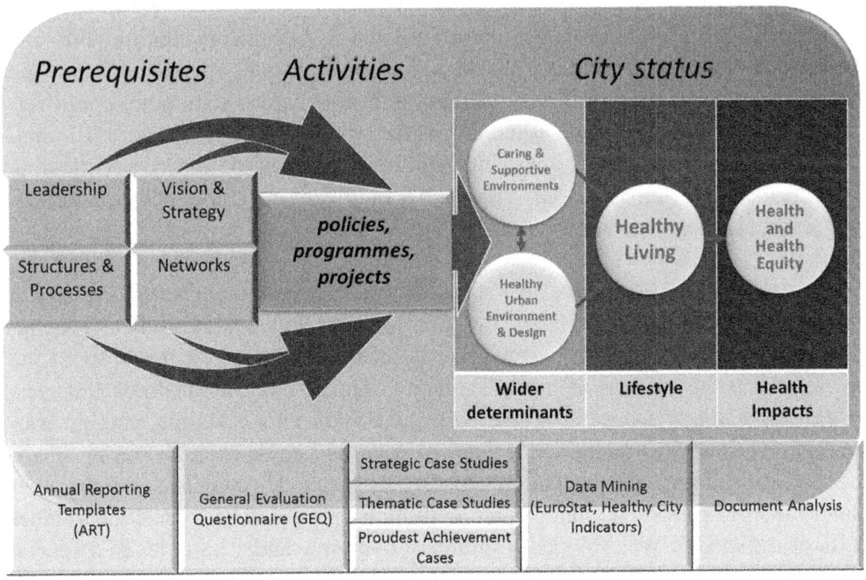

Fig. 20.3 Framework for the evaluation of phase V

To be designated a healthy city in each phase, city administrations take an official position and express commitment to various attributes of local action for health such as governance, equity, working in partnership, assuming leadership for health, engaging stakeholders and communities to participate, and move toward sustainable policy development (Table 20.1). These attributes are found in different mixes and balances, stages of evolution and implementation; the unique context and trajectory of each healthy city must be acknowledged. We developed an evaluative approach that values and celebrates the distinct contribution of each city.

This is why case studies were the most important of our research instruments. Case study research in itself is a powerful tool to generate insights with good validity and generalizability (Flyvbjerg 2006). Case material captures local stories for local conditions, uncovers continuity in corporate and community history, showcases unique and rewarding experiences, and generates a breadth and depth of data and information that might not otherwise be captured. A good case study for research purposes is not a simple story. Our template was highly structured, leading respondents through a development process with the option of focusing on one of four core themes. Key questions were about context, what prompted action, the rationale behind the choice of intervention, whether it was a project, programme or policy, what happened where and over what time frame, whether there was evidence that the intervention made a difference, and finally, how things might have been done differently or what recommendations might be made for extending the activity to other cities.

The other key research instrument was a general evaluation questionnaire, electronically administered through Surveymonkey. The conceptual approach again followed the logic of the model and asked for a self-assessment of prerequisites, activities, and city status for 3 years: 2009 (phase V start), 2014 (phase V end) and informed guesstimates for 2019 (phase VI end), addressing all the components in our model (Fig. 20.3). The questionnaire was developed in consultation with, and pre-tested by, the Healthy Cities coordination team. Questions aligned with strategic attributes as well as the four thematic areas. Innovatively (and sometimes controversially), city coordinators were asked to rate the situation in their city on a 10-point scale at each of these three points. In part this was a heuristic device encouraging cities to focus on outcomes. The approach would capture in a condensed format, progress that cities noted in the annual reporting exercise (Table 20.1) developed for phase III.

The exciting and innovative next stage for the evaluation team was to synthesize the evidence from the case studies, the questionnaire, city health profiles, designation portfolios and secondary data sources such as Eurostat. At the heart of the operation was the ordering of case study data to determine commonalities, linkages, patterns and trends using the NVivo software package for labelling, coding, connecting and assessing qualitative text-based data (de Leeuw et al. 2015a, b). It was the responsibility of each evaluation team to triangulate this evidence with the other data sources to provide a rounded picture of their theme and its connection to other parts of the model. We provided a summary overview and 12 articles in a special supplement of *Health Promotion International* (2015) which assessed the degree to which Healthy Cities have made a difference.

Discussion

This is not the place to provide a systemic review of the findings of our various efforts at Healthy Cities evaluation in Europe—the remainder of this book will provide an account of the particular areas where evidence has been accumulated; but we do want to reflect on the development of our logic of method over the 30 years in which Healthy Cities has had a formal (and codified) existence in Europe. Our account of its evolution above shows that the demands on evaluation, and the tools that we have chosen to deploy, have become increasingly sophisticated.

Evaluations of the earlier phases of the European Network were found to have drawbacks. The first phase harvested commonalities between a small group of highly committed cities, and mapped the conditions for becoming a Healthy City (Ashton et al. 1986; Draper et al. 1993). This research was neither evidence- nor theory-based, but produced some of the seminal texts used globally in Healthy Cities development (e.g. Tsouros 1991; WHO 1992). The research collaboration funded by the European Commission for phase II limited the geographical scope mainly to Western Europe (Berkeley and Humphreys 1998; Capello 2000; de Leeuw et al. 1998). Price and Tsouros (1996) compiled a collection of case studies that seems to have had more impact than the scholarly enterprise. A more integrated, mixed-method approach was advocated from phase III onwards (de Leeuw 2009), but the proposed MARI framework (Table 20.1) stretched local research commitments beyond their limits, and resourcing such a comprehensive inquiry into a multinational socioecological health paradigm was virtually impossible. With the highly condensed approach to MARI through annual reporting templates (Green and Tsouros 2013) a methodology started to crystallize in which research questions and approaches were defined and refined in collaboration with all relevant stakeholders, reflecting the naturalistic fourth generation evaluation advocated by Lincoln and Guba (1986).

With advances in methodological approaches to real-world policy research and development called realist evaluation (Pawson and Tilley 1997) and the increased computing power of software managing qualitative data (Bazeley and Jackson 2013) it was felt that responding to the needs, requirements and assets of local governments in health development in evaluations for phase V onwards was feasible and appropriate. Adding to this conviction was the emergence of the realist synthesis approach, pioneered in Britain (Pawson et al. 2004) and implemented as the European Union-funded DECiPHEr project on 'Developing an Evidence-Based Approach to City Level Public Health Planning and Investment in Europe' (Whitfield et al. 2013). In such a vision it is not necessary to investigate anew the full logical causal and final sequence of proximal and distal (social, political and commercial) determinants of health to demonstrate the evidence of effectiveness of systems interventions on health, well-being and equity; rather, evidence available from other sources may be inserted in a programme logic to approximate the likelihood of effectiveness under certain (complex) conditions that such systems' interventions work.

From the start of phase V we negotiated this approach with stakeholders in the European Network. More than any other group, the designated cities and their coordinators were quickly convinced of the legitimacy and feasibility of the novel approach. As a programme essentially driven by local government, WHO adopted the logic required for the implementation of a realist synthesis approach, although one of the drawbacks of earlier phase evaluations remained: full resourcing of a comprehensive socioecological research framework extending across dozens of highly diverse Healthy Cities would require a long-term investment in a broad and (virtually permanently available) group of institutional research actors—a situation not within the governance prerogative and funding capabilities of WHO (e.g. Bernier and Clavier 2011).

Our realist synthesis approach has yielded a number of insights that go beyond what has been found earlier and elsewhere. First there is a complete absence of any research (either in the peer-reviewed scholarly literature or outside it) that collects, compiles and analyses a diversity of qualitative and quantitative data from a set of 99 local government areas across any other WHO region as heterogeneous as the European one. The analyses carried out on, in and with local governments in this enormously heterogeneous part of the world are testament to the power of the methodology deployed. The cities, and their actions, constitute a unique living laboratory of health innovation at the local level, and the research builds on that exceptional accomplishment.

What remains is the obstinate, and partly justified, critique that answering the question 'Do Healthy Cities make a difference?' can only happen when applying a methodology that allows for a case-control or pseudo-experimental design. To do this we would have had to recruit a matched set of 99 non-Healthy Cities (or, within designated Healthy Cities, communities or neighbourhoods that were explicitly excluded from Healthy City-like strategies and actions). Apart from logistical and possible moral–ethical barriers, it would have been operationally impossible, within the resource limitations, to do this. But proponents of the realist evaluation paradigm would claim that it is indeed the very purpose of this methodological approach to demonstrate that, under varying contexts, different policy expressions can yield different yet effective outcomes.

This brings us to some of the drawbacks of our realist synthesis methodology. Although the 'negotiated reality' with city representatives early in phase V demonstrated a deep commitment of all Healthy City coordinators and their political representatives, the actual deployment of the range of research tools caused substantial research fatigue and responses that might have had considerable social desirability bias. Through triangulation between different data sources, researchers and theories, we have attempted to control for such bias (Creswell and Clark 2007); we also contrasted our findings with non-response characteristics to ascertain that responses fell within credible levels of reliability and validity. Biased self-reporting may have led cities to give some degree of positive spin on their performance in the general evaluation questionnaire. They were asked to rate their own performance on issues such as health equity and the ability to deal with non-communicable disease. Not surprisingly, most respondents suggested that membership of the Network enhanced the designation prerequisites and improved the health of their city. A more sophisticated survey design would control for this pattern.

Some features of the Healthy Cities realist synthesis in action must be described as substandard. The development of the research tools, based on the programme logic and processes of pre-testing (testing format, language, internal logic, etc.), was slow but effective. Their implementation was delayed by personnel changes, hitches in decision-making processes and formal approval procedures. As described in de Leeuw et al. (2015a, b), the response rates to each of the instruments were more than satisfactory, and the quantity of raw data was at the limits of manageable. If double-sided printed versions of all documents were piled up, the stack would tower above UN City, the office of WHO Europe in Copenhagen. In a first rough analysis of case studies and general evaluation questionnaires into 'mother reports' we compiled over 1000 pages of analysis—and these excluded designation documents, case studies submitted to WHO throughout the phase, annual report template responses, and statistics derived from OECD and EuroStat.

The overwhelming wealth of responses, combined with extremely limited resourcing provided to the research team, has meant that a full-scale synthesis across the programme logic has not been possible. As is clear from the papers in our supplement to *Health Promotion International* (de Leeuw et al. 2015a, b), the interface between the different domains are reported on remain largely unexplored: for instance, we know from our cursory analysis of data across equity, health urban planning, governance and policymaking that this challenging area of work does receive significant attention in a number of cities; and from a synthetic point of view such action should lead to radically different outcomes for age-friendliness and healthy living. For the time being it appears that these data must remain in the data morgue if not the data graveyard (Custer 2015). Putting a more optimistic spin on this, one could say that they are a sleeping beauty waiting to be kissed to vibrant life.

Conclusion

A negotiated 'logic of method' firmly rooted in an explicit conceptual framework to evaluate a network of Healthy Cities has shown to be feasible and acceptable to all stakeholders in the enterprise. However, the research management and governance structures that are required to develop and implement this approach beyond the mere operational resourcing of data collection, analysis and reporting are highly challenging. The resources required for an undertaking of this size are not only financial but infrastructural and logistical. Although the quality of the process and output of the various research endeavours described in this chapter have been acceptable (and could be rated, considering the context, exceptional) we have only scratched the surface of a rock that may well be a diamond.

The scholars and methodologists who advocate negotiated evaluation paradigms and realist (synthesis) approaches rarely discuss operational and resourcing aspects, or the temporal dimensions of these—effective—research endeavours. We suggest it is time to undertake a review of these aspects of the approach. This would inform both scholars as well as research funding bodies of the needs and feasibilities of this type of work. In the realm of Healthy Cities, we have observed that local government institutions are often not aware of the real costs of research, and too easily assume that 'a researcher' from 'a university' can easily run a bit of an evaluation project to find a few answers to a few questions.

We argue that it is time to frame a fifth generation evaluation building on the success of 4GE and realist synthesis, in which resource negotiation and allocation with research partners such as local government and stakeholders in civil society can be addressed more rigorously. This would enhance the relevance and responsiveness of the methodology, and may permit a fuller exploitation of the breadth and depth of data generated by these approaches.

References

Arendt, H. (1969). Reflections on violence. *Journal of International Affairs, 23*(1), 1–35.

Ashton, J., Grey, P., & Barnard, K. (1986). Healthy cities—WHO's new public health initiative. *Health Promotion International, 1*(3), 319–324.

Banta, H. D. (2004). Considerations in defining evidence for public health. The European Advisory Committee for health research. *International Journal of Technology Assessment in Health Care, 19*(3), 559–572.

Bazeley, P., & Jackson, K. (Eds.). (2013). *Qualitative data analysis with NVivo*. Sage.

Berkeley, D., & Humphreys, P. (1998). *In-depth evaluation of 10 EU Healthy Cities projects: Constraints and possibilities*. Luxemburg: Commission of the European Union, DGV.

Bernier, N. F., & Clavier, C. (2011). Public health policy research: Making the case for a political science approach. *Health Promotion International, 26*(1), 109–116.

Birckmayer, J. D., & Weiss, C. H. (2000). Theory-based evaluation in practice. What do we learn? *Evaluation Review, 24*(4), 407–431.

Boutilier, M., Mason, R., & Rootman, I. (1997). Community action and reflective practice in health promotion research. *Health Promotion International, 12*(1), 69–78.

Capello, R. (2000). The city-network paradigm: Measuring urban network externalities. *Urban Studies, 37*(11), 1925–1945.

CHPPHF, Cochrane Health Promotion and Public Health Field. (2002). *The Cochrane Collaboration Cochrane Health Promotion and Public Health Field start page*. VicHealth, Adelaide, Australia. Retrieved September 4, 2002, from http://www.vichealth.vic.gov.au/cochrane/welcome/index.htm.

Creswell, J. W., & Clark, V. L. P. (2007). *Designing and conducting mixed methods research*. Sage.

Critchley, R., Gilbertson, J. M., Green, G., & Grimsley, M. J. (2004). *Housing investment and health in Liverpool*. Sheffield: Centre for Regional Economic and Social Research, Sheffield Hallam University.

Cummins, S., & Macintyre, S. (2002). "Food deserts"—Evidence and assumption in health policy making. *British Medical Journal, 325*, 436–438.

Custer, S. (2015). Avoiding data graveyards: 5 questions the data revolution should be asking. *AidDataBeta*, (11), 1925–1945. Retrieved May 10, 2015, from http://aiddata.org/blog/avoiding-data-graveyards-5-questions-the-data-revolution-should-be-asking.

Davies, S., Winpenny, E., Ball, S., Fowler, T., Rubin, J., & Nolte, E. (2014). For debate: A new wave in public health improvement. *The Lancet, 384*, 1889–1895. doi:10.1016/S0140-6736(13)62341-7.

de Leeuw, E. (1989). Concepts in health promotion: The notion of relativism. *Social Science and Medicine, 29*(11), 1281–1288.

de Leeuw, E. (2009). Evidence for Healthy Cities: Reflections on practice, method and theory. *Health Promotion International, 24*(Suppl. 1), i19–i36.

de Leeuw, E., Abbema, E., & Commers, M. (1998). *Healthy Cities policy evaluation—Final report.* Maastricht: World Health Organization Collaborating Centre for Research on Healthy Cities.

de Leeuw, E., Clavier, C., & Breton, E. (2014a). Health policy—Why research it and how: Health political science. *Health Research Policy and Systems, 12*, 55.

de Leeuw, E., Green, G., Dyakova, M., Spanswick, L., & Palmer, N. (2015a). European Healthy Cities evaluation: Conceptual framework and methodology. *Health Promotion International, 30*(Suppl. 1), i8–i17.

de Leeuw, E., McNess, A., Crisp, B., & Stagnitti, K. (2008). Theoretical reflections on the nexus between research, policy and practice. *Critical Public Health, 18*(1), 5–20.

de Leeuw, E., & Skovgaard, T. (2005). Utility-driven evidence for healthy cities: Problems with evidence generation and application. *Social Science and Medicine, 61*(6), 1331–1341.

de Leeuw, E., Tsouros, A. D., Dyakova, M., & Green, G. (Eds.). (2014b). *Healthy cities. Promoting health and equity—Evidence for local policy and practice. Summary evaluation of Phase V of the WHO European Healthy Cities Network.* Copenhagen: World Health Organization Regional Office for Europe.

de Leeuw, E., Tsouros, A., & Green, G. (2015b). Intersectoral governance for health and equity in European Cities. *Healthy Cities in Europe, 30*(Suppl. 1), 1–125.

Draper, R., Curtice, L., Hopper, L., & Goumans, M. (1993). *WHO Healthy Cities project: review of the first five years (1987–1992): A working tool and reference framework for evaluating the project.* Copenhagen: WHO Regional Office for Europe. (EUR/ICP/HSC 644).

Eriksson, C. (2000). Learning and knowledge-production for public health: A review of approaches to evidence-based public health. *Scandinavian Journal of Public Health, 28*, 298–308.

Flyvbjerg, B. (2006). Five misunderstandings about case-study research. *Qualitative Inquiry, 12*(2), 219–245.

Green, G. (1998). *Health and governance in European cities. A compendium of trends and responsibilities for public health in 46 member states of the WHO European Region.* Copenhagen: World Health Organization European Regional Office and European Hospital Management Journal Ltd.

Green, G., Acres, J., Price, C., & Tsouros, A. (2009). City health development planning. *Health Promotion International, 24*(Suppl. 1), i72–i80. doi:10.1093/heapro/dap057.

Green, G., Gilbertson, J. M., & Grimsley, M. J. (2002). Fear of crime and health in residential tower blocks. A case study in Liverpool, UK. *European Journal of Public Health, 12*(1), 10–15.

Green, G., & Tsouros, A. (2007). Evaluating the impact of healthy cities in Europe. *Italian Journal of Public Health, 4*, 255–260.

Green, G., & Tsouros, A. (2013). *City leadership for health: Summary evaluation of Phase IV of the European Healthy Cities Network.* Copenhagen: WHO Regional Office for Europe.

Green, G., Jackisch, J., Zamaro, G. (2015) 'Healthy Cities as catalysts for caring and supportive environments', Health Promotion International 30: Suppl 1 i99–i107

Health Development Agency. (2002). *HDA Evidence base: Putting public health evidence into practice.* Start page, HDA, London. Retrieved September 3, 2002, from http://194.83.94.80/hda/docs/evidence/eb2000/corehtml/intro.htm; Retrieved February 14, 2016.

Hunter, D. (2013). Getting knowledge on 'wicked problems' in health promotion into action. In C. Clavier & E. de Leeuw (Eds.), *Health promotion and the policy process* (pp. 131–153). Oxford: Oxford University Press.

International Union for Health Promotion and Education. (2000a). *The Evidence of health promotion effectiveness. Shaping public health in a New Europe. Part one: Core document.* Brussels/Luxembourg: ECSC-EC-EAEC.

International Union for Health Promotion and Education. (2000b). *The evidence of health promotion effectiveness. Shaping public health in a New Europe. Part two: Evidence book.* Brussels/Luxembourg: ECSC-EC-EAEC.

Ison, E. (2013). Health impact assessment in a network of European Cities. *Healthy Promotion International, 90*(Suppl. 1), S105–S115.
Kickbusch, I. (2007). Health governance: The health society. In D. V. McQueen, I. Kickbusch, & L. Potvin (Eds.), *Health and modernity: The role of theory in health promotion* (pp. 144–161). New York: Springer.
Kickbusch, I., & Gleicher, D. (2012). *Governance for health in the 21st century*. Copenhagen: WHO Regional Office for Europe. Retrieved from http://www.euro.who.int/__data/assets/pdf_file/0019/171334/.
Kingdon, J. W. (1995). *Agendas, alternatives and public policies* (2nd ed.). New York: Harper Collins College Publishers.
Lincoln, Y. S., & Guba, E. G. (1986). But is it rigorous? Trustworthiness and authenticity in naturalistic evaluation. *New Directions for Program Evaluation, 1986*(30), 73–84.
Lindblom, C., & Cohen, D. (1979). *Usable knowledge*. New Haven: Yale University Press.
Marmot, M., Allen, J., Bell, R., Bloomer, E., & Goldblatt, P. (2012). WHO European review of social determinants of health and the health divide. *The Lancet, 380*(9846), 1011–1029.
Marmot, M., & Wilkinson, R. G. (2003a). *Social determinants of health* (2nd ed.). Oxford: Oxford University Press.
Marmot, M., & Wilkinson, R. (2003b). *Social determinants of health: The solid facts* (2nd ed.). Copenhagen: WHO Regional Office for Europe. Retrieved February 12, 2016, from http://www.euro.who.int/__data/assets/pdf_file/0005/98438/e81384.pdf.
McDougall, C., & de Leeuw, E. (2006). Healthy public policy. In: H. Keleher, C. McDougall, & B. Murphy (Eds.), *Understanding health promotion* (pp. 199–205). Melbourne: Oxford University Press.
McQueen, D. V., & Anderson, L. (2001). What counts as evidence: Issues and debates. In I. Rootman, M. Goodstadt, B. Hyndman, D. V. McQueen, L. Potvin, J. Springett, & E. Ziglio (Eds.), *Evaluation in health promotion. Principles and perspectives* (pp. 63–82). European Series, No. 92, Ch. 3. Copenhagen: WHO Regional Publications.
Milewa, T., & de Leeuw, E. (1995). Reason and protest in the new urban public health movement: An observation on the sociological analysis of political discourse in the 'healthy city'. *British Journal of Sociology, 47*(4), 657–670.
Nationaal Instituut voor Gezondheidsbevordering en Ziektepreventie (NIGZ). (2002). Effectiviteit—Preffi. De checklist voor effectievere preventieprojecten ('Effectiveness—Prevention Effectiveness Instrument. The checklist for more effective prevention projects'). Start page, NIGZ, Woerden, The Netherlands. Retrieved September 3, 2002, from http://www.nigz.nl/dossiers/index.cfm?action=dossierandvardossier=27.
Pawson, R., Greenhaigh, T., Harvey, G., & Walshe, K. (2004). *Realist Synthesis: An introduction*. Manchester, UK: University of Manchester. Retrieved February 14, 2016, from https://www.researchgate.net/profile/Gill_Harvey/publication/228855827_Realist_synthesis_an_introduction/links/0fcfd507f0b7cbb2ce000000.pdf.
Pawson, R., Greenhaigh, T., Harvey, G., & Walshe, K. (2005). Realist review—A new method of systematic review designed for complex policy interventions. *Journal of Health Services Research and Policy, 10*(Suppl. 1), 21–34.
Pawson, R., & Tilley, N. (1997). *Realist evaluation*. Thousand Oaks: Sage.
Price, C., & Tsouros, A. (1996). *Our cities, our future policies and action plans for health and sustainable development*. Copenhagen: World Health Organization European Regional Office, Healthy City Project Office. Retrieved February 12, 2015, from RC62BD01-Governance-for-Health-Web.pdf.
Sihto, M., Ollila, E., & Koivusalo, M. (2006). Principles and challenges of health in all policies. In T. Ståhl, M. Wismar, E. Ollila, E. Lahtinen, & K. Leppo (Eds.), *Health in all policies: Prospects and potential*. Helsinki: Ministrry of Social Affairs and Health.
Stone, D. (1997). *Policy Paradox. The art of political decision making*. New York/London: W.W. Norton.
Tones, K. (1997). Beyond the randomized controlled trial: A case for "judicial review". *Health Education Research, 12*(2), 1–4.

Tsouros, A. D. (1991). World Health Organization Health Cities Project: A project becomes a movement. Review of progress 1987 to 1990.

United Nations. (1992). *Rio declaration on environment and development.* New York: UN.

Vedung, E. (2000). *Public policy and programme evaluation.* London: Transaction Publishers.

Vos, T., Rob Carter, R., Barendregt, J., Mihalopoulos, C., Veerman, L., Magnus, A., … Wallace, A. (2010). *Assessing cost-effectiveness in prevention.* Brisbane: The University of Queensland.

Weiss, C. H. (1979). The many meanings of research utilization. *Public Administration Review, 39*(5), 426–431.

Weiss, C. (1998). *Evaluation: Methods for studying programs and policies* (2nd ed.). Upper Saddle River: Prentice Hall.

Whitfield, M., Machaczek, K., & Green, G. (2013). Developing a model to estimate the potential impact of municipal investment on city health. *Journal of Urban Health, 90*(1), 62–73.

WHO Centre for Urban Health. (1997). *WHO Healthy Cities Project Phase III: 1998–2002. The requirements and the designation process.* Copenhagen: WHO Regional Office for Europe.

WHO Commission on Social Determinants of Health. (2008). *Glosing the gap in a generation: Health equity through action on the social determinants of health.* Geneva: WHO.

World Health Organization. (1978). *Declaration of Alma Ata.* Geneva: WHO.

World Health Organization. (1985). *Targets for health for all: Targets in support of the European regional strategy for health for all.* Copenhagen: WHO Regional Office for Europe.

World Health Organization. (1992). *Twenty steps for developing a healthy city project.* WHO Regional Office for Europe.

World Health Organization. (2001). *City health development planning; concept, process, structure and content.* WHO Regional Office for Europe.

World Health Organization. (2005). *The health for all policy framework for the WHO European Region: 2005 Update.* Copenhagen: WHO Regional Office for Europe. Retrieved February 12, 2016, from http://www.euro.who.int/__data/assets/pdf_file/0008/98387/E87861.pdf.

World Health Organization. (2007). *Global age-friendly cities: A guide.* Geneva: WHO.

World Health Organization. (2008). *Phase V (2009–2013) of the European healthy cities network: Goals and requirements.* Copenhagen: WHO Regional Office for Europe.

World Health Organization. (2011). *Healthy urbanization: Regional framework for scaling up and expanding healthy cities in the Western Pacific 2011–2015.* Manila: WHO Regional Office for the Western Pacific.

World Health Organization. (2012). *Healthy cities tackle the social determinants of inequities in health: A framework for action.* Copenhagen: WHO Regional Office for Europe.

World Health Organization. (2013). *Phase VI (2014–2018) of the WHO European Healthy Cities Network: Goals and requirements.* Copenhagen: WHO Regional Office for Europe.

World Health Organization. (2014). *Athens declaration for Healthy Cities: Strengthening urban leadership for health, health equity and well-being for all.* Copenhagen: WHO Regional Office for Europe.

Chapter 21
Healthy Cities: A Political Project Designed to Change How Cities Understand and Deal with Health

Agis D. Tsouros

Introduction

Making change happen on the ground in the public health domain can be very complex; it is never linear. Evaluations and historical reviews often reconstruct the past to fit and mirror theory and methodical thinking, leaving out important information.

There is no shortage of rather clinical accounts that tell very little about the why and the how underlying reported successes and failures of public health policies, strategies and plans, and yet it is not uncommon to talk about the need for scientific narratives and stories to better understand innovation and implementation processes in the real world, taking into consideration sociopolitical and organizational contextual factors.

The nearly 30-year history of Healthy Cities in Europe provides an excellent example of a story that still remains to be told fully, in all its multiple facets. It is a story of the modern public health movement in action at the local level that evolved over a period of major scientific and geopolitical developments.

Healthy Cities was launched with the ambitious aim to put health high on the social and political agenda of cities. It became a thriving global movement that caught the imagination of thousands of city leaders and professionals concerned with urban health and sustainable development. It is undoubtedly one of the longest running and most successful initiatives ever introduced by WHO.

The intention in this concluding chapter is to talk about the evolution and the key attributes of Healthy Cities that made it into such a thriving and sustainable movement for health, equity and well-being.

A.D. Tsouros (✉)
International Adviser on Health Policy and Strategy for City and National Governments,
Former Director of Policy and Governance for Health and Well-being and Director of Healthy Cities at WHO, 30 Rue des Salenques, Toulouse 31000, France
e-mail: agistsouros@globalhealthycities.com

E. de Leeuw, J. Simos (eds.), *Healthy Cities*,
DOI 10.1007/978-1-4939-6694-3_21

The Dawn of the New Public Health Movement

Healthy Cities was created at a time of openness and innovation in the international public health arena. Following a decade of questioning and re-thinking health and medicine, and setting the values and principles of a new public health era, the 1980s provided the political legitimacy and the strategic means for taking forward an agenda for health for all, based on powerful concepts and ideas and engaging a wide range of new actors. Most notably, the strategy for Health for All (WHO Regional Office for Europe 1984) and the Ottawa Charter for Health Promotion (WHO Regional Office for Europe 1986) inspired new types of leadership for health that transcended traditional sectoral and professional boundaries.

Health policy analysts have not given adequate attention to the fact that inspired political leadership is driven first of all by values and visions which in turn generate the will to overcome barriers that stand in the way of progress towards a better world. The WHO Regional Office for Europe launched Healthy Cities during such a period of great openness to new ideas and concepts, and also of willingness to take risks and work with totally new partners. The creation of the WHO European Healthy Cities Network, the Regional Office's strategic vehicle to bring Health for All to the local level, was the result of several developments and initiatives in the early 1980s both at the local level and at WHO (Tsouros 2015). Asvall (1987), the WHO Regional Director for Europe, in his speech at the European Congress on Healthy Cities in 1987, said, "Building a healthy city becomes first and foremost a formidable challenge on how to create a movement for health where many players can be inspired and motivated for taking actions to think new and better solutions and to work together in new partnerships for health". The following year, in his speech at a meeting to celebrate Copenhagen joining the WHO European Healthy Cities Network (Asvall 1988), he said, "Why concentrate on cities? For two reasons: on the one hand their problems are acute and rising; and on the other hand, the city level represents a particularly interesting and promising area for action in Health for All. The mayor of the city has much more power over his area than the Prime Minister has over the country; a city administration can much more easily instruct different sectors to work together in health; and … community participation is not a theoretical issue; it is daily at the finger-tips of the whole city administration". Jo Asvall, a truly visionary Regional Director, gave Healthy Cities huge political and strategic legitimacy from the start. He established it as a cross-cutting initiative that had the strategic mandate to actively engage local governments in the implementation of Health for All. Within a very short period, the WHO Regional Office for Europe further strengthened its capacity to reaching out to new partners by establishing the health-promoting schools and hospitals settings networks and soon later the Regions for Health Network.

Creating Legitimacy to Work with Local Governments

Healthy Cities was created on the recognition of the importance of action at the local and urban levels and the key role of local governments. The fact that working directly with city governments was far beyond the mandate of WHO, whose counterparts are national governments and, in practice, mainly health ministries, has not been widely appreciated. As an organization that is uniquely accountable to the Member States, the Regional Office took the bold step to work directly with subnational agents at the frontline of creating and delivering health development strategies, where it could be directly in touch with local considerations and challenges in implementing high-level strategic ambitions. It took energy and diplomacy to address the initial concerns of and resistance to the idea of designating and working closely with municipal governments expressed by health ministries. The importance of working at the community level was reflected in many WHO resolutions but was not generally regarded as a green light to engage and work with local political leaders.

The recent history of public health is rich with examples of the enormous constitutional limitations and diplomatic obstacles WHO frequently faces in its efforts to address key issues such as inequalities, the control of non-communicable diseases, international public health emergencies or antimicrobial resistance, which require reaching out to a host of other sectors and all levels of government as well as to civil society and non-state actors. Similarly, the increasing relevance of the urban dimension in health development has not yet earned wide and appropriate programmatic status within WHO and local political leaders have not assumed, as they should, a legitimate place in country delegations to important conferences, regional committees or the world health assembly.

Concepts such as healthy public policy, intersectoral action, health in all policies and whole-of-government and whole-of-society approaches continue to be elusive for many national governments. These concepts constitute the premise on which Healthy Cities was designed: a whole-of-local-government approach to health strongly emphasizing equity and partnerships with statutory and non-statutory partners.

Key Attributes of Healthy Cities: Designing a Project for Change and Innovation

New initiatives, even if they are based on brilliant ideas, may be short-lived unless adequate thought is invested into creating conditions for a viable future. This applies especially to innovations that demand new ways of thinking and working. Healthy Cities was created and launched at a time of a strong momentum for innovation in the WHO Regional Office for Europe and with a strategic mandate that gave it enormous legitimacy from the start; the concept appealed to local politicians instantly and instinctively. WHO needed to design a project that could turn this enormous capital into an initiative that combined six essential features:

1. Local relevance and openness to innovation and a cutting-edge public health agenda.
2. Strong leadership and political commitment and a multisectoral approach to health development.
3. Partnership-based management of change, transparency and democratic governance.
4. Strategic thinking and planning and concrete deliverables and outcomes.
5. Adaptability and receptiveness to emerging needs and ideas.
6. Commitment to solidarity and international and local networking.

The project had to be underpinned by strong values, sound evidence, cutting-edge approaches and processes that were sensitive and adaptable to local socioeconomic, organizational and political contexts.

It had to be clear from the start that Healthy Cities was not about doing more on core city functions and responsibilities in the areas of health and environment. Its scope and breadth were much wider, framed on the "health is everybody's business" approach. Furthermore, Healthy Cities was geared towards experimenting with, testing and mainstreaming new approaches to dealing with health, with a strong focus on equity and the determinants of health and sustainability.

The design of Healthy Cities was not meant to be static or cast in stone either. It was launched as a value-based open system that would constantly reinvent itself and evolve, learning from practice and embracing new evidence and ideas to maintain its relevance and grounding itself on local concerns and perspectives. Healthy Cities was to be a pioneer in generating know-how for all urban communities to learn from, not an esoteric movement to benefit only its member cities. Healthy Cities was created as a real-world laboratory to generate good practices, evidence and knowledge, methods and expertise that could be used to promote health in all cities, in the European Region and beyond. Its implementation in diverse political and organizational environments provided a credible basis for the transferability of experience and know-how. All phases of Healthy Cities have been evaluated and the results published in journals and become the source of precious material for developing guidance and know-how tools and guidance documents.

Looking back through the agenda and experience of Healthy Cities in Europe since its launch, one can easily trace the history of the new public health movement in the past 30 years. There is no new concept, approach in the areas of public health and sustainable development that was not embraced and tested by Healthy Cities. Healthy Cities became on many occasions the source of innovation and leadership in areas that later gained major significance: for example, the launch of the solid facts publication on the social determinants of health in 1998 (Wilkinson and Marmot 1998) led the way to the establishment of the global Commission on Social Determinants of Health. The strategic focus of Healthy Cities work meant a focus on upstream, high-impact approaches to health development and equity. Other examples of innovative conceptual work that emerged from the WHO Healthy Cities programme include publications on home care, healthy urban planning, Agenda 21 and health at the local level, active living in the urban context, healthy ageing and health literacy.

Healthy Cities thrived during times of major changes in Europe and the world, including the fall of the Berlin Wall, the wars in the former Yugoslavia, the expansion of the European Union, globalization, the rapid expansion of the information society and austerity waves and significant changes in the social landscape of the European Region. For example, the WHO European Healthy Cities Network proved an extraordinary force of solidarity and significant material support to refugees in Croatia and a valuable network promoting peace and resilience during the horrors of the wars, showing timely adaptability and foresight that earned the respect of politicians and communities around the region. It was thanks to the fact that WHO had a direct relationship with the city of Zagreb, a member of the WHO European Healthy Cities Network before Croatia gained independence, which allowed WHO to send a mission to that city to assess the situation of the war refugees, without needing clearance from Belgrade, the capital of the former Yugoslavia. Its value-based foundations sometimes inspired and empowered communities of member cities to question and demand changes by their local leaders.

Political Prerequisites

Most cities can boast membership of several international city networks, thematic, geographical or strategic. Participation in such networks is largely based on declarations and statements of support of principles and values and much less on commitments to specific action. Membership of the WHO European Healthy Cities Network includes several commitments—political, organizational, managerial and strategic—and a commitment to network and engage internationally. This is a source of respectability and prestige for members of the networks. But more importantly each commitment is translated into a series of requirements that essentially represent preconditions and pathways to success.

Evidence of strong political commitment is a fundamental requirement of the designation process. Political commitment expressed in a letter signed by the mayor or political leader of the city is not enough to secure continuity and shield the project from changes in administration. The enormous added value of city council resolutions to participate in the WHO European Healthy Cities Network or a national network, intersectoral partnership statements by public and private stakeholders and the requirement to appoint a politically responsible person for the project in the city and to include the mayor or other senior politicians in delegations to annual meetings has been undeniable. In this context, member cities highly value the membership certificate for the WHO European Healthy Cities Network signed by the Regional Director.

Managing change and supporting innovation, especially when these imply new ways of working, must be supported by people who have the necessary knowledge, skills and seniority, to enable resources, mechanisms and processes to engage public sectors and agencies and civil society. The requirement for a coordinator and a project office that is strategically located within (or sometimes outside) the city administration is essential. The profile and seniority of the coordinator and the proximity of the project office to the political leadership of the city have proven crucial. Healthy Cities cannot reach its potential if it is reduced to a technical project far from the policy and strategy locus of the city.

The Political and Strategic Rationale of Phases

Healthy Cities evolved over 5-year phases. These phases allowed the regular renewal of goals and requirements; they were long enough to see results and to evaluate progress, and participating member cities could leave the project at the end of a phase at no political cost. Healthy Cities was also uniquely integrated and sheltered within—and beyond—the Regional Office's biennial planning process, which is based on 2-year cycles. Every phase started afresh with a newly (re-)designated group of member cities, many "old" and several "new blood". The phase approach proved most valuable in keeping Healthy Cities momentum alive and strong. There have been 6 phases in Europe, with the latest launched in 2014. The goals and themes of every phase define the priorities of work over the 5 years of the phase. Table 21.1 shows an overview of the main goals and themes in every phase. Each phase was launched with a political declaration that allowed wider engagement in the principles and goals of Healthy Cities.

Managing and Supporting an Inclusive International Initiative

Interest in Healthy Cities spread quickly. An international network, to be credible, needs to be manageable. However, the more traditional approach of piloting first a project proved inapplicable in the case of healthy cities, which generated considerable demand from cities across Europe within months from the time it was launched. As a result, WHO provided the encouragement and legitimacy to all interested cities to be associated with this initiative by creating national healthy cities networks. For reasons of limited capacity, WHO cannot realistically manage a network of more than about 100 cities across the European Region. National networks were established virtually at the same time as the WHO European Healthy Cities Network became operational. The birth of a movement was in sight (Tsouros 1990). WHO is able to work directly with the cities in the WHO European Healthy Cities Network and with the associations represented by the national network coordinators.

Table 21.1 WHO European Healthy Cities Network—six phases

Phases	Main themes	Key strategies and political statements that defined the content of Healthy Cities: global, WHO and Healthy Cities policies and declarations
I (1987–1992)	Creating new structures for and introducing new ways of working for health in cities. City health profiles—an essential tool	Health for All
		Ottawa Charter for Health Promotion
		Milan Declaration on Healthy Cities
II (1993–1997)	Emphasis on intersectoral action, community participation and comprehensive city health planning	Rio Declaration on Environment and Development
III (1997–2003)	Action on health and sustainable development and healthy urban planning. Action on key risk factors for non-communicable diseases. Addressing the social determinants of health. City health development plans—an essential tool. Partnership with other city networks in Europe	Jakarta Declaration on Leading Health Promotion into the 21st Century
		Athens Declaration for Healthy Cities
		Agenda 21 and Rio+10
		Health 21—health for all in the 21st century
		European Sustainable Cities and Towns Campaign
		Millennium Development Goals
IV (2003–2007)	Increasing emphasis on partnership-based health development plans. Core themes include healthy urban planning, health impact assessment and healthy ageing	Belfast Declaration on Healthy Cities
		Report of the WHO Commission on Social Determinants of Health (2008)
V (2008–2013)	Health and health equity in all local policies. Core thematic strands: caring and supportive environments, healthy living, healthy urban environment and design	The Tallinn Charter: Health Systems for Health and Wealth
		Zagreb Declaration on Healthy Cities
		Review of social determinants of health and the health divide in the WHO European Region
		Governance studies
VI (2014–2018)	Leadership for health	European policy and strategy for health and well-being—Health 2020
	City health diplomacy	Athens Declaration on Healthy Cities
	Applying a Health 2020 lens emphasizing life-course approaches, community resilience and health literacy	Sustainable Development Goals

National networks of healthy cities have been created in approximately 31 European countries, involving approximately 1500 cities and by and large established with legal-entity status as associations of local authorities.

National networks hold the greatest promise for the future of the Healthy Cities movement, with a key strategic role in promoting healthy cities principles and ideas, supporting member cities, organizing training and learning events, working with different ministries and participating in national programmes. National networks are accredited at the start of each phase based on explicit criteria reflecting the scope and goals of that phase as well as on the establishment of minimum managerial requirements similar to those applying to the cities in the WHO European Healthy Cities Network. National networks also represent an important public health platform at the European level. National networks have been established in several other WHO regions, most notably in the Western Pacific Region, Region of the Americas and Eastern Mediterranean Region.

One of the most important strategic mechanisms of the Healthy Cities movement in Europe is the annual business meeting and technical conference. The conferences of the WHO European Healthy Cities Network and the Network of European National Healthy Cities Networks are mainly political and strategic and always include a strong political presence from the participating cities and national networks. The main objectives of these annual meetings are to make joint decisions about strategic, managerial and technical issues regarding the running of the WHO European Healthy Cities Network, to share experiences and learn from evidence and expertise related to the themes and priorities of the phase, to learn about emerging issues and connect with new agendas when relevant and appropriate and to get to know each other. Healthy Cities in Europe has become like a family, continually welcoming newcomers. They participate in these annual meetings represented by the mayor or responsible politician, the coordinator, other city officials and delegations from national networks. Organizing the annual conference depends wholly on the generous support of the cities. The programme always includes presentations and sessions on new concepts and new developments, panels and sessions for politicians, sessions to learn from city practices on all main themes, strategic debates and teach-in and training events.

Every 5 years, at the end of each phase, a major conference (with 800–1000 participants) is organized that is open to all cities in Europe and beyond. This is an important opportunity to share the achievements and results of the evaluation of each phase and to officially launch the new phase with a new political declaration.

Shaping the Content: Thriving at the Cutting Edge of Public Health

Healthy Cities as a Vehicle and Platform for Innovation

Healthy Cities is a movement committed to change and innovation, and it needs to sustain its strategic course to fulfil its potential. It will lose its uniqueness if it is limited to the trivial or conventional initiatives and practices, however important, that comprise the day-to-day work of existing public health and environmental services. Part of its uniqueness is the political legitimacy that allows it to address challenging issues such as equity, vulnerability, the determinants of health and sustainability. It was built on a whole-of-government and whole-of-society approach at its inception almost 30 years ago. One of the greatest strengths of the movement is the diversity of political, social and organizational contexts within which it is implemented across Europe.

The agenda, themes and goals of each phase reflect WHO European priorities and strategies: global strategies and priorities and issues emerging from urban (health, social, environmental) conditions in Europe. Specific themes and deliverables are decided in close consultation with the cities in the WHO European Healthy Cities Network and national networks before the launch of every phase, generating a strong basis of ownership.

Within the dominant goals and themes of each phase, cities have the flexibility to identify and give weight to areas that are of particular relevance to local realities. However, all cities are expected to work on the overarching and innovative themes of every phase, participating in conceptual development and the brokering of new ideas at the local level. Some of the issues or themes in every phase require conceptual or operational development, appropriate for the local level and context.

The Phase VI Framework

Five phases have been completed and evaluated, and Phase VI was launched in Athens in 2014. The main documents guiding the priority themes of each phase are the document describing the goals and themes of the phase and membership requirements based on which cities apply to become members and the political declaration. Currently, these relevant documents are the Phase VI goals and requirements document and the Athens Declaration for Healthy Cities.

The goals and themes of all phases are underpinned by a constant commitment to address equity, the determinants of health and sustainable development, as well as a commitment to intersectoral action and democratic governance.

As was the case 27 years ago when Healthy Cities was launched as a vehicle of Health for All at the local level, Phase VI was launched as an official WHO strategic vehicle for implementing the European policy for health and well-being—Health 2020, at the local level. The overarching goals and themes of Phase VI (WHO European Healthy Cities Network 2014) and of the Athens Declaration for Healthy Cities were therefore based on Health 2020 (which was heavily informed by the experience of Healthy Cities).

The cities of the WHO European Healthy Cities Network and national networks are strategic vehicles for delivering Health 2020 at the local level, with opportunities for increased collaborative leadership across levels of government.

Health 2020 provides a timely and strong unifying framework for the context of Phase VI of the WHO European Healthy Cities Network. It builds on and reinforces the fundamental values, principles and work that have been at the core of Healthy Cities since its inception. Cities are uniquely placed to provide leadership for health and well-being. In the complex world of multiple tiers of government, numerous sectors and both public and private stakeholders, local governments have the capacity to influence the determinants of health and well-being and of inequities through whole-of-local-government approaches and health in all policies as well as in regulation, integrated strategies and plans and partnerships across society.

Phase VI is an adaptable and practical framework for delivering Health 2020 at the local level. It recognizes that each city is unique and will pursue the overarching goals and core themes of Phase VI according to its needs. Cities can use different entry points and approaches but will remain united in achieving the dominant goals and core themes of the phase. The new WHO European Healthy Cities Network brings together many old and several new cities, an important consideration when advising and supporting member cities.

Political commitment has always been fundamental to implementation. Cities are encouraged to strengthen leadership and participatory governance for health. Phase VI will explore and promote innovative action for whole-of-government and whole-of-society approaches. The concept of a city health development plan (or the equivalent) remains valid and desirable, encompassing the emphasis on such approaches and strategic thinking. Work on health in all policies will be taken forward, and a new major focus is on building community resilience and health literacy. City health profiles, integrated planning for health and sustainable development remain at the heart of urban health work. The sustainable development goals provide additional legitimacy and reinforce the need for working across government and society for better health. The core themes in Phase VI will be based on a local adaptation of the four priorities for policy action of Health 2020.

There is a healthy tension and debate within the movement between the need to be generic and the drive to be specific. Both generic goal setting and detailed target specifications have been deployed in different phases, each offering challenges and benefits. A clear strategic framework incorporating both perspectives seems most responsive to cities' needs.

Both overarching strategic goals (Table 21.2) reinforce the strong and longstanding commitment of the Regional Office to address equity and the social determinants of health and to strive to improve governance for health and promote health in all policies.

Healthy Cities has extensive experience and made significant progress in its efforts to address inequalities and the social determinants of health. There is scope to do much more, but this goal should remain at the top of the priority list. Applying the equity and social determinants of health lens is crucial in every aspect of city policies and activities, including providing social support, caring for those who find themselves in vulnerable situations, accessing services and dealing with subgroups such as older people or migrants. Policies and interventions within a life-course approach include action on children's well-being and early childhood development, improving employment and working conditions and lifelong learning, enhancing the conditions of life for older people, improving social protection and reducing poverty, addressing community resilience, enhancing social inclusion and cohesion and mainstreaming gender equality.

Intersectoral action and participatory governance have underpinned the work of Healthy Cities from the start, and there is a lot to show, share and learn from. Again, Health 2020, the Sustainable Development Goals (especially SDG 3 and SDG 11)

Table 21.2 Overview of Phase VI overarching goals, themes and priorities based on Health 2020

<table>
<tr><td colspan="4">Overarching goals</td></tr>
<tr><td colspan="2">Tackling health inequalities and the right to health</td><td colspan="2">Promoting city leadership and participatory governance for health with special focus on whole-of-government, whole-of-society and health in all policies approaches and on city health diplomacy</td></tr>
<tr><td colspan="4">Core themes</td></tr>
<tr><td>Life-course and empowering people with special focus on the following issues: a healthy start in life; healthy ageing; vulnerability and health; and health literacy</td><td>Tackling public health priorities with special focus on chronic disease prevention and healthy living on the following issues: physical activity; nutrition and obesity; mental well-being; and alcohol and tobacco</td><td>Strengthening people-centred health systems and public health capacity with special focus on primary care and community services; social services; and capacity for population-based public health</td><td>Creating resilient communities and supportive environments with special focus on community resilience; healthy urban planning and design: environmental health; housing and regeneration; healthy transport; and dealing with challenges related to climate change</td></tr>
</table>

and several other European and global agendas relating to health (such as the epidemic of non-communicable diseases or poverty) and development can only be implemented through approaches that involve different local sectors and community partners. Cities and national networks have a good and sophisticated understanding of related terms such as whole-of-government, whole-of-society and health in all policies approaches. Governance therefore remains at the top of action priorities for all member cities in Phase VI. Recognizing the need to be effective at the international level as well, the concept of city health diplomacy has been introduced. The issue of developing the capacity for local leadership for health is key and has been welcomed by the politicians who participate in the WHO European Healthy Cities Network and national networks.

Several Phase VI priority issues continue to draw considerable interest and attention: a healthy start in life; healthy ageing; addressing the vulnerability and health of people in disadvantaged situations, such as those living in poverty and migrants, and protecting those who might become vulnerable as a result of austerity and social exclusion; health literacy; physical activity; childhood obesity; mental health, including dealing with stress, depression and alcohol abuse; developing the capacity of local public health to promote health and prevent ill health, and in particular adopting population-based approaches; community resilience in dealing with both social and physical environmental challenges; healthy urban planning and design; and climate change, especially in terms of disaster preparedness and response. A topic that has attracted considerable attention in recent years is tourism and health, which is based on the application of Healthy Cities values and principles to traditional models of tourism.

Connecting with Other WHO Regions, WHO Headquarters and Other Partners and Networks

Healthy Cities was first launched as a WHO initiative in the European Region and soon spread to other regions. Variations of Healthy Cities initiatives (including healthy municipalities, towns, communities, villages, islands, territories, and districts) are today well established and several of them thriving in the WHO Region of the Americas, Western Pacific Region, Eastern Mediterranean Region, African Region and, to some extent, South-East Asia Region. In the 1990s, there was an active focal point for Healthy Cities at WHO headquarters. Unfortunately, this important role has been missing for almost two decades. There is a chronic need for dialogue and experience sharing on Healthy Cities between the WHO regions. Bilateral exchanges between regions in this area are not uncommon. The WHO Centre for Health Development in Kobe, Japan has strongly focused on urban health for some time, in particular on the pursuit of equity, through an instrument called Urban Heart. The Kobe Centre has also been a focal point for relations and collaboration with the United Nations Human Settlements Programme (UN-Habitat) and has led the production of the *Global report on urban health.*

The WHO European Healthy Cities Network over the years has cultivated and supported a wide range of partnerships and collaborations with other global or European networks, including ICLEI—Local Governments for Sustainability, the Council of European Municipalities and Regions, Energy Cities and Mediterranean Cities as well as other organizations, including the European Commission, OECD, Council of Europe and UNICEF. Worth mentioning are a number of European Union-funded projects on sustainable development and physical activity, the European Sustainable Cities and Towns Campaign, of which the WHO European Healthy Cities Network was a founding member, and a global conference co-organized with the OECD in Madrid. There is a need to re-think and invigorate partnerships with relevant agencies and established networks of cities but also explore broader links and synergy with others such as Smart Cities, the C40 Cities Climate Leadership Group, Resilient Cities, Tobacco or Smoke Free Cities and Child-Friendly Cities. The Age-friendly Cities initiative led by WHO headquarters deserves special mention here. This initiative became an integral part of the WHO European Healthy Cities Network for three reasons: healthy ageing was already a priority of Healthy Cities in Europe; Healthy Cities had well-established links with city leaders and promoted comprehensive and integrated approaches to promoting health in cities; and integration would avoid competition (for resources and attention) and potential confusion.

Over the years, efforts have been made in exploring synergy and long-term strategic collaboration with healthy settings networks such as health-promoting schools, health-promoting hospitals, healthy prisons and health-promoting universities, but these were short-lived mainly because of the different ways of working of these networks. Such collaboration is easier at the local level. There was always more scope to work closely with the Regions for Health Network, which also addresses a level of government and thus is more compatible with Healthy Cities.

In recent years, the connection of both networks with the Health 2020 agenda opened up excellent opportunities for closer collaboration. Differences in the political decision-making capacity between regions (with independently elected or centrally appointed leaders) and cities (always with elected leaders) have always been in the way of aligning the commitments made by the members of the two networks. Increasingly, however, it is important to systematically promote national and subnational coordination, dialogue, synergy and coherence in addressing the public health challenges of our time.

Reflections and Recommendations for Strengthening Healthy Cities

The success and relevance of Healthy Cities cannot be taken for granted. It is important to strengthen, reinforce and continually explore the role of local governments in addressing health and its determinants. WHO should develop a strong focus for urban health and Healthy Cities in all regional offices and at headquarters. Most programmatic areas are relevant to the urban environment and require active local leadership, and this needs to be acknowledged and reflected in each country's implementation plans. It makes no sense to pursue health in all policies or whole-of-government and whole-of-society approaches in countries without explicit and systematic local political involvement.

Countries should also be encouraged to use the expertise and connections of national networks, which can be a valuable interface between local and national efforts for health development as well as platforms for democratic dialogue. Mayors are emerging as powerful and influential agents for change, locally, nationally and internationally. City health diplomacy can make a true difference, but this implies coherent and strategic thinking. The local voice is essential in the decision-making work of WHO's governing bodies and can also be helpful in discussions regarding the engagement of non-state actors. The sustainable development agenda provides a new opportunity to strengthen health and equity in our cities and communities. Health 2020 and the Sustainable Development Goals agenda are mutually reinforcing and provide enormous legitimacy for strong leadership and action. Healthy Cities is ready now to embrace and integrate the Sustainable Development Goals agenda, which goes hand in hand with priority areas such as equity; vulnerability, including poverty and migrants' health; community resilience; climate change; and the whole determinants of health agenda. Priority should be given to strengthening leadership and local capacity for managing change. Political presence in Healthy Cities should be increased internationally and in countries. Now is the time to scale up Healthy Cities as an important global force for health and equity. This means greater dialogue between regions and a more unified and coherent voice in global forums. There is scope to establish new Healthy Cities collaboration centres on issues such as community resilience and whole-of-society governance for health;

or on migrants' health; but it is imperative to preserve the strategic attributes and uniqueness of the Healthy Cities movement.

Priority should also be given to reviving old and exploring new partnerships with various city networks and relevant agencies and non-governmental organizations. Taking advantage of the enormous potential of the web and social media is essential. For instance, investing in developing well-designed, interactive resource websites may be extremely cost-effective. There are still several countries in the European Region without involvement in Healthy Cities: there is a need to develop a systematic plan to reach out to these countries.

A vast amount of knowledge has been generated by the WHO European Healthy Cities Network that remains invisible, undocumented and undervalued. The stories and achievements of cities and networks need to be systematically explored and documented. This knowledge needs to be continually translated into tools and resource materials to benefit all cities. Training local professionals in this exercise will also increase ownership. Healthy Cities must also continue to identify emerging issues and innovative concepts that hold promise for public health development. It is this anticipatory quality of Healthy Cities that has made it attractive to cities in all countries, even those with a very advanced public health presence. The solid facts series, which has covered several topics, is a case in point and has inspired a host of global audiences.

Long-term strategic alliances with healthy settings networks are challenging to forge because of differences in ways of working and because the main ones, health-promoting schools and health-promoting hospitals, are managed outside WHO. There is scope, however, to work with them in the context of WHO's effort to reach out to other sectors such as health-promoting schools and education ministries. More effort needs to be invested in strengthening the committal base and strategic outlook of the Regions for Health Network, which then could pave the way to a more meaningful and coherent collaboration with the WHO European Healthy Cities Network.

A key aspect of the success on the ground and sustainable evolution of the WHO European Healthy Cities Network has been its ability to connect with other local strategies and programmes, to be a convener and facilitator of intersectoral and community dialogue and cooperation, to be continually open to new concepts and to be sensitive to needs and emerging priorities. Healthy Cities in Europe would lose its relevance if it were, for instance, to not include on today's agenda such issues as the health of migrants, the needs of an ageing population, the challenge of obesity and a physically inactive population or the implications of implementing the sustainable development goals locally.

The 9th Global Conference on Health Promotion (Shanghai, 21–24 November 2016) provides an excellent opportunity to showcase Healthy Cities achievements. The time is right to revisit Healthy Cities indicators and consider evaluation outcomes and deliverables as benchmarks in formulating membership or certification criteria of cities and in setting priorities for future Healthy Cities work.

A Personal Note and Acknowledgement I joined the WHO Regional Office for Europe to run the WHO European Healthy Cities programme from the start. I was responsible for its strategic and technical development and management throughout its 27-year history in Europe. Passionate and convinced about the tremendous value of this programme, I was able to keep this responsibility as I moved on to various senior positions in the Organization until I retired in 2015 from the European Office as Director of Policy and Governance for Health and Well-being. Continually arguing for and positioning Healthy Cities as a WHO cross-cutting policy programme that addresses and engages local governments has always been a major priority in my efforts during periods of significant organizational changes. I will always feel immense gratitude for Jo Asvall, my first WHO Regional Director for Europe, and Ilona Kickbusch, Director of Health Promotion at the Regional Office in the late 1980s, for their trust in me, huge support and inspiration. I should also like to remember with great gratitude and affection my mentor Ron Draper, a man of amazing intellect and wisdom, who as a consultant in my WHO unit during Phase I, was a source of priceless advice and support. During all these years, I have had the privilege to work with local political leaders and interdisciplinary professionals from hundreds of cities across the European Region. It is their leadership, commitment and innovative work that made Healthy Cities into such a prestigious and sustainable public health movement. The WHO Healthy Cities programme over the years has been blessed by many incredibly committed colleagues and advisers, some of them are unfortunately no longer with us. It would be impossible to attempt to mention them all, but it would be equally a missed opportunity to not acknowledge the unique roles and contributions of some of them: Eric Giroult, Xavier Bonnefoy, Linda Petersen, Jill Farrington, Connie Petersen, Evelyne de Leeuw, Geoff Green and Hugh Barton. Lastly, I wish to express deep gratitude to June Crown, Director of Public Health in Bloomsbury District Health Authority (Central London), who offered me space and encouragement in 1983–1984 to develop an initiative that aimed to implement the WHO strategy for Health for All at the local level. This initiative, Health for All in Bloomsbury, attracted the attention of the WHO Regional Office for Europe and the Regional Director, Jo Asvall, and became one of the strong influences that led to the creation and shaping of the WHO European Healthy Cities Network.

References

Asvall, J. E. (1987). *Address at opening session.* European Congress on Healthy Cities, Düsseldorf, Germany, 14–18 June 1987. Copenhagen: WHO Regional Office for Europe (unpublished original speech from Healthy Cities archives).

Asvall, J. E. (1988). Copenhagen—A Healthy City. *Opening statement at a meeting to celebrate Copenhagen joining the World Health Organization project, 14 June, 1988.* Copenhagen: WHO Regional Office for Europe (unpublished original speech from Healthy Cities archives).

Tsouros, A. D. (1990). *WHO Healthy Cities project: A project becomes a movement.* Review of progress 1987 to 1990. Copenhagen: WHO Regional Office for Europe. Retrieved from http://www.euro.who.int/en/health-topics/environment-and-health/urban-health/activities/healthy-cities/who-european-healthy-cities-network/phases-ivi-of-the-who-european-healthy-cities-network/phases-i-and-ii/who-healthy-cities-project-a-project-becomes-a-movement.

Tsouros, A. D. (2015). Twenty-seven years of the WHO European Healthy Cities movement: A sustainable movement for change and innovation at the local level. *Health Promotion International, 30*(S1), i3–i7.

WHO European Healthy Cities Network (2014). *Phase VI (2014–2018) of the WHO European Healthy Cities Network goals and requirements.* Copenhagen: WHO Regional Office for Europe. Retrieved April 28, 2016, from http://www.euro.who.int/__data/assets/pdf_file/0017/244403/Phase-VI-20142018-of-the-WHO-European-Healthy-Cities-Network-goals--and-requirements-Eng.pdf.

WHO Regional Office for Europe. (1984). *Health for all targets*. Copenhagen: WHO Regional Office for Europe.

WHO Regional Office for Europe. (1986). Ottawa charter for health promotion. *First international conference on health promotion, Ottawa, Canada, 17–21 November, 1986*. Copenhagen: WHO Regional Office for Europe. Retrieved April 28, 2016, from http://www.euro.who.int/__data/assets/pdf_file/0004/129532/Ottawa_Charter.pdf?ua=1.

Wilkinson, R., & Marmot, M. (Eds.). (1998). *The social determinants of health. The solid facts*. Copenhagen: WHO Regional Office for Europe. Retrieved April 28, 2016, from http://apps.who.int/iris/handle/10665/108082.

Index

E. de Leeuw, J. Simos (eds.), *Healthy Cities*,
DOI 10.1007/978-1-4939-6694-3

Q

R

S

Printed by Printforce, the Netherlands